Introduction to Classical Electrodynamics 2

Dedicated to the memory of my parents
Dedicated to the memory of Daniel Arnaudon
Dedicated to Prof. Amitabha Chakrabarti

Series Editor
Frédérique de Fornel

Introduction to Classical Electrodynamics 2

Boucif Abdesselam

iSTE

WILEY

First published 2025 in Great Britain and the United States by ISTE Ltd and John Wiley & Sons, Inc.

ISTE Ltd
27-37 St George's Road
London SW19 4EU
UK

www.iste.co.uk

John Wiley & Sons, Inc.
111 River Street
Hoboken, NJ 07030
USA

www.wiley.com

Library of Congress Control Number: 2025940803

British Library Cataloguing-in-Publication Data
A CIP record for this book is available from the British Library
ISBN 978-1-83669-001-6

The manufacturer's authorized representative according to the EU General Product Safety Regulation is Wiley-VCH GmbH, Boschstr. 12, 69469 Weinheim, Germany, e-mail: Product_Safety@wiley.com.

Table of Contents

Preface

There is no place in this new kind of physics both
for the field and matter, for the field is the only reality.

Albert Einstein

An *electric charge* is an abstract concept, comparable to the concept of *mass*, introduced to explain certain behaviors in physics. However, unlike mass, an electric charge can take two experimental forms that are arbitrarily called *positive* and *negative*. Two charges of the same nature, both positive or both negative, for instance, repel each other, while two opposite charges attract each other. This phenomenon is called *electrostatic interaction*.

The history of electric charges goes back to the ancient Greeks who observed that rubbing certain objects against fur produced an imbalance in the total electric charge. These "electrified" objects would also attract other light objects, like hair. What is more, if the objects are rubbed against each other for long enough, it is even possible to create a spark.

Electric charges can be experimentally measured. Particles observed in nature have charges that are integer multiples of an *elementary charge*, which is a *fundamental physical constant*. An elementary charge is the electric charge on a proton or, equivalently, the opposite of the electric charge on an electron. It is expressed in *Coulombs* (C), or in $A \times s$ in the international system of units. The value of an electric charge, considered to be invisible, and its *discrete nature* were first discovered in 1909 by Robert Andrews Millikan.

The elementary charge, denoted by e, has an approximate value $e = 1.602176634 \times 10^{-19}$C. The exact value of the elementary charge is currently defined by: $e = \sqrt{\frac{2h\alpha}{\mu_0 c}}$, where h is *Planck's constant*, with $h = 6.626070040 \times 10^{-34}$Js; α is the *fine structure constant*, with $\alpha = 7.2973525664 \times 10^{-3}$; μ_0 is the *vacuum magnetic permeability*, with $\mu_0 = 4\pi \times 10^{-7}$Hm^{-1}; and c is the speed of light in a vacuum, with $c = 299,792,458$ m/s. The existence of quarks was postulated in the beginning of 1960 and they are believed to possess a fractional electric charge $(\pm\frac{1}{3}e)$ or $(\pm\frac{2}{3}e)$, but they are confined within hadrons, whose charge and may be zero, and may be equal to the elementary charge, or may be an integer multiple of the elementary charge. Quarks have not yet been separately detected, but they are assumed to have existed in a free state in the very first instants of the Universe (the quark epoch). Charge is an invariant in the theory of special relativity. A particle of charge q, whatever its velocity, will always retain the same charge q.

There are two branches of study:

– *Electrostatics* is the branch of physics that studies the phenomena created by static electric charges for the observer. In physics, an electrically charged particle creates a vector field that is called an *electric field*. In the presence of a charged particle, the local properties of a space are therefore modified. This is therefore an accurate translation of the concept of a field. If there is another charge in this same field, it will be subject to the action of the electric force (one of the four forces of nature) exerted at a distance by the particle: the electric field is in a way the mediator of this distant action. If there is a movement of electric charges, generally electrons, within a conductive material under the effect of a potential difference at the ends of the material, an electric current is produced. The concept of electric current was first introduced by Benjamin Franklin. He was the first to imagine electricity as a type of invisible fluid present in all matter. He postulated that friction of an insulating surface causes the fluid to move and that a flow of this fluid constitutes an electric current. A material is made up of a large number of electric charges, but these charges balance out among themselves, that is, there is the same number of electrons (negative) as protons (positive). At normal temperatures, the material is electrically neutral. When a static electricity effect is produced, this means that there has been a displacement of charges from material A to material B. This is the *electrification phenomenon*. Excess or missing charges, that is, *uncompensated charges*, are responsible for electrical effects on a body (the example of the rubbed stick). There are two kinds of materials:

a) A material is said to be a *perfect conductor* if, when electrified, the uncompensated charges move freely through the material.

b) If the uncompensated charges do not move freely and remain fixed in the place where they were deposited, we say that it is a *perfect insulator (or dielectric)*.

In reality, a real material is obviously lies between these two limit states:

– *Magnetism* represents all physical phenomena in which objects exert attractive or repulsive forces on other materials. The electric currents and magnetic moments of the fundamental elementary particles are at the origin of the magnetic field generated by these forces. The magnetic field is therefore a vector quantity (characterized by an intensity and a direction) defined for each point in space. It is determined by the position and orientation of magnets, electromagnets and the displacement of electric charges. The presence of this field is seen in the existence of a force acting upon the moving electric charges (called the *Lorentz force*), and various effects that affect certain materials (paramagnetism, diamagnetism, or ferromagnetism, depending on the case). Magnetic susceptibility is the quantity that determines how much a magnetic field interacts with a material.

Electricity and magnetism are intimately related. In 1820, Ørsted discovered the relationship between electricity and magnetism: *a wire carrying an electric current is capable of deflecting the magnetized needle of a compass*. Subsequently, Ampère, based on Ørsted's work, discovered and formulated a number of laws governing the relationship between magnetism and electrodynamics. In 1831, Faraday discovered that if an electric current produces a magnetic field, the reverse is also true. An electric current can be produced by setting a magnetic field in motion, according to *Lenz's law*.

This marked the birth of electromagnetism. It is therefore the field of physics that studies the interactions between electrically charged particles, whether at rest or in motion, using the concept of the electromagnetic field. Electromagnetism can be defined as the branch that studies the electromagnetic field and its interactions with charged particles. The electric field and the magnetic field are the two components of the electromagnetic field described by electromagnetism. The equations that describe the evolution of the electromagnetic field are called *Maxwell equations*. Electric and magnetic fields waves can propagate freely in space and in most materials. These are called *electromagnetic waves* and correspond to all manifestations of light across wavelengths (radio waves, microwaves, infrared, visible domain, ultraviolet rays, X-rays and gamma rays). Maxwell wrote: *The agreement of the results seems to show that light and magnetism are affections of the same substance, and that light is an electromagnetic disturbance propagated through the field according to electromagnetic laws*. From this point of view, optics can be seen as an application of electromagnetism. The electromagnetic interaction is also one of the four fundamental interactions. Together with quantum mechanics, it enables us to understand the existence, cohesion and stability of atoms and molecules.

From the point of view of fundamental physics, the theoretical development of classical electromagnetism led to the theory of special relativity at the beginning of the 20th century. The need to reconcile electromagnetic theory and quantum mechanics led to the construction of quantum electrodynamics, in which electromagnetic interaction is interpreted as an exchange of particles called *photons*. In particle physics, *electromagnetic interaction* and *weak interaction* are unified within the *electroweak theory*.

Note to readers

Electromagnetic field theory is often poorly or insufficiently taught in university physics programs. The heavy dependence on mathematics, especially vector calculus, integral calculus, complex functions and special functions, etc., add to the difficulty in teaching this subject and may obscure certain phenomena such that students get mired in mathematical difficulties and lose sight of practical applications and vice versa. This modest contribution aims to lay out the essential ideas of the theory in the most logical order. Special importance is given to the mathematical developments of this theory. The experimental side of the theory is only lightly approached. We will not insist on units of measurement. These three volumes will contain many examples (the most classic). I have tried to give as much detail as possible in the calculations. Across the three volumes, the discussion is essentially divided into three main topics: (1) charges and/or dipoles, as the source of the electrostatic field; (2) currents and/or magnetization as sources of the magnetic field; (3) electrodynamics, where the electric field and magnetic field are of equal importance. This book discusses many applications of electromagnetic theory. For instance, there are the solutions of Laplace and Poisson equations in various coordinate systems, dielectric medias, magnetized medias, waveguides, transmission lines, electromagnetic radiation, radiation by point-like charges, Kirchhoff diffraction, special relativity, etc. For the reader's convenience, we have listed all of the mathematical tools used in the book into six appendices, which will be available to download online: vector analysis, tensors, special/hypergeometric functions, etc. References are presented at the end of each volume. Formulas are given here in the international system of units. We have tried to adopt the notations most commonly used in the literature. Graphs and 3D representations have been produced using Maple.

This three-volume set is meant for physics students, engineering students, as well as mathematics students who are interested in problems in electromagnetic theory. It can also be used by anyone interested in physics in general, or who simply wants to learn more out of intellectual curiosity. There are many books on the same subject and therefore it seems reasonable to repeat some classical developments that we find excellent, rather than striving for originality at any cost.

Volume 1 will cover the following topics:

– electrostatic field in vacuum (Gauss' theorem, etc.);

– conductor at equilibrium;

– electrokinetics;

– magnetic field in vacuum (Laplace force, Biot–Savart law, the Ampère theorem, etc.);

– electric and magnetic moments;

– boundary problems for static potentials in vacuum.

Volume 2 will study the following points:

– static potential with a source – Poisson equation (image method, Green function, etc.);

– electromagnetic induction phenomena;

– electromagnetic waves in vacuum;

– electromagnetic energies;

– polarization, polarized media (dielectric);

– magnetization, magnetized media;

– electromagnetic fields in matter.

Volume 3 will analyze the following concepts:

– waveguides;

– transmission lines;

– electromagnetic radiation;

– radiation from point-like charges;

– Kirchhoff diffraction;

– theory of special relativity;

– covariant formulation.

The online appendices will be available at the following URL:
www.iste.co.uk/abdesselam/electrodynamics1.zip

The appendices will cover:

– some vector analysis results (Appendix 1);

– tensors (Appendix 2);

– the Dirac function (Appendix 3);

– orthogonal functions (Appendix 4);

– beta and gamma functions (Appendix 5);

– special functions (Appendix 6).

Acknowledgments

This work would never have seen the light of day without the encouragement of many people. I am particularly grateful to my students, colleagues and many others. I thank Frédérique de Fornel for her patience and advice. Finally, I dedicate this book to my wife, Nadia, and my children, Abir, Ahmed Nour, Mohammed Sami and Samah. I would like to express my gratitude to them.

Boucif Abdesselam
July 2025

Chapter 1

Static Potentials with Sources

1.1. Poisson equation

Before going into the details, let us start by recalling a few basic but important notions that have already been covered in Volume 1. Consider the laws of electrostatics $\vec{\nabla} \cdot \vec{\mathbf{E}} = \frac{\rho}{\varepsilon_0}$ and $\vec{\nabla} \times \vec{\mathbf{E}} = 0$, where $\vec{\mathbf{E}}$ and ρ are the electric field and the local charge density, respectively. Since $\vec{\nabla} \times \vec{\mathbf{E}} = 0$, it follows that $\vec{\mathbf{E}}$ can be expressed as the gradient of a scalar function: the electric potential is defined by $\vec{\mathbf{E}} = -\vec{\nabla} V$ (the circulation of a gradient field along a closed loop is zero). We thus obtain the *Laplace equation* $\Delta V = 0$. When sources are present, the electric field $\vec{\mathbf{E}}$ is no longer a loop. The combination of equations $\vec{\nabla} \cdot \vec{\mathbf{E}} = \frac{\rho}{\varepsilon_0}$ and $\vec{\mathbf{E}} = -\vec{\nabla} V$ yields the *Poisson equation* (also sometimes called the *inhomogeneous Laplace equation*):

$$\Delta V\left(\vec{\mathbf{r}}\right) = -\frac{\rho\left(\vec{\mathbf{r}}\right)}{\varepsilon_0} \qquad [1.1]$$

Consider a point charge located at the position $\vec{\mathbf{r}}'$ ($\vec{\mathbf{r}}'$ is called the *point of charge*). We then write $\rho\left(\vec{\mathbf{r}}\right) = q\delta^3\left(\vec{\mathbf{r}} - \vec{\mathbf{r}}'\right)$. The presence of the Dirac distribution indicates that there is no charge at any other position except $\vec{\mathbf{r}}'$ (which clearly expresses the concept of a *point charge*). Furthermore, the total charge contained in the point of charge is $\int_\tau \rho\left(\vec{\mathbf{r}}\right) d\tau = q\int_\tau \delta^3\left(\vec{\mathbf{r}} - \vec{\mathbf{r}}'\right) d\tau = q$, where $d\tau$ is an infinitesimal volume element surrounding the point $\vec{\mathbf{r}}$, and τ is any finite volume that contains the point $\vec{\mathbf{r}}'$. The corresponding electric field created at the position $\vec{\mathbf{r}}$ by the charge q is equal to $\vec{\mathbf{E}}\left(\vec{\mathbf{r}}\right) = \frac{q\left(\vec{\mathbf{r}} - \vec{\mathbf{r}}'\right)}{4\pi\varepsilon_0\left|\vec{\mathbf{r}} - \vec{\mathbf{r}}'\right|^3} = -\frac{q}{4\pi\varepsilon_0}\vec{\nabla}\left(\frac{1}{\left|\vec{\mathbf{r}} - \vec{\mathbf{r}}'\right|}\right)$. We therefore obtain the potential:

$$V\left(\vec{\mathbf{r}}\right) = \frac{q}{4\pi\varepsilon_0\left|\vec{\mathbf{r}} - \vec{\mathbf{r}}'\right|} \qquad [1.2]$$

The *inverse distance* between the vectors $\overrightarrow{\mathbf{r}}$ and $\overrightarrow{\mathbf{r}}'$ is given by the function $d\left(\overrightarrow{\mathbf{r}}\right) = \frac{1}{\left|\overrightarrow{\mathbf{r}}-\overrightarrow{\mathbf{r}}'\right|}$. Here, $\overrightarrow{\mathbf{r}}'$ is a fixed position (the position of the point charge), and $\overrightarrow{\mathbf{r}}$ is the observer position. Let us calculate the Laplacian of the inverse distance $\Delta d\left(\overrightarrow{\mathbf{r}}\right) = \Delta\left(\frac{1}{\left|\overrightarrow{\mathbf{r}}-\overrightarrow{\mathbf{r}}'\right|}\right)$. Since $\overrightarrow{\mathbf{r}}'$ is fixed, it is more practical to define a new variable $\overrightarrow{\mathbf{R}} = \overrightarrow{\mathbf{r}} - \overrightarrow{\mathbf{r}}'$. As a result, $\Delta_{\overrightarrow{\mathbf{R}}} = \Delta_{\overrightarrow{\mathbf{r}}}$, where the index indicates the variable used in the derivations. Therefore, we see that $\Delta\left(\frac{1}{R}\right) = \frac{1}{R^2}\frac{d}{dR}\left[R^2\frac{d}{dR}\left(\frac{1}{R}\right)\right] = \frac{1}{R^2}\frac{d}{dR}\left(-1\right) = 0$ (in spherical coordinates). The angular derivatives disappear here, as the function is radial. We therefore arrive at the following result:

$$\Delta\left(\frac{1}{\left|\overrightarrow{\mathbf{r}}-\overrightarrow{\mathbf{r}}'\right|}\right) = 0 \qquad [1.3]$$

However, this equation does not always seem to be true. We have already seen that a point charge produces an electric potential $V = -\frac{q}{4\pi\varepsilon_0\left|\overrightarrow{\mathbf{r}}-\overrightarrow{\mathbf{r}}'\right|}$ and Poisson's equation indicates that $\Delta V = -\frac{\rho}{\varepsilon_0} \neq 0$. Given a point charge located at $\overrightarrow{\mathbf{r}}'$, $\Delta V = \frac{q}{4\pi\varepsilon_0}\Delta\left(\frac{1}{\left|\overrightarrow{\mathbf{r}}-\overrightarrow{\mathbf{r}}'\right|}\right) = -\frac{\rho\left(\overrightarrow{\mathbf{r}}\right)}{\varepsilon_0} = -\frac{q}{\varepsilon_0}\delta^3\left(\overrightarrow{\mathbf{r}}-\overrightarrow{\mathbf{r}}'\right)$, it can be derived that:

$$\Delta\left(\frac{1}{\left|\overrightarrow{\mathbf{r}}-\overrightarrow{\mathbf{r}}'\right|}\right) = -4\pi\delta^3\left(\overrightarrow{\mathbf{r}}-\overrightarrow{\mathbf{r}}'\right) \qquad [1.4]$$

A comparison of [1.3] and [1.4] shows that the previous calculation is correct only for $\overrightarrow{\mathbf{r}} \neq \overrightarrow{\mathbf{r}}'$. Indeed, for $\overrightarrow{\mathbf{r}} = \overrightarrow{\mathbf{r}}'$, the inverse distance is singular, and consequently, the explicit derivation of the Laplacian given earlier to establish equation [1.3] is not valid in this case. We can confirm the accuracy of formula [1.4] as follows: it is easier to work with the variable $\overrightarrow{\mathbf{R}}$. In this case, we find $\Delta\left(\frac{1}{R}\right) = -4\pi\delta^3\left(\overrightarrow{\mathbf{R}}\right)$, where $R = \left|\overrightarrow{\mathbf{R}}\right|$. Let us integrate $\Delta\left(\frac{1}{R}\right)$ on any volume τ that contains the origin. This volume is divided into two parts $\tau = \tau_a + \overline{\tau_a}$, where τ_a is a sphere of radius a and center O (entirely contained within τ). $\overline{\tau_a}$ is the remaining part. It therefore follows that:

$$\int_{\tau}\Delta\left(\frac{1}{R}\right)d\tau = \int_{\tau_a}\Delta\left(\frac{1}{R}\right)d\tau + \int_{\overline{\tau_a}}\Delta\left(\frac{1}{R}\right)d\tau = \int_{\tau_a}\Delta\left(\frac{1}{R}\right)d\tau \qquad [1.5]$$

since $\Delta\left(\frac{1}{R}\right) = 0$ at any point of $\overline{\tau_a}$ $(0 \notin \overline{\tau_a})$. Using the divergence theorem, it comes:

$$\int_{\tau_a}\Delta\left(\frac{1}{R}\right)d\tau = \int_{\tau_a}\overrightarrow{\nabla}\cdot\left[\overrightarrow{\nabla}\left(\frac{1}{R}\right)\right]d\tau = \oint_{S_a}\overrightarrow{\nabla}\left(\frac{1}{R}\right)\cdot\overrightarrow{\mathbf{e}}_R dS \qquad [1.6]$$

where S_a is the closed spherical surface enclosing τ_a. The normal to S_a directed toward the exterior is the unit radial vector $\vec{\mathbf{e}}_R \equiv \frac{\vec{\mathbf{R}}}{R}$. The infinitesimal surface element is $dS = a^2 d\Omega$, where $d\Omega = \sin\theta d\theta d\varphi$ is the differential solid angle. Since $\vec{\nabla}\left(\frac{1}{R}\right)\Big|_{R=a} = \vec{\mathbf{e}}_R \frac{d}{dR}\left(\frac{1}{R}\right)\Big|_{R=a} = -\frac{1}{a^2}\vec{\mathbf{e}}_R$, we find:

$$\int_{\tau_a} \Delta\left(\frac{1}{R}\right) d\tau = -\oint_{S_a} \vec{\mathbf{e}}_R \cdot \vec{\mathbf{e}}_R \frac{1}{a^2} a^2 d\Omega = -\oint_{S_a} d\Omega = -4\pi \qquad [1.7]$$

We have therefore shown that $\Delta\left(\frac{1}{R}\right) = 0$ for $R \neq 0$ and $\int_\tau \Delta\left(\frac{1}{R}\right) d\tau = -4\pi$, for any volume τ containing the origin. The only function that satisfies these relations is $\Delta\left(\frac{1}{R}\right) = -4\pi\delta^3\left(\vec{\mathbf{R}}\right)$, since $\delta^3\left(\vec{\mathbf{R}}\right) = 0$ for $\vec{\mathbf{R}} \neq 0$ and $\int_\tau \delta^3\left(\vec{\mathbf{R}}\right) d\tau = 1$ for any volume containing the origin. This confirms the validity of equation [1.4]. Let us conclude by recalling an important property of the function Δ, $\int_\tau f\left(\vec{\mathbf{r}}'\right)\delta^3\left(\vec{\mathbf{r}} - \vec{\mathbf{r}}'\right) d\tau = f\left(\vec{\mathbf{r}}'\right)$, where $\vec{\mathbf{r}}$ is within volume τ. This result is easy to understand, as the function δ would vanish at any point, except the point $\vec{\mathbf{r}}' = \vec{\mathbf{r}}$.

The situation is different for the induction magnetic field, as current sources must necessarily produce a closed-loop magnetic field. If, on the other hand, the magnetic field originates from a point magnetic moment, the loop would disappear at any point and, in this case, it would be possible to express $\vec{\mathbf{B}}$ as the gradient of a scalar potential. As there is no equivalent for a conductor in magnetostatics, this problem will be examined later.

EXAMPLE 1.1. (Point charge within a hollow spherical conductor).– Consider a sphere of radius R and center O. Suppose a point charge q is placed at point O. In addition, the potential is $V = V_0 \cos\theta$ on the surface of the sphere. Let us determine the electric potential inside and outside of the sphere. The potential is given by the Poisson equation:

$$\Delta V = \begin{cases} -\frac{q}{\varepsilon_0}\delta\left(\vec{\mathbf{r}}\right) & r \leq R \\ 0 & r > R \end{cases} \qquad [1.8]$$

The finite general solutions within and outside the sphere are as follows (taking into account the symmetry of the problem):

$$V(r,\theta) = \begin{cases} \dfrac{q}{4\pi\varepsilon_0 r} + \displaystyle\sum_{n=0}^{\infty} A_n r^n P_n(\cos\theta) & r \leq R \\ \displaystyle\sum_{n=0}^{\infty} \frac{B_n}{r^{n+1}} P_n(\cos\theta) & r > R \end{cases} \qquad [1.9]$$

From the condition of continuity of the potential on the sphere, that is, $V(R^-,\theta) = V(R^+,\theta)$, we obtain $A_0 = -\frac{q}{4\pi\varepsilon_0 R}$, $A_1 = \frac{V_0}{R}$, $B_0 = 0$, $A_n = B_n = 0$ for $n \geq 2$, and, consequently:

$$V(r,\theta) = \begin{cases} \dfrac{q}{4\pi\varepsilon_0 r} - \dfrac{q}{4\pi\varepsilon_0 R} + \dfrac{V_0 \cos\theta}{R} r & r \leq R \\ \dfrac{V_0 R^2}{r^2} \cos\theta & r > R \end{cases} \qquad [1.10]$$

EXAMPLE 1.2. (Point charge inside an infinite square tube maintained at zero potential).– Let us study the solutions to the Poisson equation in the case of a point charge located within a hollow infinite tube grounded to the earth. Consider a very long conductive tube with a square cross section and sides D, as shown in Figure 1.1(a). A point charge Q is suspended in the center of the square section. We wish to calculate the electric potential within the tube in the form of an infinite series. In this case, the Poisson equation and the boundary conditions are written as follows:

$$\begin{aligned} &\Delta V(x,y,z) = -\frac{Q}{\varepsilon_0}\delta(x)\,\delta(y)\,\delta(z) \\ &V\left(\pm\frac{D}{2},y,z\right) = V\left(x,\pm\frac{D}{2},z\right) = 0 \end{aligned} \qquad [1.11]$$

Through the Fourier transform $\mathsf{V}(x,y,k) = \int_{-\infty}^{\infty} V(x,y,z)\, e^{ikz} dz$, equation [1.11] becomes:

$$\begin{aligned} &\left[\frac{\partial^2}{\partial x^2} + \frac{\partial^2}{\partial y^2} - k^2\right]\mathsf{V}(x,y,k) = -\frac{Q}{\varepsilon_0}\delta(x)\,\delta(y) \\ &\mathsf{V}\left(\pm\frac{D}{2},y,k\right) = \mathsf{V}\left(x,\pm\frac{D}{2},k\right) = 0 \end{aligned} \qquad [1.12]$$

Let $F(\Omega)$ be the space of functions that cancel for $x = \pm\frac{D}{2}$ and $y = \pm\frac{D}{2}$. This functional space admits the basis $\left\{\frac{2}{D}\cos\frac{(2m+1)\pi x}{D}\cos\frac{(2m'+1)\pi y}{D}, \frac{2}{D}\cos\frac{(2m+1)\pi x}{D}\sin\frac{2n'\pi y}{D}, \frac{2}{D}\sin\frac{2n\pi x}{D}\cos\frac{(2m'+1)\pi y}{D}, \frac{2}{D}\sin\frac{2n\pi x}{D}\sin\frac{2n'\pi y}{D}\right\}$, where $m, m' \geq 0$ and $n, n' \geq 1$. $\frac{2}{D}$ is the *normalization constant*. In this basis, the term $\delta(x)\,\delta(y)$ can be expanded as follows:

$$\delta(x)\,\delta(y) = \left(\frac{2}{D}\right)^2 \sum_{m,m'=0}^{\infty} \cos\frac{(2m+1)\pi x}{D}\cos\frac{(2m'+1)\pi y}{D} \qquad [1.13]$$

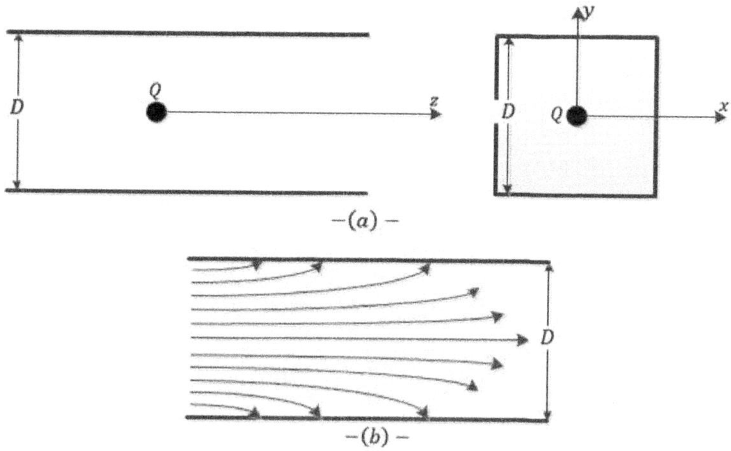

Figure 1.1. *(a) Charge suspended in the center of the square, transversal section of an infinite tube; (b) field lines generated within the tube*

To solve the Poisson equation, we propose a solution in the form of a series $\mathsf{V}(x, y, k) = \sum_{m,m'=0}^{\infty} \mathsf{V}_{mm'}(k) \cos\frac{(2m+1)\pi x}{D} \cos\frac{(2m'+1)\pi y}{D}$. Substituting this expression in [1.12], we obtain:

$$\mathsf{V}_{mm'}(k) = \frac{4Q}{\varepsilon_0 D^2 \left[k^2 + \left(\frac{(2m+1)\pi}{D}\right)^2 + \left(\frac{(2m'+1)\pi}{D}\right)^2\right]} \quad [1.14]$$

and, consequently:

$$\mathsf{V}(x, y, k) = \frac{4Q}{\varepsilon_0 D^2} \sum_{m,m'=0}^{\infty} \frac{\cos\frac{(2m+1)\pi x}{D} \cos\frac{(2m'+1)\pi y}{D}}{k^2 + \left(\frac{(2m+1)\pi}{D}\right)^2 + \left(\frac{(2m'+1)\pi}{D}\right)^2} \quad [1.15]$$

To return to the function V, let us use the inverse Fourier transform: $V(x, y, z) = \frac{1}{2\pi}\int_{-\infty}^{\infty} \mathsf{V}(x, y, k)\, e^{-ikz} dk$ and the integral formula $\int_{-\infty}^{\infty} \frac{e^{ikz}}{k^2+\lambda^2} dk = \frac{\pi}{\lambda} e^{-\lambda|z|}$, where $\lambda > 0$. We therefore obtain:

$$V(x, y, z) = \frac{2Q}{\pi\varepsilon_0 D} \sum_{m,m'=0}^{\infty} \frac{\cos\frac{(2m+1)\pi x}{D} \cos\frac{(2m'+1)\pi y}{D}}{\sqrt{(2m+1)^2 + (2m'+1)^2}} \times e^{-\frac{\sqrt{(2m+1)^2+(2m'+1)^2}\,\pi|z|}{D}} \quad [1.16]$$

For points far from the point charge, we keep the term with $m = m' = 0$, as an approximation, in the expression of the potential, that is:

$$V(x, y, z) = \frac{\sqrt{2}Q}{\pi\varepsilon_0 D} \cos\frac{\pi x}{D} \cos\frac{\pi y}{D} e^{-\frac{\sqrt{2}\pi|z|}{D}} \tag{1.17}$$

In the region $z \gg 0$, the electric field has approximately the following values:

$$\begin{aligned} E_x &= -\frac{\partial V}{\partial x} = \frac{\sqrt{2}Q}{\varepsilon_0 D^2} \sin\frac{\pi x}{D} \cos\frac{\pi y}{D} e^{-\frac{\sqrt{2}\pi z}{D}} \\ E_y &= -\frac{\partial V}{\partial y} = \frac{\sqrt{2}Q}{\varepsilon_0 D^2} \cos\frac{\pi x}{D} \sin\frac{\pi y}{D} e^{-\frac{\sqrt{2}\pi z}{D}} \\ E_z &= -\frac{\partial V}{\partial z} = \frac{2Q}{\varepsilon_0 D^2} \cos\frac{\pi x}{D} \cos\frac{\pi y}{D} e^{-\frac{\sqrt{2}\pi z}{D}} \end{aligned} \tag{1.18}$$

The field lines are described by equations $y = x$ and $z = \frac{\sqrt{2}D}{\pi} \ln\left|\sin\left(\frac{\pi x}{D}\right)\right| + u$, where u is a constant. They are depicted in Figure 1.1(b). ■

Generally speaking, the direct solution of the Poisson equation is rather complicated, so we have to introduce new methods to calculate the potential V. We will look at solutions to the Poisson equation [1.1] using two methods, namely, the image method and Green's function method. Let us formulate the problem for a point charge, q, placed near a conductor. The presence of charge induces a charge density $\rho\left(\vec{\mathbf{r}}'\right)$ on the conductor surface (the readjusted charge density on the conductor surface). Consequently, the resulting potential at any point $\vec{\mathbf{r}}$ is given by:

$$V\left(\vec{\mathbf{r}}\right) = \frac{q}{4\pi\varepsilon_0\left|\vec{\mathbf{r}} - \vec{\mathbf{r}}_q\right|} + \frac{1}{4\pi\varepsilon_0}\int\frac{\sigma\left(\vec{\mathbf{r}}'\right)dS'}{\left|\vec{\mathbf{r}} - \vec{\mathbf{r}}'\right|} \tag{1.19}$$

The problem is then to determine the charge density induced at the surface of the conductor by the introduction of known charges or by the presence of another conductor in the vicinity. In this case, it is easy to solve the Poisson equation [1.1] and then determine the charge density induced by the field on the conductor's surface.

1.2. The image charge method

1.2.1. *General properties*

The *image method*, also known as the mirror charge method, is a basic tool used to solve electrostatic problems when charges are in the presence of one or more conductors. When a conductor is close to a charge, a charge distribution is induced on the conductor's surface such that the electric field inside the equipotential surface is canceled out. This charge distribution in turn contributes to the external electric field, with the boundary condition that the conductor is an equipotential surface (i.e. that the field ends perpendicular to its surface). These are the conditions of the Dirichlet and Neumann problems described above. In general, it is difficult to obtain a solution because the charge distribution on the surface cannot be known as long as the field is unknown, so we use boundary conditions or, more precisely, the discontinuity equation for the field normal to the surface.

Note that for some simple geometries, the electric field or electric potential can be found by replacing the conducting surface with equivalent charges located inside the conductor. These are called *image charges*. Once the images are known, the problem can be solved as if the conductor were not present. The charge–conductor system is replaced by the original (real) charge plus one or more imaginary charges (images). The validity of the image method is based on a corollary of the *uniqueness theorem*, which states that the electric potential in a volume τ is univocally determined if the charge density in the entire region and the value of the electric potential at the boundaries are specified (i.e. the Dirichlet or Neumman boundary conditions). The resultant potential, at any point $\vec{\mathbf{r}}$, is then given by:

$$V\left(\vec{\mathbf{r}}\right) = \frac{1}{4\pi\varepsilon_0}\frac{q}{\left|\vec{\mathbf{r}} - \vec{\mathbf{r}}_q\right|} + \frac{1}{4\pi\varepsilon_0}\sum_i \frac{Q_i}{\left|\vec{\mathbf{r}} - \vec{\mathbf{r}}_{Q_i}\right|} \qquad [1.20]$$

where Q_i are the image charges.

1.2.2. *Point charge in proximity to a conducting plane*

A charge q is placed at a distance a from a grounded conducting plane (i.e. the plane is at zero potential). It can be shown that the potential created in the half-space on the charge side is the same as that resulting from the presence of a charge $-q$ positioned symmetrically with respect to the plane. In the reference frame $Oxyz$, we assume a conducting plane that coincides with the Oxy plane, that is, with equation $z = 0$, and a point charge q located at position $A\,(0, 0, a)$. The half-space on the charge side therefore satisfies the equation $z \geq 0$. The position B of the charge $-q$, placed symmetrically with respect to the conducting plane, is $B\,(0, 0, -a)$. Since the charge q is very close to the conductive plane, it can be approximated as an infinite conductive plane (even along $z < 0$), as shown in Figure 1.2. The potential, at a point $M\,(x, y, z)$, created by the two charges (the conducting plane is removed), is:

$$V\,(M) = \frac{q}{4\pi\varepsilon_0}\left(\frac{1}{\sqrt{x^2 + y^2 + (z-a)^2}} - \frac{1}{\sqrt{x^2 + y^2 + (z+a)^2}}\right) \qquad [1.21]$$

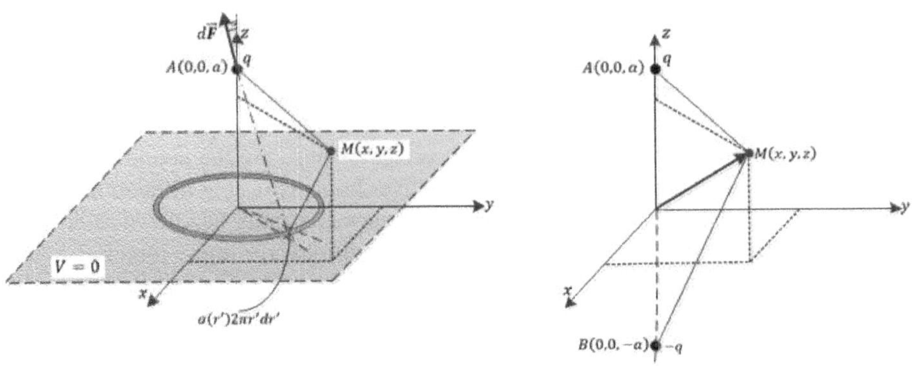

Figure 1.2. *Image method for a point charge q located in proximity to an infinite conducting plane in the Oxy plane. For a color version of this figure, see www.iste.co.uk/abdesselam/electrodynamics2.zip*

It can be easily verified that $V(x, y, z)$ cancels out at $z = 0$, that is, at any point on the median plane of AB. This plane is therefore an equipotential of $V = 0$. Consequently, if we re-establish this equipotential surface (bring the conductor back and ground it) and remove the $-q$ image, then nothing changes above the plane in terms of electric field values and potential values, since the boundary conditions of the conducting planes are the same.

The electric field is zero for $z < 0$. For $z > 0$, the field induced by potential [1.21] is given by:

$$\vec{\mathbf{E}}(M) = \frac{q}{4\pi\varepsilon_0}\left(\frac{x\,\vec{\mathbf{i}} + y\,\vec{\mathbf{j}} + (z-a)\,\vec{\mathbf{k}}}{\left(x^2 + y^2 + (z-a)^2\right)^{3/2}} - \frac{x\,\vec{\mathbf{i}} + y\,\vec{\mathbf{j}} + (z+a)\,\vec{\mathbf{k}}}{\left(x^2 + y^2 + (z+a)^2\right)^{3/2}}\right) \quad [1.22]$$

If $\left|\vec{\mathbf{AB}}\right|$ is small compared to the lengths $\left|\vec{\mathbf{AM}}\right|$ and $\left|\vec{\mathbf{BM}}\right|$, the electric field is approximately that of a dipole. The total electric field, produced at the level of the Oxy plane by the charge q and its image $-q$, is:

$$\vec{\mathbf{E}}(x, y, 0) = \frac{-2qa\,\vec{\mathbf{k}}}{4\pi\varepsilon_0\left(x^2 + y^2 + a^2\right)^{3/2}} \quad [1.23]$$

and, consequently, the charge density induced on the conductor's surface is obtained from the formula:

$$\sigma(x, y) = \varepsilon_0 E_n(x, y, 0) = \varepsilon_0\,\vec{\mathbf{E}}(x, y, 0)\cdot\vec{\mathbf{k}} = -\frac{2qa}{4\pi\left(r^2 + a^2\right)^{3/2}} \quad [1.24]$$

The charge Q_t distributed on the plane is obtained simply by calculating the integral $Q_t = \int_S \sigma dS$. For the elementary surface dS, we can take the area between the radius circles r and $r + dr$ (i.e. $2\pi r dr$), carrying the charge $2\pi\sigma r dr$. It thus follows that:

$$Q_t = -\frac{2qa}{4\pi}\int_{r=0}^{\infty}\frac{2\pi r dr}{\left(r^2 + a^2\right)^{3/2}} = -q \qquad [1.25]$$

Since $Q_t = -q$, we conclude that there is a total influence between the charge q and the conductor. Let us now calculate the force exerted by the plane on the point charge. The ring mentioned above, dS, exerts the following force on the charge q: $dF_z = \frac{q}{4\pi\varepsilon_0}\frac{\sigma dS}{(r^2+a^2)}\sin\alpha$ where $\sin\alpha = \frac{a}{\sqrt{r^2+a^2}}$. We therefore obtain $F_z = -\frac{a^2q^2}{4\pi\varepsilon_0}\int_0^{\infty}\frac{rdr}{(r^2+a^2)^3} = -\frac{q^2}{4\pi\varepsilon_0}\frac{1}{(2a)^2}$. As a result:

$$\overrightarrow{\mathbf{F}} = -\frac{q^2}{4\pi\varepsilon_0}\frac{1}{(2a)^2}\overrightarrow{\mathbf{k}} \qquad [1.26]$$

This is the same force as that exerted by a charge $-q$ on another charge $+q$, separated by a distance $2a$.

EXAMPLE 1.3. (A point charge in proximity to two intersecting conducting planes).– Now consider a charge q placed close to two intersecting grounded conductor planes. The charge q not only has images behind each conducting plane, but each image, in turn, has an image on the other conductor, etc., much like an image in optics. Consequently, when two plates meet at right angles, an image of the charge $-q$ appears behind each plate, and an image q lies on the diagonal, relative to the point of intersection of the two planes, as shown in Figure 1.3(a). If the two conducting planes lie along Oxz and Oyz, then the two planes will exert the following force on a charge q located at point $(a, b, 0)$:

$$\begin{aligned}\overrightarrow{\mathbf{F}} &= \frac{1}{4\pi\varepsilon_0}\left[\frac{(-q^2)\,2a\,\overrightarrow{\mathbf{i}}}{(4a^2)^{3/2}} + \frac{q^2\left(2a\,\overrightarrow{\mathbf{i}} + 2b\,\overrightarrow{\mathbf{j}}\right)}{(4a^2 + 4b^2)^{3/2}} + \frac{(-q^2)\,2b\,\overrightarrow{\mathbf{j}}}{(4b^2)^{3/2}}\right] \\ &= \frac{q^2}{16\pi\varepsilon_0}\left[\frac{a\,\overrightarrow{\mathbf{i}} + b\,\overrightarrow{\mathbf{j}}}{(a^2 + b^2)^{3/2}} - \frac{\overrightarrow{\mathbf{i}}}{a^2} - \frac{\overrightarrow{\mathbf{j}}}{b^2}\right]\end{aligned} \qquad [1.27]$$

In the region of the real charge, the potential is: $V(x,y) = \frac{q}{4\pi\varepsilon_0}\left[\frac{1}{\sqrt{(x-a)^2+(y-b)^2}} - \frac{1}{\sqrt{(x+a)^2+(y-b)^2}} - \frac{1}{\sqrt{(x-a)^2+(y+b)^2}} + \frac{1}{\sqrt{(x+a)^2+(y+b)^2}}\right]$. For $a = b$, the equipotential surfaces $V(x,y) = \frac{qv}{4\pi\varepsilon_0}$, where $v = \frac{1}{100}, \frac{1}{10}, \frac{1}{5}, \frac{1}{4}, \frac{1}{2},$

are shown in Figure 1.4. When the planes cross at $\frac{\pi}{3}$, five images with alternating signs appear, as in Figure 1.3(b). For $\frac{\pi}{4}$ and $\frac{\pi}{6}$, we obtain, respectively, 7 and 11 images, as shown in Figures 1.3(c) and (d). It is also clear that when the angle is not a fraction of π, an infinite number of images appear, and consequently the force experienced by the charge q contains an infinite number of terms.

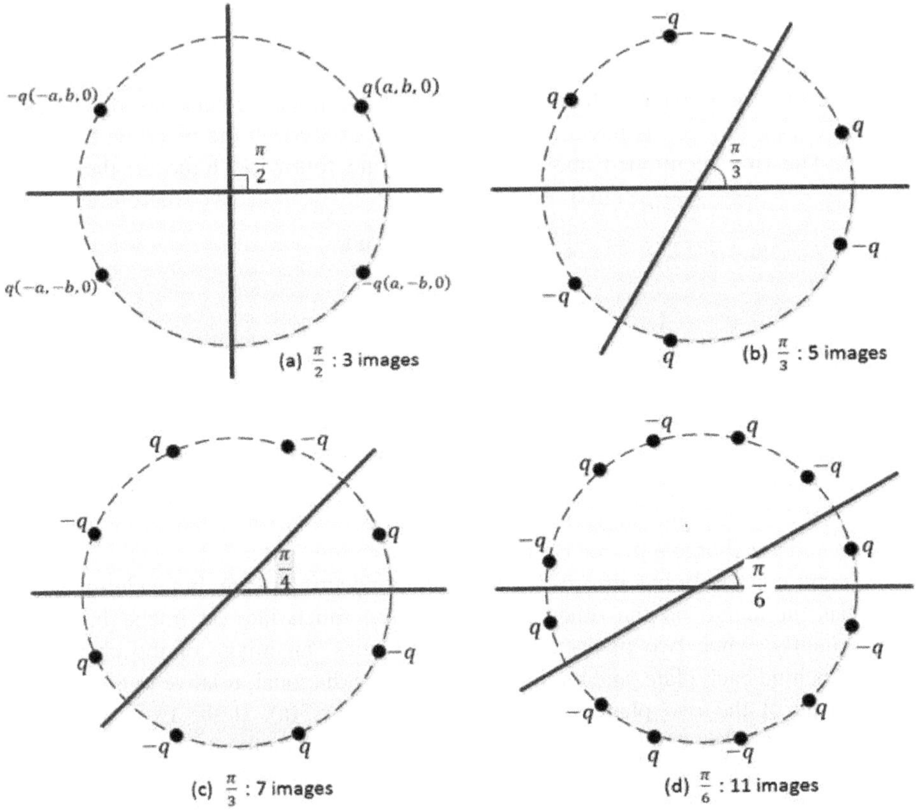

Figure 1.3. *Number of images, depending on the angle of inclination between the two conducting planes (for $\frac{\pi}{N}$, we obtain $2N-1$ images)*

EXAMPLE 1.4. (A point charge between two parallel conducting planes).– Two parallel conducting planes, separated by a distance D, are connected by a wire and maintained at a potential of zero. A point charge is placed between the two planes at an equal distance from them, as shown in Figure 1.5(a). Let us find the expression for the surface charge density induced in the lower plane using the image method. The positions of the images are shown in Figure 1.5(b). The resulting potential between the conducting planes, that is, for $-\frac{D}{2} \leq z \leq \frac{D}{2}$, is as follows:

$$V(x,y,z) = \sum_{n=-\infty}^{\infty} \frac{(-1)^n Q}{4\pi\varepsilon_0 \sqrt{x^2+y^2+(z-nD)^2}} \quad [1.28]$$

and consequently:

$$\vec{\mathbf{E}}(x,y,z) = \sum_{n=-\infty}^{\infty} \frac{(-1)^n Q\left[x\vec{\mathbf{i}} + y\vec{\mathbf{j}} + (z-nD)\vec{\mathbf{k}}\right]}{4\pi\varepsilon_0\left[x^2+y^2+(z-nD)^2\right]^{3/2}} \quad [1.29]$$

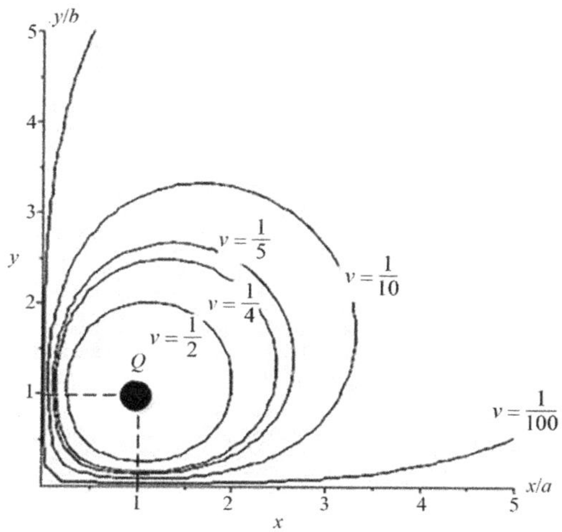

Figure 1.4. *Equipotentials generated by a real charge in proximity to two conducting planes (one perpendicular to the other and connected to earth)*

The charge density induced at point A on plane (1) is obtained from the expression for the normal component of the field, $\vec{\mathbf{E}}$, to the plane (the field to the right of plane (1) is, of course, zero):

$$\sigma(x,y) = -\varepsilon_0 \left.E_z\right|_{z=\frac{D}{2}} = \frac{QD}{8\pi}\sum_{n=-\infty}^{\infty}\frac{(-1)^n(2n-1)}{\left[x^2+y^2+\frac{1}{4}(2n-1)^2 D^2\right]^{3/2}} \quad [1.30]$$

To obtain the total induced charge on plane (1), let us integrate the previous experience over the total surface of the plane:

$$Q_t = \int_{-\infty}^{\infty}\int_{-\infty}^{\infty} \sigma(x,y)\,dxdy = \frac{QD}{8\pi}\sum_{n=-\infty}^{\infty}(-1)^n(2n-1)$$
$$\times \int_0^{\infty} \frac{2\pi r dr}{\left[r^2 + \frac{1}{4}(2n-1)^2 D^2\right]^{3/2}} \qquad [1.31]$$
$$= \frac{Q}{2}\sum_{n=-\infty}^{\infty} \frac{(-1)^n(2n-1)}{|2n-1|} = \frac{Q}{2}(\cdots -1+1-1-1+1+\cdots)$$

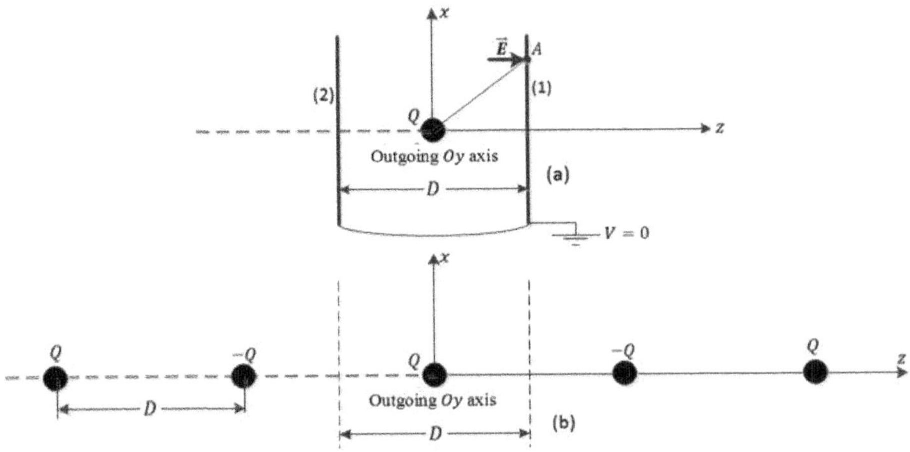

Figure 1.5. *Point charge Q placed halfway between two (infinite) conducting planes grounded to the earth*

This series is divergent and therefore does not yield any result. In a slightly artificial (but accurate) manner, it is seen that $\sum_{n=-\infty}^{\infty}\frac{(-1)^n(2n-1)}{|2n-1|} = \sum_{n=-\infty}^{-1}(-1)^{n+1} - 1 + \sum_{n=1}^{\infty}(-1)^n = \sum_{n=1}^{\infty}(-1)^{n-1} - 1 + \sum_{n=1}^{\infty}(-1)^n = -1$, which gives $Q_t = -\frac{Q}{2}$. This result is expected, as only half of the field lines leaving the charge Q reach plane (1). This explains why the induced charge is indeed $-\frac{Q}{2}$. To solve this problem and ensure the convergence of the series, the two planes can be considered to be two concentric spheres with radius R and $R + D$, with R tending to infinity. In this case, we also obtain an infinite series of images. However, the series is convergent as each successive image decreases in magnitude. This result was established by Jackson (1962).

EXAMPLE 1.5. (Two point charges between two parallel conducting planes).– Two parallel conducting planes (connected to the earth) are separated by a distance of $4a$. Two charges, Q and $-Q$, are placed at distances a and $3a$, respectively, from the left-hand plane, as shown in Figure 1.6(a). Let us calculate the electrostatic energy of the system. The potential is determined by the image method. The two planes can be replaced by the images $+Q$ in $\ldots, -9a, -5a, 3a, 7a, \ldots$ and the images $-Q$ in $\ldots, -7a, -3a, 5a, 9a, \ldots$ along the Ox axis, as shown in Figure 1.6(b). The electrostatic energy is:

$$W = \frac{1}{2}QV_+ - \frac{1}{2}QV_- \qquad [1.32]$$

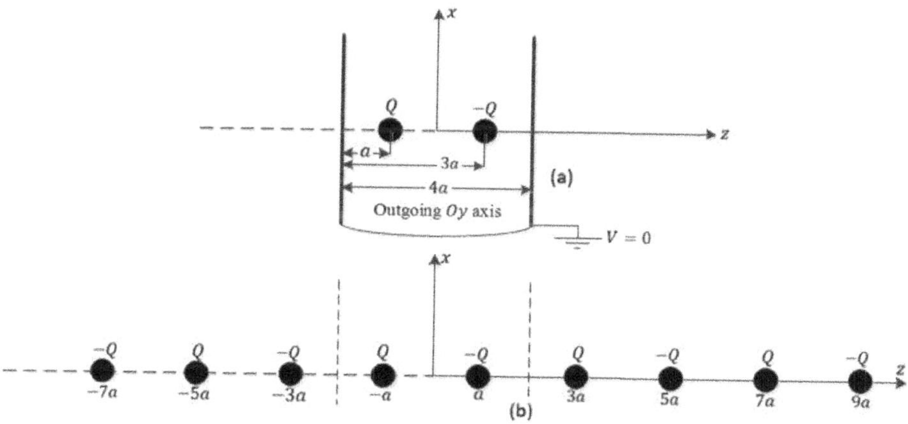

Figure 1.6. *Two point charges, $+Q$ and $-Q$, placed between two conducting planes grounded to the earth*

V_+ and V_- are, respectively, the potentials at the positions of the real charges $+Q$ and $-Q$:

$$\begin{aligned} V_+ &= \frac{1}{4\pi\varepsilon_0}\left[-\frac{2Q}{2a} + \frac{2Q}{4a} - \cdots\right] = -\frac{Q}{4\pi\varepsilon_0 a}\sum_{n=1}^{\infty}\frac{(-1)^{n+1}}{n} = -\frac{Q\ln 2}{4\pi\varepsilon_0 a} \\ V_- &= \frac{1}{4\pi\varepsilon_0}\left[\frac{2Q}{2a} - \frac{2Q}{4a} + \cdots\right] = \frac{Q}{4\pi\varepsilon_0 a}\sum_{n=1}^{\infty}\frac{(-1)^{n+1}}{n} = \frac{Q\ln 2}{4\pi\varepsilon_0 a} \end{aligned} \qquad [1.33]$$

We thus obtain:

$$W = -\frac{Q^2\ln 2}{4\pi\varepsilon_0 a} \qquad [1.34]$$

Therefore, the energy required to move the two charges an infinite distance away from the planes and from each other will be $-W$. Note that the force acting on $+Q$ is just the force exerted by the field created by the other real charge, $-Q$, and the fields of the images produced by the system. Similarly, the force acting on $-Q$ is also zero. Calculate the force exerted on the conductive plane on the left. This is the result of all of the forces acting on the images in the left-hand plane (i.e. the images to the left of the left-hand plane) due to the real charge $+Q$, $-Q$ and all of the images in the right-hand plane (i.e. the images to the right of the right-hand plane). Let us first consider the force F_1 acting on the images in the left-hand plane due to the real charge $+Q$:

$$F_1 = \frac{Q^2}{4\pi\varepsilon_0 (2a)^2} - \frac{Q^2}{4\pi\varepsilon_0 (4a)^2} + \cdots = \frac{Q^2}{16\pi\varepsilon_0 a^2} \sum_{n=1}^{\infty} \frac{(-1)^{n-1}}{n^2} \quad [1.35]$$

This force is directed in the positive direction of the Ox axis. Next, let us calculate the force F_2 between the real charge $-Q$ and the images of the left-hand side plane:

$$F_2 = -\frac{Q^2}{4\pi\varepsilon_0 (4a)^2} + \frac{Q^2}{4\pi\varepsilon_0 (6a)^2} - \cdots = \frac{Q^2}{16\pi\varepsilon_0 a^2} \sum_{n=2}^{\infty} \frac{(-1)^{n-1}}{n^2} \quad [1.36]$$

Finally, the force F_3 exerted upon the images of the left-hand side plane by the images of the right-hand side plane is equal to:

$$\begin{aligned} F_3 &= \frac{Q^2}{4\pi\varepsilon_0 (6a)^2} - \frac{Q^2}{4\pi\varepsilon_0 (8a)^2} - \cdots - \frac{Q^2}{4\pi\varepsilon_0 (8a)^2} + \frac{Q^2}{4\pi\varepsilon_0 (10a)^2} - \cdots \\ &+ \frac{Q^2}{4\pi\varepsilon_0 (10a)^2} - \frac{Q^2}{4\pi\varepsilon_0 (12a)^2} + \cdots = \frac{Q^2}{16\pi\varepsilon_0 a^2} \sum_{m=3}^{\infty} \sum_{n=m}^{\infty} \frac{(-1)^{n-1}}{n^2} \end{aligned} \quad [1.37]$$

Therefore, the total force exerted on the left-hand side plane is equal to:

$$\begin{aligned} F = F_1 + F_2 + F_3 &= \frac{Q^2}{16\pi\varepsilon_0 a^2} \left[\sum_{n=1}^{\infty} \frac{(-1)^{n-1}}{n^2} + \sum_{n=2}^{\infty} \frac{(-1)^{n-1}}{n^2} \right. \\ &\left. + \sum_{m=3}^{\infty} \sum_{n=m}^{\infty} \frac{(-1)^{n-1}}{n^2} \right] = \frac{Q^2}{16\pi\varepsilon_0 a^2} \left(1 - \frac{2}{2^2} + \frac{3}{3^2} - \frac{4}{4^2} + \cdots \right) \\ &= \frac{Q^2}{16\pi\varepsilon_0 a^2} \sum_{n=1}^{\infty} \frac{(-1)^{n-1}}{n} = \frac{Q^2}{16\pi\varepsilon_0 a^2} \ln 2 \end{aligned} \quad [1.38]$$

This force is directed to the right. In the same way, it can be shown that the force exerted on the right-hand side plane is also equal to $F = -\frac{Q^2}{16\pi\varepsilon_0 a^2} \ln 2$ (its direction being toward the left). Now, let us look at the charges induced in the conducting planes. Note that the potential over the $x = 0$ plane is zero. This means that only half of the field lines produced by the $-Q$ charge reach the right-hand plane, while none of the field lines produced by the $+Q$ charge reach this plane. Consequently, the total charge induced on the right-hand plane is $+\frac{Q}{2}$, and similarly the total charge induced on the left-hand plane is $-\frac{Q}{2}$. To verify this, let us give the expressions for the electric potential and electric field in the space between the two planes:

$$V\left(\vec{\mathbf{r}}\right) = \sum_{n=-\infty}^{\infty} \frac{(-1)^{n+1} Q}{4\pi\varepsilon_0 \sqrt{x^2 + y^2 + (z - (2n+1)\,a)^2}}$$
$$\vec{\mathbf{E}}\left(\vec{\mathbf{r}}\right) = \sum_{n=-\infty}^{\infty} \frac{(-1)^{n+1} Q \left[x\,\vec{\mathbf{i}} + y\,\vec{\mathbf{j}} + (z - (2n+1)\,a)\,\vec{\mathbf{k}}\right]}{4\pi\varepsilon_0 \left[x^2 + y^2 + (z - (2n+1)\,a)^2\right]^{3/2}} \qquad [1.39]$$

The surface charge density:

$$\sigma(x, y) = -\varepsilon_0 E_z(x, y, 2a) = \frac{Qa}{4\pi} \sum_{n=-\infty}^{\infty} \frac{(-1)^{n+1}(2n-1)}{\left[x^2 + y^2 + (2n-1)^2 a^2\right]^{3/2}} \qquad [1.40]$$

will therefore appear on the right-hand side plane. Consequently, the total charge induced on the plane is:

$$\begin{aligned} Q_t &= \frac{Qa}{4\pi} \sum_{n=-\infty}^{\infty} (-1)^{n+1}(2n-1) \int_0^{\infty} \frac{2\pi r dr}{\left[r^2 + (2n-1)^2 a^2\right]^{3/2}} \\ &= \frac{Q}{2} \sum_{n=-\infty}^{\infty} \frac{(-1)^{n+1}(2n-1)}{|2n-1|} = \frac{Q}{2}(\cdots + 1 - 1 + \mathbf{1} + 1 - 1 + \cdots) \end{aligned} \qquad [1.41]$$

This series is divergent and gives no results. However, a $\frac{Q}{2}$ factor appears in the middle of the series. The problem is also circumvented by assimilating the planes to spheres of radius R and $R + 4a$ and making $R \longrightarrow \infty$.

1.2.3. *Point charge in proximity to a conducting sphere*

Let us find the charge q' that will imitate the surface charge density over a conducting sphere of radius R and center O, induced by a point charge q located at a distance h from the point O, as shown in Figure 1.7. The potential resulting from the two charges must be constant on the surface of the sphere. We will take it to be zero, but it is always possible to take it to be non-zero by placing another charge, q'', at the center of the sphere. By symmetry, if q lies on the Oz axis, then q' must also lie on the same axis. We place q' at a distance b from the center O. Consequently, the potential on the surface of the sphere is given by:

$$V(R,\theta) = \frac{1}{4\pi\varepsilon_0}\left(\frac{q}{r_1}+\frac{q'}{r_2}\right) = \frac{1}{4\pi\varepsilon_0}\left(\frac{q}{\sqrt{h^2+R^2-2Rh\cos\theta}} + \frac{q'}{\sqrt{b^2+R^2-2Rb\cos\theta}}\right) = 0 \qquad [1.42]$$

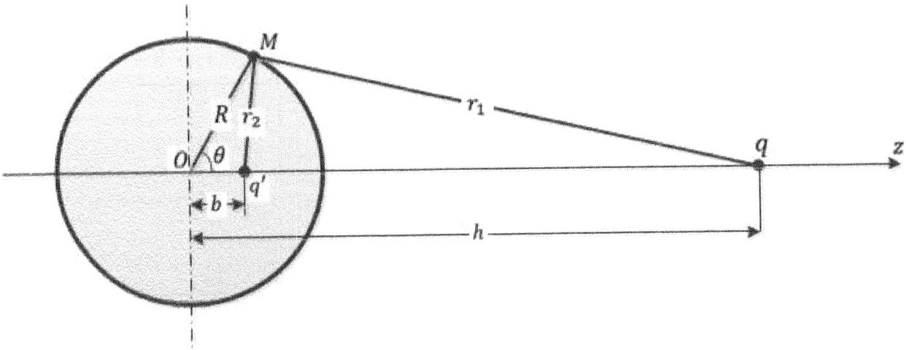

Figure 1.7. *Point charge in the neighborhood of a spherical conductor*

The distance b is chosen so that $q^2\left(b^2+R^2-2Rb\cos\theta\right) = q'^2(h^2+R^2-2Rh\cos\theta)$. Since this equation must be valid for any θ, it can be derived that:

$$q^2\left(b^2+R^2\right) = q'^2\left(h^2+R^2\right) \qquad [1.43]$$

$$Rbq^2 = Rhq'^2 \qquad [1.44]$$

Equation [1.44] immediately yields:

$$\frac{q'^2}{q^2} = \frac{b}{h} \qquad [1.45]$$

Substituting [1.45] into [1.43], we find $b^2+R^2 = \frac{b}{h}\left(h^2+R^2\right)$. This is a second-order equation in b that admits two different solutions $b = h$ and $b = \frac{R^2}{h}$. We will clearly choose:

$$b = \frac{R^2}{h} \qquad [1.46]$$

and therefore:

$$q' = -\frac{R}{h}q \tag{1.47}$$

The other solution corresponds to a trivial situation. It is important to note that when the charge q is inside the sphere (i.e. $h < R$) and we search for the potential inside the sphere, the image q' passes outside the sphere (i.e. $b = \frac{R^2}{h} > R$).

Let us generalize this result. A charge q located at $\overrightarrow{\mathbf{r}}'$ has an image $q' = -\frac{R}{r'}q$ located at $\overrightarrow{\mathbf{r}}'_2 = \frac{R^2}{r'^2}\overrightarrow{\mathbf{r}}'$. Consequently, the potential at any point $\overrightarrow{\mathbf{r}}$ is:

$$V\left(\overrightarrow{\mathbf{r}}\right) = \frac{q}{4\pi\varepsilon_0}\left(\frac{1}{\left|\overrightarrow{\mathbf{r}} - \overrightarrow{\mathbf{r}}'\right|} - \frac{\frac{R}{r'}}{\left|\overrightarrow{\mathbf{r}} - \frac{R^2}{r'^2}\overrightarrow{\mathbf{r}}'\right|}\right) \tag{1.48}$$

This expression can be easily extended to a continuous distribution (of charges) $\rho\left(\overrightarrow{\mathbf{r}}'\right)$ located in the neighborhood of a conducting sphere:

$$V\left(\overrightarrow{\mathbf{r}}\right) = \frac{1}{4\pi\varepsilon_0}\int_{\tau'} \rho\left(\overrightarrow{\mathbf{r}}'\right)\left(\frac{1}{\left|\overrightarrow{\mathbf{r}} - \overrightarrow{\mathbf{r}}'\right|} - \frac{\frac{R}{r'}}{\left|\overrightarrow{\mathbf{r}} - \frac{R^2}{r'^2}\overrightarrow{\mathbf{r}}'\right|}\right) d\tau' \tag{1.49}$$

for $r > R$. We can see that this potential cancels out for $r = R$.

The force perceived by the sphere (grounded) is then just the force felt by the image q' due to the field of q:

$$F_z = \frac{qq'}{4\pi\varepsilon_0 (h-b)^2} = -\frac{q^2 R}{4\pi\varepsilon_0 h (h-b)^2} = -\frac{q^2 Rh}{4\pi\varepsilon_0 \left(h^2 - R^2\right)^2} \tag{1.50}$$

The electric field outside the sphere is given by:

$$\begin{aligned}\overrightarrow{\mathbf{E}} = \frac{1}{4\pi\varepsilon_0}\bigg(&\frac{q}{s_1^3}\left[(r - h\cos\theta)\,\overrightarrow{\mathbf{e}}_r + h\sin\theta\,\overrightarrow{\mathbf{e}}_\theta\right] \\ &+\frac{q'}{s_2^3}\left[(r - b\cos\theta)\,\overrightarrow{\mathbf{e}}_r + b\sin\theta\,\overrightarrow{\mathbf{e}}_\theta\right]\bigg)\end{aligned} \tag{1.51}$$

where $s_1 = \sqrt{r^2 + h^2 - 2rh\cos\theta}$ and $s_2 = \sqrt{r^2 + b^2 - 2rb\cos\theta}$. On the sphere, that is, when $r_2 = \frac{R}{h}r_1$, the charge distribution is found from the normal field discontinuity equation:

$$\sigma(R) = \varepsilon_0 E_r(R) = -\frac{q\left(h^2 - R^2\right)}{4\pi R\left[R^2 + h^2 - 2Rh\cos\theta\right]^{3/2}} \tag{1.52}$$

The total charge on the sphere:

$$Q_t = \int_0^{\pi} \sigma 2\pi R^2 \sin\theta d\theta = -\frac{q\left(h^2 - R^2\right)}{4h} \int_0^{\pi} \frac{\sin\theta d\theta}{\left[R^2 + h^2 - 2Rh\cos\theta\right]^{3/2}} \quad [1.53]$$

is evaluated by introducing the new variable $u = R^2 + h^2 - 2Rh\cos\theta$:

$$Q_t = -\frac{q\left(h^2 - R^2\right)}{4h} \int_{(h-R)^2}^{(h+R)^2} \frac{du}{u^{3/2}} = -\frac{qR}{h} \quad [1.54]$$

It is precisely equal to the charge of the image q'.

EXAMPLE 1.6. (Sphere with constant charge Q_0).–

1) When the point charge q is outside a conducting sphere $(h > R)$ carrying a constant total charge Q_0, the induced charge is always $q' = -\frac{R}{h}q$. Since the total charge on the sphere is Q_0, we need to add another image that keeps the sphere on an equipotential surface. This new image has a value of $Q_0 + \frac{R}{h}q$ and must be placed in the center of the sphere, as shown in Figure 1.8. The charge q and its image $q' = -\frac{R}{h}q$ place the sphere at zero potential. The new additional image at the center of the sphere raises the sphere's potential to:

$$V = \frac{Q_0 + \frac{R}{h}q}{4\pi\varepsilon_0 R} \quad [1.55]$$

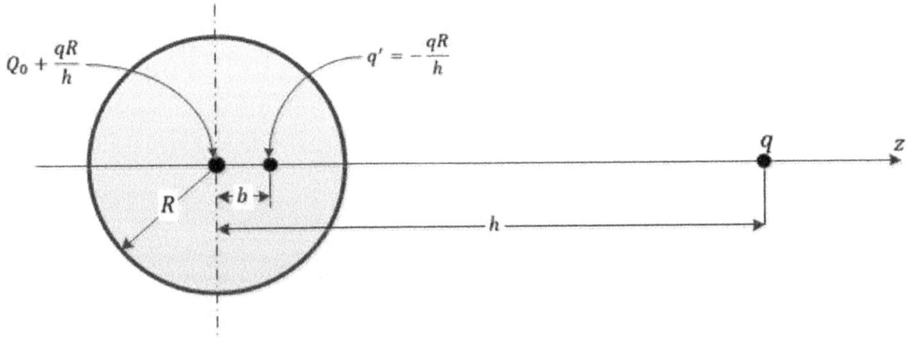

Figure 1.8. *Charge q in proximity to a conducting sphere of charge Q_0*

The force on the sphere is now due to the charge field q acting on the two images:

$$
\begin{aligned}
F_z &= \frac{q}{4\pi\varepsilon_0}\left[-\frac{qR}{h\,(h-b)^2}+\frac{Q_0+\frac{R}{h}q}{h^2}\right] \\
&= \frac{q}{4\pi\varepsilon_0}\left[-\frac{qRh}{\left(h^2-R^2\right)^2}+\frac{Q_0+\frac{R}{h}q}{h^2}\right]
\end{aligned}
\qquad [1.56]
$$

2) Let us calculate the minimum charge that must be given to a spherical conductor with radius R, which is isolated and influenced by an external point charge $+q$ at a distance $h > R$ from its center, so that the surface charge density over the sphere can be positive everywhere. To do this, let us use Cartesian coordinates with the origin at the center of the sphere and the axis Oz along the line that joins the center of the sphere and the charge q. Clearly, the maximum indicated charge density (which is negative) on the sphere will occur at $(0, 0, R)$. The effect of the conductive spherical surface can be replaced by that of a point charge $-\frac{qR}{h}$ at $\left(0, 0, \frac{R^2}{h}\right)$ and a point charge $\frac{qR}{h}$ at the center of the sphere $(0, 0, 0)$. The electric field at the point $(0, 0, R^+)$ is then:

$$
\begin{aligned}
\vec{\mathbf{E}} &= \frac{1}{4\pi\varepsilon_0}\left[\frac{\frac{qR}{h}}{R^2}-\frac{q}{(h-R)^2}-\frac{\frac{qR}{h}}{\left(R-\frac{R^2}{h}\right)^2}\right]\vec{\mathbf{k}} \\
&= \frac{q}{4\pi\varepsilon_0}\left[\frac{1}{hR}-\frac{1+\frac{h}{R}}{(h-R)^2}\right]\vec{\mathbf{k}}
\end{aligned}
\qquad [1.57]
$$

Consequently, the maximum induced (negative) charge density is given as:

$$
\sigma = \varepsilon_0 E_n = \frac{q}{4\pi}\left[\frac{1}{hR}-\frac{1+\frac{h}{R}}{(h-R)^2}\right] \qquad [1.58]
$$

If a positive charge Q is given to the sphere, it will be distributed uniformly over the spherical surface with a density $\frac{Q}{4\pi R^2}$. For the (total) surface density of charges to be positive everywhere, $\frac{Q}{4\pi R^2} + \sigma \geq 0$ is required, which means that:

$$
Q \geq -\sigma 4\pi R^2 = R^2 q\left[\frac{1+\frac{h}{R}}{(h-R)^2}-\frac{1}{hR}\right] = \frac{R^2\,(3h-R)}{h\,(h-R)^2}q \qquad [1.59]
$$

Note that the field at the point $(0, 0, -R^-)$ is:

$$
\vec{\mathbf{E}} = -\frac{q}{4\pi\varepsilon_0}\left[\frac{1}{hR}+\frac{1-\frac{h}{R}}{(h+R)^2}\right]\vec{\mathbf{k}} \qquad [1.60]
$$

If we replace q by $-q$, the maximum induced negative charge density will occur at $(0,0,-R)$. As before, the positive charge required is:

$$Q \geq -\sigma 4\pi R^2 = -\varepsilon_0 4\pi R^2 \left(-\frac{q}{4\pi\varepsilon_0}\right)\left[\frac{1}{hR} + \frac{1-\frac{h}{R}}{(h+R)^2}\right] = \frac{qR^2(3h+R)}{h(h+R)^2} \qquad [1.61]$$

EXAMPLE 1.7. (Sphere maintained at a constant potential).– If the sphere is held at a constant potential V_0, the image $q' = -\frac{qR}{h}$ at distance $b = \frac{R^2}{h}$ from the center of the sphere always holds the sphere at zero potential. To increase the potential of the sphere to V_0, another image $Q_0 = 4\pi\varepsilon_0 R V_0$ is placed in the center O, as shown in Figure 1.9. To obtain the capacitance of the sphere, let us divide the total charge accumulated inside the sphere by V_0:

$$C = \frac{Q_0 - \frac{qR}{h}}{V_0} = \frac{4\pi\varepsilon_0 R V_0 - \frac{qR}{h}}{V_0} = C_0 - \frac{qR}{hV_0} \qquad [1.62]$$

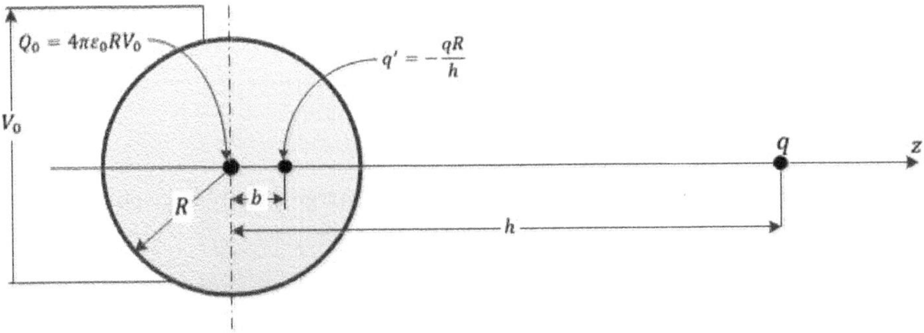

Figure 1.9. *Charge q near a conducting sphere maintained at a constant potential V_0*

where C_0 is the capacity of the isolated sphere. Clearly, if $q > 0$ (respectively, $q < 0$), then $C < C_0$ (respectively, $C > C_0$). The force exerted on the sphere is therefore:

$$F_z = \frac{q}{4\pi\varepsilon_0}\left[-\frac{qR}{h(h-b)^2} + \frac{Q_0}{h^2}\right] = \frac{q}{4\pi\varepsilon_0}\left[-\frac{qRh}{(h^2-R^2)^2} + \frac{Q_0}{h^2}\right] \qquad [1.63]$$

EXAMPLE 1.8. (Capacitance of two conducting spheres in contact).– Let us calculate the capacitance of two identical conducting spheres in direct contact and maintained to a potential V_0, as shown in Figure 1.10. To do this, we first need to find the total charge accumulated on the two spheres. We can use the image method by first placing an image $q_1 = Q_0 = 4\pi\varepsilon_0 RV_0$ in the center of each sphere to bring each surface to potential V_0. Each of these charges will induce an image q_2 in the other sphere at the distance b_2 from the center:

$$\begin{aligned} q_2 &= -\frac{Rq_1}{h} = -\frac{Q_0}{2} \\ b_2 &= \frac{R^2}{h} = \frac{R}{2} \end{aligned} \qquad [1.64]$$

because the distance between the charge q_1 and the center of the other sphere is $h = 2R$. The new charge q_2 (for each sphere) produces another image q_3 in the opposite sphere at distance b_3:

$$\begin{aligned} q_3 &= -\frac{Rq_2}{h - b_2} = \frac{Q_0}{3}, \\ b_3 &= \frac{R^2}{h - b_2} = \frac{2R}{3} \end{aligned} \qquad [1.65]$$

which, in turn, induces a new image q_4 (in the opposite sphere). This operation takes place in both directions. An infinite number of images appear in each sphere. The nth image (charge q_n and distance b_n) is linked to the $(n-1)$th image by the equations:

$$q_n = -\frac{Rq_{n-1}}{h - b_{n-1}} = -\frac{Rq_{n-1}}{2R - b_{n-1}} \qquad [1.66]$$

$$b_n = \frac{R^2}{h - b_{n-1}} = \frac{R^2}{2R - b_{n-1}} \qquad [1.67]$$

The relation [1.66] immediately gives $2R - b_{n-1} = -\frac{q_{n-1}}{q_n}R$ and, therefore, $b_n = \frac{q_n}{q_{n+1}}R + 2R$. Substituting this result into [1.67], we can deduce that $\frac{q_n}{q_{n+1}}R + 2R = -\frac{q_n}{q_{n-1}}R$, and, consequently, $\frac{q_{n+1}}{R} + 2R = \frac{q_{n+1}}{R} + 2R$:

$$\frac{1}{q_{n+1}} + \frac{2}{q_n} + \frac{1}{q_{n-1}} = 0 \qquad [1.68]$$

To solve the difference equation [1.68], let us introduce the reciprocal charges $p_n = \frac{1}{q_n}$. This gives $p_{n+1} + 2p_n + p_{n-1} = 0$. This is a linear and homogeneous differential equation with constant coefficients. To find non-vanishing solutions of this equation, we form the characteristic equation by writing $p_n = \lambda^n$. Thus, we obtain $\lambda^{n+1} + 2\lambda^n + \lambda^{n-1} = 0$, or, again, $\lambda^2 + 2\lambda + 1 = 0$. This equation admits a double root $\lambda = -1$. The solution is therefore of the form $p_n = (A_1 + A_2 n)(-1)^n$, where A_1 and A_2 are arbitrary constants determined from the initial conditions. Since $p_1 = -\frac{1}{q_1} = -\frac{1}{Q_0} = -A_1 - A_2$ and $p_2 = -\frac{1}{q_2} = -\frac{2}{Q_0} = A_1 + 2A_2$, it follows that $A_1 = 0$ and $A_2 = -\frac{1}{Q_0}$. Consequently, the charge on the nth image will be:

$$q_n = \frac{1}{p_n} = \frac{1}{-\frac{(-1)^n n}{Q_0}} = \frac{(-1)^{n-1} Q_0}{n} \qquad [1.69]$$

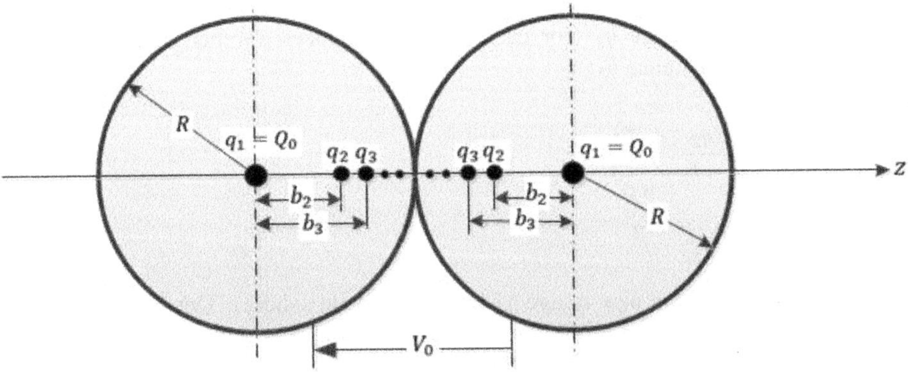

Figure 1.10. *Two identical spheres in direct contact taken to a potential* V_0

The capacitance of the system can therefore be determined by a simple addition of all charges present in the two spheres and dividing the result by V_0:

$$C = \frac{2\sum_{n=1}^{\infty} q_n}{V_0} = \frac{2Q_0}{V_0}\sum_{n=1}^{\infty}\frac{(-1)^{n-1}}{n} = 8\pi\varepsilon_0 R \ln 2 \qquad [1.70]$$

The distances b_n are obtained by substituting [1.69] in [1.66]:

$$b_n = \left[\frac{(-1)^{n-1}(n+1)}{(-1)^n n} + 2\right] R = \left(\frac{n-1}{n}\right) R \qquad [1.71]$$

To calculate the force exerted between the spheres, we need to add all of the forces applied to the images on the right side by all of the images on the left sphere and vice versa. This is because the force applied to the mth image of the right sphere by the nth image of the left sphere is $F_{nm} = \frac{q_n q_m}{4\pi\varepsilon_0[2R-b_n-b_m]^2} = \frac{Q_0^2(-1)^{n+m}mn}{4\pi\varepsilon_0 R^2(m+n)^2}$. Thus, the total force exerted between the two spheres is obtained by summing all possible values of m and n:

$$F = \frac{Q_0^2}{4\pi\varepsilon_0 R^2}\sum_{m=1}^{\infty}\sum_{n=1}^{\infty}\frac{(-1)^{n+m}\,mn}{(m+n)^2} = \frac{Q_0^2}{4\pi\varepsilon_0 R^2}\frac{1}{6}\left[\ln 2 - \frac{1}{4}\right] \qquad [1.72]$$

where the following identity was used: $\sum_{m=1}^{\infty}\sum_{n=1}^{\infty}\frac{(-1)^{n+m}mn}{(m+n)^2} = \frac{1}{6}\left[\ln 2 - \frac{1}{4}\right]$.

1.2.4. *Dipole in proximity to a conducting sphere*

An electric dipole $\vec{\mathbf{p}}$ is placed in proximity to a conducting sphere of radius R, maintained at zero potential (earthed). The dipole is randomly oriented with respect to the sphere. Let us find the image dipole whose field is equivalent to that created by the surface charge distribution induced on the sphere. Replace the dipole with two charges separated by a distance a, as shown in Figure 1.11. The dipole charges $+q$ and $-q$ are located at $\vec{\mathbf{r}}_+$ and $\vec{\mathbf{r}}_-$, respectively. The position of the dipole, that is, the midpoint of the dipole is indexed by $\vec{\mathbf{r}}$. The dipole moment of these charges is $q\vec{\mathbf{p}} = q\vec{\mathbf{r}}_+ - q\vec{\mathbf{r}}_- = q\vec{\mathbf{a}}$. The $+q$ charge has an image $-q\frac{R}{r_+}$ located at $\frac{R^2}{r_+^2}\vec{\mathbf{r}}_+$, while the $-q$ charge has an image $+q\frac{R}{r_-}$ located at $\frac{R^2}{r_-^2}\vec{\mathbf{r}}_-$. It is now easy to calculate the dipole moment $\vec{\mathbf{p}}'$ of the image dipole. In fact, we write:

$$\vec{\mathbf{p}}' = q\frac{R}{r_-}\frac{R^2}{r_-^2}\vec{\mathbf{r}}_- - q\frac{R}{r_+}\frac{R^2}{r_+^2}\vec{\mathbf{r}}_+ = qR^3\left(\frac{\vec{\mathbf{r}}_-}{r_-^3} - \frac{\vec{\mathbf{r}}_+}{r_+^3}\right) \qquad [1.73]$$

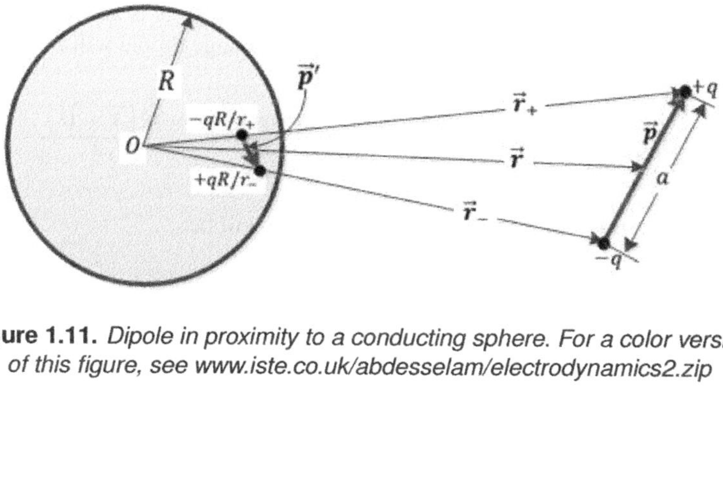

Figure 1.11. *Dipole in proximity to a conducting sphere. For a color version of this figure, see www.iste.co.uk/abdesselam/electrodynamics2.zip*

We can find the limit of this expression when $a \longrightarrow 0$ (the case of a point dipole) simply by writing $\vec{\mathbf{r}}_+ = \vec{\mathbf{r}} + \frac{1}{2}\vec{\mathbf{a}}$ and $\vec{\mathbf{r}}_- = \vec{\mathbf{r}} - \frac{1}{2}\vec{\mathbf{a}}$. It therefore follows that:

$$\begin{aligned} \frac{1}{r_\pm^3} &= \frac{1}{\left(r^2 + \vec{\mathbf{r}} \cdot \vec{\mathbf{a}} + \frac{1}{4}a^2\right)^{3/2}} \simeq \frac{1}{r^3}\left(1 \pm \frac{\vec{\mathbf{r}} \cdot \vec{\mathbf{a}}}{r^2}\right)^{-3/2} \\ &\simeq \frac{1}{r^3}\left[1 \mp \frac{3}{2}\frac{\vec{\mathbf{r}} \cdot \vec{\mathbf{a}}}{r^2} + O\left(a^2\right)\right] \end{aligned} \qquad [1.74]$$

Consequently, the induced dipole moment is:

$$\begin{aligned} \vec{\mathbf{p}}' &= \frac{qR^3}{r^3}\left[\vec{\mathbf{r}}_-\left(1 + \frac{3}{2}\frac{\vec{\mathbf{r}} \cdot \vec{\mathbf{a}}}{r^2}\right) - \vec{\mathbf{r}}_+\left(1 - \frac{3}{2}\frac{\vec{\mathbf{r}} \cdot \vec{\mathbf{a}}}{r^2}\right)\right] \\ &= \frac{qR^3}{r^3}\left[\left(\vec{\mathbf{r}}_- - \vec{\mathbf{r}}_+\right) + \frac{3}{2}\frac{\vec{\mathbf{r}} \cdot \vec{\mathbf{a}}}{r^2}\left(\vec{\mathbf{r}}_- + \vec{\mathbf{r}}_+\right)\right] \\ &= \frac{R^3}{r^3}\left[-\vec{\mathbf{p}} + \frac{3\left(\vec{\mathbf{r}} \cdot \vec{\mathbf{p}}\right)\vec{\mathbf{r}}}{r^2}\right] \end{aligned} \qquad [1.75]$$

It is located $\vec{\mathbf{r}}' = \frac{R^2}{r^2}\vec{\mathbf{r}}$ from the center of the sphere. If θ (respectively, θ') represents the angle measured between $\vec{\mathbf{p}}$ and $\vec{\mathbf{r}}$ (respectively, $\vec{\mathbf{p}}'$ and $\vec{\mathbf{r}}$), then:

$$\cos\theta' = \frac{2\cos\theta}{\sqrt{1 + 3\cos^2\theta}} \qquad [1.76]$$

because $p' = \frac{R^3}{r^3}\sqrt{1 + 3\cos^2\theta}p$. In conclusion, note that the image charges leave a net charge inside the sphere, which was initially neutral and isolated. This charge can be detected by Gauss's theorem, unless a compensating charge is added to the center of the sphere to keep it neutral. The net charge of the image dipole will be:

$$\Delta q' = \frac{qR}{r_-} - \frac{qR}{r_+} = qR\left(\frac{r_+ - r_-}{r_+ r_-}\right) \simeq \frac{qRa\cos\theta}{r^2} \equiv \frac{R\left(\vec{\mathbf{r}} \cdot \vec{\mathbf{p}}\right)}{r^3} \qquad [1.77]$$

Consequently, the required compensation charge will be:

$$q_c = -\frac{R\left(\vec{\mathbf{r}} \cdot \vec{\mathbf{p}}\right)}{r^3} \qquad [1.78]$$

EXAMPLE 1.9. (Dipole oriented perpendicular to the conducting sphere).– Let us now try to write the expression for the electrostatic potential of a dipole $\vec{\mathbf{p}}$ located along the axis Oz, at a distance D, from the center of a grounded conducting sphere of radius R. To simplify the problem, we assume that the axis of the dipole is oriented in the positive direction of Oz. The system is equivalent to a dipole $p\overrightarrow{\mathbf{k}}$ located at position $D\overrightarrow{\mathbf{k}}$, an image dipole $p\overrightarrow{\mathbf{p}}' = 2\frac{R^3}{D^3}p\overrightarrow{\mathbf{k}}$ located at position $p\overrightarrow{\mathbf{r}}' = 2\frac{R^2}{D}\overrightarrow{\mathbf{k}}$, and a compensating point charge $q_c = -\frac{Rp}{D^2}$ located in the center of the sphere, as shown in Figure 1.12. The electrostatic potential of the system, measured at a point outside the sphere, indexed by $\overrightarrow{\mathbf{r}}$ with respect to the center O, is equal to:

$$\begin{aligned}V\left(\overrightarrow{\mathbf{r}}\right) &= \frac{\left(\overrightarrow{\mathbf{r}} - D\overrightarrow{\mathbf{k}}\right)\cdot\overrightarrow{\mathbf{p}}}{4\pi\varepsilon_0\left|\overrightarrow{\mathbf{r}} - D\overrightarrow{\mathbf{k}}\right|^3} + \frac{\left(\overrightarrow{\mathbf{r}} - \frac{R^2}{D}\overrightarrow{\mathbf{k}}\right)\cdot\overrightarrow{\mathbf{p}}'}{4\pi\varepsilon_0\left|\overrightarrow{\mathbf{r}} - \frac{R^2}{D}\overrightarrow{\mathbf{k}}\right|^3} \\ &+ \frac{q_c}{4\pi\varepsilon_0\left|\overrightarrow{\mathbf{r}} - D\overrightarrow{\mathbf{k}}\right|}\frac{p}{4\pi\varepsilon_0}\left[\frac{(r\cos\theta - D)}{\left[r^2 + D^2 - 2rD\cos\theta\right]^{3/2}}\right. \\ &\left.+\frac{2\left(r\cos\theta - \frac{R^2}{D}\right)}{\left[\frac{r^2D^2}{R^2} + R^2 - 2rD\cos\theta\right]^{3/2}} - \frac{R}{D^2 r}\right]\end{aligned} \qquad [1.79]$$

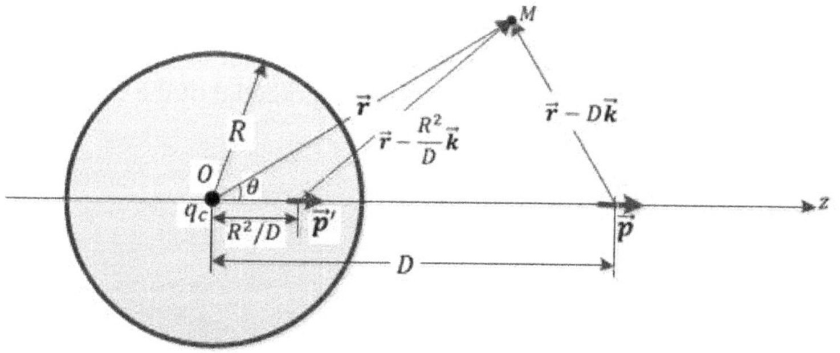

Figure 1.12. *Electric dipole with moment $\overrightarrow{\mathbf{p}}$ oriented perpendicular to a perfectly conducting sphere that is grounded to the earth. For a color version of this figure, see www.iste.co.uk/abdesselam/electrodynamics2.zip*

In place of the real dipole, the resulting electric field produced by the image dipole and the compensation charge will be:

$$\begin{aligned}\overrightarrow{\mathbf{E}} &= \frac{q_c\overrightarrow{\mathbf{k}}}{4\pi\varepsilon_0 R^2} + \frac{3\left[\left(R - \frac{R^2}{D}\right)\overrightarrow{\mathbf{k}}\cdot\overrightarrow{\mathbf{p}}'\right]\left(R - \frac{R^2}{D}\right)\overrightarrow{\mathbf{k}} - \left(R - \frac{R^2}{D}\right)^2\overrightarrow{\mathbf{p}}'}{4\pi\varepsilon_0\left(R - \frac{R^2}{D}\right)^5} \\ &= \frac{p}{4\pi\varepsilon_0}\left[-\frac{1}{RD^2} + \frac{4}{(D-R)^3}\right]\overrightarrow{\mathbf{k}}\end{aligned} \qquad [1.80]$$

and consequently, the energy of the real dipole is:

$$W = -\vec{\mathbf{p}} \cdot \vec{\mathbf{E}} = -\frac{p^2}{4\pi\varepsilon_0}\left[-\frac{1}{RD^2} + \frac{4}{(D-R)^3}\right] \qquad [1.81]$$

Thus, the force experienced by the real dipole is equal to:

$$\vec{\mathbf{F}} = -\vec{\nabla} W = -\frac{dW}{dD}\vec{\mathbf{k}} = \frac{p^2}{2\pi\varepsilon_0}\left[\frac{1}{RD^3} - \frac{6}{(D-R)^4}\right]\vec{\mathbf{k}} \qquad [1.82]$$

1.2.5. *Infinite line charge in proximity to a plane*

Consider a line of charge of density λ located at a distance a from a plane maintained at zero potential. To simplify the problem, we assume that this plane is confounded with the Oyz plane and that the line is directed along Oz (passing through $x = a$), as shown in Figure 1.13. To maintain this plane at zero potential, simply place an image charge line of density $-\lambda$, parallel to the previous line, and passing through $x = -a$. In this case, the resulting potential of the two lines is:

$$V(x, y) = \frac{\lambda}{2\pi\varepsilon_0} \ln\left[\frac{(x+a)^2 + y^2}{(x-a)^2 + y^2}\right] \qquad [1.83]$$

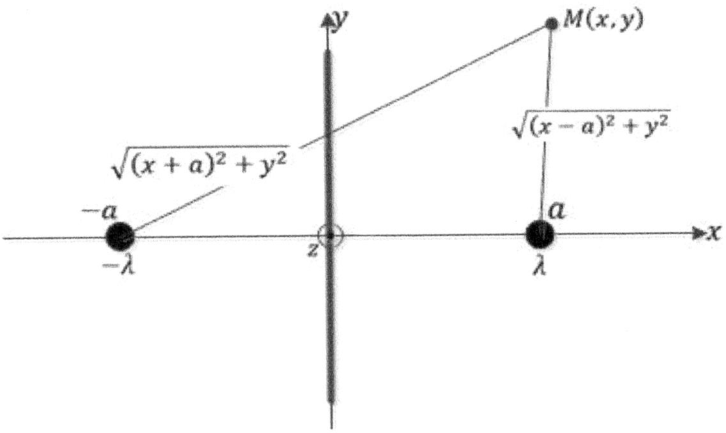

Figure 1.13. *Line charge in proximity to a grounded plane*

($V(x, y)$ is, of course, zero for $x = 0$) and consequently:

$$\vec{\mathbf{E}}(x, y) = -\vec{\nabla} V = \frac{\lambda a}{\pi \varepsilon_0} \frac{\left(x^2 - y^2 - a^2\right) \vec{\mathbf{i}} + 2xy \vec{\mathbf{j}}}{\left[(x-a)^2 + y^2\right]\left[(x+a)^2 + y^2\right]} \qquad [1.84]$$

The force per unit length experienced on the original line is only due to the field created by the image line:

$$\vec{\mathbf{F}} = \lambda \left(-\frac{\lambda}{2\pi\varepsilon_0 (2a)}\right) \vec{\mathbf{i}} = -\frac{\lambda^2}{4\pi\varepsilon_0 a} \vec{\mathbf{i}} \qquad [1.85]$$

Similarly, the induced charge density on the surface of the Oyz plane is given by the discontinuity of the component normal to the electric field:

$$\sigma(y) = \varepsilon_0 E_x(0, y) = -\frac{\lambda a}{\pi\left(y^2 + a^2\right)} \qquad [1.86]$$

where the field inside the conductor is zero. The total charge induced on the plane per unit length (along Oz) is obtained by integrating [1.86] along the Oy axis:

$$\lambda_T = \int_{-\infty}^{\infty} \sigma(y)\, dy = -\frac{\lambda a}{\pi} \int_{-\infty}^{\infty} \frac{dy}{y^2 + a^2} = -\lambda \qquad [1.87]$$

We thus arrive at the density of the image line introduced earlier.

1.2.6. *Infinite line charge in proximity to a conducting cylinder*

A line charge of density λ is located at a distance h from the axis of a perfectly conducting cylinder of radius R, as shown in Figure 1.14.

We would like to find an image line of charge density λ' that will imitate the cylinder, that is, make the cylinder's surface equipotential. We assume that the λ' image line lies at a distance b from the axis. At any point $\vec{\mathbf{r}}$ outside the cylinder, the potential is given by:

$$\begin{aligned} V &= -\frac{\lambda}{2\pi\varepsilon_0} \ln r_1 - \frac{\lambda'}{2\pi\varepsilon_0} \ln r_2 \\ &= -\frac{1}{2\pi\varepsilon_0} \left[\lambda \ln \sqrt{r^2 + h^2 - 2rh\cos\theta} + \lambda' \ln \sqrt{r^2 + b^2 - 2rb\cos\theta}\right] \\ &= -\frac{1}{4\pi\varepsilon_0} \left[\lambda \ln\left(r^2 + h^2 - 2rh\cos\theta\right) + \lambda' \ln\left(r^2 + b^2 - 2rb\cos\theta\right)\right] \end{aligned} \qquad [1.88]$$

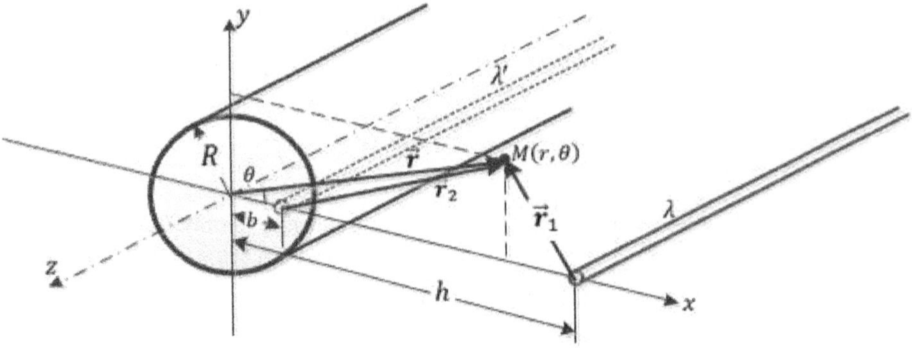

Figure 1.14. *Line charge λ placed in proximity to a conducting cylinder*

We must therefore have $\frac{\partial V}{\partial \theta}\Big|_{r=R} = 0$, that is:

$$\frac{\partial V}{\partial \theta}\Big|_{r=R} = -\frac{1}{2\pi\varepsilon_0}\left(\frac{\lambda R h \sin\theta}{r_1^2} + \frac{\lambda' R b \sin\theta}{r_2^2}\right)\Big|_{r=R} = 0 \qquad [1.89]$$

We can therefore conclude that $\left(\frac{\lambda h}{r_1^2} + \frac{\lambda' b}{r_2^2}\right)\Big|_{r=R} = 0$. It follows that $\frac{r_1^2}{r_2^2}\Big|_{r=R} = -\frac{h}{b}\frac{\lambda}{\lambda'}$ and, consequently, $R^2 + h^2 - 2Rh\cos\theta = -\frac{h}{b}\frac{\lambda}{\lambda'}\left(R^2 + b^2 - 2Rb\cos\theta\right)$. This last equation must be independent of θ. This gives $2Rh = -\frac{h}{b}\frac{\lambda}{\lambda'}2Rb$ and $R^2 + h^2 = \frac{h}{b}\left(R^2 + b^2\right)$. Thus, we obtain:

$$\lambda' = -\lambda$$
$$b = \frac{R^2}{h} \qquad [1.90]$$

It can be seen that the expression for the position of the image line relative to the cylinder axis is identical to the expression for the image of the point charge relative to the center of the conducting sphere. On the other hand, the density value of the image line is opposite to the value of the density of the real line. According to [1.90], the expression for the potential outside of the cylinder [1.88] reduces to:

$$V(r,\theta) = -\frac{\lambda}{4\pi\varepsilon_0}\ln\left[\frac{r^2 + h^2 - 2rh\cos\theta}{r^2 + \frac{R^4}{h^2} - 2r\frac{R^2}{h}\cos\theta}\right] \qquad [1.91]$$

and consequently, the value of the corresponding electric field will be:

$$\begin{aligned}\vec{\mathbf{E}} &= \frac{\lambda}{2\pi\varepsilon_0} \frac{r\left(r^4-h^4\right)+h\left(h^2-R^2\right)\left(r^2+R^2\right)\cos\theta}{h^2\left(r^2+h^2-2rh\cos\theta\right)\left(r^2+\frac{R^4}{h^2}-2r\frac{R^2}{h}\cos\theta\right)}\vec{\mathbf{e}}_r \\ &+ \frac{\lambda\sin\theta}{2\pi\varepsilon_0} \frac{\left(h^2-R^2\right)\left(r^2-R^2\right)}{h\left(r^2+h^2-2rh\cos\theta\right)\left(r^2+\frac{R^4}{h^2}-2r\frac{R^2}{h}\cos\theta\right)}\vec{\mathbf{e}}_\theta\end{aligned} \quad [1.92]$$

The induced charge density on the cylinder is equal to:

$$\sigma(\theta) = \varepsilon_0 E_r(R,\theta) = -\frac{\lambda\left(h^2-R^2\right)}{2\pi R\left(R^2+h^2-2Rh\cos\theta\right)} \quad [1.93]$$

To calculate the total charge accumulated on the cylinder per unit length, simply integrate the previous expression around the cylinder, that is, integrate between 0 and π and then multiply the result by two, because the integral is symmetrical:

$$\lambda_t = -\frac{\lambda\left(h^2-R^2\right)}{\pi R}\int_0^\pi \frac{Rd\theta}{R^2+h^2-2Rh\cos\theta} = -\lambda \quad [1.94]$$

The force exerted on the cylinder, per unit length, is therefore just the force exerted on the image line:

$$\vec{\mathbf{F}} = -\lambda\left(\frac{\lambda}{2\pi\varepsilon_0\left(h-\frac{R^2}{h}\right)}\right)\vec{\mathbf{e}}_r = -\frac{\lambda^2 h}{2\pi\varepsilon_0\left(h^2-R^2\right)}\vec{\mathbf{e}}_r \quad [1.95]$$

Let us generalize these results. For a charge line λ located at $\vec{\mathbf{r}}'$ (outside the cylinder), there must be an image line $\lambda' = -\lambda$ at $\frac{R^2}{r'^2}\vec{\mathbf{r}}'$ (inside the cylinder). These two lines induce the following potential outside of the cylinder:

$$V\left(\vec{\mathbf{r}}\right) = -\frac{\lambda}{2\pi\varepsilon_0}\ln\frac{\left|\vec{\mathbf{r}}-\vec{\mathbf{r}}'\right|}{\left|\vec{\mathbf{r}}-\frac{R^2}{r'^2}\vec{\mathbf{r}}'\right|} \quad [1.96]$$

and the corresponding field:

$$\vec{\mathbf{E}}\left(\vec{\mathbf{r}}\right) = \frac{\lambda}{2\pi\varepsilon_0}\left[\frac{\vec{\mathbf{r}}-\vec{\mathbf{r}}'}{\left|\vec{\mathbf{r}}-\vec{\mathbf{r}}'\right|^2} - \frac{\vec{\mathbf{r}}-\frac{R^2}{r'^2}\vec{\mathbf{r}}'}{\left|\vec{\mathbf{r}}-\frac{R^2}{r'^2}\vec{\mathbf{r}}'\right|^2}\right] \quad [1.97]$$

For a continuous charge distribution (invariant along Oz), it is possible to generalize the above results as follows:

$$\begin{aligned} V\left(\vec{\mathbf{r}}\right) &= -\frac{1}{2\pi\varepsilon_0}\int_{\tau'}\rho\left(\vec{\mathbf{r}}'\right)\ln\frac{\left|\vec{\mathbf{r}}-\vec{\mathbf{r}}'\right|}{\left|\vec{\mathbf{r}}-\frac{R^2}{r'^2}\vec{\mathbf{r}}'\right|}d\tau' \\ \vec{\mathbf{E}}\left(\vec{\mathbf{r}}\right) &= \frac{1}{2\pi\varepsilon_0}\int_{\tau'}\rho\left(\vec{\mathbf{r}}'\right)\left[\frac{\vec{\mathbf{r}}-\vec{\mathbf{r}}'}{\left|\vec{\mathbf{r}}-\vec{\mathbf{r}}'\right|^2}-\frac{\vec{\mathbf{r}}-\frac{R^2}{r'^2}\vec{\mathbf{r}}'}{\left|\vec{\mathbf{r}}-\frac{R^2}{r'^2}\vec{\mathbf{r}}'\right|^2}\right]d\tau' \end{aligned} \quad [1.98]$$

Clearly, the equipotentials of two equal linear charges of opposite signs are non-concentric cylinders. An equipotential plane lies between the two linear charges. This means that, in general, parallel conducting cylinders, or cylinders parallel to a plane, can be imitated by linear charges.

EXAMPLE 1.10. (Parallel cylinders with opposite charges).– The image method can also be used in the case of two parallel cylinders of opposite charges, of radius R_1 and R_2, respectively, whose centers are separated by a distance D, as shown in Figure 1.15(a) when the cylinders are adjacent; and as shown in Figure 1.15(b) when the smaller cylinder is inside the other. The two cylinders can be considered as the equipotentials of the two image lines $+\lambda$ and $-\lambda$ located, respectively, at a distance b_1 from the center of the cylinder of radius R_1 and at a distance b_2 from the center of the cylinder of radius R_2. The two linear charges are located along the line connecting the two centers. The positions of the images can be found from relation [1.90], realizing that the distance between each image and the center of the opposite cylinder is $D - b$:

$$\begin{aligned} b_1 &= \frac{R_1^2}{D \mp b_2} \\ b_2 &= \pm\frac{R_2^2}{D - b_1} \end{aligned} \quad [1.99]$$

where the upper signs are used when the cylinders are adjacent, and the lower signs are used when the smaller cylinder is inside the larger one. To solve the system [1.99], the two coupled equations must be separated into a quadratic equation in b_1 or b_2:

$$b_1^2 - \frac{\left[D^2 - R_2^2 + R_1^2\right]}{D}b_1 + R_1^2 = 0 \quad [1.100]$$

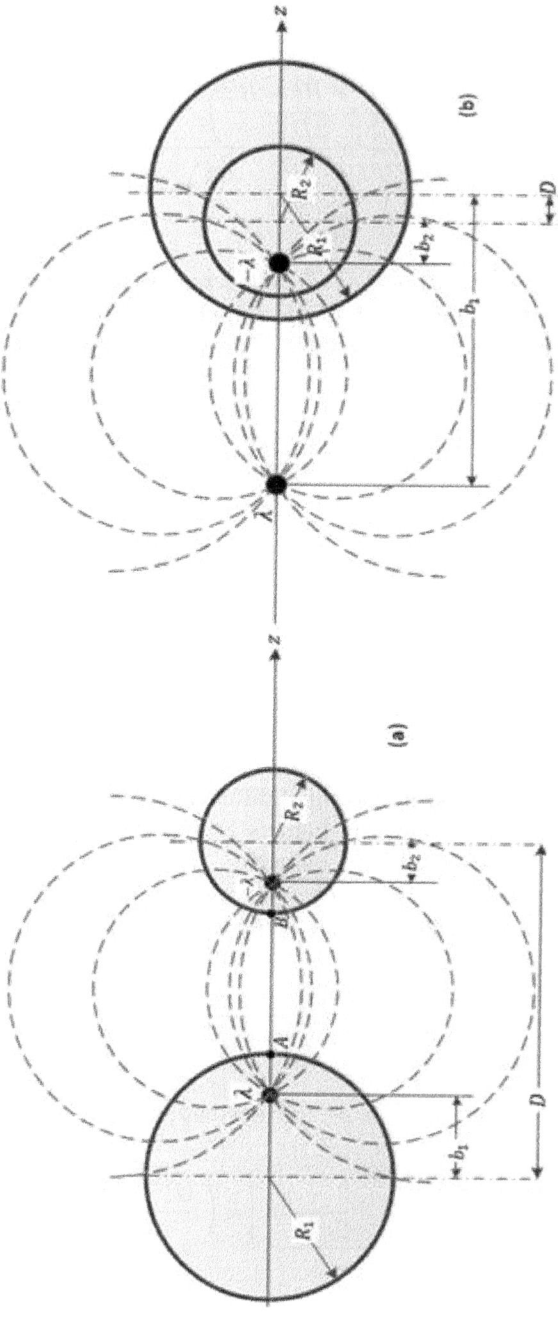

Figure 1.15. *(a) Two adjacent cylinders with opposite charges (the surface charge density is high at the level of the closest points A and B). This is called the proximity effect; (b) cylindrical conductors with opposite charges, one inside the other*

The value of b_2 is obtained using the second equation in system [1.99]. The solutions are as follows:

$$b_1 = \frac{\left[D^2 + R_1^2 - R_2^2\right]}{2D} \mp \sqrt{\left(\frac{D^2 + R_1^2 - R_2^2}{2D}\right)^2 - R_1^2}$$
$$b_2 = \pm \frac{\left[D^2 - R_1^2 + R_2^2\right]}{2D} - \sqrt{\left(\frac{D^2 + R_1^2 - R_2^2}{2D}\right)^2 - R_1^2} \qquad [1.101]$$

where solutions in the region between the two cylinders have been eliminated. The force of attraction exerted per unit length of the cylinder of radius R_1 is the force exerted on the image line $+\lambda$ by the electric field of the opposite image line $-\lambda$:

$$\vec{\mathbf{F}} = \frac{\lambda^2}{4\pi\varepsilon_0 \left[\pm(D - b_1) - b_2\right]} \vec{\mathbf{k}} = \frac{\lambda^2}{4\pi\varepsilon_0 \sqrt{\left(\frac{D^2 + R_1^2 - R_2^2}{2D}\right)^2 - R_1^2}} \vec{\mathbf{k}} \qquad [1.102]$$

The potential in the region lying between the two cylinders depends on the distances between the point being considered, M, and the image lines:

$$V(M) = -\frac{\lambda}{2\pi\varepsilon_0} \ln \frac{r_1}{r_2} \qquad [1.103]$$

where r_1 (respectively, r_2) is the distance measured between the image line $+\lambda$ (respectively, $-\lambda$) and M. To find the potential difference between these cylinders, we choose the points A and B, shown in Figures 1.15(a) and (b), to compute the potential on the surface of each cylinder, $A[r_1 = \pm(R_1 - b_1)$, $r_2 = \pm(D \mp b_2 - R_1)]$ and $B\left[r_1 = \pm(D - b_1 \mp R_2), r_2 = R_2 - R_2\right]$. It is also possible to select other points on the cylinders. The potential difference is then:

$$V_1 - V_2 = -\frac{\lambda}{2\pi\varepsilon_0} \ln \left[\pm \frac{(R_1 - b_1)(R_2 - b_2)}{(D \mp b_2 - R_1)(D - b_1 \mp R_2)}\right] \qquad [1.104]$$

Since $D \mp b_2 = \frac{R_1^2}{b_1}$ and $D - b_1 = \pm\frac{R_2^2}{b_2}$, the previous expression reduces to:

$$V_1 - V_2 = -\frac{\lambda}{2\pi\varepsilon_0} \ln \frac{b_1 b_2}{R_1 R_2} = -\frac{\lambda}{2\pi\varepsilon_0} \ln \left[\pm\left(\frac{D^2 - R_1^2 - R_2^2}{2R_1 R_2}\right) + \sqrt{\left(\frac{D^2 - R_1^2 - R_2^2}{2R_1 R_2}\right)^2 - 1}\right] \qquad [1.105]$$

The potential difference $V_1 - V_2$ is linearly related to the linear charge λ by a factor that depends only on the geometry of the conductors. This factor is defined as the capacitance per length (equal to the ratio between the linear load and the potential difference):

$$C = \frac{\lambda}{V_1 - V_2} = \frac{2\pi\varepsilon_0}{\ln\left[\pm\left(\frac{D^2-R_1^2-R_2^2}{2R_1R_2}\right) + \sqrt{\left(\frac{D^2-R_1^2-R_2^2}{2R_1R_2}\right)^2 - 1}\right]}$$
$$= \frac{2\pi\varepsilon_0}{\cosh^{-1}\left[\pm\frac{D^2-R_1^2-R_2^2}{2R_1R_2}\right]} \quad [1.106]$$

where we used the identity $\ln\left[y + \sqrt{y^2-1}\right] = \cosh^{-1} y$. Let us look at some special cases:

1) We first consider the case of adjacent cylinders $(D > R_1 + R_2)$. When the distance D is much greater than their radii, that is, $D \gg R_1$ and R_2, then:

$$C \simeq \frac{2\pi\varepsilon_0}{\cosh^{-1}\left[\frac{D^2}{2R_1R_2}\right]} \quad [1.107]$$

2) The capacitance between a cylinder and an infinite plane can be obtained by tending R_1 toward infinity, but keeping the nearest distance $h = D - R_1 - R_2$ between the cylinders finite. If R_1 is allowed to become infinite in [1.106], the capacitance becomes:

$$C = \frac{2\pi\varepsilon_0}{\cosh^{-1}\left[\frac{h+R_2}{R_2}\right]} \quad [1.108]$$

3) If the two cylinders are identical, that is, $R_1 = R_2 = R$, the capacitance per unit length is reduced to:

$$C = \frac{2\pi\varepsilon_0}{\cosh^{-1}\left[\frac{D^2}{2R^2} - 1\right]} = \frac{\pi\varepsilon_0}{\cosh^{-1}\left[\frac{D}{2R}\right]} \quad [1.109]$$

We have used the identity $2\cosh^{-1} y = \cosh^{-1}\left(2y^2 - 1\right)$;

4) When the two cylinders are concentric, that is, $D = 0$, the capacitance per unit length is:

$$C = \frac{2\pi\varepsilon_0}{\cosh^{-1}\left(\frac{R_1^2+R_2^2}{2R_1R_2}\right)} = \frac{\pi\varepsilon_0}{\cosh^{-1}\left(\frac{R_1+R_2}{2\sqrt{R_1R_2}}\right)} \quad [1.110]$$

1.3. Green's function

1.3.1. *Position of the problem*

In the previous examples, we saw that the use of image charges (to satisfy the prescribed boundary conditions) allows us to see that the potential is simply the integral of the density $\rho(\vec{\mathbf{r}})$ of the source multiplied by a function that varies with the boundary conditions (see [1.49]). This function is called *Green's function* and appears to be the solution of the Poisson equation with only one point source. Green's function, denoted here by $G(\vec{\mathbf{r}}, \vec{\mathbf{r}}')$, is a function of two variables, $\vec{\mathbf{r}}$ and $\vec{\mathbf{r}}'$, where $\vec{\mathbf{r}}'$ denotes the position of the point source. It must be noted that the utility of Green's functions goes well beyond solving the Poisson equation. We will try to give a simple idea of how Green functions can be used to solve linear and inhomogeneous partial differential equations. Consider a function $G(\vec{\mathbf{r}}, \vec{\mathbf{r}}'; t, t')$, which appears as the solution of the linear differential equation:

$$DG(\vec{\mathbf{r}}, \vec{\mathbf{r}}'; t, t') = -\delta^3(\vec{\mathbf{r}} - \vec{\mathbf{r}}')\,\delta(t - t') \qquad [1.111]$$

where D is a linear operator acting on $\vec{\mathbf{r}}$ and t. The function:

$$\psi(\vec{\mathbf{r}}, t) = \int d\tau' dt' f(\vec{\mathbf{r}}', t')\, G(\vec{\mathbf{r}}, \vec{\mathbf{r}}'; t, t') \qquad [1.112]$$

is a solution of the equation:

$$D\psi(\vec{\mathbf{r}}, t) = -f(\vec{\mathbf{r}}, t) \qquad [1.113]$$

because $D\psi(\vec{\mathbf{r}}, t) = D\int d\tau' dt' f(\vec{\mathbf{r}}', t')\, G(\vec{\mathbf{r}}, \vec{\mathbf{r}}'; t, t') = \int d\tau' dt' f(\vec{\mathbf{r}}', t')$ $DG(\vec{\mathbf{r}}, \vec{\mathbf{r}}'; t, t') = -\int d\tau' dt' f(\vec{\mathbf{r}}', t')\,\delta^3(\vec{\mathbf{r}} - \vec{\mathbf{r}}')\,\delta(t - t') = -f(\vec{\mathbf{r}}, t)$. The function $\psi(\vec{\mathbf{r}}, t)$ and the Green's function $(\vec{\mathbf{r}}, \vec{\mathbf{r}}'; t, t')$ satisfy the same boundary conditions. In fact, if $\psi(\vec{\mathbf{r}}, t)$ cancels on a given boundary, then $G(\vec{\mathbf{r}}, \vec{\mathbf{r}}'; t, t')$ also cancels on the same boundary.

1.3.2. *Poisson equation solutions*

Consider a known non-zero charge distribution $\rho(\vec{\mathbf{r}})$, on a finite volume of space. We wish to solve the Poisson equation [1.1] with boundary conditions typical of a Dirichlet problem (i.e. the potential is specified on a closed surface) or a Neumann problem (i.e. $\vec{\mathbf{E}} = -\vec{\nabla}V$ is specified on a closed surface). The solution takes the following form:

$$V(\vec{\mathbf{r}}) = V_p(\vec{\mathbf{r}}) + V_c(\vec{\mathbf{r}}) \qquad [1.114]$$

where $V_p\left(\vec{\mathbf{r}}\right)$ is the particular solution of the Poisson equation and $V_c\left(\vec{\mathbf{r}}\right)$ is the complementary solution of the Laplace equation, $\Delta V_c\left(\vec{\mathbf{r}}\right) = 0$. Of course, the following condition must be imposed on the particular solution:

$$\lim_{r \longrightarrow \infty} V_p\left(\vec{\mathbf{r}}\right) = 0 \qquad [1.115]$$

which can be seen as a boundary condition indicating that $V_p\left(\vec{\mathbf{r}}\right)$ vanishes over very large distances. Thus, the function $V_p\left(\vec{\mathbf{r}}\right)$ is:

$$V_p\left(\vec{\mathbf{r}}\right) = \frac{1}{4\pi\varepsilon_0}\int_{\tau'} \frac{\rho\left(\vec{\mathbf{r}}'\right) d\tau'}{\left|\vec{\mathbf{r}} - \vec{\mathbf{r}}'\right|} \qquad [1.116]$$

where integration takes place in $3D$ space. To prove that this solution obeys the Poisson equation and satisfies the boundary condition [1.115], first note that $\left|\vec{\mathbf{r}} - \vec{\mathbf{r}}'\right| = r\left[1 + O\left(\frac{1}{r}\right)\right]$ for $r \longrightarrow \infty$ and, consequently, $\lim_{r\longrightarrow\infty} V_p\left(\vec{\mathbf{r}}\right) = \lim_{r\longrightarrow\infty} \frac{1}{4\pi\varepsilon_0 r} \times \int_{r\tau'} \rho\left(\vec{\mathbf{r}}'\right) d\tau' + O\left(\frac{1}{r^2}\right) = \lim_{r\longrightarrow\infty} \frac{q}{4\pi\varepsilon_0 r} + O\left(\frac{1}{r^2}\right) = 0$. Next, let us calculate the Laplacian of $V_p\left(\vec{\mathbf{r}}\right)$:

$$\begin{aligned} \Delta V_p &= \frac{1}{4\pi\varepsilon_0}\Delta \int_{\tau'} \frac{\rho\left(\vec{\mathbf{r}}'\right) d\tau'}{\left|\vec{\mathbf{r}} - \vec{\mathbf{r}}'\right|} = \frac{1}{4\pi\varepsilon_0}\int_{\tau'} \rho\left(\vec{\mathbf{r}}'\right) \Delta\left(\frac{1}{\left|\vec{\mathbf{r}} - \vec{\mathbf{r}}'\right|}\right) d\tau' \\ &= -\frac{4\pi}{4\pi\varepsilon_0}\int_{\tau'} \rho\left(\vec{\mathbf{r}}'\right) \delta^3\left(\vec{\mathbf{r}} - \vec{\mathbf{r}}'\right) d\tau' = -\frac{\rho\left(\vec{\mathbf{r}}\right)}{\varepsilon_0} \end{aligned} \qquad [1.117]$$

We can therefore conclude that $V_p\left(\vec{\mathbf{r}}\right)$ is indeed a solution of the Poisson equation. $V_p\left(\vec{\mathbf{r}}\right)$ is the only solution of the Poisson equation, valid at any point in space, which verifies the boundary conditions [1.115].

If we now want to solve the Poisson equation in a restricted region of space, we need to specify the boundary conditions on the closed surface surrounding this region. Consequently, the general solution is given by:

$$V\left(\vec{\mathbf{r}}\right) = V_c\left(\vec{\mathbf{r}}\right) + \frac{1}{4\pi\varepsilon_0}\int_{\tau'} \frac{\rho\left(\vec{\mathbf{r}}'\right) d\tau'}{\left|\vec{\mathbf{r}} - \vec{\mathbf{r}}'\right|} \qquad [1.118]$$

where $V_c\left(\vec{\mathbf{r}}\right)$ is a particular solution of the Laplace equation chosen so that the complete solution satisfies the problem's pre-established boundary conditions.

1.3.3. *Green's theorem*

Green's theorem, also known as *Green's second identity*, converts the Poisson differential equation into an integral equation. This theorem is derived from the divergence theorem. In particular, when $\overrightarrow{\mathbf{F}} = f\overrightarrow{\nabla}g$, we obtain $\overrightarrow{\nabla}\cdot\overrightarrow{\mathbf{F}} = \overrightarrow{\nabla}f\cdot\overrightarrow{\nabla}g + f\Delta g$. It can thus be derived that:

$$\int_{\tau}\left(\overrightarrow{\nabla}f\cdot\overrightarrow{\nabla}g + f\Delta g\right)d\tau = \oint_{S} f\left(\overrightarrow{\nabla}g\cdot\overrightarrow{\mathbf{n}}\right)dS = \oint f\frac{\partial g}{\partial n}dS \qquad [1.119]$$

This equation is called *Green's first identity*. By permuting f and g, it comes:

$$\int_{\tau}\left(\overrightarrow{\nabla}g\cdot\overrightarrow{\nabla}f + g\Delta f\right)d\tau = \oint_{S} g\frac{\partial f}{\partial n}dS \qquad [1.120]$$

Subtracting [1.120] from [1.119], we obtain Green's second identity:

$$\int_{\tau}(f\Delta g - g\Delta f)\,d\tau = \oint_{S}\left(f\frac{\partial g}{\partial n} - g\frac{\partial f}{\partial n}\right)dS \qquad [1.121]$$

If τ is the whole space, then the surface S is at infinity. It therefore follows that when the term $f\frac{\partial g}{\partial n} - g\frac{\partial f}{\partial n}$ cancels out at infinity, the right-hand side of equation [1.121] disappears and, consequently:

$$\int_{\text{Wholespace}} f\Delta g d\tau = \int_{\text{Wholespace}} g\Delta f d\tau \qquad [1.122]$$

1.3.4. *Inverse Laplacian and Green's function*

Let us start with the Poisson equation [1.1], valid at any point in space, where $\rho\left(\overrightarrow{\mathbf{r}}\right)$ is not zero, except in a finite region of space. This restriction is stronger than necessary. It suffices to assume that $\rho\left(\overrightarrow{\mathbf{r}}\right) \longrightarrow 0$ quite rapidly when $r \longrightarrow \infty$ so that the volume integral of $\rho\left(\overrightarrow{\mathbf{r}}\right)$ over the entire space converges. Under this hypothesis, the solution of the Poisson equation is given by [1.118] with $V_c\left(\overrightarrow{\mathbf{r}}\right) = 0$. In other words, we have: $V\left(\overrightarrow{\mathbf{r}}\right) = V_p\left(\overrightarrow{\mathbf{r}}\right)$, where $V_p\left(\overrightarrow{\mathbf{r}}\right)$ is calculated from equation [1.116]. Let us derive the solution of the Poisson equation under these conditions by introducing the inverse Laplacian operator as follows:

$$\Delta^{-1}\Delta V\left(\overrightarrow{\mathbf{r}}\right) = -\frac{1}{\varepsilon_0}\Delta^{-1}\rho\left(\overrightarrow{\mathbf{r}}\right) \qquad [1.123]$$

By construction, it is clear that $\Delta^{-1}\Delta$ must be equal to the identity operator. It therefore comes that:

$$V\left(\vec{\mathbf{r}}\right) = -\frac{1}{\varepsilon_0}\Delta^{-1}\rho\left(\vec{\mathbf{r}}\right) \qquad [1.124]$$

By comparing this result with [1.116], we derive the identity:

$$\Delta^{-1}\rho\left(\vec{\mathbf{r}}\right) = -\frac{1}{4\pi}\int_{\tau'}\frac{\rho\left(\vec{\mathbf{r}}'\right)d\tau'}{\left|\vec{\mathbf{r}}-\vec{\mathbf{r}}'\right|} \qquad [1.125]$$

Inserting [1.125] into [1.124], we find $V = \frac{1}{4\pi\varepsilon_0}\int_{\tau'}\frac{\rho\left(\vec{\mathbf{r}}'\right)d\tau'}{\left|\vec{\mathbf{r}}-\vec{\mathbf{r}}'\right|}$, as predicted. The definition of the inverse Laplacian operator [1.124] shows that this operator acts non-locally, that is, the value of $\Delta^{-1}\rho\left(\vec{\mathbf{r}}\right)$ at the point $\vec{\mathbf{r}}$ depends on the value of $\rho\left(\vec{\mathbf{r}}'\right)$ for other points in space.

To verify that the definition of the inverse Laplacian operator [1.125] makes sense, let us perform the following calculations:

$$\begin{aligned}
\Delta\Delta^{-1}\rho\left(\vec{\mathbf{r}}\right) &= -\frac{1}{4\pi}\Delta\int_{\tau'}\frac{\rho' d\tau'}{\left|\vec{\mathbf{r}}-\vec{\mathbf{r}}'\right|} = -\frac{1}{4\pi}\int_{\tau'}\rho'\Delta\left(\frac{1}{\left|\vec{\mathbf{r}}-\vec{\mathbf{r}}'\right|}\right)d\tau' \\
&= \int_{\tau'}\rho'\delta^3\left(\vec{\mathbf{r}}-\vec{\mathbf{r}}'\right)d\tau' = \rho\left(\vec{\mathbf{r}}\right) \\
\Delta^{-1}\Delta\rho\left(\vec{\mathbf{r}}\right) &= -\frac{1}{4\pi}\int_{\tau'}\frac{\Delta'\rho' d\tau'}{\left|\vec{\mathbf{r}}-\vec{\mathbf{r}}'\right|} = -\frac{1}{4\pi}\int_{\tau'}\rho'\Delta\left(\frac{1}{\left|\vec{\mathbf{r}}-\vec{\mathbf{r}}'\right|}\right)d\tau' \\
&= \int_{\tau'}\rho'\delta^3\left(\vec{\mathbf{r}}-\vec{\mathbf{r}}'\right)d\tau' = \rho\left(\vec{\mathbf{r}}\right)
\end{aligned} \qquad [1.126]$$

where Δ' acts here on $\vec{\mathbf{r}}'$ and $\rho' \equiv \rho\left(\vec{\mathbf{r}}'\right)$. We used [1.122] to obtain the second equation. In fact, let us take $f = \rho\left(\vec{\mathbf{r}}'\right)$ and $g = \frac{1}{\left|\vec{\mathbf{r}}-\vec{\mathbf{r}}'\right|}$ in [1.121]. Since $\rho\left(\vec{\mathbf{r}}'\right)$ and $\frac{1}{\left|\vec{\mathbf{r}}-\vec{\mathbf{r}}'\right|}$ both vanish at infinity, the right-hand term in [1.121] is zero and, hence, $\int_{\tau'}\frac{\Delta'\rho\left(\vec{\mathbf{r}}'\right)d\tau'}{\left|\vec{\mathbf{r}}-\vec{\mathbf{r}}'\right|} = \int_{\tau'}\rho\left(\vec{\mathbf{r}}'\right)\Delta\left(\frac{1}{\left|\vec{\mathbf{r}}-\vec{\mathbf{r}}'\right|}\right)d\tau'$. We therefore conclude that:

$$\Delta\Delta^{-1}\rho\left(\vec{\mathbf{r}}\right) = \Delta^{-1}\Delta\rho\left(\vec{\mathbf{r}}\right) = \rho\left(\vec{\mathbf{r}}\right) \qquad [1.127]$$

on the condition that $\rho\left(\vec{\mathbf{r}}\right)$ is canceled for $r \longrightarrow \infty$. This confirms that $\Delta\Delta^{-1}$ and $\Delta^{-1}\Delta$ are both equivalent to the identity operator.

The inverse Laplacian operator can also be used to determine Green's function for the Poisson equation. Let us first assume that the potential vanishes rapidly at infinity, as indicated by equation [1.115]. The Green's function $G\left(\vec{\mathbf{r}}, \vec{\mathbf{r}}'\right)$ is defined as the solution of equation:

$$\Delta G\left(\vec{\mathbf{r}}, \vec{\mathbf{r}}'\right) = -\delta^3\left(\vec{\mathbf{r}} - \vec{\mathbf{r}}'\right) \qquad [1.128]$$

It can be derived from this that:

$$\begin{aligned} G\left(\vec{\mathbf{r}}, \vec{\mathbf{r}}'\right) &= -\Delta^{-1}\delta^3\left(\vec{\mathbf{r}} - \vec{\mathbf{r}}'\right) \\ &= \frac{1}{4\pi}\int_{\tau''} \frac{\delta^3\left(\vec{\mathbf{r}}'' - \vec{\mathbf{r}}'\right)}{\left|\vec{\mathbf{r}} - \vec{\mathbf{r}}''\right|} d\tau'' = \frac{1}{4\pi}\frac{1}{\left|\vec{\mathbf{r}} - \vec{\mathbf{r}}'\right|} \end{aligned} \qquad [1.129]$$

The Green's function is symmetric with respect to $\vec{\mathbf{r}}$ and $\vec{\mathbf{r}}'$, that is:

$$G\left(\vec{\mathbf{r}}, \vec{\mathbf{r}}'\right) = G\left(\vec{\mathbf{r}}', \vec{\mathbf{r}}\right) \qquad [1.130]$$

In addition, we have also:

$$G\left(\vec{\mathbf{r}}, \vec{\mathbf{r}}'\right) = G\left(\vec{\mathbf{r}} - \vec{\mathbf{r}}, 0\right) \qquad [1.131]$$

To conclude, the inverse Laplacian operator allows a quick and simple derivation of the function $G\left(\vec{\mathbf{r}}, \vec{\mathbf{r}}'\right)$. At first sight, Green's function seems to be something complicated, but we can see that Green's functions have been used since our first problems in electrostatics. Green's function can be clearly interpreted as referring to the potential resulting from the presence of a point charge located at $\vec{\mathbf{r}}'$. Green's function is useful because it can be used to construct the potential of an arbitrary charge density through the equation:

$$V\left(\vec{\mathbf{r}}\right) = \frac{1}{\varepsilon_0}\int G\left(\vec{\mathbf{r}}, \vec{\mathbf{r}}'\right)\rho\left(\vec{\mathbf{r}}'\right) d\tau' \qquad [1.132]$$

We can easily verify that equation [1.132] satisfies the Poisson equation, that is:

$$\Delta V = \frac{1}{\varepsilon_0}\int \rho'\Delta G\left(\vec{\mathbf{r}}, \vec{\mathbf{r}}'\right) d\tau' = -\frac{1}{\varepsilon_0}\int \rho'\delta^3\left(\vec{\mathbf{r}} - \vec{\mathbf{r}}'\right) d\tau' = \frac{\rho}{\varepsilon_0} \qquad [1.133]$$

There is also another interpretation of Green's function, which can be obtained directly from equation [1.128]. The Dirac distribution is analogous to the Kronecker symbol δ_{ij} when the indices i and j become continuous. The Dirac distribution can therefore be considered as an infinite matrix corresponding to the identity matrix, where $\delta^3\left(\vec{\mathbf{r}} - \vec{\mathbf{r}}'\right)$ are the elements of this matrix. Consequently, the $G\left(\vec{\mathbf{r}}, \vec{\mathbf{r}}'\right)$ values will appear as the elements of an infinite matrix associated with the inverse Laplacian operator.

1.3.5. *Derivation of the Green's function by the Fourier transform*

Note that it is also possible to directly obtain the $G\left(\vec{\mathbf{r}}, \vec{\mathbf{r}}'\right)$ function from equation [1.128] using the $3D$ Fourier transform. The $3D$ Fourier transform $F\left(\vec{\mathbf{r}}\right)$ is defined by:

$$\mathsf{F}\left(\vec{\mathbf{q}}\right) = \int f\left(\vec{\mathbf{r}}\right) e^{\mathrm{i}\vec{\mathbf{q}}\cdot\vec{\mathbf{r}}} dxdydz \qquad [1.134]$$

(to within one factor) where $\vec{\mathbf{q}}$ and $\vec{\mathbf{r}}$ are $3D$ vectors. When $\mathsf{F}\left(\vec{\mathbf{q}}\right)$ is known, it is possible to calculate $F\left(\vec{\mathbf{r}}\right)$ using the inverse Fourier transform:

$$F\left(\vec{\mathbf{r}}\right) = \frac{1}{(2\pi)^3} \int \mathsf{F}\left(\vec{\mathbf{q}}\right) e^{-\mathrm{i}\vec{\mathbf{q}}\cdot\vec{\mathbf{r}}} dq_x dq_y dq_z \qquad [1.135]$$

Let us take the Fourier transform on both sides in [1.128] with respect to $\vec{\mathbf{r}}$:

$$\begin{aligned}\int \Delta G\left(\vec{\mathbf{r}}, \vec{\mathbf{r}}'\right) e^{\mathrm{i}\vec{\mathbf{q}}\cdot\vec{\mathbf{r}}} dxdydz &= -\int \delta^3\left(\vec{\mathbf{r}} - \vec{\mathbf{r}}'\right) e^{\mathrm{i}\vec{\mathbf{q}}\cdot\vec{\mathbf{r}}} dxdydz \\ &= -e^{\mathrm{i}\vec{\mathbf{q}}\cdot\vec{\mathbf{r}}'}\end{aligned} \qquad [1.136]$$

To evaluate the left-hand side of equation [1.136], we need to use Green's second identity. We choose $f = G\left(\vec{\mathbf{r}}, \vec{\mathbf{r}}'\right)$ and $g = e^{i\vec{\mathbf{q}}\cdot\vec{\mathbf{r}}}$. Since $G\left(\vec{\mathbf{r}}, \vec{\mathbf{r}}'\right)$ and $\vec{\nabla} G\left(\vec{\mathbf{r}}, \vec{\mathbf{r}}'\right)$ cancel out at infinity, then:

$$\int \Delta G\left(\vec{\mathbf{r}}, \vec{\mathbf{r}}'\right) e^{\mathrm{i}\vec{\mathbf{q}}\cdot\vec{\mathbf{r}}} dxdydz = \int G\left(\vec{\mathbf{r}}, \vec{\mathbf{r}}'\right) \Delta e^{\mathrm{i}\vec{\mathbf{q}}\cdot\vec{\mathbf{r}}} dxdydz \qquad [1.137]$$

Noting that $\Delta e^{i\vec{\mathbf{q}}\cdot\vec{\mathbf{r}}} = -q^2 e^{\mathrm{i}\vec{\mathbf{q}}\cdot\vec{\mathbf{r}}}$, it can thus be concluded that:

$$\begin{aligned}\int \Delta G\left(\vec{\mathbf{r}}, \vec{\mathbf{r}}'\right) e^{\mathrm{i}\vec{\mathbf{q}}\cdot\vec{\mathbf{r}}} dxdydz &= -q^2 \int G\left(\vec{\mathbf{r}}, \vec{\mathbf{r}}'\right) e^{\mathrm{i}\vec{\mathbf{q}}\cdot\vec{\mathbf{r}}} dxdydz \\ &= -q^2 \mathsf{G}\left(\vec{\mathbf{q}}, \vec{\mathbf{r}}'\right)\end{aligned} \qquad [1.138]$$

where $\mathsf{G}\left(\vec{\mathbf{q}}, \vec{\mathbf{r}}'\right)$ represents the Fourier transform of $G\left(\vec{\mathbf{r}}, \vec{\mathbf{r}}'\right)$. By inserting [1.138] into [1.136], we get:

$$\mathsf{G}\left(\vec{\mathbf{q}}, \vec{\mathbf{r}}'\right) = \frac{e^{\mathrm{i}\vec{\mathbf{q}}\cdot\vec{\mathbf{r}}'}}{q^2} \qquad [1.139]$$

To once again arrive at $G\left(\vec{\mathbf{r}},\vec{\mathbf{r}}'\right)$, we just calculate the inverse Fourier transform:

$$G\left(\vec{\mathbf{r}},\vec{\mathbf{r}}'\right)=\frac{1}{(2\pi)^3}\int\frac{e^{-i\vec{\mathbf{q}}\cdot\left(\vec{\mathbf{r}}-\vec{\mathbf{r}}'\right)}}{q^2}dq_xdq_ydq_z \tag{1.140}$$

To perform the integration, let us switch to spherical coordinates. Choosing the axis Oz along the vector $\vec{\mathbf{r}}-\vec{\mathbf{r}}'$, we will have: $\vec{\mathbf{q}}\cdot\left(\vec{\mathbf{r}}-\vec{\mathbf{r}}'\right)=q\left|\vec{\mathbf{r}}-\vec{\mathbf{r}}'\right|\cos\theta$ and $dq_xdq_ydq_z=q^2\sin\theta dqd\theta d\varphi$. It follows that:

$$\begin{aligned}G\left(\vec{\mathbf{r}},\vec{\mathbf{r}}'\right)&=\frac{1}{(2\pi)^3}\int_0^\infty dq\int_0^\pi d\theta\int_0^{2\pi}d\varphi\sin\theta e^{-iq\left|\vec{\mathbf{r}}-\vec{\mathbf{r}}'\right|\cos\theta}\\&=\frac{1}{2\pi^2}\int_0^\infty\frac{\sin q\left|\vec{\mathbf{r}}-\vec{\mathbf{r}}'\right|}{q\left|\vec{\mathbf{r}}-\vec{\mathbf{r}}'\right|}dq\\&=\frac{1}{2\pi^2}\frac{1}{\left|\vec{\mathbf{r}}-\vec{\mathbf{r}}'\right|}\int_0^\infty\frac{\sin q}{q}dq=\frac{1}{4\pi}\frac{1}{\left|\vec{\mathbf{r}}-\vec{\mathbf{r}}'\right|}\end{aligned} \tag{1.141}$$

where we used the result $\int_0^\infty\frac{\sin q}{q}dq=\frac{\pi}{2}$.

1.4. Integral form of the Poisson equation

When solving the Poisson equation in a restricted region τ with clearly prescribed boundary conditions on the closed surface S surrounding τ, the corresponding Green function is still a solution of equation [1.128], but now it must also satisfy the boundary conditions on S. Therefore, by analogy with equation [1.118], Green's function must take the following form:

$$G\left(\vec{\mathbf{r}},\vec{\mathbf{r}}'\right)=\frac{1}{4\pi\left|\vec{\mathbf{r}}-\vec{\mathbf{r}}'\right|}+F\left(\vec{\mathbf{r}},\vec{\mathbf{r}}'\right) \tag{1.142}$$

where $F\left(\vec{\mathbf{r}},\vec{\mathbf{r}}'\right)$ is a solution of the Laplace equation, adjusted so that $G\left(\vec{\mathbf{r}},\vec{\mathbf{r}}'\right)$ verifies the boundary conditions of the problem. Under these conditions, equation [1.132] gives:

$$\begin{aligned}V\left(\vec{\mathbf{r}}\right)&=\frac{1}{\varepsilon_0}\int_\tau G\left(\vec{\mathbf{r}},\vec{\mathbf{r}}'\right)\rho\left(\vec{\mathbf{r}}'\right)d\tau'\\&=\frac{1}{4\pi\varepsilon_0}\int_\tau\frac{\rho\left(\vec{\mathbf{r}}'\right)}{\left|\vec{\mathbf{r}}-\vec{\mathbf{r}}'\right|}d\tau'+\frac{1}{\varepsilon_0}\int_\tau F\left(\vec{\mathbf{r}},\vec{\mathbf{r}}'\right)\rho\left(\vec{\mathbf{r}}'\right)d\tau'\end{aligned} \tag{1.143}$$

By comparing this result with [1.118], it can be derived that:

$$V_c\left(\vec{\mathbf{r}}\right)=\frac{1}{\varepsilon_0}\int_\tau F\left(\vec{\mathbf{r}},\vec{\mathbf{r}}'\right)\rho\left(\vec{\mathbf{r}}'\right)d\tau' \tag{1.144}$$

Poisson's equation [1.1] can be converted into an integral equation by choosing, in Green's second identity, $f = G(\vec{\mathbf{r}}, \vec{\mathbf{r}}')$ and $g = V(\vec{\mathbf{r}}')$:

$$\int_\tau \left[G(\vec{\mathbf{r}}, \vec{\mathbf{r}}')\,\Delta' V(\vec{\mathbf{r}}') - V(\vec{\mathbf{r}}')\,\Delta' G(\vec{\mathbf{r}}, \vec{\mathbf{r}}')\right] d\tau' = \oint_S \left[G(\vec{\mathbf{r}}, \vec{\mathbf{r}}')\frac{\partial V}{\partial n} - V(\vec{\mathbf{r}}')\frac{\partial G}{\partial n}\right] dS' \qquad [1.145]$$

Since $\Delta' V(\vec{\mathbf{r}}') = -\frac{\rho(\vec{\mathbf{r}}')}{\varepsilon_0}$ and $\Delta' G(\vec{\mathbf{r}}, \vec{\mathbf{r}}') = -\delta^3(\vec{\mathbf{r}} - \vec{\mathbf{r}}')$, it follows that:

$$\int_\tau \left[-G(\vec{\mathbf{r}}, \vec{\mathbf{r}}')\frac{\rho(\vec{\mathbf{r}}')}{\varepsilon_0} + V(\vec{\mathbf{r}}')\,\delta(\vec{\mathbf{r}} - \vec{\mathbf{r}}')\right] d\tau' = \oint_S \left[G(\vec{\mathbf{r}}, \vec{\mathbf{r}}')\frac{\partial V}{\partial n} - V(\vec{\mathbf{r}}')\frac{\partial G}{\partial n}\right] dS' \qquad [1.146]$$

or again:

$$V(\vec{\mathbf{r}}) = \frac{1}{\varepsilon_0}\int_\tau G(\vec{\mathbf{r}}, \vec{\mathbf{r}}')\,\rho(\vec{\mathbf{r}}')\,d\tau' + \oint_S \left[G(\vec{\mathbf{r}}, \vec{\mathbf{r}}')\frac{\partial V}{\partial n} - V(\vec{\mathbf{r}}')\frac{\partial G}{\partial n}\right] dS' \qquad [1.147]$$

This is the *integral form of the Poisson equation.*

Note that if no condition is specified on the surface S with volume τ, the equation $\Delta G(\vec{\mathbf{r}}, \vec{\mathbf{r}}') = -\delta^3(\vec{\mathbf{r}} - \vec{\mathbf{r}}')$ admits the solution $G(\vec{\mathbf{r}}, \vec{\mathbf{r}}') = \frac{1}{4\pi|\vec{\mathbf{r}} - \vec{\mathbf{r}}'|}$ and, consequently, equation [1.147] becomes:

$$V(\vec{\mathbf{r}}) = \frac{1}{4\pi\varepsilon_0}\int_{\tau'} \frac{\rho(\vec{\mathbf{r}}')}{|\vec{\mathbf{r}} - \vec{\mathbf{r}}'|} d\tau' + \frac{1}{4\pi}\oint_{S'} \left[\frac{\partial V/\partial n}{|\vec{\mathbf{r}} - \vec{\mathbf{r}}'|} - V(\vec{\mathbf{r}}')\frac{\partial}{\partial n}\left(\frac{1}{|\vec{\mathbf{r}} - \vec{\mathbf{r}}'|}\right)\right] dS' \qquad [1.148]$$

The first integral, called the *Coulomb integral*, relates to charge density. The surface integrals reflect the contribution of charges or induced fields at the boundaries of the τ volume. The $\frac{\partial V}{\partial n}$ element of the first term of the surface integral is none other than the normal component of the field at the boundary. The potential resulting from this term is the same as the potential produced by the surface density $\sigma = -\varepsilon_0 \frac{\partial V}{\partial n}$. The last term is the potential corresponding to a dipolar layer $\vec{\mathbf{D}} = \varepsilon_0 V \vec{\mathbf{n}}$ on the surface. If $S \longrightarrow \infty$, the surface integral disappears and we return to the familiar Coulomb result.

If the boundary conditions prescribe V over S (Dirichlet problem),we will choose $G_D\left(\vec{\mathbf{r}}, \vec{\mathbf{r}}'\right) = 0$ when $\vec{\mathbf{r}}' \in S$. In this case, formula [1.147] reduces to:

$$V\left(\vec{\mathbf{r}}\right) = \frac{1}{\varepsilon_0} \int_\tau \rho\left(\vec{\mathbf{r}}'\right) G_D\left(\vec{\mathbf{r}}, \vec{\mathbf{r}}'\right) d\tau' - \oint_S V\left(\vec{\mathbf{r}}'\right) \frac{\partial G_D\left(\vec{\mathbf{r}}, \vec{\mathbf{r}}'\right)}{\partial n} dS' \quad [1.149]$$

Solving the Poisson equation therefore reduces to evaluating two integrals (of course, only if $G_D\left(\vec{\mathbf{r}}, \vec{\mathbf{r}}'\right)$ is known). $G_D\left(\vec{\mathbf{r}}, \vec{\mathbf{r}}'\right)$ is called the *Green–Dirichlet function*.

If the boundary conditions now prescribe the normal derivative $\frac{\partial V}{\partial n}$ over S (Neumann problem), we could consider choosing $\frac{\partial G_N}{\partial n} = 0$; however, this leads to an inconsistency with respect to the definition of $G_N\left(\vec{\mathbf{r}}, \vec{\mathbf{r}}'\right)$. Indeed, after integration over $\vec{\mathbf{r}}'$, the definition $\Delta' G_N\left(\vec{\mathbf{r}}, \vec{\mathbf{r}}'\right) = \vec{\nabla}' \cdot \left(\vec{\nabla}' G_N\left(\vec{\mathbf{r}}, \vec{\mathbf{r}}'\right)\right) = -\delta^3\left(\vec{\mathbf{r}} - \vec{\mathbf{r}}'\right)$ yields:

$$\int_\tau \Delta G_N\left(\vec{\mathbf{r}}, \vec{\mathbf{r}}'\right) d\tau' = -1 = \oint_S \vec{\nabla}' G_N\left(\vec{\mathbf{r}}, \vec{\mathbf{r}}'\right) \cdot d\vec{\mathbf{S}}' = \oint_S \frac{\partial G_N}{\partial n} dS' \quad [1.150]$$

This equation can only be satisfied if $\frac{\partial G_N}{\partial n} = 0$ on S'. In fact, we must choose:

$$\frac{\partial G_N}{\partial n} = -\frac{1}{S} \quad [1.151]$$

Inserting [1.151] into [1.147], we arrive at:

$$\begin{aligned} V\left(\vec{\mathbf{r}}\right) = \frac{1}{\varepsilon_0} \int_\tau G_N\left(\vec{\mathbf{r}}, \vec{\mathbf{r}}'\right) \rho\left(\vec{\mathbf{r}}'\right) d\tau' + \oint_S \left[G_N\left(\vec{\mathbf{r}}, \vec{\mathbf{r}}'\right) \frac{\partial V\left(\vec{\mathbf{r}}'\right)}{\partial n} \right. \\ \left. + V\left(\vec{\mathbf{r}}'\right) \frac{1}{S} \right] dS' = \langle V \rangle + \frac{1}{\varepsilon_0} \int_\tau G_N\left(\vec{\mathbf{r}}, \vec{\mathbf{r}}'\right) \rho\left(\vec{\mathbf{r}}'\right) d\tau' \\ + \oint_S G_N\left(\vec{\mathbf{r}}, \vec{\mathbf{r}}'\right) \frac{\partial V\left(\vec{\mathbf{r}}'\right)}{\partial n} dS' \end{aligned} \quad [1.152]$$

where $\langle V \rangle$ is the mean potential over the boundary:

$$\langle V \rangle = \frac{1}{S} \oint_S V\left(\vec{\mathbf{r}}'\right) dS' \quad [1.153]$$

$G_N\left(\vec{\mathbf{r}},\vec{\mathbf{r}}'\right)$ is called the *Green–Neumann function*. Once the Green–Neumann function is known, the solution to the Poisson equation can be reduced to the calculation of a few integrals. In the following sections, let us try to find Green's function for the special case of the Dirichlet problem with a spherical boundary (internal or external).

1.5. Expansion of Green's function on spherical harmonics

The potential of a point charge q located at $\vec{\mathbf{r}}'$, in the neighborhood of a conducting sphere grounded to the earth, is the solution of the Poisson equation $\Delta V = -\frac{q}{\varepsilon_0}\delta^3\left(\vec{\mathbf{r}}-\vec{\mathbf{r}}'\right)$, with boundary condition $V = 0$ on the surface of the sphere. According to [1.48], the Green's function (outside the sphere) associated with the problem (satisfying the boundary condition) takes the value:

$$G_D\left(\vec{\mathbf{r}},\vec{\mathbf{r}}'\right) = \frac{1}{4\pi}\left(\frac{1}{\left|\vec{\mathbf{r}}-\vec{\mathbf{r}}'\right|} - \frac{\frac{R}{r'}}{\left|\vec{\mathbf{r}}-\frac{R^2}{r'^2}\vec{\mathbf{r}}'\right|}\right) \qquad [1.154]$$

The first term in this expression can be rewritten in the form:

$$\frac{1}{\left|\vec{\mathbf{r}}-\vec{\mathbf{r}}'\right|} = \frac{1}{r_>}\sum_{l=0}^{\infty}\left(\frac{r_<}{r_>}\right)^l P_l\left(\cos\gamma\right) \qquad [1.155]$$

where γ is the angle measured between $\vec{\mathbf{r}}$ and $\vec{\mathbf{r}}'$. For the second term, we obtain the following (the charge q and the position indicated by the vector $\vec{\mathbf{r}}$ are outside the sphere, that is, $r, r' > R$):

$$\begin{aligned}\frac{R}{r'\left|\vec{\mathbf{r}}-\frac{R^2}{r'^2}\vec{\mathbf{r}}'\right|} &= \frac{R}{rr'\sqrt{1+\frac{R^4}{r^2r'^2}-\frac{2R^2}{rr'}\cos\gamma}} \\ &= \frac{R}{rr'}\sum_{l=0}^{\infty}\left(\frac{R^2}{rr'}\right)^l P_l\left(\cos\gamma\right)\end{aligned} \qquad [1.156]$$

It follows therefore that:

$$\begin{aligned}G_D &= \frac{1}{4\pi}\sum_{l=0}^{\infty}\left(\frac{r_<^l}{r_>^{l+1}} - \frac{R^{2l+1}}{r_>^{l+1}r_<^{l+1}}\right)P_l\left(\cos\gamma\right) \\ &= \frac{1}{4\pi}\sum_{l=0}^{\infty}\frac{1}{r_>^{l+1}}\left(r_<^l - \frac{R^{2l+1}}{r_<^{l+1}}\right)P_l\left(\cos\gamma\right) \\ &= \sum_{l=0}^{\infty}\frac{1}{2l+1}\sum_{m=-l}^{l}\frac{1}{r_>^{l+1}}\left(r_<^l - \frac{R^{2l+1}}{r_<^{l+1}}\right)Y_l^m\left(\theta,\varphi\right)\overline{Y_l^m\left(\theta',\varphi'\right)}\end{aligned} \qquad [1.157]$$

where we used the formula $P_l(\cos\gamma) = \frac{4\pi}{2l+1}\sum_{m=-l}^{l} Y_l^m(\theta,\varphi)\overline{Y_l^m(\theta',\varphi')}$. This is the *Green–Dirichlet function outside the sphere*. We can clearly see that $G_D(\vec{\mathbf{r}}, \vec{\mathbf{r}}')$ cancels out on the boundary, that is, for $r_< = R$.

If the positions r and r' are now inside the sphere, we must write (to ensure convergence):

$$\frac{R}{r'\left|\vec{\mathbf{r}} - \frac{R^2}{r'^2}\vec{\mathbf{r}}'\right|} = \frac{1}{R\sqrt{1+\frac{r^2r'^2}{R^4} - \frac{2rr'}{R^2}\cos\gamma}} = \frac{1}{R}\sum_{l=0}^{\infty}\left(\frac{rr'}{R^2}\right)^l P_l(\cos\gamma) \quad [1.158]$$

In this case, the Green–Dirichlet function is:

$$G_D = \sum_{l=0}^{\infty}\frac{1}{2l+1}\sum_{m=-l}^{l} r_<^l\left(\frac{1}{r_>^{l+1}} - \frac{r_>^l}{R^{2l+1}}\right)Y_l^m(\theta,\varphi)\overline{Y_l^m(\theta',\varphi')} \quad [1.159]$$

The Green function vanishes on the surface of the sphere $r_> = R$.

EXAMPLE 1.11. (Circular disk placed within a conducting sphere).– Let us calculate the potential induced by a uniformly charged circular disk of radius a (with total charge Q) placed at the center of a conducting sphere of radius b, grounded to earth, as shown in Figure 1.16. The charge density on the disk is:

$$\rho(\vec{\mathbf{r}}') = \frac{Q\delta(\cos\theta)}{\pi a^2 r'}, \qquad r' \leq a \quad [1.160]$$

as can be easily verified by integrating this density over the volume of a fictitious sphere containing the disk:

$$\begin{aligned}\int_{\tau'}\rho(\vec{\mathbf{r}}')\,d\tau' &= \int_0^a r'^2dr'\int_0^{\pi}d(\cos\theta)\int_0^{2\pi}d\varphi\frac{Q\delta(\cos\theta)}{\pi a^2 r'}\\ &= \frac{2Q}{a^2}\int_0^a r'dr' = Q\end{aligned} \quad [1.161]$$

Since $V = 0$ on the sphere S, the potential inside the sphere is obtained from [1.132], where the Green function is given by [1.159], that is:

$$\begin{aligned}V(\vec{\mathbf{r}}) = \frac{1}{\varepsilon_0}\int_{\tau}\rho(\vec{\mathbf{r}}')\sum_{l=0}^{\infty}\sum_{m=-l}^{l}\frac{1}{2l+1}r_<^l\left(\frac{1}{r_>^{l+1}} - \frac{r_>^l}{b^{2l+1}}\right)\\ \times Y_l^m(\theta,\varphi)\overline{Y_l^m(\theta',\varphi')}d\tau'\end{aligned} \quad [1.162]$$

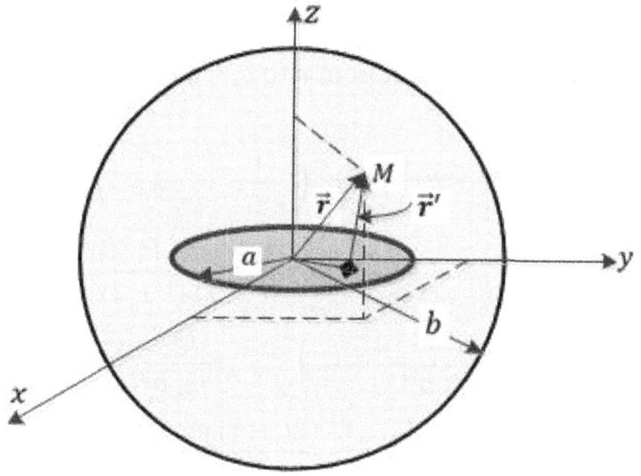

Figure 1.16. *A uniformly charged disk of radius a enclosed within a conducting sphere of radius b maintained at zero potential*

The potential V does not depend on the variable φ (for reasons of symmetry). We can therefore conclude that only $Y_l^0(\theta,\varphi) = \sqrt{\frac{2l+1}{4\pi}} P_l(\cos\theta)$ can contribute to the previous sum. It therefore follows that:

$$\begin{aligned} V &= \frac{1}{4\pi\varepsilon_0}\sum_{l=0}^{\infty}\int_\tau P_l(\cos\theta)\,P_l(\cos\theta')\,\rho\left(\vec{\mathbf{r}}'\right) r_<^l\left(\frac{1}{r_>^{l+1}}-\frac{r_>^l}{b^{2l+1}}\right)d\tau' \\ &= \frac{Q}{4\pi\varepsilon_0\pi a^2}\sum_{l=0}^{\infty}P_l(\cos\theta)\int_\tau\frac{1}{r'}\delta(\cos\theta)\,P_l(\cos\theta')\,r_<^l \\ &\quad\times\left(\frac{1}{r_>^{l+1}}-\frac{r_>^l}{b^{2l+1}}\right)r'^2dr'd(\cos\theta')\,d\varphi' \\ &= \frac{Q}{2\pi\varepsilon_0 a^2}\sum_{l=0}^{\infty}P_l(\cos\theta)\,P_l(0)\int_0^a r_<^l\left(\frac{1}{r_>^{l+1}}-\frac{r_>^l}{b^{2l+1}}\right)r'dr' \end{aligned} \quad [1.163]$$

Two cases can be distinguished:

1) If $r \geq a$, then $r_> = r$ and $r_< = r'$. We therefore obtain:

$$\begin{aligned} V(r>a) &= \frac{Q}{2\pi\varepsilon_0 a^2}\sum_{l=0}^{\infty}P_l(\cos\theta)\,P_l(0)\left(\frac{1}{r^{l+1}}-\frac{r^l}{b^{2l+1}}\right)\int_0^a r'^{l+1}dr' \\ &= \frac{Q}{2\pi\varepsilon_0 a^2}\sum_{l=0}^{\infty}P_l(\cos\theta)\,P_l(0)\,\frac{a^{l+2}}{l+2}\left(\frac{1}{r^{l+1}}-\frac{r^l}{b^{2l+1}}\right) \\ &= \frac{Q}{2\pi\varepsilon_0}\sum_{l=0}^{\infty}\frac{(-1)^l(2l-1)!!a^{2l}}{2^{l+1}(l+1)!}\left(\frac{1}{r^{l+1}}-\frac{r^l}{b^{2l+1}}\right)P_{2l}(\cos\theta) \end{aligned} \quad [1.164]$$

since $P_{2l+1}(0) = 0$ and $P_{2l}(0) = \frac{(-1)^l(2l-1)!!}{2^l l!}$.

2) If $r < a$, the integral $\int_0^a r_<^l \left(\frac{1}{r_>^{l+1}} - \frac{r_>^l}{b^{2l+1}}\right) r'dr'$ is separately calculated over two segments, that is, from 0 to r then from r to a. Consequently, we get:

$$V(r<a) = \frac{Q}{2\pi\varepsilon_0 a^2}\sum_{l=0}^{\infty} P_l(\cos\theta) P_l(0)\left[\left(\frac{1}{r^{l+1}} - \frac{r^l}{b^{2l+1}}\right)\int_0^r r'^{l+1} dr' + r^l \int_r^a \left(\frac{1}{r'^l} - \frac{r'^{l+1}}{b^{2l+1}}\right) dr'\right] = \frac{Q}{2\pi\varepsilon_0 a^2}\sum_{l=0}^{\infty}\frac{P_l(\cos\theta)P_l(0)}{(l-1)(l+2)}\times \left[(2l+1)r - \frac{r^l}{a^{l-1}}\left(l+2-\frac{(l-1)a^{2l+1}}{b^{2l+1}}\right)\right] = \frac{Q}{2\pi\varepsilon_0 a^2}\sum_{l=0}^{\infty}\frac{(-1)^l}{2^{l+1}}\times \frac{(2l-3)!!}{(l+1)!}\left[(4l+1)r - \frac{r^{2l}}{a^{2l-1}}\left(2l+2-\frac{(2l-1)a^{4l+1}}{b^{4l+1}}\right)\right] P_{2l}(\cos\theta) \qquad [1.165]$$

In particular, when $b \longrightarrow \infty$, the previous results [1.164] and [1.165] reduce to:

$$V(r>a) = \frac{Q}{2\pi\varepsilon_0}\sum_{l=0}^{\infty}\frac{(-1)^l(2l-1)!!}{2^{l+1}(l+1)!}\frac{a^{2l}}{r^{l+1}} P_{2l}(\cos\theta) \qquad [1.166]$$

$$V(r<a) = \frac{Q}{2\pi\varepsilon_0 a^2}\sum_{l=0}^{\infty}\frac{(-1)^l(2l-3)!!}{2^{l+1}(l+1)!}\times \left[(4l+1)r - 2(l+1)\frac{r^{2l}}{a^{2l-1}}\right] P_{2l}(\cos\theta) \qquad [1.167]$$

The charge density induced on the surface of the sphere (the inner surface) is:

$$\sigma(\theta) = -\varepsilon_0 E_n(b) = \varepsilon_0 \lim_{r\longrightarrow b}\frac{dV(r\geq a)}{dr} = -\frac{Q}{2\pi}\sum_{l=0}^{\infty}\frac{(-1)^l(2l+1)!!a^{2l}}{2^{l+1}(l+1)!b^{l+2}} P_{2l}(\cos\theta) \qquad [1.168]$$

Consequently, the total charge accumulated on the inner surface of the sphere is equal to:

$$Q_T = 2\pi b^2\int_0^{\pi}\sigma(\theta)\sin\theta d\theta = -\frac{2\pi b^2 Q}{2\pi}\sum_{l=0}^{\infty}\frac{(-1)^l(2l+1)!!a^{2l}}{2^{l+1}(l+1)!b^{l+2}}\times \int_0^{\pi} P_{2l}(\cos\theta)\sin\theta d\theta = -Q\sum_{l=0}^{\infty}\frac{(-1)^l(2l+1)!!a^{2l}}{2^{l+1}(l+1)!b^l}\frac{2}{4l+1}\delta_{l0} = -Q \qquad [1.169]$$

since $\int_0^{\pi} P_l(\cos\theta)\sin\theta d\theta = \frac{2}{2l+1}\delta_{l0}$.

EXAMPLE 1.12. (Ring placed within a conducting sphere).– Let us find the potential induced by a uniformly charged ring of radius a (total charge Q) enclosed in a grounded conducting sphere of radius b (held at zero potential). The centers of the ring and the sphere coincide, as shown in Figure 1.17. The charge density observed on the ring is:

$$\rho\left(\vec{\mathbf{r}}'\right) = \frac{Q\delta\left(r' - a\right)\delta\left(\cos\theta'\right)}{2\pi a r'} \qquad [1.170]$$

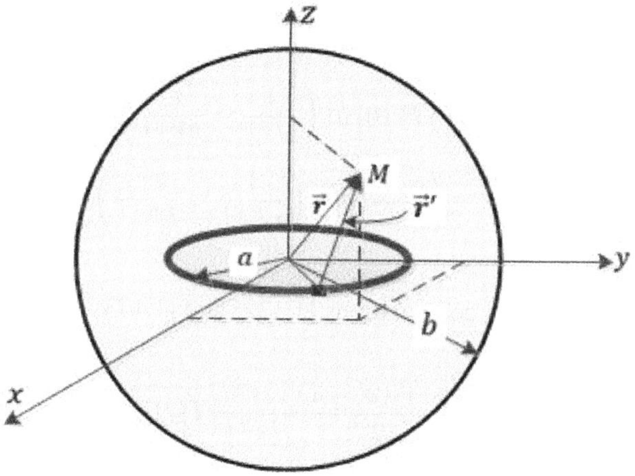

Figure 1.17. *Uniformly charged ring of radius a enclosed in a grounded conducting sphere of radius b*

Since the potential cancels out on the sphere and cannot depend on the variable φ, the functions $Y_l^m\left(\theta,\varphi\right)$ cannot contribute to the integral expression of $V\left(\vec{\mathbf{r}}\right)$ given by [1.162] unless $m = 0$. Thus, we can write that:

$$\begin{aligned} V &= \frac{1}{4\pi\varepsilon_0}\sum_{l=0}^{\infty}\int_{\tau} P_l\left(\cos\theta\right)P_l\left(\cos\theta'\right)\rho\left(\vec{\mathbf{r}}'\right)r_<^l\left(\frac{1}{r_>^{l+1}} - \frac{r_>^l}{b^{2l+1}}\right)d\tau' \\ &= \frac{Q}{8\pi^2\varepsilon_0 a}\sum_{l=0}^{\infty}P_l\left(\cos\theta\right)\int_0^a\int_0^{\pi}\int_0^{\pi}\delta\left(r'-a\right)\delta\left(\cos\theta'\right)P_l\left(\cos\theta'\right)\times \\ &\quad r_<^l\left(\frac{1}{r_>^{l+1}} - \frac{r_>^l}{b^{2l+1}}\right)r'dr'd\left(\cos\theta'\right)d\varphi' \\ &= \frac{Q}{4\pi\varepsilon_0}\sum_{l=0}^{\infty}P_l\left(\cos\theta\right)P_l\left(0\right)\int_0^a\delta\left(r'-a\right)r_<^l\left(\frac{1}{r_>^{l+1}} - \frac{r_>^l}{b^{2l+1}}\right)dr' \end{aligned} \qquad [1.171]$$

As in the previous example, two cases are considered:

1) For $r < a$, we see that $r_< = r$ and $r_> = r'$. In this case, the integral [1.171] is:

$$\begin{aligned} V(r \le a) &= \frac{Q}{4\pi\varepsilon_0} \sum_{l=0}^{\infty} P_l(\cos\theta) P_l(0) r^l \left(\frac{1}{a^{l+1}} - \frac{a^l}{b^{2l+1}} \right) \\ &= \frac{Q}{4\pi\varepsilon_0} \sum_{l=0}^{\infty} \frac{(-1)^l (2l-1)!!}{2^l l!} r^{2l} \left(\frac{1}{a^{2l+1}} - \frac{a^{2l}}{b^{4l+1}} \right) P_{2l}(\cos\theta) \end{aligned} \quad [1.172]$$

2) For $r > a$, we have: $r_< = r'$ and $r_> = r$. It follows that:

$$\begin{aligned} V(r > a) &= \frac{Q}{4\pi\varepsilon_0} \sum_{l=0}^{\infty} P_l(\cos\theta) P_l(0) a^l \left(\frac{1}{r^{l+1}} - \frac{r^l}{b^{2l+1}} \right) \\ &= \frac{Q}{4\pi\varepsilon_0} \sum_{l=0}^{\infty} \frac{(-1)^l (2l-1)!!}{2^l l!} a^{2l} \left(\frac{1}{r^{2l+1}} - \frac{r^{2l}}{b^{4l+1}} \right) P_{2l}(\cos\theta) \end{aligned} \quad [1.173]$$

In particular, when $b \longrightarrow \infty$, equations [1.172] and [1.173] reduce to:

$$V(r \le a) = \frac{Q}{4\pi\varepsilon_0} \sum_{l=0}^{\infty} \frac{(-)^l (2l-1)!!}{2^l l!} \frac{r^{2l}}{a^{2l+1}} P_{2l}(\cos\theta) \quad [1.174]$$

$$V(r > a) = \frac{Q}{4\pi\varepsilon_0} \sum_{l=0}^{\infty} \frac{(-)^l (2l-1)!!}{2^l l!} \frac{a^{2l}}{r^{2l+1}} P_{2l}(\cos\theta) \quad [1.175]$$

The charge density induced on the inner surface of the sphere is:

$$\sigma(\theta) = -\varepsilon_0 E_n(b) = -\frac{Q}{4\pi} \sum_{l=0}^{\infty} \frac{(-)^l (2l+1)!!}{2^l l!} \frac{a^{2l}}{b^{2l+2}} P_{2l}(\cos\theta) \quad [1.176]$$

and, consequently, the accumulated charge is equal to:

$$\begin{aligned} Q_T = 2\pi b^2 \int_0^{\pi} \sigma(\theta) \sin\theta d\theta = -\frac{2\pi b^2 Q}{4\pi} \sum_{l=0}^{\infty} \frac{(-1)^l (2l+1)!!}{2^l l!} \frac{a^{2l}}{b^{2l+1}} \times \\ \int_0^{\pi} P_{2l}(\cos\theta) \sin\theta d\theta = -\frac{Q}{2} \sum_{l=0}^{\infty} \frac{(-1)^l (2l+1)!!}{2^l l!} \frac{a^{2l}}{b^{2l}} \frac{2}{4l+1} \delta_{l0} = -Q \end{aligned} \quad [1.177]$$

1.6. Green–Dirichlet function from its differential equation

When there is both an inner and an outer spherical boundary, we obtain an infinite number of images, as the image produced by each surface is reflected in the other, and so on. In this case, it is simpler to solve the differential equation governing the Green–Dirichlet function $\Delta' G_D\left(\vec{\mathbf{r}},\vec{\mathbf{r}}'\right) = -\delta^3\left(\vec{\mathbf{r}}-\vec{\mathbf{r}}'\right)$. In spherical coordinates, this differential equation takes the form:

$$\begin{aligned}\frac{1}{r'}\frac{\partial^2\left(r'G_D\right)}{\partial r'^2}+\frac{1}{r'^2\sin\theta'}\frac{\partial}{\partial\theta'}\left(\sin\theta'\frac{\partial G_D}{\partial\theta'}\right)&\\+\frac{1}{r'^2\sin^2\theta'}\frac{\partial^2 G_D}{\partial\varphi'^2}=-\delta^3\left(\vec{\mathbf{r}}-\vec{\mathbf{r}}'\right)&\end{aligned} \quad [1.178]$$

Since the Green–Dirichlet function $G_D\left(\vec{\mathbf{r}},\vec{\mathbf{r}}'\right)$ depends on $\vec{\mathbf{r}}'$, it can always be developed on spherical harmonic functions as follows:

$$G_D\left(\vec{\mathbf{r}},\vec{\mathbf{r}}'\right)=\sum_{l=0}^{\infty}\sum_{m=-l}^{l}A_{l,m}\left(\theta,\varphi\right)R_l\left(r,r'\right)Y_l^m\left(\theta',\varphi'\right) \quad [1.179]$$

Let us substitute [1.179] into [1.178]:

$$\begin{aligned}\sum_{l=0}^{\infty}\sum_{m=-l}^{l}\left[\frac{1}{r'}\frac{\partial^2\left(r'R_l\right)}{\partial r'^2}-\frac{l\left(l+1\right)R_l}{r'^2}\right]A_{l,m}\left(\theta,\varphi\right)Y_l^m\left(\theta',\varphi'\right)&\\=-\delta^3\left(\vec{\mathbf{r}}-\vec{\mathbf{r}}'\right)&\end{aligned} \quad [1.180]$$

where $\left[\frac{1}{\sin\theta'}\frac{\partial}{\partial\theta'}\left(\sin\theta'\frac{\partial}{\partial\theta'}\right)+\frac{1}{\sin^2\theta'}\frac{\partial^2}{\partial\varphi'^2}\right]Y_l^m\left(\theta',\varphi'\right) = -l\left(l+1\right)Y_l^m\left(\theta',\varphi'\right)$ (properties of harmonic functions).

On the other hand, the second term in [1.180] can also be written as:

$$\begin{aligned}\delta^3\left(\vec{\mathbf{r}}-\vec{\mathbf{r}}'\right)&=\frac{\delta\left(r-r'\right)}{r'^2}\delta\left(\cos\theta-\cos\theta'\right)\delta\left(\varphi-\varphi'\right)\\&=\sum_{l=0}^{\infty}\sum_{m=-l}^{l}\frac{\delta\left(r-r'\right)}{r'^2}Y_l^m\left(\theta',\varphi'\right)\overline{Y_l^m\left(\theta,\varphi\right)}\end{aligned} \quad [1.181]$$

We therefore deduce that:

$$\begin{aligned}\sum_{l,m}\left[\frac{1}{r'}\frac{\partial^2\left(r'R_l\right)}{\partial r'^2}-\frac{l\left(l+1\right)R_l}{r'^2}\right]A_{l,m}\left(\theta,\varphi\right)Y_l^m\left(\theta',\varphi'\right)&\\=-\frac{\delta\left(r-r'\right)}{r'^2}\sum_{l,m}\overline{Y_l^m\left(\theta,\varphi\right)}Y_l^m\left(\theta',\varphi'\right)&\end{aligned} \quad [1.182]$$

It is easy to see that we can take $A_{l,m}(\theta,\varphi) = \overline{Y_l^m(\theta,\varphi)}$ and, consequently, we arrive at:

$$\frac{1}{r'}\frac{\partial^2(r'R_l)}{\partial r'^2} - \frac{l(l+1)R_l}{r'^2} = -\frac{\delta(r-r')}{r'^2} \qquad [1.183]$$

When $r \neq r'$, the right-hand side in expression [1.183] disappears. As a result, it is easy to solve the left-hand side, that is, to obtain the function $R_l(r,r')$ in terms of r'. The general solution is $R_l(r,r') = A_l r'^l + \frac{B_l}{r'^{l+1}}$. It follows that the differential equation [1.183] has the following solutions on either side of r:

$$R_l(r'<r) = A_l(r)\,r'^l + \frac{B_l(r)}{r'^{l+1}} \qquad [1.184]$$

$$R_l(r'>r) = C_l(r)\,r'^l + \frac{D_l(r)}{r'^{l+1}} \qquad [1.185]$$

The function G_D (and, therefore, R_l) must disappear on the boundaries. In the inner sphere of radius a, we must have $R_l(a) = A_l(r)\,a^l + \frac{B_l(r)}{a^{l+1}} = 0$; which means that $B_l(r) = -A_l(r)\,a^{2l+1}$. Similarly, for the outer spherical boundary of radius b, we must also obtain $R_l(b) = C_l(r)\,b^l + \frac{D_l(r)}{b^{l+1}} = 0$, which gives $C_l(r) = -\frac{D_l(r)}{b^{2l+1}}$. We therefore conclude that:

$$R_l(r'<r) = A_l(r)\left(r'^l - \frac{a^{2l+1}}{r'^{l+1}}\right) \qquad [1.186]$$

$$R_l(r'>r) = D_l(r)\left(\frac{1}{r'^{l+1}} - \frac{r'^l}{b^{2l+1}}\right) \qquad [1.187]$$

Since $G_D(\vec{\mathbf{r}},\vec{\mathbf{r}}') = G_D(\vec{\mathbf{r}}',\vec{\mathbf{r}})$, let us permute r and r' in [1.186], that is, $R_l(r'>r) = A_l(r')\left(r^l - \frac{a^{2l+1}}{r^{l+1}}\right)$, and compare with [1.187], $A_l(r')\left(r^l - \frac{a^{2l+1}}{r^{l+1}}\right)$ $= D_l(r)\left(\frac{1}{r'^{l+1}} - \frac{r'^l}{b^{2l+1}}\right)$. Therefore, it is obvious that $A_l(r') = A_l'\left(\frac{1}{r'^{l+1}} - \frac{r'^l}{b^{2l+1}}\right)$ and $D_l(r) = D_l'\left(r^l - \frac{a^{2l+1}}{r^{l+1}}\right)$. Finally, we obtain the following:

$$R_l(r'<r) = A_l\left(\frac{1}{r^{l+1}} - \frac{r^l}{b^{2l+1}}\right)\left(r'^l - \frac{a^{2l+1}}{r'^{l+1}}\right) \qquad [1.188]$$

$$R_l(r'>r) = A_l\left(r^l - \frac{a^{2l+1}}{r^{l+1}}\right)\left(\frac{1}{r'^{l+1}} - \frac{r'^l}{b^{2l+1}}\right) \qquad [1.189]$$

Clearly, the function R_l is continuous at $r' = r$. These two equations can be combined into a single one of the form:

$$R_l\left(r,r'\right)=A_l\left(r_<^l-\frac{a^{2l+1}}{r_<^{l+1}}\right)\left(\frac{1}{r_>^{l+1}}-\frac{r_>^l}{b^{2l+1}}\right) \qquad [1.190]$$

The constant A_l can be evaluated just by integrating [1.183] over an infinitesimal interval around r such that:

$$\lim_{\epsilon\longrightarrow 0}\int_{r-\epsilon}^{r+\epsilon}\left(\frac{d^2\left(r'R_l\right)}{dr'^2}-\frac{l\left(l+1\right)R_l}{r'}\right)dr'=-\lim_{\epsilon\longrightarrow 0}\int_{r-\epsilon}^{r+\epsilon}\frac{\delta\left(r-r'\right)}{r'}dr' \qquad [1.191]$$

or, again:

$$\left.\frac{d\left(r'R_l\right)}{dr'}\right|_{r'=r_+}-\left.\frac{d\left(r'R_l\right)}{dr'}\right|_{r'=r_-}=-\frac{1}{r} \qquad [1.192]$$

The integral $\lim_{\epsilon\longrightarrow 0}\int_{r-\epsilon}^{r+\epsilon}\frac{l(l+1)R_l}{r'}dr'=0$, since R_l is continuous at the point $r'=r$. Let us evaluate the terms $\left.\frac{d(r'R_l)}{dr'}\right|_{r'=r_+}$ and $\left.\frac{d(r'R_l)}{dr'}\right|_{r'=r_-}$:

$$\begin{aligned}\left.\frac{d\left(r'R_l\right)}{dr'}\right|_{r'=r_+}&=A_l\left(r^l-\frac{a^{2l+1}}{r^{l+1}}\right)\left(-\frac{l}{r^{l+1}}-\frac{\left(l+1\right)r^l}{b^{2l+1}}\right)\\&=A_l\left(\frac{la^{2l+1}}{r^{2l+2}}-\frac{\left(l+1\right)r^{2l}}{b^{2l+1}}-\frac{l}{r}+\left(l+1\right)\frac{a^{2l+1}}{b^{2l+1}}\frac{1}{r}\right)\end{aligned} \qquad [1.193]$$

$$\begin{aligned}\left.\frac{d\left(r'R_l\right)}{dr'}\right|_{r'=r_-}&=A_l\left(\frac{1}{r^{l+1}}-\frac{r^l}{b^{2l+1}}\right)\left(\left(l+1\right)r^l+\frac{la^{2l+1}}{r^{l+1}}\right)\\&=A_l\left(\frac{la^{2l+1}}{r^{2l+2}}-\frac{\left(l+1\right)r^{2l}}{b^{2l+1}}+\frac{l+1}{r}-l\frac{a^{2l+1}}{b^{2l+1}}\frac{1}{r}\right)\end{aligned} \qquad [1.194]$$

By inserting [1.193] and [1.194] into [1.192], we obtain $A_l=\frac{1}{(2l+1)\left[1-\left(\frac{a}{b}\right)^{2l+1}\right]}$. According to the results obtained, the Green–Dirichlet function is finally equal to:

$$\begin{aligned}G&=\sum_{l=0}^{\infty}\sum_{m=-l}^{l}\frac{Y_l^m\left(\theta',\varphi'\right)\overline{Y_l^m\left(\theta,\varphi\right)}}{\left(2l+1\right)\left[1-\left(\frac{a}{b}\right)^{2l+1}\right]}\left(r_<^l-\frac{a^{2l+1}}{r_<^{l+1}}\right)\left(\frac{1}{r_>^{l+1}}-\frac{r_>^l}{b^{2l+1}}\right)\\&=\sum_{l=0}^{\infty}\sum_{m=-l}^{l}\frac{Y_l^m\left(\theta',\varphi'\right)\overline{Y_l^m\left(\theta,\varphi\right)}}{\left(2l+1\right)\left[a^{2l+1}-b^{2l+1}\right]}\left(r_<^l-\frac{a^{2l+1}}{r_<^{l+1}}\right)\left(r_>^l-\frac{b^{2l+1}}{r_>^{l+1}}\right)\end{aligned} \qquad [1.195]$$

In particular, if $a\longrightarrow 0$, that is, the system has a single external boundary located at b, equation [1.195] reduces to:

$$G=\sum_{l=0}^{\infty}\sum_{m=-l}^{l}\frac{Y_l^m\left(\theta',\varphi'\right)\overline{Y_l^m\left(\theta,\varphi\right)}r_<^l}{\left(2l+1\right)}\left(\frac{1}{r_>^{l+1}}-\frac{r_>^l}{b^{2l+1}}\right) \qquad [1.196]$$

Similarly, when $b \longrightarrow \infty$ (a single internal boundary located at a for the system), we get:

$$G = \sum_{l=0}^{\infty} \sum_{m=-l}^{l} \frac{Y_l^m(\theta', \varphi') \overline{Y_l^m(\theta, \varphi)}}{(2l+1)\, r_>^{l+1}} \left(r_<^l - \frac{a^{2l+1}}{r_<^{l+1}} \right) \qquad [1.197]$$

EXAMPLE 1.13. (Cylindrical coordinates).– Let us now try to solve the Green equation in cylindrical coordinates:

$$\begin{aligned} \frac{1}{r'} \frac{\partial}{\partial r'} \left(r' \frac{\partial G_D}{\partial r'} \right) &+ \frac{1}{r'^2} \frac{\partial^2 G_D}{\partial \varphi'^2} \\ &+ \frac{\partial^2 G_D}{\partial z'^2} = -\frac{1}{r'} \delta(r - r')\, \delta(\varphi - \varphi')\, \delta(z - z') \end{aligned} \qquad [1.198]$$

where the delta function is given here in cylindrical coordinates. The terms $\delta(z - z')$ and $\delta(\varphi - \varphi')$ can be expressed as follows:

$$\begin{aligned} \delta(z - z') &= \frac{1}{2\pi} \int_{-\infty}^{\infty} d\mathrm{k}\, e^{\mathrm{ik}(z - z')} = \frac{1}{\pi} \int_0^{\infty} d\mathrm{k} \cos\left[\mathrm{k}(z - z')\right] \\ \delta(\varphi - \varphi') &= \frac{1}{2\pi} \sum_{m=-\infty}^{\infty} e^{im(\varphi - \varphi')} \end{aligned} \qquad [1.199]$$

Let us expand the Green function in the same way:

$$G(\vec{\mathbf{r}}, \vec{\mathbf{r}}') = \frac{1}{2\pi^2} \sum_{m=-\infty}^{\infty} \int_0^{\infty} d\mathrm{k}\, e^{im(\varphi - \varphi')} \cos\left[\mathrm{k}(z - z')\right] R_m(r, r') \qquad [1.200]$$

By substituting [1.200] into [1.198] and taking into account the expressions [1.199], we obtain the radial equation:

$$\frac{1}{r'} \frac{d}{dr'} \left(r' \frac{dR_m}{dr'} \right) - \left(\mathrm{k}^2 + \frac{m^2}{r'^2} \right) R_m = -\frac{1}{r'} \delta(r - r') \qquad [1.201]$$

For $r \neq r'$, this last equation reduces to the modified Bessel equation. The general solution is a linear combination of the functions $I_m(\mathrm{k}r')$ and $K_m(\mathrm{k}r')$. Consequently, equation [1.201] admits the solutions on either side of r: $R_m(r' < r) = A_m(r) I_m(\mathrm{k}r') + B_m(r) K_m(\mathrm{k}r')$ and $R_m(r' > r) = C_m(r) I_m(\mathrm{k}r') + D_m(r) K_m(\mathrm{k}r')$. Since G_D cancels on the boundaries, we must have the following restrictions on the inner (respectively, outer) boundary of radius a

(respectively, b): $A_m(r) I_m(\mathrm{k}a) + B_m(r) K_m(\mathrm{k}a) = 0$ and $C_m(r) I_m(\mathrm{k}b) + D_m(r) K_m(\mathrm{k}b) = 0$. This gives $B_m(r) = -A_m(r) \frac{I_m(\mathrm{k}a)}{K_m(\mathrm{k}a)}$ and $C_m(r) = -D_m(r) \frac{K_m(\mathrm{k}b)}{I_m(\mathrm{k}b)}$. So, we see that $R_m(r' < r) = A_m(r) \left[I_m(\mathrm{k}r') - \frac{I_m(\mathrm{k}a)}{K_m(\mathrm{k}a)} K_m(\mathrm{k}r')\right]$ and $R_m(r' > r) = D_m(r) \left[K_m(\mathrm{k}r') - \frac{K_m(\mathrm{k}b)}{I_m(\mathrm{k}b)} I_m(\mathrm{k}r')\right]$. The symmetry of the Green function in r and r' imposes that:

$$
\begin{aligned}
R_m|_{r'<r} &= A_m \left[I_m(\mathrm{k}r') - \frac{I_m(\mathrm{k}a)}{K_m(\mathrm{k}a)} K_m(\mathrm{k}r')\right]\left[K_m(\mathrm{k}r) - \frac{K_m(\mathrm{k}b)}{I_m(\mathrm{k}b)} I_m(\mathrm{k}r)\right] \\
R_m|_{r'>r} &= A_m \left[I_m(\mathrm{k}r) - \frac{I_m(\mathrm{k}a)}{K_m(\mathrm{k}a)} K_m(\mathrm{k}r)\right]\left[K_m(\mathrm{k}r') - \frac{K_m(\mathrm{k}b)}{I_m(\mathrm{k}b)} I_m(\mathrm{k}r')\right]
\end{aligned} \quad [1.202]
$$

or, again:

$$
\begin{aligned}
R_m(r, r') = A_m &\left[I_m(\mathrm{k}r_<) - \frac{I_m(\mathrm{k}a)}{K_m(\mathrm{k}a)} K_m(\mathrm{k}r_<)\right] \times \\
&\left[K_m(\mathrm{k}r_>) - \frac{K_m(\mathrm{k}b)}{I_m(\mathrm{k}b)} I_m(\mathrm{k}r_>)\right] = A_m \psi_1(\mathrm{k}r_<) \psi_2(\mathrm{k}r_>)
\end{aligned} \quad [1.203]
$$

The normalization factor A_m of the product $\psi_1(\mathrm{k}r_<) \psi_2(\mathrm{k}r_>)$ is determined from an equation that is analogous to [1.192]:

$$
\left.\frac{dR_m}{dr'}\right|_{r'=r_+} - \left.\frac{dR_m}{dr'}\right|_{r'=r_-} = -\frac{1}{r} \quad [1.204]
$$

where $|_{r'=r_\pm}$ signifies that the term is evaluated at $r' = r \pm \epsilon$. From [1.204], it is clear that:

$$
\left.\frac{dR_m}{dr'}\right|_{r'=r_+} - \left.\frac{dR_m}{dr'}\right|_{r'=r_-} = \mathrm{k}\left(\psi_1 \psi_2' - \psi_1' \psi_2\right) = \mathrm{k} W[\psi_1, \psi_2] = -\frac{1}{r} \quad [1.205]
$$

where $W[\psi_1, \psi_2]$ is the Wronskian of ψ_1 and ψ_2. If there are no boundaries, that is, when $a \longrightarrow 0$ and $b \longrightarrow \infty$, the functions $\psi_1(\mathrm{k}r_<)$ and $\psi_2(\mathrm{k}r_>)$ reduce to $I_m(\mathrm{k}r_<)$ and $K_m(\mathrm{k}r_>)$, respectively, because $\lim_{a \longrightarrow 0} \frac{I_m(\mathrm{k}a)}{K_m(\mathrm{k}a)} = 0$ and $\lim_{b \longrightarrow \infty} \frac{K_m(\mathrm{k}b)}{I_m(\mathrm{k}b)} = 0$. In fact, we have $I_m(z) \sim \frac{1}{\Gamma(m+1)} \left(\frac{z}{2}\right)^m$ and $K_m(z) \sim \begin{cases} \frac{\Gamma(m)}{2} \left(\frac{2}{z}\right)^m & m \neq 0 \\ -\left[\ln(x/2) + \gamma\right] & m = 0 \end{cases}$, for $z \ll 1$. Similarly, we also have $I_m(z) \sim \sqrt{\frac{1}{2\pi z}} e^z \left[1 + \mathcal{O}\left(\frac{1}{z}\right)\right]$ and $K_m(z) \sim \sqrt{\frac{\pi}{2z}} e^{-z} \left[1 + \mathcal{O}\left(\frac{1}{z}\right)\right]$, for $z \gg 1$.

The constant A_m is determined from [1.205]. Using the approximate expressions for $I_m(z)$ and $K_m(z)$, for $z \ll 1$, we find $A_m = 1$. Therefore, the expansion of $\frac{1}{|\vec{\mathbf{r}} - \vec{\mathbf{r}}'|}$ becomes:

$$\frac{1}{|\vec{\mathbf{r}} - \vec{\mathbf{r}}'|} = \frac{2}{\pi} \sum_{m=-\infty}^{\infty} \int_0^\infty dk e^{im(\varphi - \varphi')} \cos[\mathrm{k}(z - z')] \times I_m(\mathrm{k}r_<) K_m(\mathrm{k}r_>) \quad [1.206]$$

This can also be written entirely in terms of real functions as follows:

$$\frac{1}{|\vec{\mathbf{r}} - \vec{\mathbf{r}}'|} = \frac{4}{\pi} \int_0^\infty \cos[\mathrm{k}(z - z')]\, dk \left[\frac{1}{2} I_0(\mathrm{k}r_<) K_0(\mathrm{k}r_>) + \sum_{m=1}^{\infty} \cos m(\varphi - \varphi') I_m(\mathrm{k}r_<) K_m(\mathrm{k}r_>)\right] \quad [1.207]$$

Some useful mathematical results can be obtained from the last expression.

1) By making $\vec{\mathbf{r}}' \longrightarrow 0$, only the term in $m = 0$ remains. The result is as follows:

$$\frac{1}{\sqrt{r^2 + z^2}} = \frac{2}{\pi} \int_0^\infty \cos \mathrm{k}z K_0(\mathrm{k}r)\, dk \quad [1.208]$$

2) By replacing r^2 by $r^2 + r'^2 - 2rr' \cos(\varphi - \varphi')$ in [1.208], the left-hand term becomes the inverse distance $\frac{1}{|\vec{\mathbf{r}} - \vec{\mathbf{r}}'|}$ with $z' = 0$. Taking $z' = 0$ from [1.206] and comparing the two results, we deduce:

$$K_0\left(\mathrm{k}\sqrt{r^2 + r'^2 - 2rr' \cos(\varphi - \varphi')}\right) = I_0(\mathrm{k}r_<) K_0(\mathrm{k}r_>) + 2\sum_{m=1}^{\infty} \cos m(\varphi - \varphi') I_m(\mathrm{k}r_<) K_m(\mathrm{k}r_>) \quad [1.209]$$

3) Now taking $\mathrm{k} \longrightarrow \infty$, we obtain an expansion of the Green function in polar coordinates (in $2D$):

$$\ln\left(\frac{1}{\sqrt{r^2 + r'^2 - 2rr' \cos(\varphi - \varphi')}}\right) = \ln\left(\frac{1}{r_>}\right) + \sum_{m=1}^{\infty} \frac{1}{m}\left(\frac{r_<}{r_>}\right)^m \cos m(\varphi - \varphi') \quad [1.210]$$

1.7. Eigenfunctions method

Another method used to obtain an expansion of the Green function is to use eigenfunctions.

1) To clarify what is meant by eigenfunctions, let us consider the elliptic differential equation:

$$\Delta\psi\left(\overrightarrow{\mathbf{r}}\right)+\left[f\left(\overrightarrow{\mathbf{r}}\right)+\lambda\right]\psi\left(\overrightarrow{\mathbf{r}}\right)=0 \qquad [1.211]$$

where $f\left(\overrightarrow{\mathbf{r}}\right)$ and λ are real. If the solutions $\psi\left(\overrightarrow{\mathbf{r}}\right)$ must satisfy certain boundary conditions on the boundaries of a volume τ (medium of study), then equation [1.211] will only have solutions for certain values of λ. These values of λ, denoted by λ_n, are called *eigenvalues* and the associated solutions, $\psi_n\left(\overrightarrow{\mathbf{r}}\right)$, are called *eigenfunctions*. The differential equation associated with the λ_n eigenvalue is written as:

$$\Delta\psi_n\left(\overrightarrow{\mathbf{r}}\right)+\left[f\left(\overrightarrow{\mathbf{r}}\right)+\lambda_n\right]\psi_n\left(\overrightarrow{\mathbf{r}}\right)=0 \qquad [1.212]$$

Using methods that are similar to those used to prove the orthogonality of the Legendre or Bessel functions, it can be shown that the eigenfunctions are orthogonal, that is, they satisfy the following relations:

$$\int_\tau \psi_m^\star\left(\overrightarrow{\mathbf{r}}\right)\psi_n\left(\overrightarrow{\mathbf{r}}\right)d\tau=\delta_{mn} \qquad [1.213]$$

where the eigenfunctions are assumed to be normalized. In fact, it is easy to check that $\psi_m^\star\Delta\psi_n+(f+\lambda_n)\psi_m^\star\psi_n=0$ and $\psi_n\Delta\psi_m^\star+(f+\lambda_m)\psi_n\psi_m^\star=0$. This shows that $(\lambda_m-\lambda_n)\int_\tau\psi_m^\star\psi_n d\tau=\int_\tau\overrightarrow{\nabla}\cdot\left(\psi_m^\star\overrightarrow{\nabla}\psi_n-\psi_n^\star\overrightarrow{\nabla}\psi_m\right)d\tau=\oint_S\left(\psi_m^\star\overrightarrow{\nabla}\psi_n-\psi_n^\star\overrightarrow{\nabla}\psi_m\right)\cdot d\overrightarrow{\mathbf{S}}$. Since ψ_m and ψ_n disappear on the boundary, we can deduce that $(\lambda_m-\lambda_n)\int_\tau\psi_m^\star\psi_n d\tau=0$ and, consequently, $\int_\tau\psi_m^\star\psi_n d\tau=\delta_{mn}$. The spectrum of eigenvalues, λ_n, can be a discrete set or a continuum, or both. It is assumed that all eigenfunctions form a complete set.

2) Let us now assume that we wish to find the Green function that satisfies the equation:

$$\Delta_{\overrightarrow{\mathbf{r}}'}G\left(\overrightarrow{\mathbf{r}},\overrightarrow{\mathbf{r}}'\right)+\left[f\left(\overrightarrow{\mathbf{r}}'\right)+\lambda\right]G\left(\overrightarrow{\mathbf{r}},\overrightarrow{\mathbf{r}}'\right)=-\delta^3\left(\overrightarrow{\mathbf{r}}-\overrightarrow{\mathbf{r}}'\right) \qquad [1.214]$$

where λ is any value (different from the eigenvalues λ_n). Additionally, let us assume that the Green's function satisfies the same boundary conditions as the eigenfunctions

[1.212]. Consequently, the Green function can be expanded over the eigenfunctions as follows:

$$G\left(\vec{\mathbf{r}}, \vec{\mathbf{r}}'\right) = \sum_n a_n\left(\vec{\mathbf{r}}\right)\psi_n\left(\vec{\mathbf{r}}'\right) \qquad [1.215]$$

The substitution in the differential equation [1.214] leads to:

$$\sum_m a_m\left(\vec{\mathbf{r}}\right)(\lambda - \lambda_m)\psi_m\left(\vec{\mathbf{r}}'\right) = -\delta^3\left(\vec{\mathbf{r}} - \vec{\mathbf{r}}'\right) \qquad [1.216]$$

By multiplying both sides by $\psi_n^\star\left(\vec{\mathbf{r}}'\right)$ and integrating over the volume τ, the orthogonality condition [1.213] reduces the left side to a single term:

$$a_n\left(\vec{\mathbf{r}}\right) = \frac{\psi_n^\star\left(\vec{\mathbf{r}}\right)}{\lambda_n - \lambda} \qquad [1.217]$$

Therefore, the Green functions [1.215] is written as:

$$G\left(\vec{\mathbf{r}}, \vec{\mathbf{r}}'\right) = \sum_n \frac{\psi_n^\star\left(\vec{\mathbf{r}}\right)\psi_n\left(\vec{\mathbf{r}}'\right)}{\lambda_n - \lambda} \qquad [1.218]$$

For a continuous spectrum, the sum is replaced by an integral. To apply what we have just seen to the Green function of the Poisson equation, let us choose $f\left(\vec{\mathbf{r}}'\right) = 0$ and $\lambda = 0$ in [1.214]. Therefore, the Green function is written as:

$$G\left(\vec{\mathbf{r}}, \vec{\mathbf{r}}'\right) = \sum_n \lambda_n^{-1}\psi_n^\star\left(\vec{\mathbf{r}}\right)\psi_n\left(\vec{\mathbf{r}}'\right) \qquad [1.219]$$

where $\psi_n\left(\vec{\mathbf{r}}\right)$ are the solutions to the wave equations in the volume τ:

$$(\Delta + \lambda_n)\psi_n\left(\vec{\mathbf{r}}\right) = 0 \qquad [1.220]$$

3) *Green function in infinite space*: as a first, essentially trivial example, let us give the development of the Green function in all space, that is, that of $\frac{1}{\left|\vec{\mathbf{r}} - \vec{\mathbf{r}}'\right|}$. This expansion is carried out in terms of the solutions of the wave equation (over all space):

$$\left(\Delta + \mathsf{k}^2\right)\psi_{\vec{\mathsf{k}}}\left(\vec{\mathbf{r}}\right) = 0 \qquad [1.221]$$

where k^2 are the continuous eigenvalues and the eigenfunctions:

$$\psi_{\vec{\mathsf{k}}}\left(\vec{\mathbf{r}}\right) = \frac{1}{(2\pi)^{3/2}} e^{\mathrm{i}\,\vec{\mathsf{k}}\cdot\vec{\mathbf{r}}} \qquad [1.222]$$

These eigenfunctions are normalized as follows:

$$\int_{\tau} \psi^{\star}_{\vec{\mathsf{k}}'}\left(\vec{\mathbf{r}}\right) \psi_{\vec{\mathsf{k}}}\left(\vec{\mathbf{r}}\right) d\tau = \delta^3\left(\vec{\mathsf{k}} - \vec{\mathsf{k}}'\right) \qquad [1.223]$$

According to [1.219], the Green function (in infinite space) admits the following expansion:

$$\frac{1}{\left|\vec{\mathbf{r}} - \vec{\mathbf{r}}'\right|} = \frac{1}{2\pi^2} \int \frac{e^{-\mathrm{i}\,\vec{\mathsf{k}}\cdot\left(\vec{\mathbf{r}} - \vec{\mathbf{r}}'\right)}}{\mathsf{k}^2} d\mathsf{k}_x d\mathsf{k}_y d\mathsf{k}_z \qquad [1.224]$$

This is exactly the three-dimensional Fourier integral representation of $\frac{1}{\left|\vec{\mathbf{r}} - \vec{\mathbf{r}}'\right|}$.

4) *Green function in a rectangular box*: as a second example, let us consider the Green function for a Dirichlet problem inside a rectangular box bounded by the planes $x = y = z = 0$, $x = a$, $y = b$ and $z = c$. The expansion must be carried out in terms of the eigenfunctions of the wave equation:

$$\left(\Delta + \mathsf{k}^2_{lmn}\right) \psi_{lmn}\left(x, y, z\right) = 0 \qquad [1.225]$$

where the eigenfunctions (which disappear on the boundary surfaces) are:

$$\psi_{lmn}\left(x, y, z\right) = \sqrt{\frac{8}{abc}} \sin\left(\frac{l\pi x}{a}\right) \sin\left(\frac{m\pi y}{b}\right) \sin\left(\frac{n\pi z}{c}\right) \qquad [1.226]$$

with:

$$\mathsf{k}^2_{lmn} = \pi^2\left(\frac{l^2}{a^2} + \frac{m^2}{b^2} + \frac{n^2}{c^2}\right) \qquad [1.227]$$

This gives us the Green function (in the rectangular box):

$$G\left(\vec{\mathbf{r}}, \vec{\mathbf{r}}'\right) = \frac{8}{\pi^2 abc} \times \sum_{l,m,n=1}^{\infty} \frac{\sin\left(\frac{l\pi x}{a}\right) \sin\left(\frac{l\pi x'}{a}\right) \sin\left(\frac{m\pi y}{b}\right) \sin\left(\frac{m\pi y'}{b}\right) \sin\left(\frac{n\pi z}{c}\right) \sin\left(\frac{n\pi z'}{c}\right)}{\frac{l^2}{a^2} + \frac{m^2}{b^2} + \frac{n^2}{c^2}} \qquad [1.228]$$

The expansion [1.197], for the spherical coordinates, and [1.206] for the cylindrical coordinates, can be reproduced using the same method.

Chapter 2

Electromagnetic Induction Phenomena

2.1. Experimental observations

Let us examine the effect of a *slow* variation in fields $\vec{\mathbf{E}}$ and $\vec{\mathbf{B}}$. By slow, we mean that the dimensions of the electrical circuits are very small compared with the wavelengths of the fields involved in propagation. In this case, we can consider that the electric current is, at a given instant, the same throughout the circuit. We have already seen that charges experience a force when moving in a magnetic field. Conversely, it is clear that a moving magnetic field source exerts a force on a stationary charge. At the charge location, a moving magnetic field source is perceived as a time-varying magnetic field, which implies that the force must depend on $\frac{\partial \vec{\mathbf{B}}}{\partial t}$. This phenomenon is known as the *electromagnetic induction effect*. In addition, this force felt by the stationary charge must also be interpreted as resulting from an electric field, since the particle is immobile.

To illustrate the effect of electromagnetic induction, consider a circuit (1) through which a current of intensity I flows. The solenoid in circuit (1) creates a magnetic field $\vec{\mathbf{B}}$ in space. A second circuit (2) with no current source (no generator) connected to a current meter is placed close to circuit (1). Circuit (2) is therefore immersed in the magnetic field $\vec{\mathbf{B}}$ of (1), as shown in Figure 2.1. The following can be seen:

– *First experiment*: the circuit (2) is deformed or moved. An *induced current* then appears in circuit (2).

– *Second experiment*: the circuit (2) is held stationary. The circuit (1) through which current I flows is moved. An induced current appears in circuit (2). We therefore conclude that only the relative movement of the two circuits is involved.

– *Third experiment*: both circuits are stationary and the current I in circuit (1) is varied. Once again, an induced current appears in circuit (2).

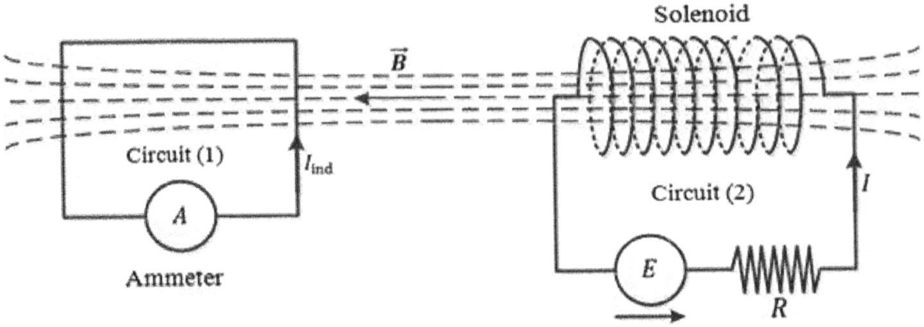

Figure 2.1. *Electromagnetic induction due to the relative movement of the two circuits (Experiments 1 and 2) and the magnetic field created by the solenoid (Experiment 3)*

For an observer attached to the laboratory, the first experiment corresponds to the displacement of a circuit in a magnetic field $\vec{\mathbf{B}}$ constant at any point in space, that is, $\frac{\partial \vec{\mathbf{B}}}{\partial t} = 0$. The other two experiments correspond to an immobile circuit in a region of space, where $\vec{\mathbf{B}}$ varies with time, that is, $\frac{\partial \vec{\mathbf{B}}}{\partial t} \neq 0$. In all three cases, an induced electromotive force (EMF) appears, which we will try to define in the following examples.

EXAMPLE 2.1.– Let us consider a closed circuit consisting of a conducting bar MN running on two parallel horizontal metal rails connected by a conducting wire PQ, as shown in Figure 2.2.

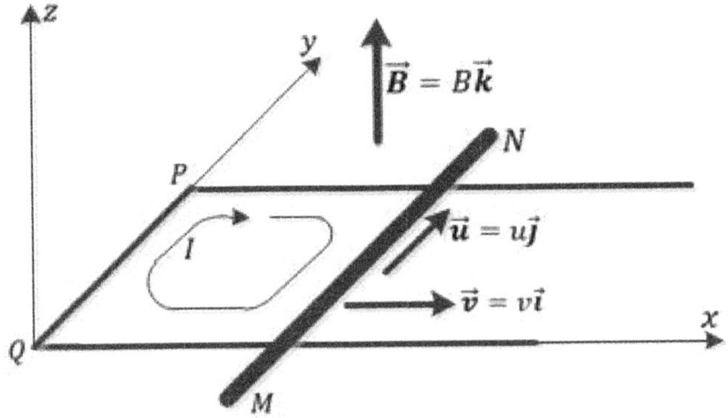

Figure 2.2. *Induction of an electric current in the $MNPQM$ circuit by the movement of the bar and the presence of a uniform magnetic field perpendicular to the plane of the two rails*

The whole system is immersed in a vertical, static and uniform magnetic field $\overrightarrow{\mathbf{B}} = B\,\overrightarrow{\mathbf{k}}$. The MN bar is assumed to move at constant speed $\overrightarrow{\mathbf{v}} = v\,\overrightarrow{\mathbf{i}}$. As a result, electric charges (in this case, electrons) move along the bar at a speed $\overrightarrow{\mathbf{u}} = u\,\overrightarrow{\mathbf{j}}$. The speed of an $-e$ electron relative to the field $\overrightarrow{\mathbf{B}}$ is $\overrightarrow{\mathbf{V}} = v\,\overrightarrow{\mathbf{i}} + u\,\overrightarrow{\mathbf{j}}$. Each electron is therefore subject to the Lorentz force:

$$\overrightarrow{\mathbf{F}}_L = -e\left(\overrightarrow{\mathbf{V}} \times \overrightarrow{\mathbf{B}}\right) = -euB\,\overrightarrow{\mathbf{i}} + evB\,\overrightarrow{\mathbf{j}} \quad [2.1]$$

During a dy displacement of the electron through the bar, the force $\overrightarrow{\mathbf{F}}_L$ performs the elementary work:

$$dW = (F_L)_y\, dy = evBdy \quad [2.2]$$

This work is equivalent to that produced by an elementary EMF, called the *induced EMF*, $d\mathcal{E}_{\text{ind}}$ defined as follows:

$$dW = -ed\mathcal{E}_{\text{ind}} \quad [2.3]$$

Comparing [2.2] with [2.3], we obtain:

$$d\mathcal{E}_{\text{ind}} = -vBdy \quad [2.4]$$

Consequently, the total EMF generated between the two ends M and N of the movable bar is:

$$\mathcal{E}_{\text{ind}} = -\int_0^l vBdy = -vBl \quad [2.5]$$

where l is the distance measured between the two rails. From an electrical point of view, the bar MN will behave exactly like a generator and will produce a current in the closed circuit $MNPQM$, which, according to Ohm's law, is equal to:

$$I_{\text{ind}} = \frac{\mathcal{E}_{\text{ind}}}{R} = -\frac{vBl}{R} \quad [2.6]$$

where R is the circuit resistance. Clearly, I_{ind} depends on the speed v of the bar and the induced field B (the two causes of the electromagnetic induction effect). If one of the two causes disappears, so does the induced current. Since $v = \frac{dx}{dt}$, where dx is the elementary displacement of the bar along the rails, equation [2.5] becomes:

$$\mathcal{E}_{\text{ind}} = -Bl\frac{dx}{dt} \quad [2.7]$$

By introducing the *flux cut* by the bar during its displacement $d\Phi_c = Bldx$, we deduce, finally, that:

$$\mathcal{E}_{\text{ind}} = -\frac{d\Phi_c}{dt} \tag{2.8}$$

The cut flux $d\Phi_c$ is also written $d\Phi_c = Bldx = Bl\,(x + dx) - Blx = \Phi\,(t + dt) - \Phi\,(t) = d\Phi\,(t)$, where $\Phi\,(t)$ is the flux passing through the circuit at time t, that is:

$$\Phi\,(t) = \int_{\text{surf.}MNPQ} \vec{\mathbf{B}} \cdot d\vec{\mathbf{S}} \tag{2.9}$$

It therefore follows that:

$$\mathcal{E}_{\text{ind}} = -\frac{d\Phi}{dt} \tag{2.10}$$

This relation is quite general and constitutes the *Faraday law*: *a closed circuit (here $MNPQM$) traversed by a variable magnetic flux is the seat of an electromotive induction force that opposes the flux variation that causes it.* We will come back to this later. The force $\vec{\mathbf{F}} = -e\left(\vec{\mathbf{v}} \times \vec{\mathbf{B}}\right) = evB\,\vec{\mathbf{j}}$ that sets the electrons in motion is also written as:

$$\vec{\mathbf{F}} = -e\vec{\mathbf{E}}_{\text{ind}} \tag{2.11}$$

where $\vec{\mathbf{E}}_{\text{ind}} = -vB\,\vec{\mathbf{j}}$ is the *electromotive field*. Let us calculate the circulation of this electromotive field along the closed contour $MNPQM$:

$$\begin{aligned} \oint_{MNPQM} \vec{\mathbf{E}}_{\text{ind}} \cdot d\vec{\mathbf{l}} &= -\oint_{MNPQM} \left(vB\,\vec{\mathbf{j}}\right) \cdot d\vec{\mathbf{l}} \\ &= -\int_M^N vBdy - \int_{NPQM} \left(vB\,\vec{\mathbf{j}}\right) \cdot d\vec{\mathbf{l}} \end{aligned} \tag{2.12}$$

The last integral is zero, because the velocity is zero. It follows that:

$$\oint_{MNPQM} \vec{\mathbf{E}}_{\text{ind}} \cdot d\vec{\mathbf{l}} = -vBl = \mathcal{E}_{\text{ind}} \tag{2.13}$$

The circulation of the electromotive field along the closed circuit $MNPQM$ is equal to $\mathcal{E}_{\text{ind}}$: it is not zero. We therefore conclude that the electromotive field does not derive from a potential. On the other hand, we know that any electrostatic field $\vec{\mathbf{E}}_{\text{elec}}$ that is added to $\vec{\mathbf{E}}_{\text{ind}}$ derives from a potential, that is, $\oint_{MNPQM} \vec{\mathbf{E}}_{\text{elec}} \cdot d\vec{\mathbf{l}} = 0$. Considering the total electric field $\vec{\mathbf{E}} = \vec{\mathbf{E}}_{\text{elec}} + \vec{\mathbf{E}}_{\text{ind}}$, we see that:

$$\oint_{MNPQM} \vec{\mathbf{E}} \cdot d\vec{\mathbf{l}} = \mathcal{E}_{\text{ind}} = -\frac{d\Phi}{dt} \qquad [2.14]$$

From [2.9], we get:

$$\oint_{MNPQM} \vec{\mathbf{E}} \cdot d\vec{\mathbf{l}} = -\frac{d}{dt}\left(\int_{\text{surf}.MNPQ} \vec{\mathbf{B}} \cdot d\vec{\mathbf{S}}\right) \qquad [2.15]$$

This is the *Maxwell–Faraday equation*. ■

We will come back to this equation later [2.15]. Once again, the following important remarks should be made:

1) The variation in magnetic flux through a closed circuit can have two causes. Either the circuit is subjected to a fixed field and is deformed or forced to move (in rotation, translation, etc.). This is known as the *Lorentz induction effect*. Or, the circuit is stationary, and it is the inducing magnetic field that varies over time. This is known as the *Neumann induction effect*. More generally, both phenomena can occur simultaneously.

2) To create an induction effect in a conductor, it is not even necessary to have a closed circuit. For example, if the MN bar moves in a stationary external magnetic field, the Lorentz force will cause the electrons to move until an electrostatic field is established that compensates for the electromotive field, as in the *Hall effect*.

3) It can happen that there is no apparent variation in flow through a closed circuit, as we will see with the *Barlow wheel*.

EXAMPLE 2.2.– Now let us suppose that the bar MN is launched at speed v_0, then subjected to the action of the external magnetic field. There are two equations to describe its motion:

$$\mathcal{E}_{\text{ind}} = -\frac{d\Phi}{dt} = -Blv = RI_{\text{ind}} \qquad \text{(electrical equation)} \qquad [2.16]$$

$$m\frac{dv}{dt} = -euB = I_{\text{ind}}lB \qquad \text{(mechanical equation)} \qquad [2.17]$$

because $I_{\text{ind}} = -\frac{eu}{l}$. m represents here the mass of the bar and v its instantaneous velocity. Taking the expression of I_{ind} as a function of v from [2.16] and inserting it into [2.17], we obtain:

$$\frac{dv}{dt} = -\frac{v}{\tau} \qquad [2.18]$$

where $\tau = \frac{mR}{l^2B^2}$. This equation has the solution:

$$v(t) = v_0 e^{-t/\tau} \qquad [2.19]$$

The expression for the induced current intensity follows immediately from this:

$$I_{\text{ind}}(t) = -\frac{Blv_0}{R} e^{-t/\tau} \qquad [2.20]$$

This solution indicates that after a few seconds the bar will stop and the induced current will disappear. The variation per unit time of the kinetic energy of the bar is equal to:

$$\frac{d}{dt}\left(\frac{1}{2}mv^2\right) = mv\frac{dv}{dt} = I_{\text{ind}}lBv = -I_{\text{ind}}\mathcal{E}_{\text{ind}} = -RI_{\text{ind}}^2 \qquad [2.21]$$

This relation clearly shows the conversion of the bar's kinetic energy into electrical energy, which in turn is transformed into heat by the *Joule effect*. Movement stops when all initial kinetic energy has been consumed.

EXAMPLE 2.3.– The conducting wire PQ contains a voltage generator E, as shown in Figure 2.3. At time $t = 0$, the current in the circuit is $I_0 = \frac{E}{R}$, where R is the total resistance of the circuit and assumed constant. At this point, the Laplace force $I_0lB\,\overrightarrow{\mathrm{i}}$ is exerted on the bar, and the mobile bar MN starts moving. At this point, the electromagnetic induction effect starts. Let I and v denote, respectively, the values of the current in the circuit and the speed of the bar at $t > 0$. The equations governing the evolution of the system are given as:

$$\begin{aligned} \mathcal{E}_{\text{ind}} &= -\frac{d\Phi}{dt} = -Blv \\ m\frac{dv}{dt} &= IlB \end{aligned} \qquad [2.22]$$

where $E + \mathcal{E}_{\text{ind}} = RI$. This implies that:

$$\frac{dv}{dt} = \frac{ElB}{mR} - \frac{v}{\tau} \qquad [2.23]$$

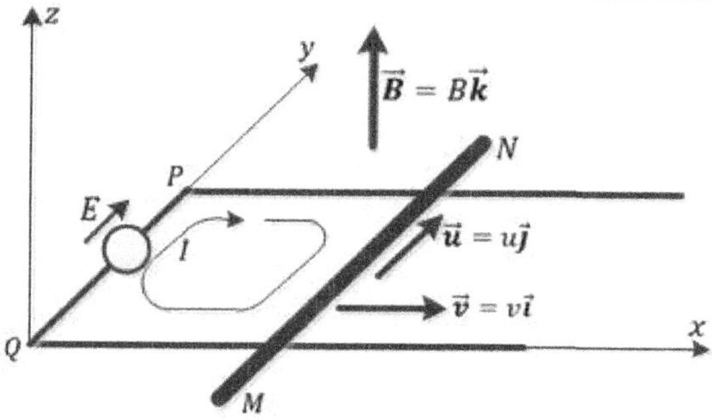

Figure 2.3. *Adding a voltage generator E to the line PQ creates an electromotive force that opposes E until it cancels it out. The bar MN will acquire a uniform velocity v_{lim} in the direction Ox*

Taking into account the initial condition $v(0) = 0$, we derive the following expression for $v(t)$:

$$v(t) = \frac{E}{Bl}\left(1 - e^{-t/\tau}\right) \qquad [2.24]$$

We can see that after a few τ the speed reaches a limit value equal to $v_{\text{lim}} = \frac{E}{Bl}$. In this case, the induced EMF is given, as a function of time, by:

$$\mathcal{E}_{\text{ind}}(t) = -Blv(t) = -E\left(1 - e^{-t/\tau}\right) \qquad [2.25]$$

The EMF will very quickly reach the limit value $-E$, progressively opposing the generator voltage, that is, the cause of the bar's movement, until it cancels it out. The current intensity is:

$$I(t) = \frac{E + \mathcal{E}_{\text{ind}}}{R} = \frac{E}{R}e^{-t/\tau} \qquad [2.26]$$

The initial current $\frac{E}{R}$ is gradually compensated for by the induced current, which opposes it. To calculate the energy balance, let us multiply the equation $E + \mathcal{E}_{\text{ind}} = RI$ by I:

$$EI = -\mathcal{E}_{\text{ind}}I + RI^2 \qquad [2.27]$$

This relation shows that the energy delivered by the generator is partly used by the induction effect and partly dissipated by the Joule effect. Thus, the generator will produce the energy:

$$W_G = \int_0^{\infty} EIdt = \frac{E^2}{R}\int_0^{\infty} e^{-t/\tau}dt = \frac{E^2\tau}{R} = \frac{mE^2}{B^2l^2} \quad [2.28]$$

which is independent of R. The energy absorbed in the induction process is given as:

$$\begin{aligned} W_{\text{ind}} &= -\int_0^{\infty} \mathcal{E}_{\text{ind}} I dt = \frac{E^2}{R}\int_0^{\infty}\left(e^{-t/\tau} - e^{-2t/\tau}\right)dt \\ &= \frac{E^2\tau}{2R} - = \frac{mE^2}{2B^2l^2} = \frac{W_G}{2} \end{aligned} \quad [2.29]$$

The energy dissipated by the Joule effect is equal to:

$$W_{\text{J}} = \int_0^{\infty} RI^2dt = \frac{E^2}{R}\int_0^{\infty} e^{-2t/\tau}dt = \frac{E^2\tau}{2R} - = \frac{mE^2}{2B^2l^2} = \frac{W_G}{2} \quad [2.30]$$

Clearly, half of the energy developed by the generator is used by induction, while the other half is dissipated by resistance.

2.2. EMF for static fields (Lorentz case)

Consider a closed filiform circuit $\Gamma(t)$, free to move, relative to a Galilean reference frame $Oxyz$, in an electrostatic field $\vec{\mathbf{E}}$ and a stationary magnetic field $\vec{\mathbf{B}}$. In addition, this circuit can be deformed without changing its composition, that is, it always remains composed of the same elements. This is known as the *Lorentz case*. Let $\vec{\mathbf{v}}_e$ denote the entrained velocity of an infinitesimal element $d\vec{\mathbf{l}}$ of the circuit with respect to $Oxyz$. A point charge q moving in this element at speed $\vec{\mathbf{v}}$ relative to $Oxyz$ is subject, in addition to the frictional force $\vec{\mathbf{F}}_{\text{col}}$ due to collisions with ions in the crystal lattice and the force $\vec{\mathbf{F}}_{\text{gen}}$ due to any generators (e.g. a battery), to the electromagnetic force $\vec{\mathbf{F}}_{EB} = q\left(\vec{\mathbf{E}} + \vec{\mathbf{v}} \times \vec{\mathbf{B}}\right)$. Let $\vec{\mathbf{f}}_{EB} = \frac{\vec{\mathbf{F}}_{EB}}{q} = \vec{\mathbf{E}} + \vec{\mathbf{v}} \times \vec{\mathbf{B}}$, $\vec{\mathbf{f}}_{\text{col}} = \frac{\vec{\mathbf{F}}_{\text{col}}}{q}$ and $\vec{\mathbf{f}}_{\text{gen}} = \frac{\vec{\mathbf{F}}_{\text{gen}}}{q}$ the forces per unit charge. By definition, the *EMF* of the entire circuit is equal to:

$$\mathcal{E} = \oint_{\Gamma(t)} \vec{\mathbf{f}} \cdot d\vec{\mathbf{l}} \quad [2.31]$$

where:

$$\vec{\mathbf{f}} = \vec{\mathbf{f}}_{EB} + \vec{\mathbf{f}}_{\text{gen}} \quad [2.32]$$

This name is somewhat incorrect, as EMF is not a force, but rather the work that would be done on a unit charge as it moves through the closed circuit. It has the dimension of a voltage.

In particular, let us focus on the so-called *induced EMF*:

$$\mathcal{E}_{\text{ind}} = \oint_{\Gamma(t)} \vec{\mathbf{f}}_{EB} \cdot d\,\vec{\mathbf{l}} \qquad [2.33]$$

Let $\vec{\mathbf{v}}_r$ be the velocity of the charge relative to the $d\,\vec{\mathbf{l}}$ portion of the circuit in which it is located. Here, $\vec{\mathbf{v}}_r$ is referred to as the *relative velocity* of the charge with respect to the reference frame $O'x'y'z'$ attached to $d\,\vec{\mathbf{l}}$. As $\vec{\mathbf{E}} = -\vec{\nabla}V$ and $\vec{\mathbf{v}} = \vec{\mathbf{v}}_e + \vec{\mathbf{v}}_r$, we obtain:

$$\mathcal{E}_{\text{ind}} = -\oint_{\Gamma(t)} \vec{\nabla}V \cdot d\,\vec{\mathbf{l}} + \oint_{\Gamma(t)} \left(\vec{\mathbf{v}}_e \times \vec{\mathbf{B}}\right) \cdot d\,\vec{\mathbf{l}} + \oint_{\Gamma(t)} \left(\vec{\mathbf{v}}_r \times \vec{\mathbf{B}}\right) \cdot d\,\vec{\mathbf{l}} \qquad [2.34]$$

The first integral is zero, because $\oint_{\Gamma(t)} \vec{\nabla}V \cdot d\,\vec{\mathbf{l}} = \oint_{\Gamma(t)} dV = 0$. Similarly $\left(\vec{\mathbf{v}}_r \times \vec{\mathbf{B}}\right) \cdot d\,\vec{\mathbf{l}} = 0$, since $\vec{\mathbf{v}}_r$ is parallel to $d\,\vec{\mathbf{l}}$ in a filiform circuit. It follows that:

$$\mathcal{E}_{\text{ind}} = \oint_{\Gamma(t)} \left(\vec{\mathbf{v}}_e \times \vec{\mathbf{B}}\right) \cdot d\,\vec{\mathbf{l}} \qquad [2.35]$$

$\vec{\mathbf{v}}_e \times \vec{\mathbf{B}}$ is called the *electromotive induction field*. We use the notation:

$$\begin{aligned} \vec{\mathbf{E}}_{\text{ind}} &= \vec{\mathbf{v}}_e \times \vec{\mathbf{B}} \\ \mathcal{E}_{\text{ind}} &= \oint_{\Gamma(t)} \vec{\mathbf{E}}_{\text{ind}} \cdot d\,\vec{\mathbf{l}} \end{aligned} \qquad [2.36]$$

Let us multiply $\left(\vec{\mathbf{v}}_e \times \vec{\mathbf{B}}\right) \cdot d\,\vec{\mathbf{l}}$ by dt:

$$\left(\vec{\mathbf{v}}_e \times \vec{\mathbf{B}}\right) \cdot d\,\vec{\mathbf{l}}\,dt = \left(d\,\vec{\mathbf{l}} \times \vec{\mathbf{v}}_e dt\right) \cdot \vec{\mathbf{B}} = -\left(d\,\vec{\mathbf{r}} \times d\,\vec{\mathbf{l}}\right) \cdot \vec{\mathbf{B}} \qquad [2.37]$$

where $d\,\vec{\mathbf{r}} = d\vec{\mathbf{v}}_e dt$ represents the displacement of the element $d\,\vec{\mathbf{l}}$ of the circuit during the time interval dt. Consequently, the quantity $d^2\vec{\mathbf{S}}_c = d\,\vec{\mathbf{r}} \times d\,\vec{\mathbf{l}}$ can be interpreted as the surface swept or simply cut by $d\,\vec{\mathbf{l}}$ when it moves by $d\,\vec{\mathbf{r}}$, as shown in Figure 2.4. The infinitesimal element $d^2\vec{\mathbf{S}}_c$ is called the *cut surface element*. $d^2\Phi_c = \vec{\mathbf{B}} \cdot d^2\vec{\mathbf{S}}_c$ is called the *cut flux*. It therefore follows that:

$$\mathcal{E}_{\text{ind}} dt = -\oint_{\Gamma(t)} d^2\Phi_c = -d\Phi_c \qquad [2.38]$$

where $d\Phi_c$ is the *flux cut by the entire circuit*:

$$d\Phi_c = \oint_{\Gamma(t)} \vec{\mathbf{B}} \cdot d^2\vec{\mathbf{S}}_c \qquad [2.39]$$

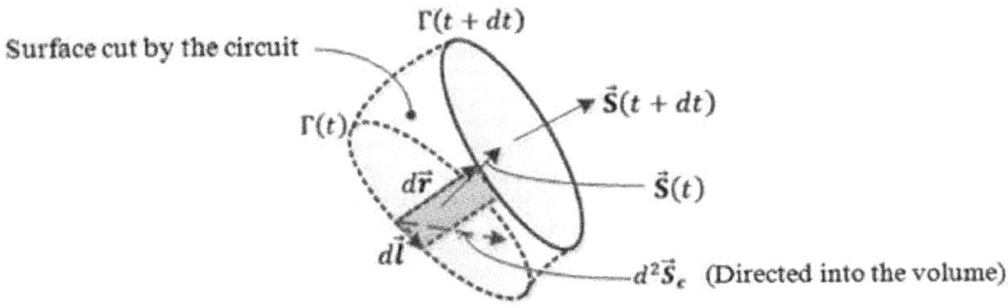

Figure 2.4. *Closed circuit moving in a stationary magnetic field (since the magnetic field flux through any closed surface is zero, the sum of the fluxes entering through $\Gamma(t)$ and through the total cut surface is equal to the flux leaving through $\Gamma(t+dt)$, that is, $-\Phi(t) - d\Phi_c + \Phi(t+dt) = 0$; which gives $d\Phi_c = d\Phi$). For a color version of this figure, see www.iste.co.uk/abdesselam/electrodynamics2.zip*

In the Lorentz case, the induction EMF is written, according to [2.38], as:

$$\mathcal{E}_{\text{ind}} = -\frac{d\Phi_c}{dt} \qquad [2.40]$$

Since the magnetic field is stationary, the flux $d\Phi_c$ cut by the circuit during its motion is exactly equal to the variation of the total flux $d\Phi = \Phi(t+dt) - \Phi(t)$, where $\Phi(t)$ is the flux measured through any surface bounded by the circuit at time t, that is:

$$\Phi(t) = \int_{S(t)} \vec{\mathbf{B}} \cdot d\vec{\mathbf{S}} \qquad [2.41]$$

In this expression, the magnetic field appears as a *magnetic flux density*. Finally, we obtain:

$$\mathcal{E}_{\text{ind}} = -\frac{d\Phi}{dt} \qquad [2.42]$$

We know that in an ohmic conductor with conductivity γ, the current density $\vec{\mathbf{J}}$ is related to $\vec{\mathbf{f}}$ as follows:

$$\vec{\mathbf{J}} = \gamma\vec{\mathbf{f}} = \gamma\left(-\vec{\nabla}V + \vec{\mathbf{v}}_r \times \vec{\mathbf{B}} + \vec{\mathbf{f}}_{\text{total}}\right) \quad [2.43]$$

where:

$$\vec{\mathbf{f}}_{\text{total}} = \vec{\mathbf{f}}_{\text{ind}} + \vec{\mathbf{f}}_{\text{gen}} \quad [2.44]$$

is called the *total electromotive field*. For a portion AB of a thin-wire circuit with resistance R_{AB}, we deduce that:

$$\begin{aligned} R_{AB}I &= \int_A^B \left(-\vec{\nabla}V + \vec{\mathbf{v}}_r \times \vec{\mathbf{B}} + \vec{\mathbf{f}}_{\text{total}}\right) \cdot d\vec{\mathbf{l}} \\ &= V_A - V_B + \int_A^B \vec{\mathbf{f}}_{\text{total}} \cdot d\vec{\mathbf{l}} \end{aligned} \quad [2.45]$$

The integral $\int_A^B \left(\vec{\mathbf{v}}_r \times \vec{\mathbf{B}}\right) \cdot d\vec{\mathbf{l}}$ is obviously zero since $\vec{\mathbf{v}}_r$ and $d\vec{\mathbf{l}}$ are parallel in a thin-wire circuit. In particular, if the circuit is closed, that is, $A = B$, we obtain:

$$RI = \oint_{\text{closedcircuit}} \vec{\mathbf{f}}_{\text{total}} \cdot d\vec{\mathbf{l}} = \mathcal{E}_{\text{total}} \quad [2.46]$$

where R is the total resistance, $\mathcal{E}_{\text{total}} = \mathcal{E}_{\text{ind}} + \mathcal{E}_{\text{gen}}$ and $\mathcal{E}_{\text{gen}} = \oint \vec{\mathbf{f}}_{\text{gen}} \cdot d\vec{\mathbf{l}}$ is the EMF due to the generators. The electromotive field therefore produces a current, whereas the electrostatic field $-\vec{\nabla}V$ cannot do so in a closed circuit. In addition, if the circuit contains no generator ($\mathcal{E}_{\text{gen}} = 0$), the induced EMF also produces an electric current in the closed circuit. This is, of course, consistent with the first experiment.

EXAMPLE 2.4.– A ring of ionized gas emitted by a nova expands radially at velocity $\vec{\mathbf{v}}$ through a static magnetic field $\vec{\mathbf{B}} = B\vec{\mathbf{k}}$, as shown in Figure 2.5. We would like to calculate the tangential acceleration of a charged particle inside the ring. The EMF generated around a loop of radius r is:

$$\mathcal{E} = \oint \frac{\vec{\mathbf{F}}}{q} \cdot d\vec{\mathbf{l}} = -\frac{d\Phi}{dt} = -2\pi r B \frac{dr}{dt} \quad [2.47]$$

Since $\vec{\mathbf{F}} = F_r\vec{\mathbf{e}}_r + F_\varphi\vec{\mathbf{e}}_\varphi$ and $d\vec{\mathbf{l}} = rd\varphi\vec{\mathbf{e}}_\varphi$, we get $\oint \frac{\vec{\mathbf{F}}}{q} \cdot d\vec{\mathbf{l}} = \frac{2\pi r F_\varphi}{q} = \frac{2\pi r m a_\varphi}{q}$. This gives:

$$\frac{2\pi r m a_\varphi}{q} = -2\pi r B \frac{dr}{dt} = -2\pi r B v_r \quad [2.48]$$

and therefore:

$$a_\varphi = -\frac{q}{m} B v_r \qquad [2.49]$$

The negative charges are accelerated in the direction of $\overrightarrow{\mathbf{e}}_\varphi$, while the positive charges are accelerated in the opposite direction.

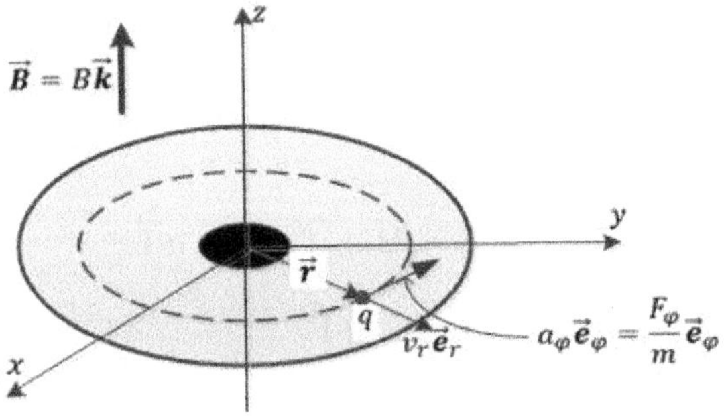

Figure 2.5. *Ionized gas emitted by a nova*

EXAMPLE 2.5. (Principle of an alternator).– An important application of the induction phenomenon is the generation of alternating current. To achieve this, a flat coil consisting of N turns of radius a (initially positioned at $t = 0$ in the Oxy plane) is placed in a uniform magnetic field $\overrightarrow{\mathbf{B}} = B\overrightarrow{\mathbf{k}}$. The coil is connected to a circuit without a source of current or voltage. The coil is rotated about the axis Ox in the anti-clockwise direction at the angular velocity ω, as shown in Figure 2.6. Due to the motion of the coil relative to the $\overrightarrow{\mathbf{B}}$ field, the flux of the latter through the coil varies over time. This flux is:

$$\Phi = NSB \cos \omega t \qquad [2.50]$$

where $S = \pi a^2$. It generates induced EMF in the coil:

$$\mathcal{E}_{\text{ind}} = -\frac{d\Phi}{dt} = \omega \Phi_0 \sin \omega t \qquad [2.51]$$

where $\Phi_0 = NSB$. If the circuit connected to the coil contains a resistor R, a sinusoidal current appears, given by:

$$I_{\text{ind}} = \frac{\mathcal{E}_{\text{ind}}}{R} = \frac{\omega \Phi_0}{R} \sin \omega t \qquad [2.52]$$

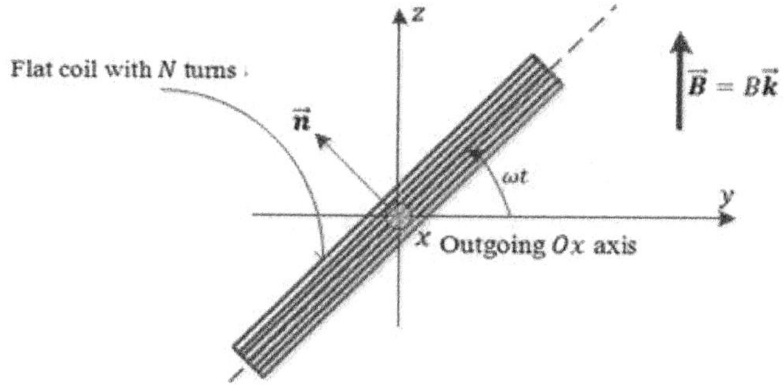

Figure 2.6. *Operating principle of an alternator*

This current is known as an *alternating current*. The above system is a simple example of the conversion of mechanical energy into electrical energy. In fact, the coil can be assimilated to a magnetic moment $\vec{m} = I_{ind} S \vec{n}$, where $\vec{n} = -\sin\omega t\, \vec{i} + \cos\omega t\, \vec{k}$ is the unit vector normal to the coil. Thus, the torque exerted by the magnetic field on the coil is given by:

$$\vec{N}_{Lap} = \vec{m} \times \vec{B} = -\vec{i}\, SNI_{ind} B \sin\omega t = -\vec{i}\, \frac{\omega\Phi_0^2}{R} \sin^2\omega t \qquad [2.53]$$

This torque is oriented along Ox, thus opposing the rotation of the coil. To maintain a constant rotational speed, the external mechanical system must exert an opposing torque $\vec{N}_{ext} = -\vec{N}_{Lap}$, meaning it must develop the power:

$$P_{ext} = \vec{N}_{ind} \cdot \omega\, \vec{i} = \frac{\omega^2\Phi_0^2}{R} \sin^2\omega t \qquad [2.54]$$

This power is exactly equal to the electrical power developed by the induced EMF (there is a complete conversion of mechanical energy into electrical energy). In fact, the electrical power developed by the induced EMF is:

$$\mathcal{E}_{ind} I_{ind} = RI_{ind}^2 = \frac{\omega^2\Phi_0^2}{R} \sin^2\omega t = P_{ext} \qquad [2.55]$$

and is dissipated by the Joule effect, causing heating of the resistor.

2.3. Variable magnetic field (Neumann case)

2.3.1. *Motion relative to the fixed reference $Oxyz$*

Consider a closed circuit (a rectangular loop), infinitesimal in size, located in the Oxy plane, with dimensions Δx and Δy. We assume that the surface remains unchanged, that is, it behaves as a rigid body, and that it is free to move within a magnetic field $\overrightarrow{\mathbf{B}}$ whose intensity varies with position. The moving sides of the circuit will naturally experience a time-dependent field. For simplicity, we also assume that the circuit moves only in the Oxy plane with uniform velocity $\overrightarrow{\mathbf{v}} = v_x \overrightarrow{\mathbf{i}} + v_y \overrightarrow{\mathbf{j}}$ (v_x and v_y are constants) through a magnetic field directed along the Oz axis, denoted as $B(x, y)\overrightarrow{\mathbf{k}}$, as shown in Figure 2.7. The induced EMF generated around the moving circuit is:

$$\begin{aligned} d\mathcal{E}_{\text{ind}} = & \left[\overrightarrow{\mathbf{v}} \times B\left(x + \Delta x, y + \frac{\Delta y}{2}\right)\overrightarrow{\mathbf{k}} - \right. \\ & \left. \overrightarrow{\mathbf{v}} \times B\left(x, y + \frac{\Delta y}{2}\right)\overrightarrow{\mathbf{k}}\right] \cdot \overrightarrow{\mathbf{j}}\,\Delta y - \\ & \left[\overrightarrow{\mathbf{v}} \times B\left(x + \frac{\Delta x}{2}, y + \Delta y\right)\overrightarrow{\mathbf{k}} - \right. \\ & \left. \overrightarrow{\mathbf{v}} \times B\left(x + \frac{\Delta y}{2}, y\right)\overrightarrow{\mathbf{k}}\right] \cdot \overrightarrow{\mathbf{i}}\,\Delta x \end{aligned} \qquad [2.56]$$

which, to the first-order approximation in $\Delta x \Delta y$, simplifies to:

$$\begin{aligned} d\mathcal{E}_{\text{ind}} & = \left(\overrightarrow{\mathbf{v}} \times \frac{\partial B}{\partial x}\Delta x\,\overrightarrow{\mathbf{k}}\right)_y \Delta y - \left(\overrightarrow{\mathbf{v}} \times \frac{\partial B}{\partial y}\Delta y\,\overrightarrow{\mathbf{k}}\right)_x \Delta x \\ & = -\left(v_x \frac{\partial B}{\partial x} + v_y \frac{\partial B}{\partial y}\right)\Delta x \Delta y = -\left(\frac{dx}{dt}\frac{\partial B}{\partial x} + \frac{dy}{dt}\frac{\partial B}{\partial y}\right)\Delta x \Delta y \\ & = -\frac{\partial B}{\partial t}\Delta x \Delta y = -\frac{\partial B}{\partial t} dS_z \end{aligned} \qquad [2.57]$$

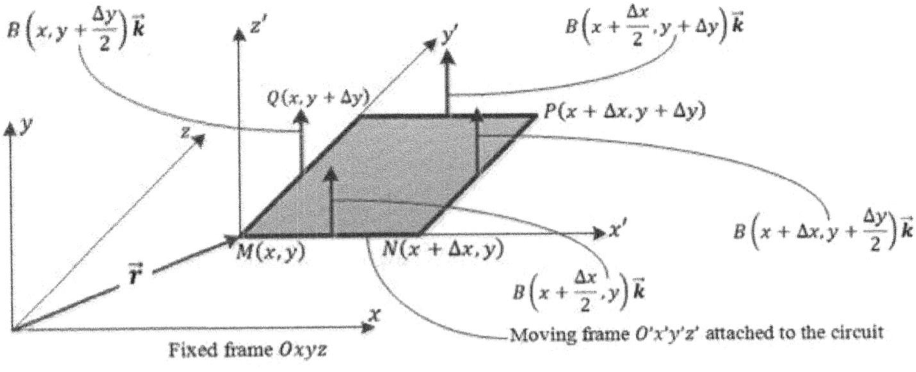

Figure 2.7. *Infinitesimal, rectangular and rigid circuit in motion within a magnetic field*

For a larger, rigid closed circuit located in the Oxy plane, we obtain the following, after integration over the entire circuit:

$$\mathcal{E}_{\text{ind}} = -\int_S \frac{\partial B}{\partial t} dS_z \qquad [2.58]$$

This result can be generalized to a closed, rigid circuit arbitrarily oriented in space and moving in an arbitrary direction through an arbitrary magnetic field $\vec{\mathbf{B}}$. The induced EMF is equal to:

$$\mathcal{E}_{\text{ind}} = -\int_S \frac{\partial \vec{\mathbf{B}}}{\partial t} \cdot d\vec{\mathbf{S}} = -\frac{d}{dt}\int_S \vec{\mathbf{B}} \cdot d\vec{\mathbf{S}} = -\frac{d\Phi}{dt} \qquad [2.59]$$

where Φ represents the magnetic field flux through an arbitrary surface S bounded by the circuit.

2.3.2. *Motion relative to the moving frame $O'x'y'z'$ attached to the circuit*

Let us now consider the reference frame $O'x'y'z'$ attached to the circuit. A unit charge inside the circuit is subjected to the electromagnetic force:

$$\vec{\mathbf{f}}_{EB} = \vec{\mathbf{E}}' + \vec{\mathbf{v}}'_r \times \vec{\mathbf{B}}' \qquad [2.60]$$

where $\vec{\mathbf{E}}'$ and $\vec{\mathbf{B}}'$ are the electric and magnetic fields measured in $O'x'y'z'$, respectively. In this reference frame, the induced EMF is given by:

$$\begin{aligned}\mathcal{E}_{\text{ind}} &= \oint_{\text{circuit}} \vec{\mathbf{f}}_{EB} \cdot d\vec{\mathbf{l}} = \oint_{\text{circuit}} \vec{\mathbf{E}}' \cdot d\vec{\mathbf{l}} + \oint_{\text{circuit}} \left(\vec{\mathbf{v}}'_r \times \vec{\mathbf{B}}'\right) \cdot d\vec{\mathbf{l}} \\ &= \oint_{\text{circuit}} \vec{\mathbf{E}}' \cdot d\vec{\mathbf{l}}\end{aligned} \qquad [2.61]$$

since $\vec{\mathbf{v}}'_r$ and $d\vec{\mathbf{l}}$ are parallel in a thin-wire circuit. This expression for the induced EMF differs from that given by [2.59], but their values must be the same because $\mathcal{E}_{\text{ind}}$ determines the motion of mobile charges relative to the conductor, not relative to $Oxyz$. We thus conclude that the electric field $\vec{\mathbf{E}}'$ must contain an additional term in addition to the $-\vec{\nabla}V'$ term, since its circulation along a closed circuit is not zero. Consequently, we can finally write:

$$\mathcal{E}_{\text{ind}} = \oint \vec{\mathbf{E}}' \cdot d\vec{\mathbf{l}} = -\frac{d}{dt}\left(\int_S \vec{\mathbf{B}} \cdot d\vec{\mathbf{S}}\right) \qquad [2.62]$$

If we now consider a closed thin-wire, rigid and stationary circuit in a time-varying magnetic field, we must have $\vec{\mathbf{E}}' = \vec{\mathbf{E}}$, since the reference frames $O'x'y'z'$ and $Oxyz$ coincide. Thus, equation [2.62] becomes:

$$\mathcal{E}_{\text{ind}} = \oint_{\text{circuit}} \vec{\mathbf{E}} \cdot d\vec{\mathbf{l}} = -\frac{d\Phi}{dt} = -\frac{d}{dt}\left(\int_S \vec{\mathbf{B}} \cdot d\vec{\mathbf{S}}\right) \qquad [2.63]$$

This is known as the *Neumann case*. It is clear that the relation $\mathcal{E}_{\text{ind}} = -\frac{d\Phi}{dt}$ also holds in this case. From [2.63], we can see that when $\vec{\mathbf{B}}$ varies with time, the relation $\vec{\nabla} \times \vec{\mathbf{E}} = 0$ is no longer valid.

EXAMPLE 2.6.– An electron moving with velocity $\vec{\mathbf{v}}$ undergoes a cyclotron motion between the parallel faces of an electromagnet, as shown in Figure 2.8. The magnetic field B, directed along Oz, varies with time. We aim to determine the tangential acceleration of the electron along the circular orbit it describes. The radius of this orbit can be obtained from the equation $F = |e|\, vB = \frac{mv^2}{r}$. This implies $r = \frac{mv}{|e|B}$. The induced EMF along this orbit is:

$$\oint_{\text{orbit}} \vec{\mathbf{E}} \cdot d\vec{\mathbf{l}} = 2\pi r E_\varphi = -\frac{d\Phi}{dt} \qquad [2.64]$$

where Φ is the magnetic field flux through any surface bounded by the orbit and $\vec{\mathbf{E}}$ is the electric field experienced by the electron. Since $\Phi = \pi r^2 B$, we can deduce that $2\pi r E_\varphi = -\pi r^2 \frac{dB}{dt}$ and consequently:

$$E_\varphi = -\frac{r}{2}\frac{dB}{dt} \qquad [2.65]$$

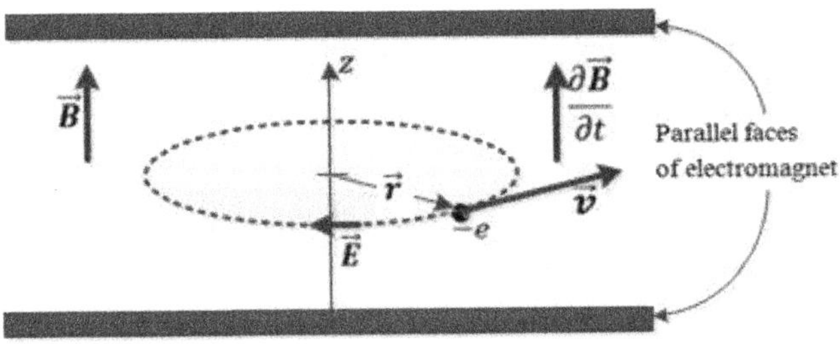

Figure 2.8. *Electron motion in a cyclotron*

Thus, the tangential acceleration is equal to:

$$a_\varphi = -\frac{|e|\, E_\varphi}{m} = \frac{|e|\, r}{2m}\frac{dB}{dt} = \frac{v}{2B}\frac{dB}{dt} \qquad [2.66]$$

The electron is therefore accelerated in the direction of its motion.

2.3.3. *Maxwell–Faraday equation*

Applying the curl theorem to the right-hand side of [2.63] and noting that $\frac{d\Phi}{dt} = \frac{d}{dt}\left(\int_S \overrightarrow{\mathbf{B}} \cdot d\overrightarrow{\mathbf{S}}\right) = \int_S \frac{\partial \overrightarrow{\mathbf{B}}}{\partial t} \cdot d\overrightarrow{\mathbf{S}}$ since the circuit is fixed, we deduce that:

$$\int_S \left(\overrightarrow{\nabla} \times \overrightarrow{\mathbf{E}}\right) \cdot d\overrightarrow{\mathbf{S}} = -\int_S \frac{\partial \overrightarrow{\mathbf{B}}}{\partial t} \cdot d\overrightarrow{\mathbf{S}} \qquad [2.67]$$

This relation is true for any surface bounded by the circuit, which implies that:

$$\overrightarrow{\nabla} \times \overrightarrow{\mathbf{E}} = -\frac{\partial \overrightarrow{\mathbf{B}}}{\partial t} \qquad [2.68]$$

This equation is called the *Maxwell–Faraday equation*. For stationary fields, it reduces to $\overrightarrow{\nabla} \times \overrightarrow{\mathbf{E}} = 0$, which is equivalent to the existence of a potential V such that $\overrightarrow{\mathbf{E}} = -\overrightarrow{\nabla} V$. The Maxwell–Faraday equation is therefore general.

Moreover, since $\overrightarrow{\nabla} \cdot \left(\overrightarrow{\nabla} \times \overrightarrow{\mathbf{E}}\right) = 0$, it follows from the Maxwell–Faraday equation that:

$$\overrightarrow{\nabla} \cdot \overrightarrow{\mathbf{B}} = 0 \qquad [2.69]$$

This is the Maxwell–Thomson equation. We can conclude that this equation always holds for the magnetic field $\overrightarrow{\mathbf{B}}$, even when it varies with time. This allows us to write:

$$\overrightarrow{\mathbf{B}} = \overrightarrow{\nabla} \times \overrightarrow{\mathbf{A}} \qquad [2.70]$$

By substituting [2.70] into the Maxwell–Faraday equation [2.68], we obtain $\overrightarrow{\nabla} \times \overrightarrow{\mathbf{E}} = -\frac{\partial \overrightarrow{\mathbf{B}}}{\partial t} = -\frac{\partial}{\partial t}\left(\overrightarrow{\nabla} \times \overrightarrow{\mathbf{A}}\right) = -\overrightarrow{\nabla} \times \frac{\partial \overrightarrow{\mathbf{A}}}{\partial t}$. This implies that:

$$\overrightarrow{\nabla} \times \left(\overrightarrow{\mathbf{E}} + \frac{\partial \overrightarrow{\mathbf{A}}}{\partial t}\right) = 0 \qquad [2.71]$$

We see that it is not $\overrightarrow{\mathbf{E}}$, but rather $\overrightarrow{\mathbf{E}} + \frac{\partial \overrightarrow{\mathbf{A}}}{\partial t}$, which can be expressed as the gradient of a scalar potential V (which can no longer be called electrostatic), that is,

$\overrightarrow{\mathbf{E}} + \frac{\partial \overrightarrow{\mathbf{A}}}{\partial t} = -\overrightarrow{\nabla} V$. Thus, we find:

$$\overrightarrow{\mathbf{E}} = -\overrightarrow{\nabla} V - \frac{\partial \overrightarrow{\mathbf{A}}}{\partial t} \qquad [2.72]$$

2.4. General case of a deformable circuit

2.4.1. *Electromotive field and induced EMF*

The Lorentz and Neumann cases are two manifestations of the same physical phenomenon, namely, the *electromagnetic induction effect*. In the general case where the circuit can move and deform in a variable magnetic field $\overrightarrow{\mathbf{B}}$, the electromagnetic force per unit charge is written as:

$$\overrightarrow{\mathbf{f}}_{EB} = -\overrightarrow{\nabla} V + \overrightarrow{\mathbf{v}}_r \times \overrightarrow{\mathbf{B}} + \overrightarrow{\mathbf{f}}_{\text{ind}} \qquad [2.73]$$

where:

$$\overrightarrow{\mathbf{f}}_{\text{ind}} = -\frac{\partial \overrightarrow{\mathbf{A}}}{\partial t} + \overrightarrow{\mathbf{v}}_e \times \overrightarrow{\mathbf{B}} \qquad [2.74]$$

is the *induced electromotive field*. Consequently, the induced EMF along a closed thin-wire circuit is:

$$\mathcal{E}_{\text{ind}} = \oint_{\text{circuit}} \overrightarrow{\mathbf{f}}_{\text{ind}} \cdot d\,\overrightarrow{\mathbf{l}} \qquad [2.75]$$

Note that the term $\overrightarrow{\mathbf{v}}_r$ does not contribute to the FEM in an thin-wire circuit, as $\overrightarrow{\mathbf{v}}_r$ and $d\,\overrightarrow{\mathbf{l}}$ are parallel. Moreover, this term is generally negligible ($\overrightarrow{\mathbf{v}}_r$ is very small) compared to the other terms of $\overrightarrow{\mathbf{f}}_{EB}$ and can therefore be ignored in the expression of the generalized local Ohm's law:

$$\overrightarrow{\mathbf{J}} \simeq \gamma \left(-\overrightarrow{\nabla} V - \frac{\partial \overrightarrow{\mathbf{A}}}{\partial t} + \overrightarrow{\mathbf{v}}_e \times \overrightarrow{\mathbf{B}} + \overrightarrow{\mathbf{f}}_{\text{gen}} \right) \qquad [2.76]$$

2.4.2. *Faraday's law*

In the following paragraph, we will show that the relation $\mathcal{E}_{\text{ind}} = -\frac{d\Phi}{dt}$, which holds in both Lorentz and Neumann cases, remains valid in the general case of a closed, deformable, thin-wire circuit whose composition does not change over time. For this purpose, let us consider a closed, deformable, thin-wire circuit $\Gamma(t)$. At the first order of approximation in dt, we can write that:

$$\begin{aligned}\Phi(t+dt)-\Phi(t) &= \int_{S(t+dt)}\left[\vec{\mathbf{B}}(t)+\frac{\partial\vec{\mathbf{B}}}{\partial t}dt\right]\cdot d\vec{\mathbf{S}}-\int_{S(t)}\vec{\mathbf{B}}(t)\cdot d\vec{\mathbf{S}}\\ &= \left[\int_{S(t+dt)}\vec{\mathbf{B}}(t)\cdot d\vec{\mathbf{S}}-\int_{S(t)}\vec{\mathbf{B}}(t)\cdot d\vec{\mathbf{S}}\right]+\int_{S(t+dt)}\frac{\partial\vec{\mathbf{B}}}{\partial t}\cdot d\vec{\mathbf{S}}\,dt\\ &\simeq \left[\int_{S(t+dt)}\vec{\mathbf{B}}(t)\cdot d\vec{\mathbf{S}}-\int_{S(t)}\vec{\mathbf{B}}(t)\cdot d\vec{\mathbf{S}}\right]+\int_{S(t)}\frac{\partial\vec{\mathbf{B}}}{\partial t}\cdot d\vec{\mathbf{S}}\,dt\end{aligned} \quad [2.77]$$

Since $\vec{\mathbf{B}}=\vec{\nabla}\times\vec{\mathbf{A}}$, we deduce that:

$$\begin{aligned}\Phi(t+dt)-\Phi(t) = &\left[\int_{S(t+dt)}\vec{\mathbf{B}}(t)\cdot d\vec{\mathbf{S}}-\int_{S(t)}\vec{\mathbf{B}}(t)\cdot d\vec{\mathbf{S}}\right]\\ &+\int_{S(t)}\frac{\partial\left(\vec{\nabla}\times\vec{\mathbf{A}}\right)}{\partial t}\cdot d\vec{\mathbf{S}}\,dt\end{aligned} \quad [2.78]$$

The first term on the right, in brackets, represents the flux cut by the circuit during the time interval dt, that is:

$$\begin{aligned}&\int_{S(t+dt)}\vec{\mathbf{B}}(t)\cdot d\vec{\mathbf{S}}-\int_{S(t)}\vec{\mathbf{B}}(t)\cdot d\vec{\mathbf{S}}=-\oint_{\Gamma(t)}\vec{\mathbf{B}}\cdot\left(d\vec{\mathbf{l}}\times\vec{\mathbf{v}}_e dt\right)\\ &=-dt\oint_{\Gamma(t)}\left(\vec{\mathbf{v}}_e\times\vec{\mathbf{B}}\right)\cdot d\vec{\mathbf{l}}\end{aligned} \quad [2.79]$$

while the second term $\int_{S(t)}\frac{\partial\left(\vec{\nabla}\times\vec{\mathbf{A}}\right)}{\partial t}\cdot d\vec{\mathbf{S}}\,dt$ can be written, according to the curl theorem, as:

$$\int_{S(t)}\frac{\partial\left(\vec{\nabla}\times\vec{\mathbf{A}}\right)}{\partial t}\cdot d\vec{\mathbf{S}}\,dt=dt\int_{S(t)}\left(\vec{\nabla}\times\frac{\partial\vec{\mathbf{A}}}{\partial t}\right)\cdot d\vec{\mathbf{S}}=dt\oint_{\Gamma(t)}\frac{\partial\vec{\mathbf{A}}}{\partial t}\cdot d\vec{\mathbf{l}} \quad [2.80]$$

It follows that:

$$\Phi(t+dt)-\Phi(t)=-dt\oint_{\Gamma(t)}\left(\vec{\mathbf{v}}_e\times\vec{\mathbf{B}}-\frac{\partial\vec{\mathbf{A}}}{\partial t}\right)\cdot d\vec{\mathbf{l}} \quad [2.81]$$

Divide by dt and then take the limit $dt \longrightarrow 0$:

$$-\frac{d\Phi}{dt} = \oint_{\Gamma(t)} \left(\overrightarrow{\mathbf{v}}_e \times \overrightarrow{\mathbf{B}} - \frac{\partial \overrightarrow{\mathbf{A}}}{\partial t} \right) \cdot d\,\overrightarrow{\mathbf{l}} \qquad [2.82]$$

We recognize in parentheses the expression of the induction form ([2.74]). The term on the right is none other than the induced EMF ([2.75]). We ultimately arrive at:

$$\mathcal{E}_{\text{ind}} = -\frac{d\Phi}{dt} \qquad [2.83]$$

This relation is known as *Faraday's law*. It is more convenient to express the time-dependent term in terms of the magnetic field $\overrightarrow{\mathbf{B}}$:

$$-\oint_{\Gamma(t)} \frac{\partial \overrightarrow{\mathbf{A}}}{\partial t} \cdot d\,\overrightarrow{\mathbf{l}} = -\int_{S(t)} \frac{\partial \overrightarrow{\mathbf{B}}}{\partial t} \cdot d\,\overrightarrow{\mathbf{S}} \qquad [2.84]$$

Thus, it follows that:

$$\mathcal{E}_{\text{ind}} = -\frac{d\Phi}{dt} = \oint_{\Gamma(t)} \left(\overrightarrow{\mathbf{v}}_e \times \overrightarrow{\mathbf{B}} \right) \cdot d\,\overrightarrow{\mathbf{l}} - \int_{S(t)} \frac{\partial \overrightarrow{\mathbf{B}}}{\partial t} \cdot d\,\overrightarrow{\mathbf{S}} \qquad [2.85]$$

Faraday's law can be used only if the circuit is closed, thin-wire and its composition does not change over time. In other cases, we must return to equation [2.75].

EXAMPLE 2.7. (Barlow wheel).– The Barlow wheel was one of the first machines used to generate electricity. This wheel is a conductive disk with center O, able to rotate (with the help of a gasoline engine, for example), in a magnetic field $\overrightarrow{\mathbf{B}}$ parallel to the disk axis (denoted here by Oz). Two points on the disk are connected to an electrical circuit. The first point is located on Oz, while the other is a sliding contact on the outer edge of the disk, as shown in Figure 2.9. Let a be the radius of the disk, $\overrightarrow{\omega} = \omega\,\overrightarrow{\mathbf{k}}$ the angular velocity of the disk, $\overrightarrow{\mathbf{B}} = B\,\overrightarrow{\mathbf{k}}$ the constant magnetic field and R the resistance of the circuit. When the wheel turns, an EMF is generated and a current of $I_{\text{ind}} = \frac{\mathcal{E}_{\text{ind}}}{R}$ travels through the circuit $OMNPO$. To calculate $\mathcal{E}_{hboxind}$, simply note that the electromotive field $\overrightarrow{\mathbf{v}}_e \times \overrightarrow{\mathbf{B}} = \omega r B\,\overrightarrow{\mathbf{i}}$ occurs only along OM. It therefore follows that:

$$\mathcal{E}_{\text{ind}} = \oint_{OMNPO} \left(\overrightarrow{\mathbf{v}}_e \times \overrightarrow{\mathbf{B}} \right) \cdot d\,\overrightarrow{\mathbf{l}} = \int_{r=0}^{a} \omega B r dr = \frac{1}{2}\omega B a^2 \qquad [2.86]$$

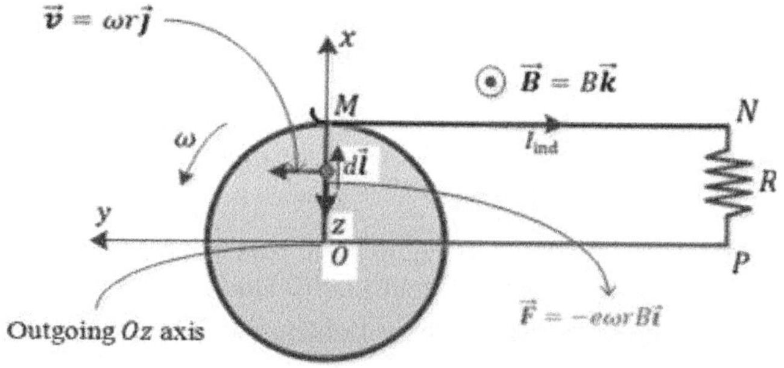

Figure 2.9. *Operating mechanism of the Barlow wheel. For a color version of this figure, see www.iste.co.uk/abdesselam/electrodynamics2.zip*

Thus, the current produced by the wheel is:

$$I_{\text{ind}} = \frac{\omega B a^3}{2R} \qquad [2.87]$$

It should be noted that the Laplace force experienced by an element $d\vec{\mathbf{l}} = dr\,\vec{\mathbf{i}}$ located along OA is equal to $d\vec{\mathbf{F}}_{\text{Lap}} = I_{\text{ind}} d\vec{\mathbf{l}} \times \vec{\mathbf{B}} = -IBdr\,\vec{\mathbf{j}}$. This element is therefore subjected to the elementary torque:

$$d\vec{\mathbf{N}}_{\text{Lap}} = \vec{\mathbf{r}} \times d\vec{\mathbf{F}}_{\text{Lap.}} = -r\,\vec{\mathbf{i}} \times I_{\text{ind}} B dr\,\vec{\mathbf{j}} = -I_{\text{ind}} B r dr\,\vec{\mathbf{k}} \qquad [2.88]$$

with respect to the point O. After integrating along OA, it becomes:

$$\vec{\mathbf{N}}_{\text{Lap}} = -I_{\text{ind}} B \int_0^a r dr\,\vec{\mathbf{k}} = -\frac{1}{2} I_{\text{ind}} B a^2\,\vec{\mathbf{k}} = -\frac{\omega B^2 a^4}{4R}\,\vec{\mathbf{k}} \qquad [2.89]$$

We can see that the torque $\vec{\mathbf{N}}_{\text{Lap}}$ applies in the opposite direction to $\vec{\omega}$. This contributes to slowing down the disc. To maintain wheel speed, an external torque $\vec{\mathbf{N}}_{\text{ext}} = -\vec{\mathbf{N}}_{\text{Lap}} = \frac{\omega B^2 a^4}{4R}\vec{\mathbf{k}}$ must be applied, that is, the external mechanical device must develop the power $\mathsf{P}_{\text{ext}} = \vec{\mathbf{N}}_{\text{ext}} \cdot \omega\,\vec{\mathbf{k}} = \frac{\omega^2 B^2 a^4}{4R}$, which will be completely converted into electrical power and then dissipated by the Joule effect. In fact, we have: $\mathsf{P}_{\text{ext}} = \frac{\omega^2 B^2 a^4}{4R} = \mathcal{E}_{\text{ind}} I_{\text{ind}} = R I_{\text{ind}}^2$.

EXAMPLE 2.8. (Eddy current).– The existence of an electromotive field resulting from time variations of a magnetic field is responsible for the appearance of an induced current in a conductor placed within this field. This current is called an *eddy current* or *Foucault current*. To illustrate this concept, let us consider a conducting sphere of radius a, with conductivity γ, immersed in a time-varying magnetic field. The sphere is stationary, and the magnetic field, oriented upwards, varies as follows:

$$\overrightarrow{\mathbf{B}} = B_0 \sin \omega t \, \overrightarrow{\mathbf{k}} \qquad [2.90]$$

where B_0 is constant, as illustrated in Figure 2.10. Let us try to determine the vector potential at the origin of the magnetic field [2.90]. Due to symmetry considerations, the vector field $\overrightarrow{\mathbf{A}}$ must not depend on the variable φ'. In addition, it must be directed along $\overrightarrow{\mathbf{e}}_{\varphi'}$, that is, it must be of the form $\overrightarrow{\mathbf{A}}\,(r', z', t) = A_{\varphi'}\,(r', z', t)\,\overrightarrow{\mathbf{e}}_{\varphi'}$. Clearly, the potential vector maintains a constant intensity along the circle passing through $M\,(r', \varphi', z')$, defined by $r' =$ constant and $z' =$ constant. Applying the curl theorem to this circle gives: $2\pi r' A_{\varphi'}\,(r', z', t) = \pi r'^2 B_0 \sin \omega t$ and consequently:

$$A_{\varphi'}\,(r', z', t) = A_{\varphi'}\,(r', t) = \frac{1}{2} r' B_0 \sin \omega t \qquad [2.91]$$

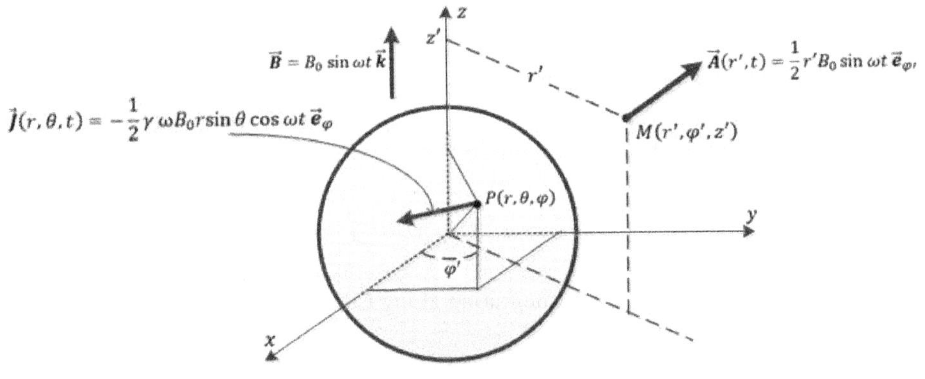

Figure 2.10. *The time variation of the magnetic field induces a current density (eddy current) on the conducting sphere, which will cause its heating*

Thus, the electromotive field is:

$$\overrightarrow{\mathbf{E}} = -\frac{\partial \overrightarrow{\mathbf{A}}}{\partial t} = -\frac{1}{2} \omega r' B_0 \cos \omega t \, \overrightarrow{\mathbf{e}}_{\varphi'} \qquad [2.92]$$

This field will induce a volume current inside the sphere, given by Ohm's law:

$$\overrightarrow{\mathbf{J}}\,(r, \theta, t) = \gamma \overrightarrow{\mathbf{E}} = -\frac{1}{2} \gamma \omega r B_0 \sin \theta \cos \omega t \, \overrightarrow{\mathbf{e}}_{\varphi} \qquad [2.93]$$

We see that $\overrightarrow{\mathbf{J}}(r,\theta,t)$ is constant along the circles defined by r = constant and θ = constant. Let us calculate the power developed by the electromotive field during the movement of charges inside the sphere:

$$\mathsf{P}=\int_{\text{sphere}}\overrightarrow{\mathbf{J}}\cdot\overrightarrow{\mathbf{E}}d\tau=\gamma\int_{\text{sphere}}\left|\overrightarrow{\mathbf{E}}\right|^2 d\tau \qquad [2.94]$$

Using spherical coordinates, we obtain:

$$\mathsf{P}=\frac{B_0^2\gamma\omega^2}{4}\cos^2\omega t\int_0^a r^4dr\int_0^\pi\sin^3\theta d\theta\int_0^{2\pi}d\varphi=\frac{2\pi B_0^2\gamma\omega^2a^5}{15}\cos^2\omega t \qquad [2.95]$$

whose average value over a period $T=\frac{2\pi}{\omega}$ is:

$$\langle\mathsf{P}\rangle=\frac{1}{T}\int_0^T\mathsf{P}dt=\frac{\pi B_0^2\gamma\omega^2a^5}{15} \qquad [2.96]$$

This power will be dissipated by Joule heating within the sphere. It is worth noting that the eddy current also induces another magnetic field, which can be evaluated using the Biot–Savart law. For example, at the center of the sphere, it is given by:

$$\begin{aligned}\overrightarrow{\mathbf{B}}_{\text{ind}}(O)&=\frac{\mu_0}{4\pi}\int_{\text{sphere}}\frac{\overrightarrow{\mathbf{J}}\times\overrightarrow{\mathbf{PO}}}{\left|\overrightarrow{\mathbf{PO}}\right|^3}d\tau=-\frac{\mu_0\gamma\omega B_0}{8\pi}\cos\omega t\int_0^a rdr\\&\times\int_0^\pi\sin^2\theta d\theta\int_0^{2\pi}d\varphi\overrightarrow{\mathbf{e}}_\theta=-\frac{\mu_0\gamma\omega B_0a^2}{16\pi}\cos\omega t\\&\times\int_0^\pi\sin^2\theta d\theta\int_0^{2\pi}d\varphi\left(\cos\theta\cos\varphi\overrightarrow{\mathbf{i}}+\cos\theta\sin\varphi\overrightarrow{\mathbf{j}}-\sin\theta\overrightarrow{\mathbf{k}}\right)\\&=\frac{\mu_0\gamma\omega B_0a^2}{8}\cos\omega t\int_0^\pi\sin^3\theta d\theta\overrightarrow{\mathbf{k}}=\frac{\mu_0\gamma\omega a^2}{6}B_0\cos\omega t\overrightarrow{\mathbf{k}}\end{aligned} \qquad [2.97]$$

2.5. Maxwell–Ampère equation and displacement current

In the following, we will see how a time-varying magnetic field can cause the curvature of the electric field, and vice versa. To demonstrate this, we consider a wire carrying a current I, with one end attached to a circular plate that becomes charged as the current passes through it, as shown in Figure 2.11. Let us draw a closed loop Γ around the wire.

$$\oint_\Gamma\overrightarrow{\mathbf{B}}\cdot d\overrightarrow{\mathbf{l}}=\int_S\left(\overrightarrow{\nabla}\times\overrightarrow{\mathbf{B}}\right)\cdot d\overrightarrow{\mathbf{S}}=\mu_0 I_{\text{enclosed}}=\mu_0\int_S\overrightarrow{\mathbf{J}}\cdot d\overrightarrow{\mathbf{S}} \qquad [2.98]$$

where S is any surface bounded by the loop Γ. We see that there are two possible choices for S: (1) $S = S_1$, S_1 is the flat surface bounded by Γ; (2) $S = S_2$, where S_2 is the sac-shaped surface, as shown in Figure 2.11. Using the surface S_1, we obtain:

$$\int_{S_1} \left(\vec{\nabla} \times \vec{\mathbf{B}}\right) \cdot d\vec{\mathbf{S}} = \mu_0 \int_{S_1} \vec{\mathbf{J}} \cdot d\vec{\mathbf{S}} = \mu_0 \int_{S_1} \vec{\mathbf{J}} \cdot d\vec{\mathbf{S}} \\ - \mu_0 \int_{S_2} \vec{\mathbf{J}} \cdot d\vec{\mathbf{S}} = -\mu_0 \oint_{S_2 - S_1} \vec{\mathbf{J}} \cdot d\vec{\mathbf{S}} \qquad [2.99]$$

where $\vec{\mathbf{J}}$ is zero on S_2. However, using the surface S_2, it follows that:

$$\int_{S_2} \left(\vec{\nabla} \times \vec{\mathbf{B}}\right) \cdot d\vec{\mathbf{S}} = 0 \qquad [2.100]$$

and, consequently, $\int_{S_1} \left(\vec{\nabla} \times \vec{\mathbf{B}}\right) \cdot d\vec{\mathbf{S}} \neq \int_{S_2} \left(\vec{\nabla} \times \vec{\mathbf{B}}\right) \cdot d\vec{\mathbf{S}}$. However, the integral $\int \left(\vec{\nabla} \times \vec{\mathbf{B}}\right) \cdot d\vec{\mathbf{S}}$ must not depend on the surface used for the calculation, since $\vec{\nabla} \cdot \left(\vec{\nabla} \times \vec{\mathbf{B}}\right) = 0$. Therefore, equation [2.100] must be adjusted. This issue can be resolved by modifying [2.100] and rewriting it in the form:

$$\int_{S_2} \left(\vec{\nabla} \times \vec{\mathbf{B}}\right) \cdot d\vec{\mathbf{S}} = -\mu_0 \oint_{S_2 - S_1} \vec{\mathbf{J}} \cdot d\vec{\mathbf{S}} \qquad [2.101]$$

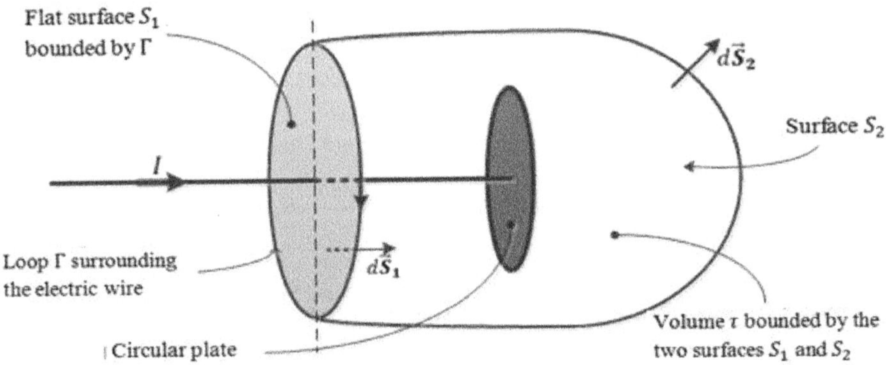

Figure 2.11. *The integral $\int_S \left(\vec{\nabla} \times \vec{\mathbf{B}}\right) \cdot d\vec{\mathbf{S}}$ over the planar surface S_1 or over the closed surface in the form of a sac S_2 yields identical values*

It follows that equation $\int_{S_1} \left(\vec{\nabla} \times \vec{\mathbf{B}}\right) \cdot d\vec{\mathbf{S}} = \int_{S_2} \left(\vec{\nabla} \times \vec{\mathbf{B}}\right) \cdot d\vec{\mathbf{S}}$ is automatically ensured. Equation [2.101] can also be written in a more convenient form. The divergence theorem and the continuity equation $\vec{\nabla} \cdot \vec{\mathbf{J}} = -\frac{\partial \rho}{\partial t}$ allow us to rewrite it as follows:

$$\int_{S_2} \left(\vec{\nabla} \times \vec{\mathbf{B}}\right) \cdot d\vec{\mathbf{S}} = -\mu_0 \int_\tau \vec{\nabla} \cdot \vec{\mathbf{J}} d\tau = \mu_0 \int_\tau \frac{\partial \rho}{\partial t} d\tau \qquad [2.102]$$

where τ is the volume enclosed by the surfaces S_1 and S_2. According to the Poisson equation $\vec{\nabla} \cdot \vec{\mathbf{E}} = \frac{\rho}{\varepsilon_0}$, the above integral can also be written as:

$$\int_{S_2} \left(\vec{\nabla} \times \vec{\mathbf{B}}\right) \cdot d\vec{\mathbf{S}} = \mu_0 \varepsilon_0 \int_\tau \frac{\partial}{\partial t} \left(\vec{\nabla} \cdot \vec{\mathbf{E}}\right) d\tau = \mu_0 \varepsilon_0 \int_\tau \vec{\nabla} \cdot \frac{\partial \vec{\mathbf{E}}}{\partial t} d\tau \qquad [2.103]$$

By subtracting equations [2.99] and [2.103] from each other, we get:

$$\oint_{S_2 - S_1} \left(\vec{\nabla} \times \vec{\mathbf{B}}\right) \cdot d\vec{\mathbf{S}} = \mu_0 \oint_{S_2 - S_1} \vec{\mathbf{J}} \cdot d\vec{\mathbf{S}} + \mu_0 \varepsilon_0 \oint_{S_2 - S_1} \frac{\partial \vec{\mathbf{E}}}{\partial t} \cdot d\vec{\mathbf{S}} \qquad [2.104]$$

where we used the divergence theorem to transform the last term of [2.103] into a surface integral. Since S_2 is arbitrary, we finally deduce:

$$\vec{\nabla} \times \vec{\mathbf{B}} = \mu_0 \vec{\mathbf{J}} + \mu_0 \varepsilon_0 \frac{\partial \vec{\mathbf{E}}}{\partial t} \qquad [2.105]$$

Equation [2.105] is known as the *Maxwell–Ampère equation*. The term $\varepsilon_0 \frac{\partial \vec{\mathbf{E}}}{\partial t}$ is called the *displacement current*. Let us note two remarks that have already been mentioned:

1) Due to charge conservation, we must have $\vec{\nabla} \cdot \vec{\mathbf{J}} + \frac{\partial \rho}{\partial t} = 0$. Moreover, since $\rho = \varepsilon_0 \vec{\nabla} \cdot \vec{\mathbf{E}}$, this shows that:

$$\vec{\nabla} \cdot \left(\vec{\mathbf{J}} + \varepsilon_0 \frac{\partial \vec{\mathbf{E}}}{\partial t}\right) = 0 \qquad [2.106]$$

This expression resembles the equation $\vec{\nabla} \cdot \vec{\mathbf{J}}' = 0$, which is valid for quasi-stationary regimes. Therefore, it is sufficient to add the term $\varepsilon_0 \frac{\partial \vec{\mathbf{E}}}{\partial t}$ (displacement current) to $\vec{\mathbf{J}}$. Thus, Ampère's equation must be generalized as follows:

$$\vec{\nabla} \times \vec{\mathbf{B}} = \mu_0 \vec{\mathbf{J}}' = \mu_0 \vec{\mathbf{J}} + \mu_0 \varepsilon_0 \frac{\partial \vec{\mathbf{E}}}{\partial t} \qquad [2.107]$$

We once again arrive at the Maxwell–Ampère equation.

2) The modification of the theorem is necessary to preserve the identity $\vec{\nabla} \cdot \left(\vec{\nabla} \times \vec{\mathbf{B}}\right) = 0$. Indeed, we verify that:

$$\vec{\nabla} \cdot \left(\vec{\nabla} \times \vec{\mathbf{B}}\right) = \mu_0 \left[\vec{\nabla} \cdot \vec{\mathbf{J}} + \varepsilon_0 \frac{\partial}{\partial t}\left(\vec{\nabla} \cdot \vec{\mathbf{E}}\right)\right] = \mu_0 \left[\vec{\nabla} \cdot \vec{\mathbf{J}} + \frac{\partial \rho}{\partial t}\right] = 0 \quad [2.108]$$

EXAMPLE 2.9. (Van de Graaff generator).– A Van de Graaff generator is a fixed hollow metallic sphere with radius R. It is supplied with a constant current I, which allows electric charges to accumulate on its surface. We aim to calculate the magnetic induction field at a distance z above the sphere, as shown in Figure 2.12. As the charge accumulates on the sphere, the electric field increases as follows:

$$\frac{\partial \vec{\mathbf{E}}}{\partial t} = \frac{1}{4\pi\varepsilon_0} \frac{dQ}{dt} \frac{\vec{\mathbf{r}}}{r^3} = \frac{I}{4\pi\varepsilon_0} \frac{\vec{\mathbf{r}}}{r^3} \quad [2.109]$$

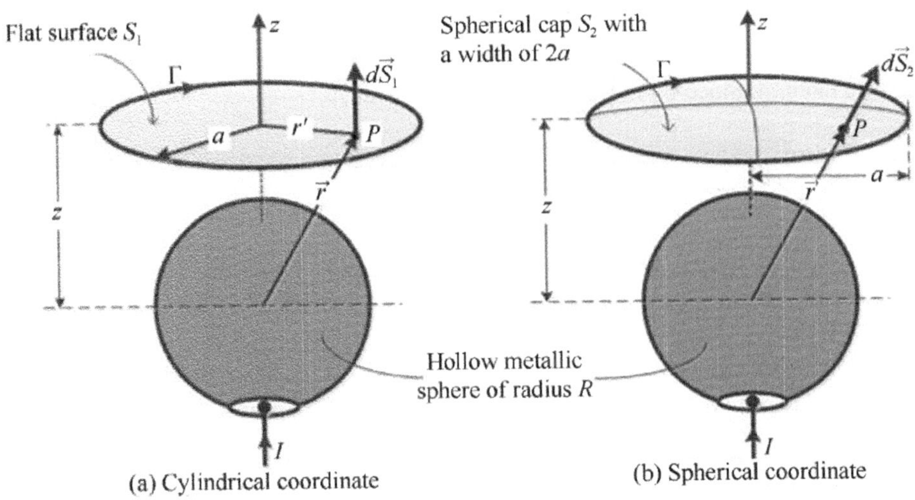

Figure 2.12. *Operating principle of a Van de Graaff generator. For a color version of this figure, see www.iste.co.uk/abdesselam/electrodynamics2.zip*

Draw a Γ loop of radius a around the axis Oz at distance z above the sphere. At this point, the current density is, of course, zero, $\vec{\mathbf{J}} = 0$. Let us integrate over any surface bounded by the loop Γ:

$$\int_S \left(\vec{\nabla} \times \vec{\mathbf{B}}\right) \cdot d\vec{\mathbf{S}} = \oint_\Gamma \vec{\mathbf{B}} \cdot d\vec{\mathbf{l}} = \mu_0\varepsilon_0 \int_S \frac{\partial \vec{\mathbf{E}}}{\partial t} \cdot d\vec{\mathbf{S}} \quad [2.110]$$

Thus, we obtain $2\pi a B_\varphi = \frac{\mu_0 I}{4\pi}\int_S \frac{\vec{\mathbf{r}}\cdot d\vec{\mathbf{S}}}{r^3} = \frac{\mu_0 I \Omega}{4\pi}$ and, therefore:

$$B_\varphi = \frac{\mu_0 I \Omega}{8\pi^2 a} \qquad [2.111]$$

where $\Omega = \int_S \frac{\vec{\mathbf{r}}\cdot d\vec{\mathbf{S}}}{r^3}$ is the solid angle under which the surface S is observed from point O. This integral can be evaluated in two ways:

– *In cylindrical coordinates*: we will choose the flat surface S_1, as shown in Figure 2.12(a). Since $d\vec{\mathbf{S}}_1 = r'dr'd\varphi\,\vec{\mathbf{k}}$ and $\vec{\mathbf{r}} = z\vec{\mathbf{k}} + r'\vec{\mathbf{e}}_r$, we obtain:

$$B_\varphi = \frac{\mu_0 I}{8\pi^2 a}\int_0^{2\pi}\int_0^a \frac{zr'dr'd\varphi}{(z^2+r'^2)^{3/2}} = \frac{\mu_0 I}{4\pi a}\left(1 - \frac{z}{\sqrt{z^2+a^2}}\right) \qquad [2.112]$$

because $z > 0$.

– *In spherical coordinates*: we will choose the spherical cap S_2 with a width of $2a$, as shown in Figure 2.12(b). Since $d\vec{\mathbf{S}}_2 = r^2 \sin\theta d\theta d\varphi\,\vec{\mathbf{e}}_r$ and $\vec{\mathbf{r}} = r\,\vec{\mathbf{e}}_r$, we obtain:

$$B_\varphi = \frac{\mu_0 I}{8\pi^2 a}\int_0^{2\pi}\int_0^{\cos^{-1}\left(\frac{z}{\sqrt{z^2+a^2}}\right)} \sin\theta d\theta d\varphi = \frac{\mu_0 I}{4\pi a}\left(1 - \frac{z}{\sqrt{z^2+a^2}}\right) \qquad [2.113]$$

We obviously obtain the same results, since the calculations must be independent of the chosen surface.

2.6. Lorentz force and canonical momentum

In classical mechanics, it is often useful to express a conservative force as the gradient of a potential energy. Although the force associated with the magnetic field $\vec{\mathbf{B}}$ is not conservative, it can still be written as the gradient of a quasipotential (or pseudopotential). To do so, let us write Newton's law for the Lorentz force:

$$\frac{d\vec{\mathbf{p}}}{dt} = q\left(\vec{\mathbf{E}} + \vec{\mathbf{v}}\times\vec{\mathbf{B}}\right) = q\left[-\vec{\nabla}V - \frac{\partial\vec{\mathbf{A}}}{\partial t} + \vec{\mathbf{v}}\times\left(\vec{\nabla}\times\vec{\mathbf{A}}\right)\right] \qquad [2.114]$$

Let us use the identity $\vec{\nabla}\left(\vec{\mathbf{v}}\cdot\vec{\mathbf{A}}\right) = \left(\vec{\mathbf{v}}\cdot\vec{\nabla}\right)\vec{\mathbf{A}} + \left(\vec{\mathbf{A}}\cdot\vec{\nabla}\right)\vec{\mathbf{v}} + \vec{\mathbf{A}}\times\left(\vec{\nabla}\times\vec{\mathbf{v}}\right) + \vec{\mathbf{v}}\times\left(\vec{\nabla}\times\vec{\mathbf{A}}\right)$. Since $\vec{\mathbf{v}}$ does not depend on position, it follows that $\vec{\mathbf{v}} \times \left(\vec{\nabla} \times \vec{\mathbf{A}}\right) = \vec{\nabla}\left(\vec{\mathbf{v}} \cdot \vec{\mathbf{A}}\right) - \left(\vec{\mathbf{v}} \cdot \vec{\nabla}\right)\vec{\mathbf{A}}$ and, consequently, equation [2.114] becomes:

$$\frac{d\vec{\mathbf{p}}}{dt} = q\left[-\vec{\nabla}V - \frac{\partial \vec{\mathbf{A}}}{\partial t} + \vec{\nabla}\left(\vec{\mathbf{v}} \cdot \vec{\mathbf{A}}\right) - \left(\vec{\mathbf{v}} \cdot \vec{\nabla}\right)\vec{\mathbf{A}}\right] \qquad [2.115]$$

The term $\frac{\partial \vec{\mathbf{A}}}{\partial t} + \left(\vec{\mathbf{v}} \cdot \vec{\nabla}\right)\vec{\mathbf{A}}$ is simply the *total derivative* of $\vec{\mathbf{A}}$:

$$\frac{\partial \vec{\mathbf{A}}}{\partial t} + \left(\vec{\mathbf{v}} \cdot \vec{\nabla}\right)\vec{\mathbf{A}} \equiv \frac{d\vec{\mathbf{A}}}{dt} \qquad [2.116]$$

In fact, we verify that $d\vec{\mathbf{A}} = \vec{\mathbf{A}}\left(\vec{\mathbf{r}}\,(t+dt)\,, t+dt\right) - \vec{\mathbf{A}}\left(\vec{\mathbf{r}}\,(t)\,, t\right) = \vec{\mathbf{A}}\left(\vec{\mathbf{r}} + \vec{\mathbf{v}}dt, t+dt\right) - \vec{\mathbf{A}}\left(\vec{\mathbf{r}} + \vec{\mathbf{v}}dt, t\right) + \vec{\mathbf{A}}\left(\vec{\mathbf{r}} + \vec{\mathbf{v}}dt, t\right) - \vec{\mathbf{A}}\left(\vec{\mathbf{r}}, t\right) = \frac{\partial \vec{\mathbf{A}}}{\partial t}dt + \left(\vec{\mathbf{v}} \cdot \vec{\nabla}\right)\vec{\mathbf{A}}dt$. Dividing by dt, we find [2.116]. Taking [2.116] into account, we obtain $\frac{d\vec{\mathbf{p}}}{dt} = q\left[-\vec{\nabla}\left(V - \vec{\mathbf{v}} \cdot \vec{\mathbf{A}}\right) - \frac{d\vec{\mathbf{A}}}{dt}\right]$, or equivalently:

$$\frac{d}{dt}\left(\vec{\mathbf{p}} + q\vec{\mathbf{A}}\right) = -q\vec{\nabla}\left(V - \vec{\mathbf{v}} \cdot \vec{\mathbf{A}}\right) \qquad [2.117]$$

The term $V - \vec{\mathbf{v}} \cdot \vec{\mathbf{A}}$ is the potential that appears in the Lagrangian of a charged particle in an electromagnetic field:

$$L = \frac{1}{2}m\vec{\mathbf{v}}^2 - q\left(V - \vec{\mathbf{v}} \cdot \vec{\mathbf{A}}\right) \qquad [2.118]$$

However, $\vec{\mathbf{p}} + q\vec{\mathbf{A}}$ is the *conjugate momentum* associated with the position $\vec{\mathbf{r}}$.

EXAMPLE 2.10. (Motion of a charged particle in a uniform magnetic field).– When the magnetic field $\vec{\mathbf{B}}$ is uniform and directed along the axis Oz, the scalar potential V and the vector potential $\vec{\mathbf{A}}$ are written as: $V = 0$ and $\vec{\mathbf{A}} = \frac{B}{2}\left(-y\,\vec{\mathbf{i}} + x\,\vec{\mathbf{j}}\right)$. Substituting these values of V and $\vec{\mathbf{A}}$ into the Lagrangian [2.118], we obtain:

$$L = \frac{1}{2}\left[m\left(v_x^2 + v_y^2 + v_z^2\right) - qB\left(v_x y - v_y x\right)\right] \qquad [2.119]$$

and, consequently, the equations of motion are:

$$\begin{aligned} &\frac{d^2x}{dt^2} - \frac{qB}{m}\frac{dy}{dt} = 0 \\ &\frac{d^2y}{dt^2} + \frac{qB}{m}\frac{dx}{dt} = 0 \\ &\frac{d^2z}{dt^2} = 0 \end{aligned} \qquad [2.120]$$

The last differential equation has the solution:

$$z(t) = (v_z)_0\, t + z_0 \qquad [2.121]$$

where $(v_z)_0$ and z_0 are constants. The charged particle thus undergoes uniform motion with velocity $(v_z)_0$ along the Oz axis from the initial position z_0. The first integration of the second differential equation gives $\frac{dy}{dt} = (v_y)_0 - \frac{qB}{m}(x - x_0)$. This result allows us to rewrite the first equation in the form: $\frac{d^2x}{dt^2} + \frac{q^2B^2}{m^2}(x - x_0) = \frac{qB}{m}(v_y)_0$. This last differential equation has the solution:

$$x(t) = x_0 + \frac{m(v_y)_0}{qB}\left(1 - \cos\frac{qBt}{m}\right) + \frac{m(v_x)_0}{qB}\sin\frac{qBt}{m} \qquad [2.122]$$

It follows that $\frac{dy}{dt} = (v_y)_0 \cos\frac{qBt}{m} - (v_x)_0 \sin\frac{qBt}{m}$ and consequently:

$$y(t) = y_0 - \frac{m(v_x)_0}{qB}\left(1 - \cos\frac{qBt}{m}\right) + \frac{m(v_y)_0}{qB}\sin\frac{qBt}{m} \qquad [2.123]$$

The trajectory of the particle is obtained by eliminating time between equations [2.122] and [2.123]:

$$\left[x - x_0 - \frac{m(v_y)_0}{qB}\right]^2 + \left[y - y_0 + \frac{m(v_x)_0}{qB}\right]^2 = \frac{m^2}{q^2B^2}\left[(v_x)_0^2 + (v_y)_0^2\right] \qquad [2.124]$$

2.7. Maxwell equations

The results obtained so far can be summarized in four coupled equations, known as *Maxwell's equations*:

$$\begin{aligned}
&\vec{\nabla}\cdot\vec{\mathbf{E}} = \frac{\rho}{\varepsilon_0} && \text{(Maxwell-Gauss equation)}\\
&\vec{\nabla}\cdot\vec{\mathbf{B}} = 0 && \text{(Maxwell-Thomson equation)}\\
&\vec{\nabla}\times\vec{\mathbf{E}} = -\frac{\partial\vec{\mathbf{B}}}{\partial t} && \text{(Maxwell-Faraday equation)}\\
&\vec{\nabla}\times\vec{\mathbf{B}} = \mu_0\vec{\mathbf{J}} + \mu_0\varepsilon_0\frac{\partial\vec{\mathbf{E}}}{\partial t} && \text{(Maxwell-Ampère equation)}
\end{aligned} \qquad [2.125]$$

These equations are linear, meaning that superimposed solutions are always solutions. However, this linearity does not always hold, especially in materials where fields cannot simply be added. Maxwell's equations, Lorentz's force equation and Newton's second law form the foundation of all classical electromagnetic interactions. The development of electromagnetic theory is entirely based on Maxwell's equations.

Substituting $\vec{\mathbf{E}}$ by $-\vec{\nabla}V - \frac{\partial\vec{\mathbf{A}}}{\partial t}$ in the Maxwell–Gauss equation, we obtain $\vec{\nabla}\cdot\vec{\mathbf{E}} = \vec{\nabla}\cdot\left(-\vec{\nabla}V - \frac{\partial\vec{\mathbf{A}}}{\partial t}\right) = \frac{\rho}{\varepsilon_0}$, which leads to:

$$-\Delta V - \frac{\partial}{\partial t}\left(\vec{\nabla}\cdot\vec{\mathbf{A}}\right) = \frac{\rho}{\varepsilon_0} \qquad [2.126]$$

Similarly, replacing $\vec{\mathbf{B}}$ by $\vec{\nabla}\times\vec{\mathbf{A}}$ and $\vec{\mathbf{E}}$ by $-\vec{\nabla}V - \frac{\partial\vec{\mathbf{A}}}{\partial t}$ in the Maxwell–Ampère equation, we find $\vec{\nabla}\times\left(\vec{\nabla}\times\vec{\mathbf{A}}\right) = \vec{\nabla}\times\left(\vec{\nabla}\cdot\vec{\mathbf{A}}\right) - \Delta\vec{\mathbf{A}} = \mu_0\vec{\mathbf{J}} - \mu_0\varepsilon_0\frac{\partial}{\partial t}\left(\vec{\nabla}V + \frac{\partial\vec{\mathbf{A}}}{\partial t}\right)$, or:

$$\Delta\vec{\mathbf{A}} - \mu_0\varepsilon_0\frac{\partial^2\vec{\mathbf{A}}}{\partial t^2} - \vec{\nabla}\left(\vec{\nabla}\cdot\vec{\mathbf{A}} + \mu_0\varepsilon_0\frac{\partial V}{\partial t}\right) = -\mu_0\vec{\mathbf{J}} \qquad [2.127]$$

2.8. Gauge transformations

We have already mentioned that the potential vector $\vec{\mathbf{A}}$ is not uniquely defined, since we can always add a gradient field $\vec{\nabla}\Lambda$ without changing physics. However, according to [2.72], we must also require a compensating modification for V in parallel with any modification of $\vec{\mathbf{A}}$ to maintain the value of the electric field $\vec{\mathbf{E}}$, ensuring that the physics remain unchanged. Indeed, if we take $\overrightarrow{\mathbf{A}}' = \overrightarrow{\mathbf{A}} + \overrightarrow{\nabla}\Lambda$ and if V' denotes the modified potential, then $\overrightarrow{\mathbf{E}}' = -\overrightarrow{\nabla}V' - \frac{\partial}{\partial t}\left(\overrightarrow{\mathbf{A}} + \overrightarrow{\nabla}\Lambda\right) = -\overrightarrow{\nabla}\left(V' + \frac{\partial\Lambda}{\partial t}\right) - \frac{\partial\overrightarrow{\mathbf{A}}}{\partial t}$. According to the original potentials, we must have $\overrightarrow{\mathbf{E}} = -\overrightarrow{\nabla}V - \frac{\partial\overrightarrow{\mathbf{A}}}{\partial t}$. Comparing these expressions, we find $V' = V - \frac{\partial\overrightarrow{\mathbf{A}}}{\partial t}$. The pair of coupled transformations:

$$\begin{aligned}\overrightarrow{\mathbf{A}}' &= \overrightarrow{\mathbf{A}} + \overrightarrow{\nabla}\Lambda \\ V' &= V - \frac{\partial\Lambda}{\partial t}\end{aligned} \qquad [2.128]$$

is called a gauge *invariance transformation.*

The invariance of fields under such a transformation is called *gauge invariance.* These transformations are of great importance, as they allow equations [2.126] and [2.127] to be written in more elegant forms. Although there are infinitely many possible gauges (i.e. an infinite number of choices for Λ), the *Coulomb* and *Lorenz gauges* are among the most widely used.

When fields are static, it is preferable to choose Λ so that:

$$\overrightarrow{\nabla} \cdot \overrightarrow{\mathbf{A}} = 0 \qquad [2.129]$$

This is the *Coulomb gauge*. With this choice of gauge, equations [2.126] and [2.127] are reduced to:

$$\begin{aligned}\Delta V &= -\frac{\rho}{\varepsilon_0} \quad \text{(i)} \\ \Delta\overrightarrow{\mathbf{A}} &= -\mu_0\overrightarrow{\mathbf{J}} \quad \text{(ii)}\end{aligned} \qquad [2.130]$$

On the other hand, if the fields are non-static and the Coulomb gauge is adopted, equation [2.130i] remains valid, while equation [2.130ii] reduces to:

$$\Delta\overrightarrow{\mathbf{A}} - \mu_0\varepsilon_0\frac{\partial^2\overrightarrow{\mathbf{A}}}{\partial t^2} = -\mu_0\overrightarrow{\mathbf{J}} + \mu_0\varepsilon_0\overrightarrow{\nabla}\left(\frac{\partial V}{\partial t}\right) \qquad [2.131]$$

The vector $\overrightarrow{\mathbf{J}}$ can always be regarded as a sum:

$$\overrightarrow{\mathbf{J}} = \overrightarrow{\mathbf{J}}_l + \overrightarrow{\mathbf{J}}_s \qquad [2.132]$$

where $\overrightarrow{\mathbf{J}}_l$ is the *longitudinal component* whose circulation along any closed curve is zero, that is, $\overrightarrow{\nabla} \times \overrightarrow{\mathbf{J}}_l = 0$, and $\overrightarrow{\mathbf{J}}_s$ is the *solenoidal component* whose flux through any closed surface is zero, that is, $\overrightarrow{\nabla} \cdot \overrightarrow{\mathbf{J}}_s = 0$. This statement holds for any vector; it is known as the *Helmholtz–Hodge theorem*. Substituting $\overrightarrow{\mathbf{J}}_l$ and $\overrightarrow{\mathbf{J}}_s$ respectively into identity $\overrightarrow{\nabla} \times \left(\overrightarrow{\nabla} \times \overrightarrow{\mathbf{J}}\right) = \overrightarrow{\nabla}\left(\overrightarrow{\nabla} \cdot \overrightarrow{\mathbf{J}}\right) - \Delta \overrightarrow{\mathbf{J}}$, we obtain:

$$\begin{aligned} \Delta \overrightarrow{\mathbf{J}}_s &= -\overrightarrow{\nabla} \times \left(\overrightarrow{\nabla} \times \overrightarrow{\mathbf{J}}_s\right) = -\overrightarrow{\nabla} \times \left(\overrightarrow{\nabla} \times \overrightarrow{\mathbf{J}}\right) \quad \text{(i)} \\ \Delta \overrightarrow{\mathbf{J}}_l &= \overrightarrow{\nabla}\left(\overrightarrow{\nabla} \cdot \overrightarrow{\mathbf{J}}_l\right) = \overrightarrow{\nabla}\left(\overrightarrow{\nabla} \cdot \overrightarrow{\mathbf{J}}\right) \quad \text{(ii)} \end{aligned} \qquad [2.133]$$

These equations obviously resemble the Poisson equation $\Delta V = -\frac{\rho}{\varepsilon_0}$, whose solution is given by $V\left(\overrightarrow{\mathbf{r}}, t\right) = \frac{1}{4\pi\varepsilon_0}\int_{\tau'} \frac{\rho\left(\overrightarrow{\mathbf{r}}', t\right) d\tau'}{\left|\overrightarrow{\mathbf{r}} - \overrightarrow{\mathbf{r}}'\right|}$, except that they are in 3D here. By analogy, the solutions of [2.133i] and [2.133ii] are written as:

$$\begin{aligned} \overrightarrow{\mathbf{J}}_s\left(\overrightarrow{\mathbf{r}}, t\right) &= \frac{1}{4\pi}\int_{\tau'} \frac{\overrightarrow{\nabla}' \times \left[\overrightarrow{\nabla}' \times \overrightarrow{\mathbf{J}}\left(\overrightarrow{\mathbf{r}}', t\right)\right] d\tau'}{\left|\overrightarrow{\mathbf{r}} - \overrightarrow{\mathbf{r}}'\right|} \quad \text{(i)} \\ \overrightarrow{\mathbf{J}}_l\left(\overrightarrow{\mathbf{r}}, t\right) &= -\frac{1}{4\pi}\int_{\tau'} \frac{\overrightarrow{\nabla}'\left[\overrightarrow{\nabla}' \cdot \overrightarrow{\mathbf{J}}\left(\overrightarrow{\mathbf{r}}', t\right)\right] d\tau'}{\left|\overrightarrow{\mathbf{r}} - \overrightarrow{\mathbf{r}}'\right|} \quad \text{(ii)} \end{aligned} \qquad [2.134]$$

Using the identity $f\overrightarrow{\nabla} g = \overrightarrow{\nabla}(fg) - g\overrightarrow{\nabla} f$, equation [2.134ii] also becomes:

$$\begin{aligned} \overrightarrow{\mathbf{J}}_l &= \frac{1}{4\pi}\int_{\tau'}\left(-\overrightarrow{\nabla}'\left[\frac{\overrightarrow{\nabla}' \cdot \overrightarrow{\mathbf{J}}'}{\left|\overrightarrow{\mathbf{r}} - \overrightarrow{\mathbf{r}}'\right|}\right] + \left[\overrightarrow{\nabla}' \cdot \overrightarrow{\mathbf{J}}'\right]\overrightarrow{\nabla}'\left(\frac{1}{\left|\overrightarrow{\mathbf{r}} - \overrightarrow{\mathbf{r}}'\right|}\right)\right) d\tau' \\ &= -\frac{1}{4\pi}\oint_{S'} \frac{\overrightarrow{\nabla}' \cdot \overrightarrow{\mathbf{J}}'}{\left|\overrightarrow{\mathbf{r}} - \overrightarrow{\mathbf{r}}'\right|} d\overrightarrow{\mathbf{S}}' - \frac{1}{4\pi}\int_{\tau'}\left[\overrightarrow{\nabla}' \cdot \overrightarrow{\mathbf{J}}'\right]\overrightarrow{\nabla}\left(\frac{1}{\left|\overrightarrow{\mathbf{r}} - \overrightarrow{\mathbf{r}}'\right|}\right) d\tau' \qquad [2.135] \\ &= -\frac{1}{4\pi}\oint_{S'} \frac{\overrightarrow{\nabla}' \cdot \overrightarrow{\mathbf{J}}'}{\left|\overrightarrow{\mathbf{r}} - \overrightarrow{\mathbf{r}}'\right|} d\overrightarrow{\mathbf{S}}' - \frac{1}{4\pi}\overrightarrow{\nabla}\left(\int_{\tau'} \frac{\overrightarrow{\nabla}' \cdot \overrightarrow{\mathbf{J}}'}{\left|\overrightarrow{\mathbf{r}} - \overrightarrow{\mathbf{r}}'\right|} d\tau'\right) \end{aligned}$$

since $\overrightarrow{\nabla}'\left(\frac{1}{\left|\overrightarrow{\mathbf{r}} - \overrightarrow{\mathbf{r}}'\right|}\right) = -\overrightarrow{\nabla}\left(\frac{1}{\left|\overrightarrow{\mathbf{r}} - \overrightarrow{\mathbf{r}}'\right|}\right)$. We have used the notation: $\overrightarrow{\mathbf{J}}\left(\overrightarrow{\mathbf{r}}', t\right) \equiv \overrightarrow{\mathbf{J}}'$. Since τ' must include all currents (i.e. no current can cross the boundaries of the volume), the integral $\oint_{S'} \frac{\overrightarrow{\nabla}' \cdot \overrightarrow{\mathbf{J}}'}{\left|\overrightarrow{\mathbf{r}} - \overrightarrow{\mathbf{r}}'\right|} d\overrightarrow{\mathbf{S}}'$ must therefore be zero. It follows that:

$$\overrightarrow{\mathbf{J}}_l = -\overrightarrow{\nabla}\left(\frac{1}{4\pi}\int_{\tau'} \frac{\overrightarrow{\nabla}' \cdot \overrightarrow{\mathbf{J}}\left(\overrightarrow{\mathbf{r}}', t\right)}{\left|\overrightarrow{\mathbf{r}} - \overrightarrow{\mathbf{r}}'\right|} d\tau'\right) \qquad [2.136]$$

Using the continuity equation $\vec{\nabla}' \cdot \vec{\mathbf{J}}\left(\vec{\mathbf{r}}', t\right) = -\frac{\partial \rho(\vec{\mathbf{r}}', t)}{\partial t}$, equation [2.136] becomes:

$$\begin{aligned} \vec{\mathbf{J}}_l &= \vec{\nabla}\left(\frac{1}{4\pi}\int_{\tau'} \frac{\frac{\partial \rho(\vec{\mathbf{r}}', t)}{\partial t}}{\left|\vec{\mathbf{r}} - \vec{\mathbf{r}}'\right|} d\tau'\right) \\ &= \vec{\nabla}\frac{\partial}{\partial t}\left(\frac{1}{4\pi}\int_{\tau'} \frac{\rho\left(\vec{\mathbf{r}}', t\right) d\tau'}{\left|\vec{\mathbf{r}} - \vec{\mathbf{r}}'\right|}\right) = \varepsilon_0 \vec{\nabla}\left(\frac{\partial V}{\partial t}\right) \end{aligned} \quad [2.137]$$

Substituting $\vec{\mathbf{J}}$ in [2.131] with $\vec{\mathbf{J}}_l + \vec{\mathbf{J}}_s$ and $\vec{\mathbf{J}}_l$ with $\varepsilon_0 \vec{\nabla}\left(\frac{\partial V}{\partial t}\right)$, we obtain:

$$\Delta \vec{\mathbf{A}} - \mu_0 \varepsilon_0 \frac{\partial^2 \vec{\mathbf{A}}}{\partial t^2} = -\mu_0 \vec{\mathbf{J}}_s \quad [2.138]$$

We conclude that in the Coulomb gauge, V satisfies a static equation giving instantaneous solutions, while the vector potential $\vec{\mathbf{A}}$ satisfies a wave equation that takes into account the finite velocity of propagation of $\vec{\mathbf{J}}_s$ perturbations.

In electrodynamics, it is also useful to choose the gauge:

$$\vec{\nabla} \cdot \vec{\mathbf{A}} = -\mu_0 \varepsilon_0 \frac{\partial V}{\partial t} \quad [2.139]$$

called the *Lorenz gauge*. With this choice, equations [2.126] and [2.127] take the form of an *inhomogeneous wave equation*:

$$\begin{aligned} \Delta V - \mu_0 \varepsilon_0 \frac{\partial^2 V}{\partial t^2} &= -\frac{\rho}{\varepsilon_0} \quad \text{(i)} \\ \Delta \vec{\mathbf{A}} - \mu_0 \varepsilon_0 \frac{\partial^2 \vec{\mathbf{A}}}{\partial t^2} &= -\mu_0 \vec{\mathbf{J}} \quad \text{(ii)} \end{aligned} \quad [2.140]$$

In the Lorenz gauge, V and $\vec{\mathbf{A}}$ satisfy similar wave equations. The solutions to these equations, with the additional conditions $V(\infty) = 0$ and $\vec{\mathbf{A}}(\infty) = 0$, which hold for a localized source distribution or one whose density decreases sufficiently rapidly with distance, are:

$$\begin{aligned} V\left(\vec{\mathbf{r}}, t\right) &= \frac{1}{4\pi\varepsilon_0}\int_{\tau'} \frac{\rho\left(\vec{\mathbf{r}}', t - \frac{\left|\vec{\mathbf{r}} - \vec{\mathbf{r}}'\right|}{c}\right) d\tau'}{\left|\vec{\mathbf{r}} - \vec{\mathbf{r}}'\right|} \quad \text{(i)} \\ \vec{\mathbf{A}}\left(\vec{\mathbf{r}}, t\right) &= \frac{\mu_0}{4\pi}\int_{\tau'} \frac{\vec{\mathbf{J}}\left(\vec{\mathbf{r}}', t - \frac{\left|\vec{\mathbf{r}} - \vec{\mathbf{r}}'\right|}{c}\right) d\tau'}{\left|\vec{\mathbf{r}} - \vec{\mathbf{r}}'\right|} \quad \text{(ii)} \end{aligned} \quad [2.141]$$

These solutions are called the *retarded potentials* because the values of the potentials at a point $\vec{\mathbf{r}}$ at time t depend on those of the sources at all points $\vec{\mathbf{r}}'$ at the earlier time $t - \frac{|\vec{\mathbf{r}} - \vec{\mathbf{r}}'|}{c}$. The quantity $\frac{|\vec{\mathbf{r}} - \vec{\mathbf{r}}'|}{c}$ corresponds to the propagation time from (x', y', z') to (x, y, z) at the speed c of the effects of a variation in the sources. There exists another class of solutions, called *advanced potentials*, for equations [2.140], where $t - \frac{|\vec{\mathbf{r}} - \vec{\mathbf{r}}'|}{c}$ is replaced by $t + \frac{|\vec{\mathbf{r}} - \vec{\mathbf{r}}'|}{c}$. However, these solutions have no physical significance, as V and $\vec{\mathbf{A}}$ would then depend on the values of the sources at a future time. They would therefore violate the *principle of causality*, which states that the cause must precede the effect.

When the sources are stationary, the retarded potentials reduce to the expressions obtained in electrostatics and magnetostatics:

$$
\begin{aligned}
V\left(\vec{\mathbf{r}}\right) &= \frac{1}{4\pi\varepsilon_0} \int_{\tau'} \frac{\rho\left(\vec{\mathbf{r}}'\right) d\tau'}{\left|\vec{\mathbf{r}} - \vec{\mathbf{r}}'\right|} \quad \text{(i)} \\
\vec{\mathbf{A}}\left(\vec{\mathbf{r}}\right) &= \frac{\mu_0}{4\pi} \int_{\tau'} \frac{\vec{\mathbf{J}}\left(\vec{\mathbf{r}}'\right) d\tau'}{\left|\vec{\mathbf{r}} - \vec{\mathbf{r}}'\right|} \quad \text{(ii)}
\end{aligned}
\qquad [2.142]
$$

and, consequently, the fields are then given by the Coulomb and Biot–Savart laws:

$$
\begin{aligned}
\vec{\mathbf{E}}\left(\vec{\mathbf{r}}\right) &= \frac{1}{4\pi\varepsilon_0} \int_{\tau'} \frac{\rho\left(\vec{\mathbf{r}}'\right)\left(\vec{\mathbf{r}} - \vec{\mathbf{r}}'\right) d\tau'}{\left|\vec{\mathbf{r}} - \vec{\mathbf{r}}'\right|^3} \quad \text{(i)} \\
\vec{\mathbf{B}}\left(\vec{\mathbf{r}}\right) &= \frac{\mu_0}{4\pi} \int_{\tau'} \frac{\vec{\mathbf{J}}\left(\vec{\mathbf{r}}'\right) \times \left(\vec{\mathbf{r}} - \vec{\mathbf{r}}'\right) d\tau'}{\left|\vec{\mathbf{r}} - \vec{\mathbf{r}}'\right|^3} \quad \text{(ii)}
\end{aligned}
\qquad [2.143]
$$

For non-stationary sources, the expressions for $\vec{\mathbf{E}}\left(\vec{\mathbf{r}}, t\right)$ and $\vec{\mathbf{B}}\left(\vec{\mathbf{r}}, t\right)$ cannot simply be obtained from the Coulomb and Biot–Savart laws by directly replacing $\rho\left(\vec{\mathbf{r}}'\right)$ with $\rho\left(\vec{\mathbf{r}}', t - \frac{|\vec{\mathbf{r}} - \vec{\mathbf{r}}'|}{c}\right)$ and $\vec{\mathbf{J}}\left(\vec{\mathbf{r}}'\right)$ with $\vec{\mathbf{J}}\left(\vec{\mathbf{r}}', t - \frac{|\vec{\mathbf{r}} - \vec{\mathbf{r}}'|}{c}\right)$. In fact, from [2.70] and [2.72], we find that:

$$
\begin{aligned}
\vec{\mathbf{E}}\left(\vec{\mathbf{r}}, t\right) &= \frac{1}{4\pi\varepsilon_0} \int_{\tau'} \left[\left(\frac{\rho'(t_r)}{\left|\vec{\mathbf{r}} - \vec{\mathbf{r}}'\right|^3} + \frac{1}{c\left|\vec{\mathbf{r}} - \vec{\mathbf{r}}'\right|^2} \frac{\partial \rho'(t_r)}{\partial t}\right) \times \right. \\
&\quad \left. \left(\vec{\mathbf{r}} - \vec{\mathbf{r}}'\right) - \frac{1}{c^2\left|\vec{\mathbf{r}} - \vec{\mathbf{r}}'\right|} \frac{\partial \vec{\mathbf{J}}'(t_r)}{\partial t}\right] d\tau' \quad \text{(i)} \\
\vec{\mathbf{B}}\left(\vec{\mathbf{r}}, t\right) &= \frac{\mu_0}{4\pi} \int_{\tau'} \left[\frac{\vec{\mathbf{J}}'(t_r)}{\left|\vec{\mathbf{r}} - \vec{\mathbf{r}}'\right|^3} + \frac{1}{c\left|\vec{\mathbf{r}} - \vec{\mathbf{r}}'\right|^2} \frac{\partial \vec{\mathbf{J}}'(t_r)}{\partial t}\right] \times \\
&\quad \left(\vec{\mathbf{r}} - \vec{\mathbf{r}}'\right) d\tau' \quad \text{(ii)}
\end{aligned}
\qquad [2.144]
$$

where $t_r = t - \frac{|\vec{\mathbf{r}} - \vec{\mathbf{r}}'|}{c}$, $\rho'(t_r) \equiv \rho\left(\vec{\mathbf{r}}', t_r\right)$ and $\vec{\mathbf{J}}'(t_r) \equiv \vec{\mathbf{J}}\left(\vec{\mathbf{r}}', t_r\right)$.

These formulas were established by Jefimenko. In addition to the usual terms in electrostatics and magnetostatics, which depend on the values of $\rho\left(\overrightarrow{\mathbf{r}}', t_r\right)$ and $\overrightarrow{\mathbf{J}}\left(\overrightarrow{\mathbf{r}}', t_r\right)$ (at retarded time) and decrease as $\frac{1}{r^2}$, the non-stationary case includes additional terms that depend on the time derivatives of $\rho\left(\overrightarrow{\mathbf{r}}', t_r\right)$ and $\overrightarrow{\mathbf{J}}\left(\overrightarrow{\mathbf{r}}', t_r\right)$ and decrease as $\frac{1}{r}$. These latter terms dominate at large distances from the sources (if the sources are not stationary) and are responsible for electromagnetic waves.

In the quasi-stationary regime, the propagation time $\frac{\left|\overrightarrow{\mathbf{r}} - \overrightarrow{\mathbf{r}}'\right|}{c}$ is negligible compared to the system's evolution time. Near the sources, the potentials $V\left(\overrightarrow{\mathbf{r}}, t\right)$ and $\overrightarrow{\mathbf{A}}\left(\overrightarrow{\mathbf{r}}, t\right)$ are then approximately given by:

$$\begin{aligned} V\left(\overrightarrow{\mathbf{r}}, t\right) &\simeq \frac{1}{4\pi\varepsilon_0} \int_{\tau'} \frac{\rho\left(\overrightarrow{\mathbf{r}}', t\right) d\tau'}{\left|\overrightarrow{\mathbf{r}} - \overrightarrow{\mathbf{r}}'\right|} \quad \text{(i)} \\ \overrightarrow{\mathbf{A}}\left(\overrightarrow{\mathbf{r}}, t\right) &\simeq \frac{\mu_0}{4\pi} \int_{\tau'} \frac{\overrightarrow{\mathbf{J}}\left(\overrightarrow{\mathbf{r}}', t\right) d\tau'}{\left|\overrightarrow{\mathbf{r}} - \overrightarrow{\mathbf{r}}'\right|} \quad \text{(ii)} \end{aligned} \qquad [2.145]$$

It follows that:

$$\overrightarrow{\mathbf{B}}\left(\overrightarrow{\mathbf{r}}, t\right) = \overrightarrow{\nabla} \times \overrightarrow{\mathbf{A}}\left(\overrightarrow{\mathbf{r}}, t\right) \simeq \frac{\mu_0}{4\pi} \int_{\tau'} \frac{\overrightarrow{\mathbf{J}}\left(\overrightarrow{\mathbf{r}}', t\right) \times \left(\overrightarrow{\mathbf{r}} - \overrightarrow{\mathbf{r}}'\right) d\tau'}{\left|\overrightarrow{\mathbf{r}} - \overrightarrow{\mathbf{r}}'\right|^3} \qquad [2.146]$$

$$\overrightarrow{\mathbf{E}}\left(\overrightarrow{\mathbf{r}}, t\right) \simeq \frac{1}{4\pi\varepsilon_0} \int_{\tau'} \frac{\rho\left(\overrightarrow{\mathbf{r}}', t\right) \left(\overrightarrow{\mathbf{r}} - \overrightarrow{\mathbf{r}}'\right) d\tau'}{\left|\overrightarrow{\mathbf{r}} - \overrightarrow{\mathbf{r}}'\right|^3} - \frac{\mu_0}{4\pi} \int_{\tau'} \frac{\frac{\partial \overrightarrow{\mathbf{J}}\left(\overrightarrow{\mathbf{r}}', t\right)}{\partial t} d\tau'}{\left|\overrightarrow{\mathbf{r}} - \overrightarrow{\mathbf{r}}'\right|} \qquad [2.147]$$

due to the presence of the term $\frac{\partial \overrightarrow{\mathbf{A}}}{\partial t}$ in [2.72].

Chapter 3

Electromagnetic Waves in a Vacuum

3.1. Wave equations in a vacuum

Consider a charged distribution $\mathcal{D}$ localized around a point O. The charge density ρ and the current density $\overrightarrow{\mathbf{J}}$ depend on time. According to the Maxwell–Gauss and Maxwell–Ampère equations, this distribution $\mathcal{D}$ acts as a source of time-varying fields $\overrightarrow{\mathbf{E}}$ and $\overrightarrow{\mathbf{B}}$, which will be created in the neighborhood of O. A point M in this region, although located outside $\mathcal{D}$, also acts as a source of fields due to the terms $\frac{\partial \overrightarrow{\mathbf{E}}}{\partial t}$ and $\frac{\partial \overrightarrow{\mathbf{B}}}{\partial t}$ (originating from O), which serve as sources in the Maxwell–Faraday and Maxwell–Ampère equations. Other points N in the neighborhood of M similarly become sources of temporally varying fields in their own surroundings. Thus, the electromagnetic field propagates like a wave on the surface of water. The coupling introduced in the Maxwell equations through the time partial derivatives $\frac{\partial \overrightarrow{\mathbf{E}}}{\partial t}$ and $\frac{\partial \overrightarrow{\mathbf{B}}}{\partial t}$ is responsible for the propagation effect of the electromagnetic field.

Mathematically, it is very easy to see that the Maxwell–Faraday and Maxwell–Ampère equations, when taken together, lead to wave equations whose solutions are well known. To decouple the Maxwell–Faraday and Maxwell–Ampère equations, we must proceed in the same way as for the pair of equations: $\frac{dx}{dt} = -ky$ and $\frac{dy}{dt} = k'x$. In fact, these equations can be decoupled by differentiating them a second time with respect to time, that is, $\frac{d^2x}{dt^2} = -k\frac{dy}{dt} = -kk'x$ and $\frac{d^2y}{dt^2} = k'\frac{dx}{dt} = -kk'y$. To apply this technique to the Maxwell equations, let us compute the curl of the Maxwell–Faraday equation:

$$\begin{aligned}\overrightarrow{\nabla} \times \left(\overrightarrow{\nabla} \times \overrightarrow{\mathbf{E}}\right) &= -\overrightarrow{\nabla} \times \left(\frac{\partial \overrightarrow{\mathbf{B}}}{\partial t}\right) = -\frac{\partial}{\partial t}\left(\overrightarrow{\nabla} \times \overrightarrow{\mathbf{B}}\right) \\ &= -\frac{\partial}{\partial t}\left(\mu_0 \overrightarrow{\mathbf{J}} + \mu_0\varepsilon_0 \frac{\partial \overrightarrow{\mathbf{E}}}{\partial t}\right)\end{aligned} \qquad [3.1]$$

Moreover, $\vec{\nabla} \times \left(\vec{\nabla} \times \vec{\mathbf{E}}\right) = \vec{\nabla}\left(\vec{\nabla} \cdot \vec{\mathbf{E}}\right) - \Delta\vec{\mathbf{E}}$ and $\vec{\nabla} \cdot \vec{\mathbf{E}} = \frac{\rho}{\varepsilon_0}$ (form the Maxwell–Gauss equation), which leads to:

$$\Delta\vec{\mathbf{E}} - \varepsilon_0\mu_0\frac{\partial^2\vec{\mathbf{E}}}{\partial t^2} = \frac{\vec{\nabla}\rho}{\varepsilon_0} + \mu_0\frac{\partial\vec{\mathbf{J}}}{\partial t} \qquad [3.2]$$

Symmetrically, let us apply the curl of the Maxwell–Ampère equation:

$$\vec{\nabla} \times \left(\vec{\nabla} \times \vec{\mathbf{B}}\right) = \mu_0\vec{\nabla} \times \vec{\mathbf{J}} + \varepsilon_0\mu_0\frac{\partial}{\partial t}\left(\vec{\nabla} \times \vec{\mathbf{E}}\right) = \mu_0\vec{\nabla} \times \vec{\mathbf{J}} - \varepsilon_0\mu_0\frac{\partial^2\vec{\mathbf{B}}}{\partial t^2} \qquad [3.3]$$

which, according to the Maxwell–Thomson equation $\vec{\nabla} \cdot \vec{\mathbf{B}} = 0$ and the identity $\vec{\nabla} \times \left(\vec{\nabla} \times \vec{\mathbf{B}}\right) = \vec{\nabla}\left(\vec{\nabla} \cdot \vec{\mathbf{B}}\right) - \Delta\vec{\mathbf{B}} = -\Delta\vec{\mathbf{B}}$, simplifies to:

$$\Delta\vec{\mathbf{B}} - \varepsilon_0\mu_0\frac{\partial^2\vec{\mathbf{B}}}{\partial t^2} = -\mu_0\vec{\nabla} \times \vec{\mathbf{J}} \qquad [3.4]$$

At a point very far from the sources, that is, in a region where $\rho = 0$ and $\vec{\mathbf{J}} = 0$, we obtain homogeneous wave equations:

$$\Delta\vec{\mathbf{E}} - \varepsilon_0\mu_0\frac{\partial^2\vec{\mathbf{E}}}{\partial t^2} = 0 \qquad [3.5]$$

$$\Delta\vec{\mathbf{B}} - \varepsilon_0\mu_0\frac{\partial^2\vec{\mathbf{B}}}{\partial t^2} = 0 \qquad [3.6]$$

These equations are the *propagation* or *wave equations of the electromagnetic field.* Similarly, equations [6.140i] and [6.140ii] (in the Lorenz gauge) also simplify to:

$$\Delta V - \varepsilon_0\mu_0\frac{\partial^2 V}{\partial t^2} = 0 \qquad [3.7]$$

$$\Delta\vec{\mathbf{A}} - \varepsilon_0\mu_0\frac{\partial^2\vec{\mathbf{A}}}{\partial t^2} = 0 \qquad [3.8]$$

in vacuum. In all cases, we arrive at the *d'Alembert equation*, which describes the three-dimensional propagation of a wave $\psi\left(\vec{\mathbf{r}}, t\right)$ at speed $c = \frac{1}{\sqrt{\varepsilon_0\mu_0}}$:

$$\left(\Delta - \frac{1}{c^2}\frac{\partial^2}{\partial t^2}\right)\psi\left(\vec{\mathbf{r}}, t\right) = 0 \qquad [3.9]$$

It is called an *unattenuated electromagnetic wave* in vacuum, where $\psi \in \left\{\vec{\mathbf{E}}, \vec{\mathbf{B}}, V, \vec{\mathbf{A}}\right\}$. This equation remains valid for the Cartesian components of $\vec{\mathbf{E}}$, $\vec{\mathbf{B}}$ and $\vec{\mathbf{A}}$, but not for cylindrical components, spherical components, etc. The speed $c = \frac{1}{\sqrt{\varepsilon_0 \mu_0}}$ of electromagnetic waves is equal to the *speed of light*. Light is therefore an electromagnetic wave. This hypothesis has been confirmed by numerous experiments.

The d'Alembert equation (the classical wave propagation equation, also known as the *vibrating string equation*) was established in the 18th century to model the vibrations of a stretched string. As we show in the next section, the solutions to this equation describe a propagation phenomenon with a speed of c. Before that, here are some basic notions about waves:

1) A wave is a disturbance propagating. It transports energy without carrying matter. For example, when the end of a string is shaken (moving it up and then down), a wave is created, consisting of upward and downward disturbances. This wave travels along the length of the string, transferring energy (kinetic energy) to the other end. It is important to note that the string itself does not move; the points on the string rise and then return to their original positions. Therefore, there is no transfer of matter.

2) There are two types of waves. When the oscillatory motion is perpendicular to the direction of propagation, the wave is called a *transverse wave*. Examples include vibrating strings, electromagnetic waves, etc. Transverse waves have two distinct characteristics: *crests* and *troughs*. Crests correspond to the maximum upward oscillations relative to the initial (equilibrium) positions, whereas troughs correspond to the maximum downward oscillations. When the oscillatory motion is parallel to the direction of wave propagation, the wave is called a *longitudinal wave*. An example of this is a sound wave. Longitudinal waves have two key features: regions of *compression* (areas with high particle concentration) and regions of *rarefaction* (areas with low particle concentration).

3) To discuss waves, we must specify the *medium*, that is, the substance through which the wave travels. For vibrating strings, the string itself serves as the medium. For sound waves, the medium is air. The air particles oscillate (constituting the matter without undergoing net movement for the sound wave), meaning they form the medium through which the sound wave propagates.

4) The *wavelength* is defined as the distance traveled by a complete wave cycle. For a transverse wave, a full cycle can be considered as the distance required for the wave to return to its initial position. Equivalently, the wavelength can also be defined as the distance from crest to crest or from trough to trough. Similarly, for a longitudinal wave, the wavelength is the distance between the centers of successive compression regions or between the centers of successive rarefaction regions.

5) The *amplitude* is defined as the maximum possible displacement of the oscillating medium. For a transverse wave, the amplitude represents the distance

between a crest or a trough and the equilibrium position. For a longitudinal wave, it corresponds to half the width of a compression or rarefaction region.

6) The *frequency* ν is defined as the number of complete wave cycles that pass through a given point per unit of time. For a transverse wave, the frequency indicates the number of crests passing through a given point per unit of time. Similarly, for a longitudinal wave, the frequency represents the number of compressions passing through a given point per unit of time. The angular frequency is defined as $\omega = 2\pi\nu$.

7) The *period* is the duration of a complete oscillation. It is given by $T = \frac{1}{\nu}$.

3.2. One-dimensional solutions

The d'Alembert equation admits *plane waves* as solutions. For simplicity, let us consider waves $\psi(x,t)$ that depend only on the coordinate x:

$$\frac{\partial^2 \psi(x,t)}{\partial x^2} - \frac{1}{c^2}\frac{\partial^2 \psi(x,t)}{\partial t^2} = 0 \qquad [3.10]$$

Set $u = t - \frac{x}{c}$ and $v = t + \frac{x}{c}$, then express the derivatives of ψ with respect to x and t in terms of the derivatives with respect to u and v:

$$\begin{aligned}\frac{\partial \psi}{\partial t} &= \frac{\partial \psi}{\partial u}\frac{\partial u}{\partial t} + \frac{\partial \psi}{\partial v}\frac{\partial v}{\partial t} = \frac{\partial \psi}{\partial u} + \frac{\partial \psi}{\partial v} \\ \frac{\partial^2 \psi}{\partial t^2} &= \frac{\partial}{\partial u}\left(\frac{\partial \psi}{\partial u} + \frac{\partial \psi}{\partial v}\right)\frac{\partial u}{\partial t} + \frac{\partial}{\partial v}\left(\frac{\partial \psi}{\partial u} + \frac{\partial \psi}{\partial v}\right)\frac{\partial v}{\partial t} \\ &= \frac{\partial^2 \psi}{\partial u^2} + 2\frac{\partial^2 \psi}{\partial u \partial v} + \frac{\partial^2 \psi}{\partial v^2}\end{aligned} \qquad [3.11]$$

and:

$$\begin{aligned}\frac{\partial \psi}{\partial x} &= \frac{\partial \psi}{\partial u}\frac{\partial u}{\partial x} + \frac{\partial \psi}{\partial v}\frac{\partial v}{\partial x} = -\frac{1}{c}\frac{\partial \psi}{\partial u} + \frac{1}{c}\frac{\partial \psi}{\partial v} \\ \frac{\partial^2 \psi}{\partial x^2} &= \frac{\partial}{\partial u}\left(-\frac{1}{c}\frac{\partial \psi}{\partial u} + \frac{1}{c}\frac{\partial \psi}{\partial v}\right)\frac{\partial u}{\partial x} + \frac{\partial}{\partial v}\left(-\frac{1}{c}\frac{\partial \psi}{\partial u} + \frac{1}{c}\frac{\partial \psi}{\partial v}\right)\frac{\partial v}{\partial x} \\ &= \frac{1}{c^2}\frac{\partial^2 \Psi}{\partial u^2} - \frac{2}{c^2}\frac{\partial^2 \psi}{\partial u \partial v} + \frac{1}{c^2}\frac{\partial^2 \psi}{\partial v^2}\end{aligned} \qquad [3.12]$$

It follows that $\frac{\partial^2 \psi}{\partial x^2} - \frac{1}{c^2}\frac{\partial^2 \psi}{\partial t^2} = -\frac{4}{c^2}\frac{\partial^2 \psi}{\partial u \partial v} = 0$, or equivalently:

$$\frac{\partial^2 \psi}{\partial u \partial v} = 0 \qquad [3.13]$$

By integrating [3.13] once with respect to u, we obtain $\frac{\partial \psi}{\partial v} = f_-(v)$. A second integration with respect to v immediately yields:

$$\psi(x,t) = F_+(u) + F_-(v) = F_+\left(t - \frac{x}{c}\right) + F_-\left(t + \frac{x}{c}\right) \quad [3.14]$$

where $F_-(v) = \int f_-(v)\,dv$. The solutions $F_+\left(t - \frac{x}{c}\right)$ and $F_-\left(t + \frac{x}{c}\right)$ correspond, as we will see, to *progressive plane waves*. We observe that $F_+\left(t - \frac{x}{c}\right) = F_+\left(t + \Delta t - \frac{x+\Delta x}{c}\right)$ for any pair Δx and Δt that satisfies $\Delta x = c\Delta t$. In other words, the value $F_+\left(t - \frac{x}{c}\right)$, observed at position x at time t, is also observed at another position $x + \Delta x$ at a later time $t + \Delta t$. If we consider $F_+\left(t - \frac{x}{c}\right)$ as a signal, we can see that it propagates from position x to position $x + \Delta x$ at velocity c. Such a signal is also called a *wave*. Equation [3.10], which governs its evolution, is called the *wave equation*. Thus, $F_+\left(t - \frac{x}{c}\right)$ represents a signal that propagates without deformation at velocity c along the Ox axis in the positive direction. The solution $F_-\left(t + \frac{x}{c}\right)$ represents a signal that propagates without deformation at velocity c along the Ox axis in the negative direction. Here are some remarks:

1) We now aim to solve the three-dimensional d'Alembert equation [3.9]. It can be verified that functions of the form $F_\pm\left(t \mp \frac{x}{c}\right)$, $G_\pm\left(t \mp \frac{y}{c}\right)$ and $H_\pm\left(t \mp \frac{z}{c}\right)$ are solutions of the three-dimensional equation. These solutions are called *plane waves*, propagating in the directions of $\overrightarrow{\mathbf{i}}$, $\overrightarrow{\mathbf{j}}$ and $\overrightarrow{\mathbf{k}}$, respectively, either in the positive or negative direction.

2) If it is a plane wave propagating in the direction of the unit vector $\pm\overrightarrow{\mathbf{n}}$, we write:

$$\psi\left(\overrightarrow{\mathbf{r}},t\right) = F_+\left(t - \frac{\overrightarrow{\mathbf{n}}\cdot\overrightarrow{\mathbf{r}}}{c}\right) + F_-\left(t + \frac{\overrightarrow{\mathbf{n}}\cdot\overrightarrow{\mathbf{r}}}{c}\right) \quad [3.15]$$

where F_+ corresponds to a progressive wave propagating in the $+\overrightarrow{\mathbf{n}}$ direction, and F_- corresponds to a progressive wave propagating in the $-\overrightarrow{\mathbf{n}}$ direction. In general, a plane wave is defined as a wave that propagates in a straight line.

3) Isotropic spherical waves are also solutions of the three-dimensional d'Alembert equation. Let us find, for example, spherically symmetric solutions $\psi(r,t)$. These solutions are particularly associated with isotropic emission in space by a point source. For this type of wave, the electromagnetic field depends only on the distance r from the source position O. Using the expression for the Laplacian in spherical coordinates, we obtain $\frac{1}{r}\frac{\partial^2(r\psi)}{\partial r^2} - \frac{1}{c^2}\frac{\partial^2\psi}{\partial t^2} = 0$, or equivalently:

$$\frac{\partial^2(r\psi)}{\partial r^2} - \frac{1}{c^2}\frac{\partial^2(r\psi)}{\partial t^2} = 0 \quad [3.16]$$

We observe that the function $r\psi(r,t)$ is a solution of the one-dimensional d'Alembert equation. It follows that $r\psi(r,t) = F_+\left(t-\frac{r}{c}\right) + F_-\left(t+\frac{r}{c}\right)$ and consequently:

$$\psi(r,t) = \frac{1}{r}F_+\left(t-\frac{r}{c}\right) + \frac{1}{r}F_-\left(t+\frac{r}{c}\right) \quad [3.17]$$

The two terms in this sum represent a divergent and a convergent spherical wave, respectively. Here, the signal does not propagate without deformation due to the presence of the factor $\frac{1}{r}$. At a large distance from the source, the spherical wave can be approximated by a plane wave.

4) *Standing plane waves*: we look for solutions to the d'Alembert equation of the form (using the method of separation of variables):

$$\psi(x,t) = X(x)\,T(t) \quad [3.18]$$

By substituting [3.18] into the d'Alembert equation [3.1], we find $X''(x)\,T(t) - \frac{1}{c^2}X(x)\,T''(t) = 0$, or equivalently, $\frac{X''(x)}{X(x)} = \frac{1}{c^2}\frac{T''(t)}{T(t)} = K$, where K is a constant. This leads to two differential equations:

$$\begin{aligned} &X''(x) - KX(x) = 0 \quad \text{(i)} \\ &T''(t) - Kc^2T(t) = 0 \quad \text{(ii)} \end{aligned} \quad [3.19]$$

If $K > 0$, equation [3.19i] has a solution of the form:

$$T(t) = Ae^{c\sqrt{K}t} + Be^{-c\sqrt{K}t} \quad [3.20]$$

where A and B are constants. This solution must be rejected because it corresponds to either a divergent solution or a transient solution. From now on, we assume that $K < 0$. By setting $-c^2K = \omega^2$, the solution can be written as:

$$T(t) = A\cos(\omega t + \theta_0) \quad [3.21]$$

where A (*amplitude*) and θ_0 (*phase shift*) are constants. In this case, the differential equation [3.19i] takes the form:

$$X''(x) + \frac{\omega^2}{c^2}X(x) = 0 \quad [3.22]$$

It has the solution:

$$X(x) = B\cos(\mathrm{k}x + \varphi_0) \quad [3.23]$$

where $\mathrm{k} = \frac{\omega}{c}$ (B and φ_0 are constants). k is called the *wavenumber*. The general solution of the d'Alembert equation is therefore:

$$\begin{aligned}\psi(x,t) &= C\cos(\mathrm{k}x+\varphi_0)\cos(\omega t+\theta_0)\\ &= \frac{C}{2}\left[\cos(\mathrm{k}x-\omega t+\varphi_0-\theta_0)+\cos(\mathrm{k}x+\omega t+\varphi_0+\theta_0)\right]\end{aligned} \quad [3.24]$$

$\psi(x,t)$ appears here as a superposition of two waves with the same amplitude, the same angular frequency and opposite wavenumbers, that is, propagating in opposite directions. In this solution, called the *standing plane wave*, the spatial and temporal dependencies appear separately. The shape of the string at different moments is shown in Figure 3.1. Some points of the string remain fixed at all times and are called *nodes* of vibration. Other points reach maximum or minimum displacement at a given instant and are called *antinodes* of vibration. The positions of the nodes are determined by setting $\cos\left(\mathrm{k}x_n^N+\varphi_0\right) = 0$, which implies that $\mathrm{k}x_n^N+\varphi_0 = (2n+1)\frac{\pi}{2}$. Thus, $x_n^N = \left(\frac{n}{2}+\frac{1}{4}\right)\lambda - \frac{\varphi_0}{\mathrm{k}}$, where $\mathrm{k} = \frac{2\pi}{\lambda}$. The distance between two successive nodes is $\frac{\lambda}{2}$. The positions of the antinodes are determined by setting $\cos\left(\mathrm{k}x_n^{AN}+\varphi_0\right) = \pm 1$, which gives $\mathrm{k}x_n^{AN}+\varphi_0 = n\pi$. Hence, $x_n^{AN} = \frac{n}{2}\lambda - \frac{\varphi_0}{\mathrm{k}}$. The distance between two successive antinodes is $\frac{\lambda}{2}$, and the distance between a node and its adjacent antinode is $\frac{\lambda}{4}$.

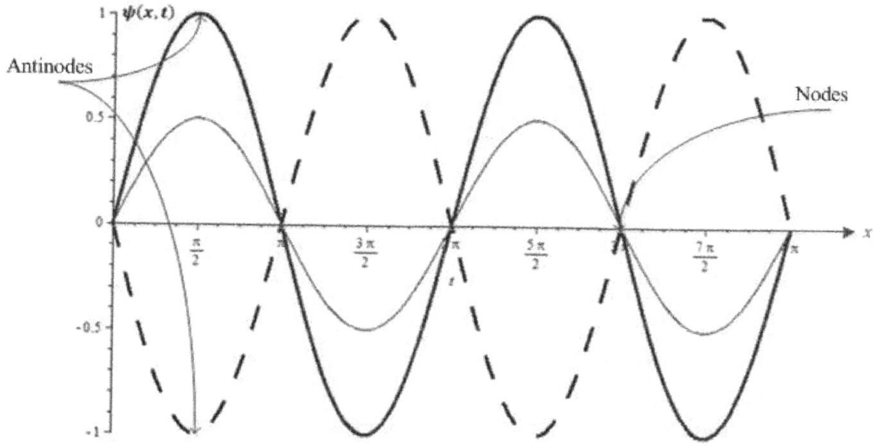

Figure 3.1. *Standing plane wave: the nodes and antinodes correspond to the fixed points and the extrema of the oscillation, respectively. For a color version of this figure, see www.iste.co.uk/abdesselam/electrodynamics2.zip*

3.3. Structure of a plane electromagnetic wave

We now focus on the d'Alembert equations associated with $\vec{\mathbf{E}}$ and $\vec{\mathbf{B}}$, which depend only on the spatial Cartesian coordinate x and time t. Based on the previous results, the corresponding electric and magnetic fields can be expressed as $\vec{\mathbf{E}}(x,t) = \vec{\mathbf{E}}_+\left(t-\frac{x}{c}\right) + \vec{\mathbf{E}}_-\left(t+\frac{x}{c}\right)$ and $\vec{\mathbf{B}}(x,t) = \vec{\mathbf{B}}_+\left(t-\frac{x}{c}\right) + \vec{\mathbf{B}}_-\left(t-\frac{x}{c}\right)$. The Maxwell equations indicate that $\vec{\nabla}\cdot\vec{\mathbf{B}} = 0$, since $\vec{\mathbf{B}}$ has no sources. Moreover, we also have $\vec{\nabla}\cdot\vec{\mathbf{E}} = 0$ because we are in a vacuum. Let us now write:

$$0 = \vec{\nabla}\cdot\vec{\mathbf{E}} = \frac{\partial E_x}{\partial x} + \frac{\partial E_y}{\partial y} + \frac{\partial E_z}{\partial z} = \frac{\partial E_x}{\partial x} = -\frac{1}{c}\frac{\partial E_{x,+}}{\partial u} + \frac{1}{c}\frac{\partial E_{x,-}}{\partial v} \quad [3.25]$$

Since u and v are independent variables, it follows that we necessarily have:

$$\frac{\partial E_{x,+}}{\partial u} = \frac{\partial E_{x,-}}{\partial v} = K \quad [3.26]$$

and the constant K must be zero because a solution of the form $E_{x,+} = Ku$, which increases with x and t, has no physical meaning. The only possible solution would therefore be constant fields $E_{x,+}$ and $E_{x,-}$. However, since static fields do not play a role in wave propagation, we must necessarily have $E_{x,+} = E_{x,-} = 0$. The same reasoning applies to the magnetic field $\vec{\mathbf{B}}$, implying that the components $B_{x,+}$ and $B_{x,-}$ must also be zero. Thus, we conclude that a plane electromagnetic wave is necessarily transverse, meaning its electric and magnetic fields are perpendicular to the direction of propagation. This result, directly derived from the Maxwell equations and experimentally verified, provides strong evidence for their validity.

The Maxwell–Faraday equation, in turn, leads to the following relationships.

$$\begin{aligned}
&\frac{\partial E_z}{\partial x} = \frac{\partial B_y}{\partial t} \Longrightarrow -\frac{1}{c}\frac{\partial E_{z,+}}{\partial u} - \frac{\partial B_{y,+}}{\partial u} = -\frac{1}{c}\frac{\partial E_{z,-}}{\partial v} + \frac{\partial B_{y,-}}{\partial v} = K_1\\
&\frac{\partial E_y}{\partial x} = -\frac{\partial B_z}{\partial t} \Longrightarrow -\frac{1}{c}\frac{\partial E_{y,+}}{\partial u} + \frac{\partial B_{z,+}}{\partial u} = -\frac{1}{c}\frac{\partial E_{y,-}}{\partial v} - \frac{\partial B_{z,-}}{\partial v} = K_2
\end{aligned} \quad [3.27]$$

By following the same reasoning, we deduce that:

$$\begin{aligned}
B_{y,+} &= -\frac{1}{c}E_{z,+} \\
B_{z,+} &= \frac{1}{c}E_{y,+}
\end{aligned} \Longrightarrow \begin{aligned}
\vec{\mathbf{B}}_+ &= \frac{1}{c}\vec{\mathbf{i}} \times \vec{\mathbf{E}}_+ \\
\vec{\mathbf{E}}_+ &= -c\,\vec{\mathbf{i}} \times \vec{\mathbf{B}}_+
\end{aligned} \quad [3.28]$$

$$\begin{aligned}
B_{y,-} &= \frac{1}{c}E_{z,-} \\
B_{z,-} &= -\frac{1}{c}E_{y,-}
\end{aligned} \Longrightarrow \begin{aligned}
\vec{\mathbf{B}}_- &= -\frac{1}{c}\vec{\mathbf{i}} \times \vec{\mathbf{E}}_- \\
\vec{\mathbf{E}}_- &= c\,\vec{\mathbf{i}} \times \vec{\mathbf{B}}_-
\end{aligned} \quad [3.29]$$

Consequently, it is very easy to observe that:

$$\vec{\mathbf{E}}_+ \cdot \vec{\mathbf{B}}_+ = \vec{\mathbf{E}}_- \cdot \vec{\mathbf{B}}_- = 0 \qquad [3.30]$$

We therefore conclude that for a plane wave propagating in a given direction, either positive or negative, the corresponding electric and magnetic fields are perpendicular to each other. Furthermore, denoting $\vec{\mathbf{n}}$ as the unit vector that defines the direction and sense of propagation, the triad $\left(\vec{\mathbf{n}}, \vec{\mathbf{E}}, \vec{\mathbf{B}}\right)$ forms a right-handed system, as shown in Figure 3.2. In fact, it is easily verified that:

$$\begin{aligned} \vec{\mathbf{B}}_+ &= \frac{1}{c}\vec{\mathbf{n}} \times \vec{\mathbf{E}}_+ \\ \vec{\mathbf{E}}_+ &= c\vec{\mathbf{B}}_+ \times \vec{\mathbf{n}} \end{aligned} \qquad [3.31]$$

holds for the wave propagating in the positive direction, and

$$\begin{aligned} \vec{\mathbf{B}}_- &= -\frac{1}{c}\vec{\mathbf{n}} \times \vec{\mathbf{E}}_- \\ \vec{\mathbf{E}}_- &= -c\vec{\mathbf{B}}_- \times \vec{\mathbf{n}} \end{aligned} \qquad [3.32]$$

for the wave propagating in the negative direction. We also note that

$$\begin{aligned} \left|\vec{\mathbf{E}}_+\right| &= c\left|\vec{\mathbf{B}}_+\right| \\ \left|\vec{\mathbf{E}}_-\right| &= c\left|\vec{\mathbf{B}}_-\right| \end{aligned} \qquad [3.33]$$

that is, the ratio of the amplitudes of the electric and magnetic fields is equal to c.

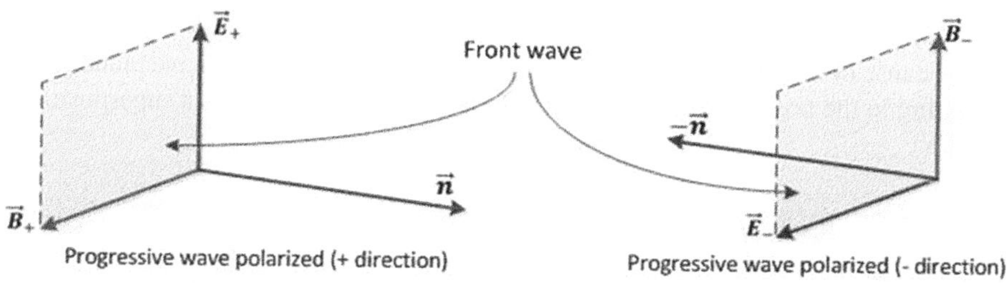

Figure 3.2. *Structure of a progressive plane electromagnetic wave: the electromagnetic field is uniform in a wavefront plane. For a color version of this figure, see www.iste.co.uk/abdesselam/electrodynamics2.zip*

3.4. Progressive plane monochromatic waves (PPMW)

3.4.1. *Sinusoidal solutions of the d'Alembert propagation equation*

There is a particularly interesting solution to the d'Alembert equation, namely, the *progressive plane monochromatic wave (PPMW)* or *harmonic plane wave* propagating along $\pm\overrightarrow{\mathbf{i}}$. This solution takes the following form:

$$\psi(x,t) = A\cos(\omega t - \mathsf{k}x + \phi_0) \qquad [3.34]$$

where (1) ω is the *angular frequency*; (2) $\nu = \frac{\omega}{2\pi}$ is the *temporal frequency*; (3) $T = \frac{2\pi}{\omega}$ is the *temporal period*; (4) $|\mathsf{k}|$ is the *wavenumber* (if $\mathsf{k} > 0$, the wave propagates in the $+\overrightarrow{\mathbf{i}}$ direction, whereas if $\mathsf{k} < 0$, it propagates in the $-\overrightarrow{\mathbf{i}}$ direction); (5) $\lambda = \frac{2\pi}{|\mathsf{k}|}$ is the *wavelength or spatial period*; (6) A is the *amplitude*; (7) $\phi_M = \omega t - \mathsf{k}x + \phi_0$ is the *phase of the wave*; and (8) ϕ_0 is the *phase shift*. The function $\psi(x,t)$ is a scalar and represents one of the Cartesian components of $\overrightarrow{\mathbf{E}}$, $\overrightarrow{\mathbf{B}}$, $\overrightarrow{\mathbf{A}}$, or the potential V. Here are some remarks:

1) By substituting [3.34] into the d'Alembert equation [3.10], we obtain $\frac{\partial^2\psi}{\partial x^2} - \frac{1}{c^2}\frac{\partial^2\psi}{\partial t^2} = A\left(-\mathsf{k}^2 + \frac{\omega^2}{c^2}\right) \times \cos(\omega t - \mathsf{k}x + \phi) = 0$, which implies that k and ω must satisfy the *dispersion relation*:

$$|\mathsf{k}| = \frac{\omega}{c} \Longleftrightarrow \lambda = cT \qquad [3.35]$$

Using [3.35], the solutions given in [3.34] can also be expressed as:

$$\psi(x,t) = \begin{cases} A_+\cos\left[\omega\left(t - \frac{x}{c}\right) + \phi_+\right] & \text{(propagation in the } \overrightarrow{\mathbf{i}} \text{ direction)} \\ A_-\cos\left[\omega\left(t + \frac{x}{c}\right) + \phi_-\right] & \text{(propagation in the } -\overrightarrow{\mathbf{i}} \text{ direction)} \end{cases} \qquad [3.36]$$

2) Because of the linearity of the d'Alembert equation, any progressive plane wave propagating in the positive direction of the Ox axis can be regarded as a superposition of PPMW:

$$\begin{aligned} F_+\left(t - \frac{x}{c}\right) &= \int_0^\infty A_+(\omega)\cos\left[\omega\left(t - \frac{x}{c}\right) + \phi_+(\omega)\right] d\omega \\ &\equiv \int_0^\infty \left(A'_+(\omega)\cos\left[\omega\left(t - \frac{x}{c}\right)\right] + A''_+(\omega)\sin\left[\omega\left(t - \frac{x}{c}\right)\right]\right) d\omega \end{aligned} \qquad [3.37]$$

where $A'_+(\omega) = A_+(\omega)\cos\phi_+(\omega)$ and $A''_+(\omega) = -A_+(\omega)\sin\phi_+(\omega)$. This is the *Fourier transform*. The coefficients $A_+(\omega)$ and $\phi_+(\omega)$ can be calculated from $F_+(u)$

using the *inverse Fourier transform*:

$$A_+(\omega)\left\{\begin{array}{c}\cos\phi_+(\omega)\\ \sin\phi_+(\omega)\end{array}\right\}=\frac{1}{\pi}\int_{-\infty}^{\infty}F_+(u)\left\{\begin{array}{c}\cos\omega u\\ -\sin\omega u\end{array}\right\}du \qquad [3.38]$$

where we used $\frac{1}{\pi}\int_{-\infty}^{\infty}\cos\omega x\cos\omega' x dx=\delta(\omega-\omega')$, $\frac{1}{\pi}\int_{-\infty}^{\infty}\sin\omega x\sin\omega' x dx=\delta(\omega-\omega')$, and $\frac{1}{\pi}\int_{-\infty}^{\infty}\cos\omega x\sin\omega' x dx=0$. This result remains valid for any progressive wave propagating in the opposite direction:

$$F_-\left(t+\frac{x}{c}\right)=\int_0^{\infty}A_-(\omega)\cos\left[\omega\left(t+\frac{x}{c}\right)+\phi_-(\omega)\right]d\omega \qquad [3.39]$$

where $A_-(\omega)$ and $\phi_-(\omega)$ are obtained from $F_-(v)$ using the integrals:

$$A_-(\omega)\left\{\begin{array}{c}\cos\phi_-(\omega)\\ \sin\phi_-(\omega)\end{array}\right\}=\frac{1}{\pi}\int_{-\infty}^{\infty}F_-(v)\left\{\begin{array}{c}\cos\omega v\\ -\sin\omega v\end{array}\right\}dv \qquad [3.40]$$

3) It is easy to verify the following equalities:

$$\frac{1}{\pi}\int_{-\infty}^{\infty}F_{\pm}^2(u_{\pm})\,du_{\pm}=\int_0^{\infty}A_{\pm}^2(\omega)\,d\omega \qquad [3.41]$$

4) If the progressive wave is a periodic function with period $T=\frac{2\pi}{\omega}$, the functions $F_{\pm}\left(t\mp\frac{x}{c}\right)$ can be written as:

$$F_{\pm}\left(t\mp\frac{x}{c}\right)=\frac{A_{0,\pm}}{2}+\sum_{n=1}^{\infty}A_{n,\pm}\cos\left[n\omega\left(t\mp\frac{x}{c}\right)+\phi_{n,\pm}\right] \qquad [3.42]$$

where $A_{n,\pm}$ and $\phi_{n,\pm}$ can be determined from $F_{\pm}\left(t\mp\frac{x}{c}\right)$ using the integral formulas:

$$\begin{aligned}A_{0,\pm}&=\frac{2}{T}\int_d^{d+T}F_{\pm}(u_{\pm})\,du_{\pm},\\ A_{n,\pm}\left\{\begin{array}{c}\cos\phi_{n,\pm}\\ \sin\phi_{n,\pm}\end{array}\right\}&=\frac{2}{T}\int_d^{d+T}F_{\pm}(u_{\pm})\left\{\begin{array}{c}\cos n\omega u_{\pm}\\ -\sin n\omega u_{\pm}\end{array}\right\}du_{\pm}\end{aligned} \qquad [3.43]$$

where $u_{\pm}=t\mp\frac{x}{c}$ $(\forall d\in\mathbb{R})$. These are the *Fourier series expansions* of the functions $F_{\pm}\left(t\mp\frac{x}{c}\right)$. It is also verified that:

$$\frac{2}{T}\int_d^{d+T}F_{\pm}^2(u_{\pm})\,du_{\pm}=\frac{A_{0,\pm}^2}{2}+\sum_{n=1}^{\infty}A_{n,\pm}^2 \qquad [3.44]$$

5) The *phase velocity* is the velocity v_ϕ at which a wavefront, defined by ϕ_M = constant, moves. It can be found by differentiating ϕ_M. For a monochromatic progressive wave propagating in the Ox direction, we get $d\phi_M = \frac{\partial \phi_M}{\partial t} dt + \frac{\partial \phi_M}{\partial x} dx = \omega dt - \mathsf{k} dx = 0$, and consequently:

$$v_\phi = \left(\frac{dx}{dt}\right)_\phi = \frac{\omega}{|\mathsf{k}|} \quad [3.45]$$

The wavelength λ corresponds to the distance that separates two successive wavefronts, corresponding to ϕ_M and $\phi_M + 2\pi$, as shown in Figure 9.3. Therefore, we have $\lambda = v_\phi T$. Finally, note that the constant k should, in the general case, be considered as a function of the angular frequency ω, that is, $\mathsf{k} = \mathsf{k}(\omega)$. The relation that expresses k as a function of ω is called the *dispersion relation*. In a vacuum, v_ϕ is independent of ω and equals c.

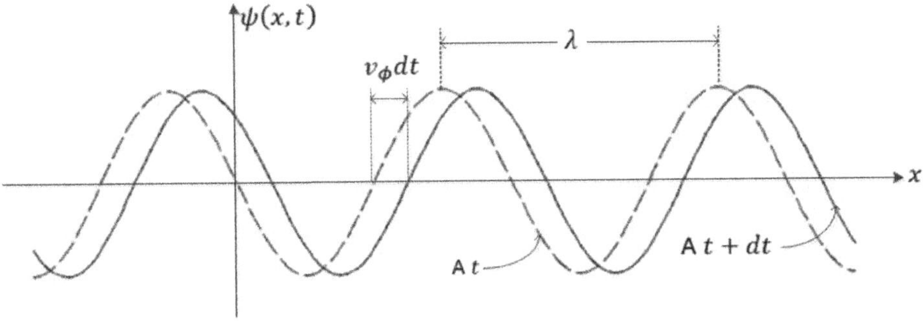

Figure 3.3. *Phase velocity*

6) For two waves with the same amplitude, with neighboring wavenumbers k_1 and k_2, neighboring angular frequencies ω_1 and ω_2, and the same phase shift (zero for simplicity), the resulting wave can be written as:

$$\begin{aligned}\psi(x,t) &= A_0 \cos(\omega_1 t - \mathsf{k}_1 x) + A_0 \cos(\omega_2 t - \mathsf{k}_2 x) \\ &= 2A_0 \cos(\Delta\omega t - \Delta\mathsf{k} x) \cos(\omega_m t - \mathsf{k}_m x)\end{aligned} \quad [3.46]$$

with $\Delta\mathsf{k} = \frac{1}{2}(\mathsf{k}_1 - \mathsf{k}_2)$, $\Delta\omega = \frac{1}{2}(\omega_1 - \omega_2)$, $\mathsf{k}_m = \frac{1}{2}(\mathsf{k}_1 + \mathsf{k}_2)$ and $\omega_m = \frac{1}{2}(\omega_1 + \omega_2)$. For small $\Delta\mathsf{k}$ and $\Delta\omega$, that is, $\Delta\omega \ll (\omega_1, \omega_2)$ and $|\Delta\mathsf{k}| \ll (|\mathsf{k}_1|, |\mathsf{k}_2|)$, a *beat phenomenon* appears, where the first wave $\cos(\Delta\omega t - \Delta\mathsf{k} x)$, called the *envelope*, produces an amplitude modulation of the average wave $\cos(\omega_m t - \mathsf{k}_m x)$, called the *carrier*, with a much longer period than the average wave, as shown in Figure 9.4. Thus, the apparent propagation velocity of the total wave is essentially equal to that of its envelope. The latter is:

$$v_{\text{env.}} = \frac{\Delta\omega}{|\Delta \mathsf{k}|} \quad [3.47]$$

while the propagation velocity of the carrier wave is:

$$v_{\text{avg.}} = \frac{\omega_1 + \omega_2}{|\mathsf{k}_1 + \mathsf{k}_2|} \quad [3.48]$$

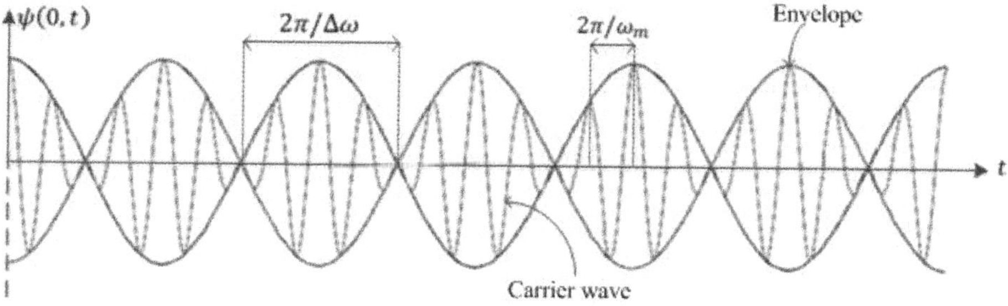

Figure 3.4. *Group velocity. For a color version of this figure, see www.iste.co.uk/abdesselam/electrodynamics2.zip*

Taking the limits $\Delta\omega \longrightarrow 0$ and $\Delta\mathsf{k} \longrightarrow 0$, the velocity of the envelope tends to the group velocity v_g, while the velocity of the average wave tends to the phase velocity v_p:

$$\begin{aligned} v_{\text{env.}} &\Rightarrow v_g = \frac{d\omega}{d\mathsf{k}} = \left(\frac{d\mathsf{k}}{d\omega}\right)^{-1} \\ v_{\text{avg.}} &\Rightarrow v_p = \frac{\omega}{\mathsf{k}} \end{aligned} \quad [3.49]$$

Note that the group velocity corresponds to the speed of propagation of the energy. In contrast to the phase velocity v_p, the group velocity v_g is a physical velocity, that is, it corresponds to the speed at which information is transported and therefore cannot exceed the speed of light. The expression for v_g can be obtained by starting from the dispersion relation $f(\mathsf{k}, \omega) = 0$ and then differentiating f:

$$df = \frac{\partial f}{\partial \mathsf{k}} d\mathsf{k} + \frac{\partial f}{\partial \omega} d\omega = 0 \Rightarrow v_g = \frac{d\omega}{d\mathsf{k}} = -\left(\frac{\partial f}{\partial \mathsf{k}}\right)\left(\frac{\partial f}{\partial \omega}\right)^{-1} \quad [3.50]$$

Another option is to start from the expression for the phase velocity $v_p = \frac{\omega}{\mathrm{k}}$:

$$v_p = \frac{\omega}{\mathrm{k}} \quad \Rightarrow \quad v_g = \frac{d\omega}{d\mathrm{k}} = v_p + \mathrm{k}\left(\frac{dv_p}{d\mathrm{k}}\right) \qquad [3.51]$$

If the medium is non-dispersive, that is, v_p does not depend on k, then we have: $v_g = v_\phi = c$. In particular, for $\Delta\omega = 0$ and $\mathrm{k}_1 = -\mathrm{k}_2$ (two waves with the same frequency and amplitude propagating in opposite directions), the superposition of the two waves creates a standing wave. This wave does not propagate, which means that it is incapable of carrying energy. In this case, it is verified that $v_g = 0$.

7) When two PPMWs with the same frequency, the same amplitude A_0, and different wavenumbers $(\mathrm{k}_1, \mathrm{k}_2)$ are superimposed, an interference effect occurs. The resulting wave is written as:

$$\begin{aligned}\psi(x,t) &= A_0 \cos(\omega t - \mathrm{k}_1 x + \phi_1) + A_0 \cos(\omega t - \mathrm{k}_2 x \phi_2) \\ &= 2A_0 \cos\left[\frac{1}{2}(\mathrm{k}_2 - \mathrm{k}_1)x + \frac{\Delta\phi}{2}\right] \cos\left[\omega t - \frac{1}{2}(\mathrm{k}_1 + \mathrm{k}_2)x + \frac{\phi_1 + \phi_2}{2}\right]\end{aligned} \qquad [3.52]$$

with $\Delta\phi = \phi_1 - \phi_2$. The structure of the resulting wave is that of a wave with pulsation ω, a wavenumber $\frac{1}{2}(\mathrm{k}_1 + \mathrm{k}_2)$, and an amplitude modulated by a sinusoidal spatial function. The wave therefore propagates along the Ox axis with a velocity $v_\phi = \frac{v_{\phi,1} v_{\phi,2}}{2(v_{\phi,1} + v_{\phi,2})}$, and it may be annihilated locally by the modulation term of the amplitude. Two particular cases can be discussed:

a) If $\mathrm{k}_1 = \mathrm{k}_2 = \mathrm{k}$, then $\psi(x,t) = 2A_0 \cos\left(\frac{\Delta\phi}{2}\right) \cos\left(\omega t - \mathrm{k}x + \frac{\phi_1+\phi_2}{2}\right)$. The resulting wave travels in the same manner as the two initial waves, but its amplitude is modulated by a constant term that depends on the value of $\Delta\phi$. In some cases, the resulting wave may have zero amplitude (for $\Delta\phi = \pi$, i.e. in phase opposition) or double amplitude (for $\Delta\phi = 0$, i.e. in phase). This is referred to as *destructive* interference and *constructive* interference, respectively.

b) If $\mathrm{k}_2 = -\mathrm{k}_1 = \mathrm{k}$ (the two incident waves travel in opposite directions with the same amplitude), then the resulting wave is given by: $\psi(x,t) = 2A_0 \cos\left(\mathrm{k}x + \frac{\Delta\phi}{2}\right) \cos\left(\omega t + \frac{\phi_1+\phi_2}{2}\right)$. We observe a decoupling between the spatial and temporal parts of the wave. This decoupling leads to the disappearance of the propagation phenomenon, resulting in a standing wave.

8) The PPMW propagating in the direction of the unit vector $\pm\vec{\mathrm{n}}$ is written as:

$$\psi(\vec{\mathrm{r}},t) = \alpha \cos\left(\omega t - \vec{\mathrm{k}} \cdot \vec{\mathrm{r}} + \phi_0\right) \qquad [3.53]$$

where $\overrightarrow{\mathsf{k}} = \pm\frac{\omega}{c}\overrightarrow{\mathbf{n}} = \pm\frac{2\pi}{\lambda}\overrightarrow{\mathbf{n}}$ ($\overrightarrow{\mathsf{k}}$ is the *wave vector*). The *phase velocity*, denoted as v_p, is the speed at which the phase $\omega t - \overrightarrow{\mathsf{k}} \cdot \overrightarrow{\mathbf{r}} + \phi_0$ propagates, that is:

$$v_p(\omega) = \frac{\omega}{\left|\overrightarrow{\mathsf{k}}\right|} \quad [3.54]$$

The *group velocity* is also defined by:

$$v_g(\omega) = \frac{d\omega}{d\left|\overrightarrow{\mathsf{k}}\right|} \quad [3.55]$$

9) Resolution of the d'Alembert equation using the Fourier transform: let us start from the three-dimensional d'Alembert equation [3.9]. The solution $\psi\left(\overrightarrow{\mathbf{r}}, t\right)$ is expressed using a Fourier transform with respect to time:

$$\psi\left(\overrightarrow{\mathbf{r}}, t\right) = \int_{-\infty}^{\infty} \frac{d\omega}{2\pi} \Psi\left(\overrightarrow{\mathbf{r}}, \omega\right) e^{-\mathrm{i}\omega t} \quad [3.56]$$

$\Psi\left(\overrightarrow{\mathbf{r}}, \omega\right)$ is said to be the *Fourier transform* of the wave $\psi\left(\overrightarrow{\mathbf{r}}, t\right)$. By inserting this expression into the wave equation [3.9], we obtain:

$$\int_{-\infty}^{\infty} \frac{d\omega}{2\pi} e^{-\mathrm{i}\omega t} \left[\Delta\Psi\left(\overrightarrow{\mathbf{r}}, \omega\right) + \frac{\omega^2}{c^2}\Psi\left(\overrightarrow{\mathbf{r}}, \omega\right)\right] = 0 \quad [3.57]$$

and consequently, the equality

$$\Delta\Psi\left(\overrightarrow{\mathbf{r}}, \omega\right) + \frac{\omega^2}{c^2}\Psi\left(\overrightarrow{\mathbf{r}}, \omega\right) = 0 \quad [3.58]$$

must be verified. This equation, which is satisfied by the Fourier transform $\Psi\left(\overrightarrow{\mathbf{r}}, \omega\right)$, is called the *Helmholtz equation*. Since the wave $\Psi(x, y, z, \omega)$ can be considered as a function of x and y in a plane $z =$ constant, let us attempt to decompose $\Psi(x, y, z, \omega)$ in terms of x and y only by writing:

$$\Psi(x, y, z, \omega) = \int_{-\infty}^{\infty} \frac{d\mathsf{k}_x}{2\pi} \int_{-\infty}^{\infty} \frac{d\mathsf{k}_y}{2\pi} \hat{\Psi}(\mathsf{k}_x, \mathsf{k}_y, z, \omega)\, e^{\mathrm{i}(\mathsf{k}_x x + \mathsf{k}_y y)} \quad [3.59]$$

The problem of determining $\Psi(x, y, z, \omega)$ is then reduced to finding the Fourier transform $\hat{\Psi}(\mathsf{k}_x, \mathsf{k}_y, z, \omega)$. The equation satisfied by $\hat{\Psi}(\mathsf{k}_x, \mathsf{k}_y, z, \omega)$ is obtained by substituting [3.59] into [3.58]. Thus, we get:

$$\int_{-\infty}^{\infty} \frac{d\mathsf{k}_x}{2\pi} \int_{-\infty}^{\infty} \frac{d\mathsf{k}_y}{2\pi} \left[\frac{\partial^2 \hat{\Psi}(\mathsf{k}_x, \mathsf{k}_y, z, \omega)}{\partial z^2} + \left(\frac{\omega^2}{c^2} - \mathsf{k}_x^2 - \mathsf{k}_y^2 \right) \hat{\Psi}(\mathsf{k}_x, \mathsf{k}_y, z, \omega) \right] e^{\mathrm{i}(\mathsf{k}_x x + \mathsf{k}_y y)} = 0 \quad [3.60]$$

which implies that:

$$\frac{\partial^2 \hat{\Psi}(\mathsf{k}_x, \mathsf{k}_y, z, \omega)}{\partial z^2} + \left(\frac{\omega^2}{c^2} - \mathsf{k}_x^2 - \mathsf{k}_y^2 \right) \hat{\Psi}(\mathsf{k}_x, \mathsf{k}_y, z, \omega) = 0 \quad [3.61]$$

The general solution of this equation can immediately be written in the form of two exponentials. Let us introduce the notation:

$$\mathsf{k}_z = \begin{cases} \sqrt{\frac{\omega^2}{c^2} - \mathsf{k}_x^2 - \mathsf{k}_y^2} & \text{for } \frac{\omega^2}{c^2} > \mathsf{k}_x^2 + \mathsf{k}_y^2 \\ \mathrm{i}\sqrt{-\frac{\omega^2}{c^2} + \mathsf{k}_x^2 + \mathsf{k}_y^2} & \text{for } \frac{\omega^2}{c^2} < \mathsf{k}_x^2 + \mathsf{k}_y^2 \end{cases} \quad [3.62]$$

The general solution of [3.61] is then given by:

$$\hat{\Psi}(\mathsf{k}_x, \mathsf{k}_y, z, \omega) = A(\mathsf{k}_x, \mathsf{k}_y, \omega) e^{\mathrm{i}\mathsf{k}_z z} + B(\mathsf{k}_x, \mathsf{k}_y, \omega) e^{-\mathrm{i}\mathsf{k}_z z} \quad [3.63]$$

If we consider a field propagating in the positive z direction, the function B is zero. The function A is determined by expressing the wave in the plane $z = 0$ using [3.59] and [9.63]. Thus, we obtain:

$$\Psi(x, y, 0, \omega) = \int_{-\infty}^{\infty} \frac{d\mathsf{k}_x}{2\pi} \int_{-\infty}^{\infty} \frac{d\mathsf{k}_y}{2\pi} A(\mathsf{k}_x, \mathsf{k}_y, \omega) e^{\mathrm{i}(\mathsf{k}_x x + \mathsf{k}_y y)} \quad [3.64]$$

Formula [3.64] shows that the complex amplitude $A(\mathsf{k}_x, \mathsf{k}_y, \omega)$ is simply the Fourier transform of the wave in the plane $z = 0$: $A(\mathsf{k}_x, \mathsf{k}_y, \omega) = \hat{\Psi}(\mathsf{k}_x, \mathsf{k}_y, 0, \omega)$. Finally, the wave is expressed at any point (x, y, z) as:

$$\Psi(x, y, z, \omega) = \int_{-\infty}^{\infty} \frac{d\mathsf{k}_x}{2\pi} \int_{-\infty}^{\infty} \frac{d\mathsf{k}_y}{2\pi} \hat{\Psi}(\mathsf{k}_x, \mathsf{k}_y, 0, \omega) e^{\mathrm{i}(\mathsf{k}_x x + \mathsf{k}_y y + \mathsf{k}_z z)} \quad [3.65]$$

and, consequently:

$$\psi(x, y, z, t) = \int_{-\infty}^{\infty} \frac{d\omega}{2\pi} \int_{-\infty}^{\infty} \frac{d\mathsf{k}_x}{2\pi} \int_{-\infty}^{\infty} \frac{d\mathsf{k}_y}{2\pi} \hat{\Psi}(\mathsf{k}_x, \mathsf{k}_y, 0, \omega) \times e^{-\mathrm{i}[\omega t - (\mathsf{k}_x x + \mathsf{k}_y y + \mathsf{k}_z z)]} \quad [3.66]$$

In this form, it is clear that the wave can be expressed as a superposition of plane waves whose wave vectors have the components $(\mathsf{k}_x, \mathsf{k}_y, \mathsf{k}_z)$ and satisfy the dispersion relation in vacuum:

$$\mathsf{k}_x^2 + \mathsf{k}_y^2 + \mathsf{k}_z^2 = \frac{\omega^2}{c^2} \qquad [3.67]$$

The complex amplitude of each plane wave is simply given by the value of the Fourier transform of the wave in the (x, y) plane. It should be noted that equation [3.66] is invertible, that is:

$$\hat{\Psi}(\mathsf{k}_x, \mathsf{k}_y, 0, \omega) = \int_{-\infty}^{\infty} dt \int_{-\infty}^{\infty} dx \int_{-\infty}^{\infty} dy \psi(x, y, z, t)\, e^{\mathrm{i}[\omega t - (\mathsf{k}_x x + \mathsf{k}_y y + \mathsf{k}_z z)]} \qquad [3.68]$$

and, consequently:

$$\int_{-\infty}^{\infty} dt \int_{-\infty}^{\infty} dx \int_{-\infty}^{\infty} dy\, |\psi(x, y, z, t)|^2 = \int_{-\infty}^{\infty} \frac{d\omega}{2\pi} \int_{-\infty}^{\infty} \frac{d\mathsf{k}_x}{2\pi} \int_{-\infty}^{\infty} \frac{d\mathsf{k}_y}{2\pi} \times \left|\hat{\Psi}(\mathsf{k}_x, \mathsf{k}_y, 0, \omega)\right|^2 \qquad [3.69]$$

3.4.2. *Complex representation of PPMWs*

1) To study PPMWs, it is simpler to use complex notation. The complex representation of a PPMW with angular frequency ω propagating in the direction $\overrightarrow{\mathsf{k}}$ is obtained as follows:

$$\begin{aligned} \psi(\overrightarrow{\mathbf{r}}, t) &= A\cos\left(\omega t - \overrightarrow{\mathsf{k}} \cdot \overrightarrow{\mathbf{r}} + \phi_0\right) = \Re\left[Ae^{-\mathrm{i}\left(\omega t - \overrightarrow{\mathsf{k}} \cdot \overrightarrow{\mathbf{r}} + \phi_0\right)}\right] \\ &= \Re\left[\hat{A}e^{-\mathrm{i}\left(\omega t - \overrightarrow{\mathsf{k}} \cdot \overrightarrow{\mathbf{r}}\right)}\right] = \Re\left[\hat{\psi}(\overrightarrow{\mathbf{r}}, t)\right] \end{aligned} \qquad [3.70]$$

where $\hat{A} = Ae^{-\mathrm{i}\phi_0}$ is the *complex amplitude*. The function:

$$\hat{\psi}(\overrightarrow{\mathbf{r}}, t) = \hat{A}e^{-\mathrm{i}\left(\omega t - \overrightarrow{\mathsf{k}} \cdot \overrightarrow{\mathbf{r}}\right)} \qquad [3.71]$$

is called the *complex representation of the PPMW*. The advantage of complex notation is that, when restricted to linear problems, the factor $e^{-\mathrm{i}\left(\omega t - \overrightarrow{\mathsf{k}} \cdot \overrightarrow{\mathbf{r}}\right)}$ appears in all terms and can therefore be eliminated, which considerably simplifies the calculations.

2) The average value of $\psi^2\left(\vec{\mathbf{r}}, t\right)$ over a period T is:

$$\begin{aligned}\langle\psi^2\rangle &= \frac{1}{T}\int_0^T \psi^2\left(\vec{\mathbf{r}}, t\right) dt = \frac{1}{T}\int_0^T A^2\cos^2\left(\omega t - \vec{\mathsf{k}}\cdot\vec{\mathbf{r}} + \phi_0\right) dt \\ &= \frac{A^2}{2} = \frac{1}{2}\hat{\psi}\hat{\psi}^\star = \frac{1}{2}\left|\hat{\psi}\right|^2\end{aligned} \qquad [3.72]$$

where $\hat{\psi}^\star$ is the complex conjugate of $\hat{\psi}$. If it is a superposition of multiple MPPWs, that is, $\psi\left(\vec{\mathbf{r}}, t\right) = \sum_l A_l \cos\left(\omega t - \vec{\mathsf{k}}\cdot\vec{\mathbf{r}} + \phi_l\right) = \Re\left[\hat{A}e^{-\mathrm{i}\left(\omega t - \vec{\mathsf{k}}\cdot\vec{\mathbf{r}}\right)}\right]$ (with the same frequency and wave vector), where $\hat{A} = \sum_l A_l e^{-\mathrm{i}\phi_l}$, then, of course, the previous formula remains valid:

$$\begin{aligned}\langle\psi^2\rangle &= \frac{1}{T}\sum_{l,l'}\int_0^T A_l A_{l'}\cos\left(\omega t - \vec{\mathsf{k}}\cdot\vec{\mathbf{r}} + \phi_l\right)\times \\ &\cos\left(\omega t - \vec{\mathsf{k}}\cdot\vec{\mathbf{r}} + \phi_{l'}\right) dt = \frac{1}{2}\sum_{l,l'} A_l A_{l'}\cos\left(\phi_l - \phi_{l'}\right) \\ &= \frac{1}{2}\Re\left[\sum_{l,l'} A_l A_{l'} e^{-\mathrm{i}\phi_l} e^{\mathrm{i}\phi_{l'}}\right] = \frac{1}{2}\hat{A}\hat{A}^\star = \frac{1}{2}\left|\hat{\psi}\right|^2\end{aligned} \qquad [3.73]$$

3) The differentiation properties of the exponential function imply the following:

$$\begin{aligned}&\frac{\partial\hat{\psi}}{\partial t} = -\mathrm{i}\omega\hat{\psi} \quad \Longrightarrow \quad \frac{\partial^2\hat{\psi}}{\partial t^2} = -\omega^2\hat{\psi} \\ &\begin{cases}\dfrac{\partial\hat{\psi}}{\partial x} = \mathrm{i}\mathsf{k}_x\hat{\psi} \\ \vec{\nabla}\hat{\psi} = \mathrm{i}\,\vec{\mathsf{k}}\,\hat{\psi}\end{cases} \quad \Longrightarrow \quad \begin{cases}\dfrac{\partial^2\hat{\psi}}{\partial x^2} = -\mathsf{k}_x^2\hat{\psi} \\ \Delta\hat{\psi} = -\left|\vec{\mathsf{k}}\right|^2\hat{\psi}\end{cases}\end{aligned} \qquad [3.74]$$

For a vectorial PPMW $\vec{\hat{\psi}}$, it is easy to verify that:

$$\begin{aligned}&\vec{\nabla}\times\vec{\hat{\psi}} = \mathrm{i}\,\vec{\mathsf{k}}\times\vec{\hat{\psi}} \\ &\vec{\nabla}\cdot\vec{\hat{\psi}} = \mathrm{i}\,\vec{\mathsf{k}}\cdot\vec{\hat{\psi}} \\ &\Delta\vec{\hat{\psi}} = -\left|\vec{\mathsf{k}}\right|^2\vec{\hat{\psi}}\end{aligned} \qquad [3.75]$$

In complex notation, the electric and magnetic fields of an MPPW can be expressed as follows:

$$\begin{aligned}&\vec{\hat{\mathbf{E}}}\left(x, t\right) = \vec{\hat{\mathbf{E}}}_0 e^{-\mathrm{i}(\omega t - \mathsf{k}x)} \\ &\vec{\hat{\mathbf{B}}}\left(x, t\right) = \vec{\hat{\mathbf{B}}}_0 e^{-\mathrm{i}(\omega t - \mathsf{k}x)}\end{aligned} \qquad [3.76]$$

The physical fields correspond to the real parts of these expressions. If the direction of the MPPW is arbitrary, the expressions used are:

$$\begin{aligned}\overrightarrow{\hat{\mathbf{E}}}\left(\overrightarrow{\mathbf{r}},t\right) &= \overrightarrow{\hat{\mathbf{E}}}_0 e^{-\mathrm{i}\left(\omega t-\overrightarrow{\mathrm{k}}\cdot\overrightarrow{\mathbf{r}}\right)}\\ \overrightarrow{\hat{\mathbf{B}}}\left(\overrightarrow{\mathbf{r}},t\right) &= \overrightarrow{\hat{\mathbf{B}}}_0 e^{-\mathrm{i}\left(\omega t-\overrightarrow{\mathrm{k}}\cdot\overrightarrow{\mathbf{r}}\right)}\end{aligned} \qquad [3.77]$$

where $\overrightarrow{\mathrm{k}} = \mathrm{k}\overrightarrow{\mathbf{n}}$. The four Maxwell equations take the form:

$$\begin{bmatrix}\overrightarrow{\nabla}\cdot\overrightarrow{\hat{\mathbf{E}}}=0\\ \overrightarrow{\nabla}\cdot\overrightarrow{\hat{\mathbf{B}}}=0\\ \overrightarrow{\nabla}\times\overrightarrow{\hat{\mathbf{E}}}=-\dfrac{\partial\overrightarrow{\hat{\mathbf{B}}}}{\partial t}\\ \overrightarrow{\nabla}\times\overrightarrow{\hat{\mathbf{B}}}=\mu_0\varepsilon_0\dfrac{\partial\overrightarrow{\hat{\mathbf{E}}}}{\partial t}\end{bmatrix} \Longrightarrow \begin{bmatrix}\overrightarrow{\nabla}\cdot\overrightarrow{\hat{\mathbf{E}}}=0\\ \overrightarrow{\nabla}\cdot\overrightarrow{\hat{\mathbf{B}}}=0\\ \overrightarrow{\nabla}\times\overrightarrow{\hat{\mathbf{E}}}=\mathrm{i}\omega\overrightarrow{\hat{\mathbf{B}}}\\ \overrightarrow{\nabla}\times\overrightarrow{\hat{\mathbf{B}}}=-\mathrm{i}\dfrac{\omega}{c^2}\overrightarrow{\hat{\mathbf{E}}}\end{bmatrix} \Longrightarrow \begin{bmatrix}\overrightarrow{\mathrm{k}}\cdot\overrightarrow{\hat{\mathbf{E}}}_0=0\\ \overrightarrow{\mathrm{k}}\cdot\overrightarrow{\hat{\mathbf{B}}}_0=0\\ \overrightarrow{\mathrm{k}}\times\overrightarrow{\hat{\mathbf{E}}}_0=\omega\overrightarrow{\hat{\mathbf{B}}}_0\\ \overrightarrow{\mathrm{k}}\times\overrightarrow{\hat{\mathbf{B}}}_0=-\dfrac{\omega}{c^2}\overrightarrow{\hat{\mathbf{E}}}_0\end{bmatrix}$$

[3.78]

We clearly recover the transverse nature of electromagnetic waves, that is, $\overrightarrow{\mathrm{k}}\cdot\overrightarrow{\hat{\mathbf{E}}}_0 = \overrightarrow{\mathrm{k}}\cdot\overrightarrow{\hat{\mathbf{B}}}_0 = 0$. We also find that $\overrightarrow{\hat{\mathbf{E}}}_0$ is perpendicular to $\overrightarrow{\hat{\mathbf{B}}}_0$, that is, $\overrightarrow{\hat{\mathbf{B}}}_0 = \frac{\overrightarrow{\mathrm{k}}}{\omega}\times\overrightarrow{\hat{\mathbf{E}}}_0 = \frac{1}{c}\overrightarrow{\mathbf{n}}\times\overrightarrow{\hat{\mathbf{E}}}_0$. These relations are, of course, only valid for PPMWs. In particular, when the wave amplitude depends on the coordinates, the relationship between the fields $\overrightarrow{\hat{\mathbf{E}}}_0$ and $\overrightarrow{\hat{\mathbf{B}}}_0$ will be different.

3.4.3. *Polarization of a PPMW*

Consider an PPMW propagating along $+\overrightarrow{\mathbf{i}}$:

$$\begin{bmatrix}E_x=0\\ E_y=A_y\cos\left(\omega t-\mathrm{k}x+\phi_y\right)\\ E_z=A_z\cos\left(\omega t-\mathrm{k}x+\phi_z\right)\end{bmatrix} \Longrightarrow \begin{array}{l}\overrightarrow{\mathbf{E}}=\Re\left[\left(A_y e^{-\mathrm{i}\phi_y}\overrightarrow{\mathbf{j}}+A_z e^{-\mathrm{i}\phi_z}\overrightarrow{\mathbf{k}}\right)\right.\\ \qquad\left. e^{-\mathrm{i}(\omega t-\mathrm{k}x)}\right]\end{array} \qquad [3.79]$$

The magnetic field is calculated using the relation $\overrightarrow{\mathbf{B}} = \frac{\overrightarrow{\mathbf{i}}\times\overrightarrow{\mathbf{E}}}{c}$:

$$\begin{bmatrix}B_x=0\\ B_y=-\dfrac{A_z}{c}\cos\left(\omega t-\mathrm{k}x+\phi_z\right)\\ B_z=\dfrac{A_y}{c}\cos\left(\omega t-\mathrm{k}x+\phi_y\right)\end{bmatrix} \Longrightarrow \begin{array}{l}\overrightarrow{\mathbf{B}}=\Re\left[\dfrac{1}{c}\left(-A_z e^{-\mathrm{i}\phi_z}\overrightarrow{\mathbf{j}}+\right.\right.\\ \qquad\left.\left. A_y e^{-\mathrm{i}\phi_y}\overrightarrow{\mathbf{k}}\right) e^{-\mathrm{i}(\omega t-\mathrm{k}x)}\right]\end{array} \qquad [3.80]$$

Setting $\phi_0 = \phi_y - \phi_z$, we obtain:

$$\frac{E_z}{A_z} = \cos(\omega t - \mathrm{k}x + \phi_y - \phi_0) = \cos(\omega t - \mathrm{k}x + \phi_y)\cos\phi_0 \\ + \sin(\omega t - \mathrm{k}x + \phi_y)\sin\phi_0 = \frac{E_y}{A_y}\cos\phi_0 + \sin(\omega t - \mathrm{k}x + \phi_y)\sin\phi_0 \qquad [3.81]$$

and, consequently:

$$\frac{E_z}{A_z} - \frac{E_y}{A_y}\cos\phi_0 = \sin(\omega t - \mathrm{k}x + \phi_y)\sin\phi_0 \\ \frac{E_y}{A_y}\sin\phi_0 = \cos(\omega t - \mathrm{k}x + \phi_y)\sin\phi_0 \qquad [3.82]$$

Squaring these last two equations and then summing them term by term:

$$\left(\frac{E_y}{A_y}\right)^2 + \left(\frac{E_z}{A_z}\right)^2 - 2\frac{E_y}{A_y}\frac{E_z}{A_z}\cos\phi_0 = \sin^2\phi_0 \qquad [3.83]$$

1) *Linear polarization*: linear polarization corresponds to the case where the electric field maintains a constant direction over time. To describe this type of polarization, we must take $\phi_0 = 0$ or π (modulo 2π), that is, $\sin\phi_0 = 0$ and $\cos\phi_0 = s = \pm 1$. In this case, equation [3.83] simplifies to:

$$\left(\frac{E_y}{A_y}\right)^2 + \left(\frac{E_z}{A_z}\right)^2 - 2s\frac{E_y}{A_y}\frac{E_z}{A_z} = \left[\frac{E_y}{A_y} - s\frac{E_z}{A_z}\right]^2 = 0 \qquad [3.84]$$

which implies:

$$E_z = \frac{sA_z}{A_y}E_y \qquad [3.85]$$

The vector $\vec{\mathbf{E}}$ thus oscillates in the plane $\left(\vec{\mathbf{i}}, \vec{\mathbf{j}}'\right)$, where $\vec{\mathbf{j}}'$ is the unit vector forming an angle $\theta = s\tan^{-1}\left(\frac{A_z}{A_y}\right)$ with $\vec{\mathbf{j}}$. The polarization of $\vec{\mathbf{E}}$ is said to be *linear* along $\vec{\mathbf{j}}'$, as illustrated in Figure 9.5. The field $\vec{\mathbf{B}}$ oscillates along the vector $\vec{\mathbf{k}}'$, which forms the same angle θ with $\vec{\mathbf{k}}$. Cases (a) and (b) are equivalent. Figure 3.6 illustrates the evolution of fields $\vec{\mathbf{E}}$ and $\vec{\mathbf{B}}$ along the Ox axis at a given instant t in this type of polarization. By setting $E_0 = \sqrt{A_y^2 + A_z^2}$, the fields $\overrightarrow{\mathbf{E}}$ and $\overrightarrow{\mathbf{B}}$ are written as:

$$\begin{bmatrix} \overrightarrow{\mathbf{E}} = E_0\,\overrightarrow{\mathbf{j}}'\cos(\omega t - \mathrm{k}x + \phi_y) \\ \overrightarrow{\mathbf{B}} = \dfrac{E_0}{c}\overrightarrow{\mathbf{k}}'\cos(\omega t - \mathrm{k}x + \phi_y) \end{bmatrix} \Longrightarrow \begin{bmatrix} \overrightarrow{\mathbf{E}} = \Re\left[\hat{E}_0\,\overrightarrow{\mathbf{j}}'e^{-\mathrm{i}(\omega t - \mathrm{k}x)}\right] \\ \overrightarrow{\mathbf{B}} = \Re\left[\dfrac{\hat{E}_0}{c}\overrightarrow{\mathbf{k}}'e^{-\mathrm{i}(\omega t - \mathrm{k}x)}\right] \end{bmatrix} \quad [3.86]$$

where $\overrightarrow{\mathbf{j}}' = \cos\theta\,\overrightarrow{\mathbf{j}} + \sin\theta\,\overrightarrow{\mathbf{k}}$, $\overrightarrow{\mathbf{j}}' = -\sin\theta\,\overrightarrow{\mathbf{j}} + \cos\theta\,\overrightarrow{\mathbf{k}}$ and $\hat{E}_0 = E_0 e^{-\mathrm{i}\phi_y}$. These are the expressions for the fields $\overrightarrow{\mathbf{E}}$ and $\overrightarrow{\mathbf{B}}$ of a linearly polarized wave.

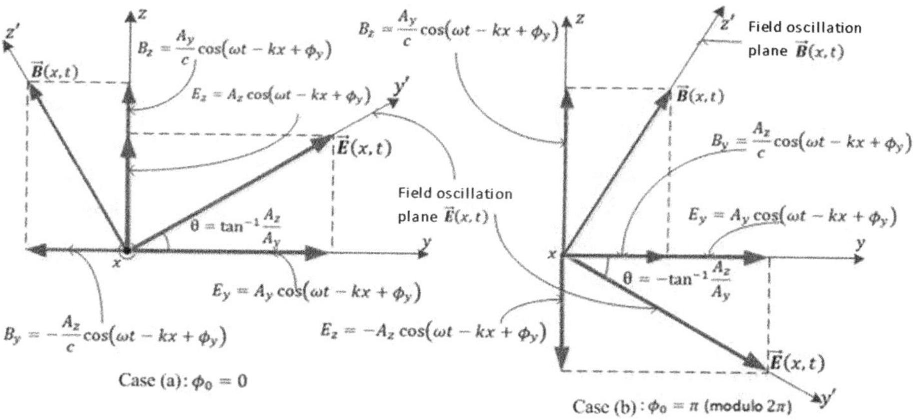

Figure 3.5. *Linearly polarized electromagnetic wave: the fields $\overrightarrow{\mathbf{E}}$ and $\overrightarrow{\mathbf{B}}$ oscillate along the x-axis without changing direction. For a color version of this figure, see www.iste.co.uk/abdesselam/electrodynamics2.zip*

2) *Elliptical polarization*: if $\phi_0 \neq 0$ or $\neq \pi$ (modulo 2π), it can be shown that [3.83] is simply the equation of an ellipse. Indeed, the equation of an ellipse, with a semi-major axis a and a semi-minor axis $b \leq a$, along the orthogonal unit vectors $\overrightarrow{\mathbf{j}}'$ and $\overrightarrow{\mathbf{k}}'$, is:

$$\left(\frac{Y}{a}\right)^2 + \left(\frac{Z}{b}\right)^2 = 1 \quad [3.87]$$

Vectors $\overrightarrow{\mathbf{j}}'$ and $\overrightarrow{\mathbf{k}}'$ are obtained from $\overrightarrow{\mathbf{j}}$ and $\overrightarrow{\mathbf{k}}$ by rotating an angle θ around the origin O, that is, $\overrightarrow{\mathbf{j}}' = \cos\theta\,\overrightarrow{\mathbf{j}} + \sin\theta\,\overrightarrow{\mathbf{k}}$ and $\overrightarrow{\mathbf{k}}' = -\sin\theta\,\overrightarrow{\mathbf{j}} + \cos\theta\,\overrightarrow{\mathbf{k}}$. In this new basis $\left(\overrightarrow{\mathbf{j}}', \overrightarrow{\mathbf{k}}'\right)$, the components Y and Z of the electric field $\overrightarrow{\mathbf{E}}$ are expressed as:

$$\begin{aligned} Y &= \vec{\mathbf{E}} \cdot \vec{\mathbf{j}}' = E_y \cos\theta + E_z \sin\theta \\ Z &= \vec{\mathbf{E}} \cdot \vec{\mathbf{k}}' = -E_y \sin\theta + E_z \cos\theta \end{aligned} \qquad [3.88]$$

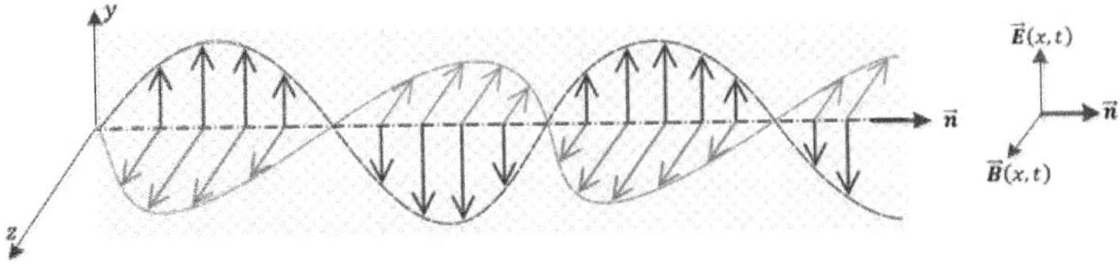

Figure 3.6. *Evolution of the fields $\vec{\mathbf{E}}$ and $\vec{\mathbf{B}}$ for a linearly polarized PPMW. For a color version of this figure, see www.iste.co.uk/abdesselam/electrodynamics2.zip*

By substituting [3.88] into [3.87], we get:

$$\begin{aligned} &E_y^2 \left(\frac{\cos^2\theta}{a^2} + \frac{\sin^2\theta}{b^2} \right) + E_z^2 \left(\frac{\sin^2\theta}{a^2} + \frac{\cos^2\theta}{b^2} \right) \\ &-2E_yE_z \sin\theta\cos\theta \left(\frac{1}{b^2} - \frac{1}{a^2} \right) = 1 \end{aligned} \qquad [3.89]$$

By comparing the result [3.89] with [3.83], we deduce that:

$$\frac{\cos^2\theta}{a^2} + \frac{\sin^2\theta}{b^2} = \frac{1}{A_y^2 \sin^2\phi_0} \qquad [3.90]$$

$$\frac{\sin^2\theta}{a^2} + \frac{\cos^2\theta}{b^2} = \frac{1}{A_z^2 \sin^2\phi_0} \qquad [3.91]$$

$$\sin\theta\cos\theta \left(\frac{1}{b^2} - \frac{1}{a^2} \right) = \frac{\cos\phi_0}{A_yA_z \sin^2\phi_0} \qquad [3.92]$$

It is clear that the tip of the electric field $\vec{\mathbf{E}}\,(x,t)$ traces an ellipse. In this case, we say that the polarization is *elliptical*, as shown in Figure 3.7(a). The case of linear polarization corresponds to a degenerate ellipse: the semi-minor axis is zero, and the ellipse reduces to a segment of length $2a$ that forms an angle θ with $\vec{\mathbf{j}}$, as illustrated in Figure 3.7(b). To determine the characteristics of this ellipse, we add equations [3.90] and [3.91], then subtract one from the other, yielding:

$$\frac{1}{a^2} + \frac{1}{b^2} = \left(\frac{1}{A_y^2} + \frac{1}{A_z^2} \right) \frac{1}{\sin^2\phi_0} \equiv \alpha \qquad [3.93]$$

$$\cos 2\theta \left(\frac{1}{b^2} - \frac{1}{a^2} \right) = \left(\frac{1}{A_z^2} - \frac{1}{A_y^2} \right) \frac{1}{\sin^2\phi_0} \equiv \beta \qquad [3.94]$$

Equation [3.92] can also be rewritten in the following form:

$$\sin 2\theta \left(\frac{1}{b^2} - \frac{1}{a^2}\right) = \frac{2\cos\phi_0}{A_y A_z \sin^2\phi_0} \equiv \gamma \qquad [3.95]$$

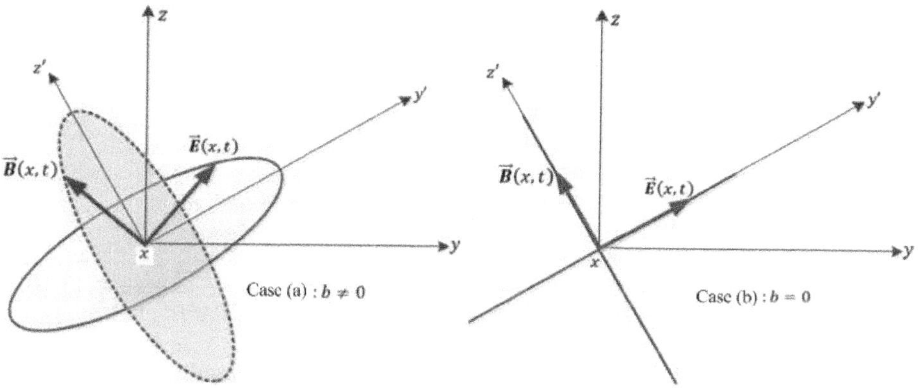

Figure 3.7. *(a) For an elliptically polarized wave, the tips of the fields $\vec{\mathbf{E}}$ and $\vec{\mathbf{B}}$ trace out ellipses. (b) When the semi-minor axis is zero, the ellipse degenerates into a line segment of length $2a$, and the fields no longer change direction; this corresponds to linear polarization. For a color version of this figure, see www.iste.co.uk/abdesselam/electrodynamics2.zip*

To eliminate the angle θ between equations [3.94] and [3.95], it is sufficient to square both equations and then sum the results term by term:

$$\frac{1}{b^2} - \frac{1}{a^2} = \sqrt{\beta^2 + \gamma^2} \qquad [3.96]$$

The solution of the systems of equations [3.93] and [3.96] directly yield:

$$a = \sqrt{\frac{2}{\alpha - \sqrt{\beta^2 + \gamma^2}}} \qquad [3.97]$$

$$b = \sqrt{\frac{2}{\alpha + \sqrt{\beta^2 + \gamma^2}}} \qquad [3.98]$$

To determine θ, we simply divide [3.94] by [3.95], which gives $\tan 2\theta = \frac{\gamma}{\beta}$ and, consequently:

$$\theta = \frac{1}{2}\tan^{-1}\frac{\gamma}{\beta}\ \left(\text{modulo } \frac{\pi}{2}\right) \qquad [3.99]$$

There is another formula to calculate θ. To find it, let us express equations [3.90]–[3.92] in terms of the parameters α, β and γ:

$$\left.\begin{array}{l}\dfrac{\cos^2\theta}{a^2}+\dfrac{\sin^2\theta}{b^2}=\dfrac{\alpha-\beta}{2}\\ \dfrac{\sin^2\theta}{a^2}+\dfrac{\cos^2\theta}{b^2}=\dfrac{\alpha+\beta}{2}\\ \sin\theta\cos\theta\left(\dfrac{1}{b^2}-\dfrac{1}{a^2}\right)=\dfrac{\gamma}{2}\end{array}\right. \Longrightarrow \left[\begin{array}{l}\beta=\left(\frac{1}{b^2}-\frac{1}{a^2}\right)\cos 2\theta=\sqrt{\beta^2+\gamma^2}\cos 2\theta\\ \\ \gamma=\left(\frac{1}{b^2}-\frac{1}{a^2}\right)\sin 2\theta=\sqrt{\beta^2+\gamma^2}\sin 2\theta\end{array}\right. \qquad [3.100]$$

Thus, we obtain:

$$\frac{\beta+\sqrt{\beta^2+\gamma^2}}{\gamma}=\frac{(\cos 2\theta+1)}{\sin 2\theta}=\frac{1}{\tan\theta} \Longrightarrow \theta=\tan^{-1}\frac{\gamma}{\beta+\sqrt{\beta^2+\gamma^2}} \qquad [3.101]$$

In particular, if $\beta=\gamma=0$, that is, $A_y=A_z$ and $\phi_0=\pm\frac{\pi}{2}$ (modulo π), the ellipse reduces to a circle. In this case, the polarization is said to be *circular*. In other words, the tips of the fields $\vec{\mathbf{E}}(x,t)$ and $\vec{\mathbf{B}}(x,t)$ describe circles, as illustrated in Figure 3.8.

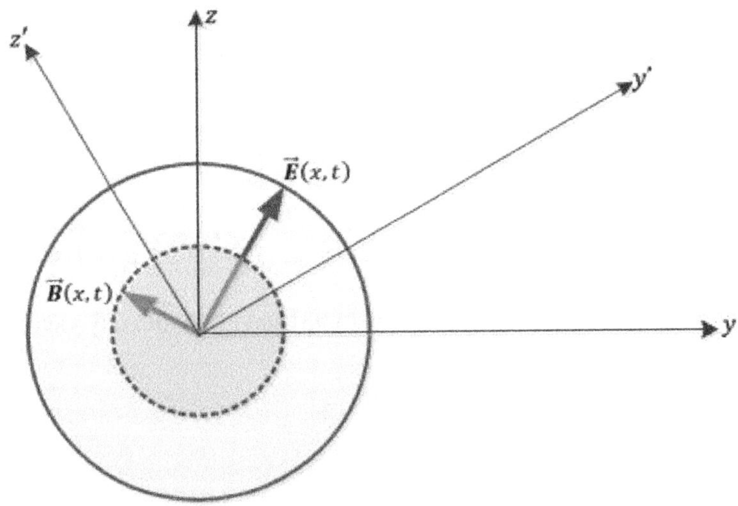

Figure 3.8. *For a circularly polarized wave, the tips of the fields $\vec{\mathbf{E}}(x,t)$ and $\vec{\mathbf{B}}(x,t)$ trace out circles. For a color version of this figure, see www.iste.co.uk/abdesselam/electrodynamics2.zip*

3) *Direction of rotation*: consider an electromagnetic wave propagating in the positive Ox direction. Polarization is said to be *left-handed* if the tip of $\vec{\mathbf{E}}$ rotates counterclockwise around the Ox axis (i.e. from Oy to Oz) over time, as seen by an observer looking at the incoming wave along Ox. In contrast, polarization is said to be *right-handed* if $\vec{\mathbf{E}}$ rotates in the opposite direction, that is, clockwise. To determine the polarization direction, we compute the vector $\vec{\mathbf{E}} \times \frac{\partial \vec{\mathbf{E}}}{\partial t}$. It is easy to verify that the polarization is left-handed if $\vec{\mathbf{E}} \times \frac{\partial \vec{\mathbf{E}}}{\partial t}$ points in the $+\vec{\mathbf{i}}$ direction for an infinitesimally small positive time interval dt. Similarly, the polarization is right-handed if $\vec{\mathbf{E}} \times \frac{\partial \vec{\mathbf{E}}}{\partial t}$ points in the $-\vec{\mathbf{i}}$ direction for $dt > 0$. In fact, we can express: $\vec{\mathbf{E}}(t) \times \vec{\mathbf{E}}(t+dt) = \vec{\mathbf{E}}(t) \times \left[\vec{\mathbf{E}}(t) + \frac{\partial \vec{\mathbf{E}}}{\partial t} dt\right] = \vec{\mathbf{E}}(t) \times \frac{\partial \vec{\mathbf{E}}}{\partial t} dt$. Since $dt > 0$, it follows that $\vec{\mathbf{E}} \times \frac{\partial \vec{\mathbf{E}}}{\partial t}$ is in the $+\vec{\mathbf{i}}$ direction for left-handed polarization and in the $-\vec{\mathbf{i}}$ direction for right-handed polarization. Now, let us proceed with the calculation:

$$\begin{aligned}
\vec{\mathbf{E}}(t) \times \frac{\partial \vec{\mathbf{E}}}{\partial t} &= \begin{vmatrix} \vec{\mathbf{i}} & \vec{\mathbf{j}} & \vec{\mathbf{k}} \\ 0 & A_y \cos(\omega t - \mathrm{k}x + \phi_y) & A_z \cos(\omega t - \mathrm{k}x + \phi_z) \\ 0 & -\omega A_y \sin(\omega t - \mathrm{k}x + \phi_y) & -\omega A_z \sin(\omega t - \mathrm{k}x + \phi_z) \end{vmatrix} \\
&= -\omega A_y A_z \left[\cos(\omega t - \mathrm{k}x + \phi_y) \sin(\omega t - \mathrm{k}x + \phi_z)\right. \\
&\quad \left. - \sin(\omega t - \mathrm{k}x + \phi_y) \cos(\omega t - \mathrm{k}x + \phi_z)\right] \vec{\mathbf{i}} = -\omega A_y A_z \sin(\phi_z - \phi_y) \vec{\mathbf{i}} \\
&= \omega A_y A_z \sin\phi_0 \vec{\mathbf{i}}
\end{aligned} \quad [3.102]$$

The wave is therefore right-hand elliptically polarized if $\phi_0 \in \left]-\pi, 0\right[$ and left-hand elliptically polarized if $\phi_0 \in \left]0, \pi\right[$.

4) Here are the expressions of the fields $\vec{\mathbf{E}}$ and $\vec{\mathbf{B}}$ for the different types of polarization:

i) *Left-handed circular polarization*:

$$\begin{cases}
\vec{\mathbf{E}}(x,t) = a \left[\cos(\omega t - \mathrm{k}x + \phi) \vec{\mathbf{j}}' + \sin(\omega t - \mathrm{k}x + \phi) \vec{\mathbf{k}}'\right] \\
\vec{\mathbf{B}}(x,t) = \frac{a}{c} \left[-\sin(\omega t - \mathrm{k}x + \phi) \vec{\mathbf{j}}' + \cos(\omega t - \mathrm{k}x + \phi) \vec{\mathbf{k}}'\right] \\
\vec{\hat{\mathbf{E}}}(x,t) = \hat{a} \left[\vec{\mathbf{j}}' + \mathrm{i}\vec{\mathbf{k}}'\right] e^{-\mathrm{i}(\omega t - \mathrm{k}x)} = \sqrt{2}\hat{a}\, \vec{\mathbf{e}}^{+} e^{-\mathrm{i}(\omega t - \mathrm{k}x)} \\
\vec{\hat{\mathbf{B}}}(x,t) = -\mathrm{i}\frac{\hat{a}}{c} \left[\vec{\mathbf{j}}' + \mathrm{i}\vec{\mathbf{k}}'\right] e^{-\mathrm{i}(\omega t - \mathrm{k}x)} = -\mathrm{i}\sqrt{2}\frac{\hat{a}}{c} \vec{\mathbf{e}}^{+} e^{-\mathrm{i}(\omega t - \mathrm{k}x)}
\end{cases} \quad [3.103]$$

where:

$$\vec{\mathbf{e}}^{+} = \frac{1}{\sqrt{2}} \left[\vec{\mathbf{j}}' + \mathrm{i}\vec{\mathbf{k}}'\right] \equiv \frac{1}{\sqrt{2}} \begin{pmatrix} 1 \\ \mathrm{i} \end{pmatrix} \quad [3.104]$$

is the *left-handed helicity vector*, and $\hat{a} = ae^{-i\phi}$. Since the Maxwell equations [3.78] are invariant under the transformation $\overrightarrow{\mathbf{B}}' \longrightarrow \frac{1}{c}\overrightarrow{\mathbf{E}}$ and $\overrightarrow{\mathbf{E}}' \longrightarrow -c\overrightarrow{\mathbf{B}}$, we can also take the solutions:

$$\begin{cases} \overrightarrow{\mathbf{E}}(x,t) = a\left[\sin(\omega t - \mathrm{k}x + \phi)\,\overrightarrow{\mathbf{j}}' - \cos(\omega t - \mathrm{k}x + \phi)\,\overrightarrow{\mathbf{k}}'\right] \\ \overrightarrow{\mathbf{B}}(x,t) = \frac{a}{c}\left[\cos(\omega t - \mathrm{k}x + \phi)\,\overrightarrow{\mathbf{j}}' + \sin(\omega t - \mathrm{k}x + \phi)\,\overrightarrow{\mathbf{k}}'\right] \\ \overrightarrow{\hat{\mathbf{E}}}(x,t) = \mathrm{i}\sqrt{2}\hat{a}\,\overrightarrow{\mathbf{e}}^{+}e^{-\mathrm{i}(\omega t - \mathrm{k}x)} \\ \overrightarrow{\hat{\mathbf{B}}}(x,t) = \sqrt{2}\frac{\hat{a}}{c}\,\overrightarrow{\mathbf{e}}^{+}e^{-\mathrm{i}(\omega t - \mathrm{k}x)} \end{cases} \quad [3.105]$$

ii) *Right-handed circular polarization*:

$$\begin{cases} \overrightarrow{\mathbf{E}}(x,t) = a\left[\cos(\omega t - \mathrm{k}x + \phi)\,\overrightarrow{\mathbf{j}}' - \sin(\omega t - \mathrm{k}x + \phi)\,\overrightarrow{\mathbf{k}}'\right] \\ \overrightarrow{\mathbf{B}}(x,t) = \frac{a}{c}\left[\sin(\omega t - \mathrm{k}x + \phi)\,\overrightarrow{\mathbf{j}}' + \cos(\omega t - \mathrm{k}x + \phi)\,\overrightarrow{\mathbf{j}}'\right] \\ \overrightarrow{\hat{\mathbf{E}}}(x,t) = \hat{a}\left[\overrightarrow{\mathbf{j}}' - \mathrm{i}\overrightarrow{\mathbf{k}}'\right]e^{-\mathrm{i}(\omega t - \mathrm{k}x)} = \sqrt{2}\hat{a}\,\overrightarrow{\mathbf{e}}^{-}e^{-\mathrm{i}(\omega t - \mathrm{k}x)} \\ \overrightarrow{\hat{\mathbf{B}}}(x,t) = \mathrm{i}\frac{\hat{a}}{c}\left[\overrightarrow{\mathbf{j}}' - \mathrm{i}\overrightarrow{\mathbf{k}}'\right]e^{-\mathrm{i}(\omega t - \mathrm{k}x)} = \mathrm{i}\sqrt{2}\frac{\hat{a}}{c}\,\overrightarrow{\mathbf{e}}^{-}e^{-\mathrm{i}(\omega t - \mathrm{k}x)} \end{cases} \quad [3.106]$$

where:

$$\overrightarrow{\mathbf{e}}^{-} = \frac{1}{\sqrt{2}}\left[\overrightarrow{\mathbf{j}}' - \mathrm{i}\overrightarrow{\mathbf{k}}'\right] \equiv \frac{1}{\sqrt{2}}\begin{pmatrix} 1 \\ -\mathrm{i} \end{pmatrix} \quad [3.107]$$

is the *right-handed helicity vector*. Similarly, we can also consider the configuration (the evolution of the field $\overrightarrow{\mathbf{E}}(x,t)$ is shown in Figure 3.9):

$$\begin{cases} \overrightarrow{\mathbf{E}}(x,t) = -c\left[\sin(\omega t - \mathrm{k}x + \phi)\,\overrightarrow{\mathbf{j}}' + \cos(\omega t - \mathrm{k}x + \phi)\,\overrightarrow{\mathbf{j}}'\right] \\ \overrightarrow{\mathbf{B}}(x,t) = \frac{a}{c}\left[\cos(\omega t - \mathrm{k}x + \phi)\,\overrightarrow{\mathbf{j}}' - \sin(\omega t - \mathrm{k}x + \phi)\,\overrightarrow{\mathbf{k}}'\right] \\ \overrightarrow{\hat{\mathbf{E}}}(x,t) = -\mathrm{i}\sqrt{2}\hat{a}\,\overrightarrow{\mathbf{e}}^{-}e^{-\mathrm{i}(\omega t - \mathrm{k}x)} \\ \overrightarrow{\hat{\mathbf{B}}}(x,t) = \sqrt{2}\frac{\hat{a}}{c}\,\overrightarrow{\mathbf{e}}^{-}e^{-\mathrm{i}(\omega t - \mathrm{k}x)} \end{cases} \quad [3.108]$$

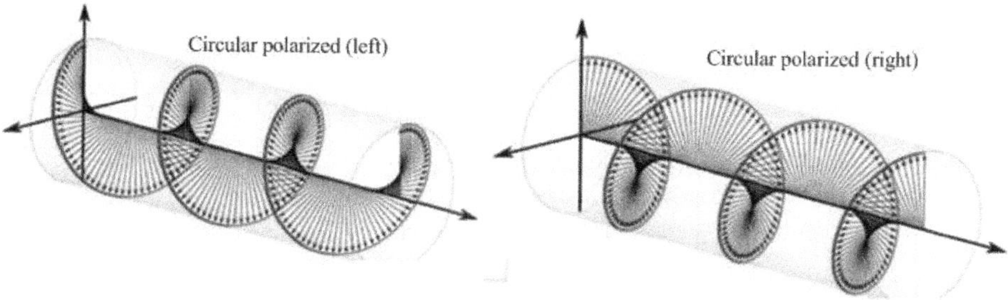

Figure 3.9. *Circularly polarized electromagnetic wave*

iii) *Left-handed elliptical polarization*:

$$\begin{cases} \overrightarrow{\mathbf{E}}(x,t) = a\cos(\omega t - \mathrm{k}x + \phi)\,\overrightarrow{\mathbf{j}}' + b\sin(\omega t - \mathrm{k}x + \phi)\,\overrightarrow{\mathbf{k}}' \\ \overrightarrow{\mathbf{B}}(x,t) = -\dfrac{b}{c}\sin(\omega t - \mathrm{k}x + \phi)\,\overrightarrow{\mathbf{j}}' + \dfrac{a}{c}\cos(\omega t - \mathrm{k}x + \phi)\,\overrightarrow{\mathbf{k}}' \end{cases}$$
$$\begin{cases} \overrightarrow{\hat{\mathbf{E}}}(x,t) = \sqrt{2}\left[\dfrac{\hat{a}+\hat{b}}{2}\overrightarrow{\mathbf{e}}^{+} + \dfrac{\hat{a}-\hat{b}}{2}\overrightarrow{\mathbf{e}}^{-}\right] e^{-\mathrm{i}(\omega t - \mathrm{k}x)} \\ \overrightarrow{\hat{\mathbf{B}}}(x,t) = -\mathrm{i}\sqrt{2}\left[\dfrac{\hat{a}+\hat{b}}{2c}\overrightarrow{\mathbf{e}}^{+} - \dfrac{\hat{a}-\hat{b}}{2c}\overrightarrow{\mathbf{e}}^{-}\right] e^{-\mathrm{i}(\omega t - \mathrm{k}x)} \end{cases} \qquad [3.109]$$

where $\hat{a} = a^{-\mathrm{i}\phi}$ and $\hat{b} = b^{-\mathrm{i}\phi}$. We can also consider:

$$\begin{cases} \overrightarrow{\mathbf{E}}(x,t) = b\sin(\omega t - \mathrm{k}x + \phi)\,\overrightarrow{\mathbf{j}}' - a\cos(\omega t - \mathrm{k}x + \phi)\,\overrightarrow{\mathbf{k}}' \\ \overrightarrow{\mathbf{B}}(x,t) = \dfrac{a}{c}\cos(\omega t - \mathrm{k}x + \phi)\,\overrightarrow{\mathbf{j}}' + \dfrac{b}{c}\sin(\omega t - \mathrm{k}x + \phi)\,\overrightarrow{\mathbf{k}}' \end{cases}$$
$$\begin{cases} \overrightarrow{\hat{\mathbf{E}}}(x,t) = \mathrm{i}\sqrt{2}\left[\dfrac{\hat{a}+\hat{b}}{2}\overrightarrow{\mathbf{e}}^{+} - \dfrac{\hat{a}-\hat{b}}{2}\overrightarrow{\mathbf{e}}^{-}\right] e^{-\mathrm{i}(\omega t - \mathrm{k}x)} \\ \overrightarrow{\hat{\mathbf{B}}}(x,t) = \sqrt{2}\left[\dfrac{\hat{a}+\hat{b}}{2c}\overrightarrow{\mathbf{e}}^{+} + \dfrac{\hat{a}-\hat{b}}{2c}\overrightarrow{\mathbf{e}}^{-}\right] e^{-\mathrm{i}(\omega t - \mathrm{k}x)} \end{cases} \qquad [3.110]$$

iv) *Right-handed elliptical polarization*:

$$\begin{cases} \overrightarrow{\mathbf{E}}(x,t) = a\cos(\omega t - \mathrm{k}x + \phi)\,\overrightarrow{\mathbf{j}}' - b\sin(\omega t - \mathrm{k}x + \phi)\,\overrightarrow{\mathbf{k}}' \\ \overrightarrow{\mathbf{B}}(x,t) = \dfrac{b}{c}\sin(\omega t - \mathrm{k}x + \phi)\,\overrightarrow{\mathbf{j}}' + \dfrac{a}{c}\cos(\omega t - \mathrm{k}x + \phi)\,\overrightarrow{\mathbf{k}}' \end{cases}$$
$$\begin{cases} \overrightarrow{\hat{\mathbf{E}}}(x,t) = \sqrt{2}\left[\dfrac{\hat{a}-\hat{b}}{2}\overrightarrow{\mathbf{e}}^{+} + \dfrac{\hat{a}+\hat{b}}{2}\overrightarrow{\mathbf{e}}^{-}\right] e^{-\mathrm{i}(\omega t - \mathrm{k}x)} \\ \overrightarrow{\hat{\mathbf{B}}}(x,t) = -\mathrm{i}\sqrt{2}\left[\dfrac{\hat{a}-\hat{b}}{2c}\overrightarrow{\mathbf{e}}^{+} - \dfrac{\hat{a}+\hat{b}}{2c}\overrightarrow{\mathbf{e}}^{-}\right] e^{-\mathrm{i}(\omega t - \mathrm{k}x)} \end{cases} \qquad [3.111]$$

We can also take:

$$\begin{cases} \overrightarrow{\mathbf{E}}(x,t) = -b\sin(\omega t - \mathrm{k}x + \phi)\,\overrightarrow{\mathbf{j}}' - a\cos(\omega t - \mathrm{k}x + \phi)\,\overrightarrow{\mathbf{k}}' \\ \overrightarrow{\mathbf{B}}(x,t) = \frac{a}{c}\cos(\omega t - \mathrm{k}x + \phi)\,\overrightarrow{\mathbf{j}}' - \frac{b}{c}\sin(\omega t - \mathrm{k}x + \phi)\,\overrightarrow{\mathbf{k}}' \\ \overrightarrow{\hat{\mathbf{E}}}(x,t) = \mathrm{i}\sqrt{2}\left[\frac{\hat{a} - \hat{b}}{2}\overrightarrow{\mathbf{e}}^{+} - \frac{\hat{a} + \hat{b}}{2}\overrightarrow{\mathbf{e}}^{-}\right] e^{-\mathrm{i}(\omega t - \mathrm{k}x)} \\ \overrightarrow{\hat{\mathbf{B}}}(x,t) = \sqrt{2}\left[\frac{\hat{a} - \hat{b}}{2c}\overrightarrow{\mathbf{e}}^{+} + \frac{\hat{a} + \hat{b}}{2c}\overrightarrow{\mathbf{e}}^{-}\right] e^{-\mathrm{i}(\omega t - \mathrm{k}x)} \end{cases} \qquad [3.112]$$

v) *Linear polarization*:

$$\begin{cases} \overrightarrow{\mathbf{E}}(x,t) = a\cos(\omega t - \mathrm{k}x + \phi)\,\overrightarrow{\mathbf{j}}' \\ \overrightarrow{\mathbf{B}}(x,t) = \frac{a}{c}\cos(\omega t - \mathrm{k}x + \phi)\,\overrightarrow{\mathbf{k}}' \end{cases}$$

$$\begin{cases} \overrightarrow{\hat{\mathbf{E}}}(x,t) = \hat{a}\,\overrightarrow{\mathbf{j}}' e^{-\mathrm{i}(\omega t - \mathrm{k}x)} = \frac{\hat{a}}{\sqrt{2}}\left[\overrightarrow{\mathbf{e}}^{+} + \overrightarrow{\mathbf{e}}^{-}\right] e^{-\mathrm{i}(\omega t - \mathrm{k}x)} \\ \overrightarrow{\hat{\mathbf{B}}}(x,t) = \frac{\hat{a}}{c}\overrightarrow{\mathbf{k}}' e^{-\mathrm{i}(\omega t - \mathrm{k}x)} = -\mathrm{i}\frac{\hat{a}}{\sqrt{2}c}\left[\overrightarrow{\mathbf{e}}^{+} - \overrightarrow{\mathbf{e}}^{-}\right] e^{-\mathrm{i}(\omega t - \mathrm{k}x)} \end{cases} \qquad [3.113]$$

This wave is linearly polarized along $\overrightarrow{\mathbf{j}}'$. Similarly, we can take (polarized along $\overrightarrow{\mathbf{k}}'$):

$$\begin{cases} \overrightarrow{\mathbf{E}}(x,t) = -a\cos(\omega t - \mathrm{k}x + \phi)\,\overrightarrow{\mathbf{k}}' \\ \overrightarrow{\mathbf{B}}(x,t) = \frac{a}{c}\cos(\omega t - \mathrm{k}x + \phi)\,\overrightarrow{\mathbf{j}}' \end{cases}$$

$$\begin{cases} \overrightarrow{\hat{\mathbf{E}}}(x,t) = \mathrm{i}\frac{\hat{a}}{\sqrt{2}}\left[\overrightarrow{\mathbf{e}}^{+} - \overrightarrow{\mathbf{e}}^{-}\right] e^{-\mathrm{i}(\omega t - \mathrm{k}x)} \\ \overrightarrow{\hat{\mathbf{B}}}(x,t) = \frac{\hat{a}}{\sqrt{2}c}\left[\overrightarrow{\mathbf{e}}^{+} + \overrightarrow{\mathbf{e}}^{-}\right] e^{-\mathrm{i}(\omega t - \mathrm{k}x)} \end{cases} \qquad [3.114]$$

5) *Decomposition*: it can be shown that an elliptically polarized wave is obtained by superimposing (i.e. adding) two linearly polarized waves along two perpendicular directions. In fact, the polarized wave [3.109] can also be written as:

$$\overrightarrow{\hat{\mathbf{E}}}(x,t) = \underbrace{\sqrt{2}\left[\frac{\hat{a} + \hat{b}}{2}\overrightarrow{\mathbf{e}}^{+} + \frac{\hat{a} - \hat{b}}{2}\overrightarrow{\mathbf{e}}^{-}\right] e^{-\mathrm{i}(\omega t - \mathrm{k}x)}}_{\text{left-handed elliptical pol.}} \qquad [3.115]$$

$$= \underbrace{\frac{\hat{a}}{\sqrt{2}}\left(\vec{\mathbf{e}}^{+}+\vec{\mathbf{e}}^{-}\right)e^{-\mathrm{i}(\omega t-\mathrm{k}x)}}_{\text{linear pol. along }\vec{\mathbf{j}}'} + \mathrm{i}\underbrace{\frac{-\mathrm{i}\hat{b}}{\sqrt{2}}\left(\vec{\mathbf{e}}^{+}-\vec{\mathbf{e}}^{-}\right)e^{-\mathrm{i}(\omega t-\mathrm{k}x)}}_{\text{linear pol. along }\vec{\mathbf{k}}'}$$

$$\vec{\hat{\mathbf{B}}}(x,t) = \underbrace{-\mathrm{i}\sqrt{2}\left[\frac{\hat{a}+\hat{b}}{2c}\vec{\mathbf{e}}^{+} - \frac{\hat{a}-\hat{b}}{2c}\vec{\mathbf{e}}^{-}\right]e^{-\mathrm{i}(\omega t-\mathrm{k}x)}}_{\text{left-handed elliptical pol.}} \qquad [3.116]$$

$$= \underbrace{-\mathrm{i}\frac{\hat{a}}{\sqrt{2}c}\left(\vec{\mathbf{e}}^{+}-\vec{\mathbf{e}}^{-}\right)e^{-\mathrm{i}(\omega t-\mathrm{k}x)}}_{\text{linear pol. along }\vec{\mathbf{j}}'} - \mathrm{i}\underbrace{\frac{\hat{b}}{\sqrt{2}c}\left(\vec{\mathbf{e}}^{+}+\vec{\mathbf{e}}^{-}\right)e^{-\mathrm{i}(\omega t-\mathrm{k}x)}}_{\text{linear pol. along }\vec{\mathbf{k}}'}$$

Similarly, a linearly polarized wave is obtained by superimposing two circularly polarized waves, rotating in opposite directions (right and left) and with the same amplitude. To prove this, it is sufficient to show that:

$$\vec{\hat{\mathbf{E}}}(x,t) = \underbrace{\frac{\hat{a}}{\sqrt{2}}\left[\vec{\mathbf{e}}^{+}+\vec{\mathbf{e}}^{-}\right]e^{-\mathrm{i}(\omega t-\mathrm{k}x)}}_{\text{linear pol. along }\vec{\mathbf{j}}'}$$

$$= \frac{1}{2}\underbrace{\sqrt{2}\hat{a}\,\vec{\mathbf{e}}^{+}e^{-\mathrm{i}(\omega t-\mathrm{k}x)}}_{\text{left-handed circularpol.}} + \frac{1}{2}\underbrace{\sqrt{2}\hat{a}\,\vec{\mathbf{e}}^{-}e^{-\mathrm{i}(\omega t-\mathrm{k}x)}}_{\text{right-handed circular pol.}} \qquad [3.117]$$

$$\vec{\hat{\mathbf{B}}}(x,t) = \underbrace{\frac{-\mathrm{i}\hat{a}}{\sqrt{2}c}\left[\vec{\mathbf{e}}^{+}-\vec{\mathbf{e}}^{-}\right]e^{-\mathrm{i}(\omega t-\mathrm{k}x)}}_{\text{linear pol. along }\vec{\mathbf{j}}'}$$

$$= \frac{1}{2}\underbrace{-\mathrm{i}\sqrt{2}\frac{\hat{a}}{c}\vec{\mathbf{e}}^{+}e^{-\mathrm{i}(\omega t-\mathrm{k}x)}}_{\text{left-handed circular pol.}} + \frac{1}{2}\underbrace{\mathrm{i}\sqrt{2}\frac{\hat{a}}{c}\vec{\mathbf{e}}^{-}e^{-\mathrm{i}(\omega t-\mathrm{k}x)}}_{\text{right-handed circular pol.}} \qquad [3.118]$$

6) *Polarizer*: a *polarizer* is a device that affects the polarization of an incident light beam. It can filter light by allowing only certain polarization components to pass through. It can also modify the polarization of an incident beam. These devices can be modeled using matrices, known as *Jones matrices*. For example, to transition from linear polarization along $\vec{\mathbf{j}}'$ to left-handed circular polarization, we simply multiply the complex representation [3.113] by the matrix $\begin{pmatrix} 1 & -\mathrm{i} \\ \mathrm{i} & 1 \end{pmatrix}$:

$$\begin{pmatrix} 1 & -\mathrm{i} \\ \mathrm{i} & 1 \end{pmatrix}\underbrace{\frac{\hat{a}}{\sqrt{2}}\left(\vec{\mathbf{e}}^{+}+\vec{\mathbf{e}}^{-}\right)e^{-\mathrm{i}(\omega t-\mathrm{k}x)}}_{\text{linear pol. along }\vec{\mathbf{j}}'} = \underbrace{\sqrt{2}\hat{a}\,\vec{\mathbf{e}}^{+}e^{-\mathrm{i}(\omega t-\mathrm{k}x)}}_{\text{left-handed circular pol.}}$$

$$\begin{pmatrix} 1 & -\mathrm{i} \\ \mathrm{i} & 1 \end{pmatrix} \underbrace{\frac{-\mathrm{i}\hat{a}}{\sqrt{2}c}\left(\vec{\mathbf{e}}^{+} - \vec{\mathbf{e}}^{-}\right) e^{-\mathrm{i}(\omega t - \mathrm{k}x)}}_{\text{linear pol. along } \vec{\mathbf{j}}'} = \underbrace{-\mathrm{i}\sqrt{2}\frac{\hat{a}}{c}\vec{\mathbf{e}}^{+} e^{-\mathrm{i}(\omega t - \mathrm{k}x)}}_{\text{left-handed circular pol.}} \quad [3.119]$$

$\begin{pmatrix} 1 & -\mathrm{i} \\ \mathrm{i} & 1 \end{pmatrix}$ is called a *left-handed circular polarizer*. Similarly, it is shown that:

$$\begin{pmatrix} 1 & \mathrm{i} \\ -\mathrm{i} & 1 \end{pmatrix} \underbrace{\frac{\hat{a}}{\sqrt{2}}\left(\vec{\mathbf{e}}^{+} + \vec{\mathbf{e}}^{-}\right) e^{-\mathrm{i}(\omega t - \mathrm{k}x)}}_{\text{linear pol. along } \vec{\mathbf{j}}'} = \underbrace{\sqrt{2}\hat{a}\vec{\mathbf{e}}^{-} e^{-\mathrm{i}(\omega t - \mathrm{k}x)}}_{\text{right-handed circular pol.}}$$

$$\begin{pmatrix} 1 & \mathrm{i} \\ -\mathrm{i} & 1 \end{pmatrix} \underbrace{\frac{-\mathrm{i}\hat{a}}{\sqrt{2}c}\left(\vec{\mathbf{e}}^{+} - \vec{\mathbf{e}}^{-}\right) e^{-\mathrm{i}(\omega t - \mathrm{k}x)}}_{\text{linear pol. along } \vec{\mathbf{j}}'} = \underbrace{\mathrm{i}\sqrt{2}\frac{\hat{a}}{c}\vec{\mathbf{e}}^{-} e^{-\mathrm{i}(\omega t - \mathrm{k}x)}}_{\text{right-handed circular pol.}} \quad [3.120]$$

$\begin{pmatrix} 1 & \mathrm{i} \\ -\mathrm{i} & 1 \end{pmatrix}$ is called a *right-handed circular polarizer*. Note that these two matrices are not invertible. By applying the first polarizer to a left-handed elliptical polarization, we obtain, according to [3.115] and [3.116]:

$$\begin{pmatrix} 1 & -\mathrm{i} \\ \mathrm{i} & 1 \end{pmatrix} \underbrace{\sqrt{2}\left[\frac{\hat{a}+\hat{b}}{2}\vec{\mathbf{e}}^{+} + \frac{\hat{a}-\hat{b}}{2}\vec{\mathbf{e}}^{-}\right] e^{-\mathrm{i}(\omega t - \mathrm{k}x)}}_{\text{left-handed elliptical pol.}}$$

$$= \underbrace{\sqrt{2}\left(\hat{a}+\hat{b}\right)\vec{\mathbf{e}}^{+} e^{-\mathrm{i}(\omega t - \mathrm{k}x)}}_{\text{left-handed circular pol.}} \quad [3.121]$$

$$\begin{pmatrix} 1 & -\mathrm{i} \\ \mathrm{i} & 1 \end{pmatrix} \underbrace{-\mathrm{i}\sqrt{2}\left[\frac{\hat{a}+\hat{b}}{2c}\vec{\mathbf{e}}^{+} - \frac{\hat{a}-\hat{b}}{2c}\vec{\mathbf{e}}^{-}\right] e^{-\mathrm{i}(\omega t - \mathrm{k}x)}}_{\text{left-handed elliptical pol.}}$$

$$= \underbrace{-\mathrm{i}\sqrt{2}\frac{\hat{a}+\hat{b}}{c}\vec{\mathbf{e}}^{+} e^{-\mathrm{i}(\omega t - \mathrm{k}x)}}_{\text{left-handed circular pol.}} \quad [3.122]$$

3.5. Interference of light waves

Let us return to the phenomenon of interference. Here, we will focus on monochromatic interference. Here are the key points to remember:

1) The wave emitted by many luminous objects is often spherical. Unlike a plane wave (which propagates in only one direction), a spherical wave propagates in all

directions in space. Therefore, it is visible to an observer at any point in space. The wave vector $\vec{k}$ has spherical symmetry and is radial. The surfaces perpendicular to the wave vectors are spheres and are called *spherical wavefronts*. A spherical wave is monochromatic if its wave vector has a constant magnitude. As the spherical wave propagates, the radius r of the wavefront increases until it becomes infinite. An infinite wavefront can be approximated as a plane wavefront. Therefore, a spherical wave can be considered as a plane wave at a point that is sufficiently distant from the source.

2) A divergent spherical wave (emitted from a source) is expressed as:

$$\hat{\psi}\left(\vec{\mathbf{r}},t\right)=\frac{\hat{\psi}_0}{r}e^{-\mathrm{i}\omega t}e^{\mathrm{i}(\mathrm{k}r-\varphi_0)} \qquad [3.123]$$

3) A converging spherical wave (toward a focal point) is expressed as:

$$\hat{\psi}\left(\vec{\mathbf{r}},t\right)=\frac{\hat{\psi}_0}{r}e^{-\mathrm{i}\omega t}e^{\mathrm{i}(-\mathrm{k}r-\varphi_0)} \qquad [3.124]$$

4) It is useful to write equation [3.123] in a Cartesian coordinate system. In Figure 3.10, the source is located at the point with coordinates $(x_S, y_S, 0)$, and the wave propagates along the positive direction of Oz. We observe that $r = \left[d^2+(x-x_S)^2+(y-y_S)^2\right]^{1/2}$. The propagated wavefront in the plane $z=d$ (located at a distance d from the source) is expressed as:

$$\hat{\psi}\left(\vec{\mathbf{r}},t\right)=\hat{\psi}_0 e^{-\mathrm{i}(\omega t+\varphi_0)}\frac{e^{\mathrm{i}\frac{2\pi}{\lambda}\sqrt{d^2+(x-x_S)^2+(y-y_S)^2}}}{\sqrt{d^2+(x-x_S)^2+(y-y_S)^2}} \qquad [3.125]$$

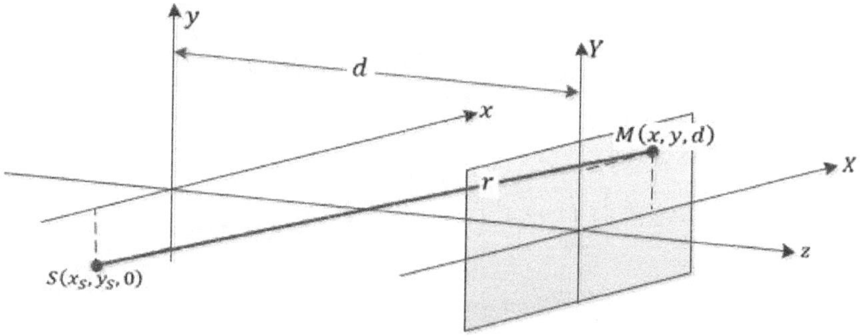

Figure 3.10. *Fresnel approximation in Cartesian coordinates. For a color version of this figure, see www.iste.co.uk/abdesselam/electrodynamics2.zip*

To simplify the previous expression, we introduce approximations for the distance d (between the source and the observed wavefront). By writing $\left[d^2 + (x - x_S)^2 + (y - y_S)^2\right]^{1/2} = d\left[1 + \frac{1}{2}\frac{(x-x_S)^2}{d^2} + \frac{1}{2}\frac{(y-y_S)^2}{d^2}\right] + \cdots$, we obtain:

$$\hat{\psi}\left(\vec{\mathbf{r}}, t\right) \simeq \frac{\hat{\psi}_0}{d} e^{-\mathrm{i}\left(\omega t - \frac{2\pi d}{\lambda} + \varphi_0\right)} e^{\frac{i\pi}{\lambda d}\left[(x-x_S)^2 + (y-y_S)^2\right]} \qquad [3.126]$$

In the denominator, we retain only d. This approximation is known as the *Fresnel approximation*. The previous expression is equivalent to replacing spherical wavefront wavelets with parabolic wavefronts. This approximation is valid on the condition that $d \gg \left(\frac{\pi d}{\lambda}\left[(x - x_S)^2 + (y - y_S)^2\right]\right)^{1/3}$.

5) The *light intensity* is defined as the power per unit area carried by the wave. Consider an electromagnetic plane wave with electric and magnetic fields $\left(\vec{\mathbf{E}}, \vec{\mathbf{B}}\right)$. Around a point M in space, defined by its position $\vec{\mathbf{r}}$, we define an oriented surface element $d\vec{\mathbf{S}} = dS\vec{\mathbf{n}}$. We will see later that the power that crosses the surface element $d\vec{\mathbf{S}}$ is the flux of the Poynting vector $\vec{\Pi}$: $\vec{\Pi} \cdot d\vec{\mathbf{S}}\,\frac{\vec{\mathbf{E}}\times\vec{\mathbf{B}}}{\mu_0} \cdot d\vec{\mathbf{S}}$. For a plane electromagnetic wave, we will see that: $\vec{\Pi} \cdot d\vec{\mathbf{S}} = \frac{|E_0||B_0|}{\mu_0}|\cos\theta|\cos^2\left(\omega t - \vec{\mathrm{k}} \cdot \vec{\mathbf{r}}\right) dS$, where θ is the angle between $\vec{\Pi}$ and $d\vec{\mathbf{S}}$. Thus, the instantaneous intensity of the plane wave, defined as positive, is: $\mathcal{I}\left(\vec{\mathbf{r}}, t\right) = \frac{|E_0|^2}{\mu_0 c}|\cos\theta|\cos^2\left(\omega t - \vec{\mathrm{k}} \cdot \vec{\mathbf{r}}\right)$. The observer, whether eye or detector, does not actually see this function $\mathcal{I}\left(\vec{\mathbf{r}}, t\right)$, but rather its statistical average over the response time τ, that is, the quantity:

$$I = \frac{1}{\tau}\int_0^\tau \frac{|E_0|^2}{\mu_0 c}|\cos\theta|\cos^2\left(\omega t - \vec{\mathrm{k}} \cdot \vec{\mathbf{r}}\right) dt \qquad [3.127]$$

For the eye, the integral covers more than 10^{12} periods of $\cos^2\left(\omega t - \vec{\mathrm{k}} \cdot \vec{\mathbf{r}}\right)$, and for high-speed cameras, the integration is calculated over approximately 10^{10} periods. The integral then becomes:

$$\begin{aligned} I &= \frac{|E_0|^2}{\mu_0 c}|\cos\theta|\frac{1}{\tau}\int_0^\tau \cos^2\left(\omega t - \vec{\mathrm{k}} \cdot \vec{\mathbf{r}}\right) dt \\ &= \frac{|E_0|^2}{\mu_0 c}|\cos\theta|\left\langle \cos^2\left(\omega t - \vec{\mathrm{k}} \cdot \vec{\mathbf{r}}\right)\right\rangle = \frac{|E_0|^2}{2\mu_0 c}|\cos\theta| \end{aligned} \qquad [3.128]$$

We obtain the result: for an electromagnetic wave, the intensity is proportional to the square of the modulus of the electric field of the wave $|E_0|^2 = \left|\vec{\mathbf{E}}\left(\vec{\mathbf{r}}, t\right)\right|^2$. This calculation easily generalizes to any monochromatic wave with complex amplitude $\hat{\psi}\left(\vec{\mathbf{r}},t\right)$, written as $\hat{\psi}\left(\vec{\mathbf{r}},t\right) = A\left(\vec{\mathbf{r}}\right)e^{\mathrm{i}\varphi\left(\vec{\mathbf{r}}\right)}e^{-\mathrm{i}\omega t}$. In fact, the wavefronts are locally assimilable to planes on the surface $d\vec{\mathbf{S}}$ defined around a point M. A first-order expansion allows us to write $\varphi\left(\vec{\mathbf{r}}\right) \simeq \varphi_0 + \vec{\nabla}\varphi \cdot \vec{\mathbf{r}}$, where ϕ_0 is a constant and $A\left(\vec{\mathbf{r}}\right) \simeq A_0$ (A_0 being a constant). Thus, the complex amplitude of the wave can be written as $\hat{\psi}\left(\vec{\mathbf{r}},t\right) = A_0 e^{-\mathrm{i}\left(\omega t - \vec{\mathsf{k}}\cdot\vec{\mathbf{r}} - \varphi_0\right)}$ (a plane wave), with $\vec{\mathsf{k}} = \vec{\nabla}\varphi$ (the local wave vector). The previous reasoning thus also applies to any wave that is locally plane. The intensity is therefore also the time-averaged square of the modulus of the complex amplitude: $I\left(\vec{\mathbf{r}}\right) = \left\langle \psi\left(\vec{\mathbf{r}},t\right)^2 \right\rangle$.

6) We consider two electromagnetic waves represented by their electric fields $\vec{\mathbf{E}}_1$ (wave vector $\vec{\mathsf{k}}_1$, frequency ω_1 and φ_1 the phase at the origin) and $\vec{\mathbf{E}}_2$ (wave vector $\vec{\mathsf{k}}_2$, frequency ω_2 and φ_2 the phase at the origin). If $\vec{\mathbf{E}}_1$ and $\vec{\mathbf{E}}_2$ are polarized in the directions $\vec{\mathbf{e}}_1$ and $\vec{\mathbf{e}}_2$, respectively, then $\vec{\mathbf{E}}_1\left(\vec{\mathbf{r}},t\right) = A_1\left(\vec{\mathbf{r}}\right) e^{-\mathrm{i}\left(\omega_1 t - \vec{\mathsf{k}}_1\cdot\vec{\mathbf{r}} + \varphi_1\right)} \times \vec{\mathbf{e}}_1$ and $\vec{\mathbf{E}}_2\left(\vec{\mathbf{r}},t\right) = A_2\left(\vec{\mathbf{r}}\right) e^{-\mathrm{i}\left(\omega_2 t - \vec{\mathsf{k}}_2\cdot\vec{\mathbf{r}} + \varphi_2\right)}\vec{\mathbf{e}}_2$. Suppose that both radiations coexist in a region of space. At any point in this region, the resulting electric field is simply the sum of the two fields: $\vec{\mathbf{E}}\left(\vec{\mathbf{r}},t\right) = \vec{\mathbf{E}}_1\left(\vec{\mathbf{r}},t\right) + \vec{\mathbf{E}}_2\left(\vec{\mathbf{r}},t\right)$. Thus, we obtain $\left|\vec{\mathbf{E}}\left(\vec{\mathbf{r}},t\right)\right|^2 = \vec{\mathbf{E}}\left(\vec{\mathbf{r}},t\right)\cdot \vec{\mathbf{E}}^\star\left(\vec{\mathbf{r}},t\right) = A_1^2\left(\vec{\mathbf{r}}\right) + A_2^2\left(\vec{\mathbf{r}}\right) + 2A_1\left(\vec{\mathbf{r}}\right)A_2\left(\vec{\mathbf{r}}\right)\cos\left[(\omega_2-\omega_1)t + \left(\vec{\mathsf{k}}_1 - \vec{\mathsf{k}}_2\right)\cdot\vec{\mathbf{r}} + \varphi_2 - \varphi_1\right]\vec{\mathbf{e}}_1\cdot\vec{\mathbf{e}}_2$ and consequently:

$$I\left(\vec{\mathbf{r}},t\right) = A_1^2\left(\vec{\mathbf{r}}\right) + A_2^2\left(\vec{\mathbf{r}}\right) + 2A_1\left(\vec{\mathbf{r}}\right)A_2\left(\vec{\mathbf{r}}\right)\langle\cos\left[(\omega_2-\omega_1)t + \left(\vec{\mathsf{k}}_1 - \vec{\mathsf{k}}_2\right)\cdot\vec{\mathbf{r}} + \varphi_2 - \varphi_1\right]\rangle\,\vec{\mathbf{e}}_1\cdot\vec{\mathbf{e}}_2 \quad [3.129]$$

The last term represents the *interference term* of the two fields; it is also called the *interference modulation term*. This term vanishes if:

- the two waves are polarized perpendicularly, that is, $\vec{\mathbf{e}}_1\cdot\vec{\mathbf{e}}_2 = 0$. This condition is known as *polarization coherence*;

- the two frequencies are different, that is, $\omega_1 \neq \omega_2$. In fact, if $\omega_1 \neq \omega_2$, the interference term oscillates very rapidly and is therefore not observable by the eye or a detector. This condition is known as *temporal coherence*;

- the two wave vectors $\vec{\mathsf{k}}_1$ and $\vec{\mathsf{k}}_2$ vary significantly in the observation region, that is, $\vec{\mathsf{k}}_1\left(\vec{\mathbf{r}}\right) \neq$ constant and $\vec{\mathsf{k}}_2\left(\vec{\mathbf{r}}\right) \neq$ constant. In this case, the interference term oscillates very rapidly in space, resulting in a zero average, making it unobservable. This condition is known as *spatial coherence*.

7) In the following, we assume that the two waves are temporally, spatially, and polarization-coherent. The intensity of the observable signal is given by $I\left(\overrightarrow{\mathbf{r}}\right) = A_1^2\left(\overrightarrow{\mathbf{r}}\right)+A_2^2\left(\overrightarrow{\mathbf{r}}\right)+2A_1\left(\overrightarrow{\mathbf{r}}\right)A_2\left(\overrightarrow{\mathbf{r}}\right)\cos\left[\left(\overrightarrow{\mathrm{k}}_1-\overrightarrow{\mathrm{k}}_2\right)\cdot\overrightarrow{\mathbf{r}}+\varphi_2-\varphi_1\right]$. By setting $I_1 = A_1^2$ and $I_2 = A_2^2$, we obtain:

$$I\left(\overrightarrow{\mathbf{r}}\right) = I_1 + I_2 + 2\sqrt{I_1 I_2}\cos\left(\frac{2\pi\delta}{\lambda}\right) \qquad [3.130]$$

where:

$$\frac{2\pi\delta}{\lambda} = \left(\overrightarrow{\mathrm{k}}_1-\overrightarrow{\mathrm{k}}_2\right)\cdot\overrightarrow{\mathbf{r}}+\varphi_2-\varphi_1 \qquad [3.131]$$

δ represents the optical path difference between the two waves. The interference phenomenon is manifested by alternating dark regions, where $\cos\left(\frac{2\pi\delta}{\lambda}\right) = -1$, and bright regions, where $\cos\left(\frac{2\pi\delta}{\lambda}\right) = +1$. Let $I_{\min}$ be the minimum intensity and $I_{\max}$ the maximum intensity, given by: $I_{\min} = I_1 + I_2 - 2\sqrt{I_1 I_2}$ and $I_{\max} = I_1 + I_2 + 2\sqrt{I_1 I_2}$. The *contrast* between these two extreme values is characterized by the *visibility factor*:

$$C = \frac{I_{\max} - I_{\min}}{I_{\max} + I_{\min}} = \frac{2\sqrt{I_1 I_2}}{I_1 + I_2} = \frac{2A_1 A_2}{A_1^2 + A_2^2} \qquad [3.132]$$

The maximum contrast is equal to 1 (achieved when $A_1 = A_2$, that is, if both waves have the same amplitude at every point in space). If $A_1 \neq A_2$, the contrast is less than 1. In particular, $A_1 \ll A_2$ or $A_2 \ll A_1$, then $C \simeq 0$. The maxima occur when $\cos\left(\frac{2\pi\delta}{\lambda}\right) = +1$, that is, $\varphi = \left(\overrightarrow{\mathrm{k}}_1-\overrightarrow{\mathrm{k}}_2\right)\cdot\overrightarrow{\mathbf{r}}+\varphi_2-\varphi_1 = 2k\pi,\ k \in \mathbb{R}$. The minima occur when $\varphi = \left(\overrightarrow{\mathrm{k}}_1-\overrightarrow{\mathrm{k}}_2\right)\cdot\overrightarrow{\mathbf{r}}+\varphi_2-\varphi_1 = (2k+1)\pi,\ k \in \mathbb{R}$. The quantity φ is called the *phase* of the interference: it represents the difference in optical phases between the two waves and is proportional to the difference in their optical paths. The distance in the observation plane that separates two consecutive regions of the same nature is called the *fringe spacing* (interfringe).

8) Case of plane waves: consider two plane waves propagating in the positive z-direction, with wave vectors that are not collinear with the Oz-axis but lie in the Oxz-plane. Figure 9.11 illustrates the geometry of the problem. In this case, we have: $\overrightarrow{\mathbf{E}}_1\left(\overrightarrow{\mathbf{r}},t\right) = A_1 e^{-\mathrm{i}\omega t} e^{\mathrm{i}\left(\overrightarrow{\mathrm{k}}_1\cdot\overrightarrow{\mathbf{r}}-\varphi_1\right)}\overrightarrow{\mathbf{j}}$ and $\overrightarrow{\mathbf{E}}_2\left(\overrightarrow{\mathbf{r}},t\right) = A_2 e^{-\mathrm{i}\omega t} e^{\mathrm{i}\left(\overrightarrow{\mathrm{k}}_2\cdot\overrightarrow{\mathbf{r}}-\varphi_2\right)}\overrightarrow{\mathbf{j}}$, with $\overrightarrow{\mathrm{k}}_1 = \frac{2\pi}{\lambda}\cos\theta_1\overrightarrow{\mathbf{k}} - \frac{2\pi}{\lambda}\sin\theta_1\overrightarrow{\mathbf{i}}$ and

$\overrightarrow{k}_2 = \frac{2\pi}{\lambda}\cos\theta_2\,\overrightarrow{\mathbf{k}} + \frac{2\pi}{\lambda}\sin\theta_2\,\overrightarrow{\mathbf{i}}$. Substituting these results into [3.130], we obtain:

$$I(x,y,z) = I_1 + I_2 + 2\sqrt{I_1 I_2}\cos\left[\frac{2\pi}{\lambda}(\cos\theta_1 - \cos\theta_2)\,z - \frac{2\pi}{\lambda}(\sin\theta_1 + \sin\theta_2)\,x + \varphi_2 - \varphi_1\right] \quad [3.133]$$

The distance separating two consecutive zones of the same nature is the interfringe. Consider the bright fringe passing through the coordinates (x_k, z_k), defined by $(\cos\theta_1 - \cos\theta_2)\,z_k - \frac{2\pi}{\lambda}(\sin\theta_1 + \sin\theta_2)\,x_k + \varphi_2 - \varphi_1 = 2k\pi$. In the Ox-direction, the next consecutive fringe is defined by the equation $(\cos\theta_1 - \cos\theta_2)\,z_k - \frac{2\pi}{\lambda}(\sin\theta_1 + \sin\theta_2)\,x_{k+1} + \varphi_2 - \varphi_1 = 2k\pi + 2\pi$. Therefore, the distance separating two consecutive bright fringes (in the Ox-direction) is given by:

$$\Delta x = \frac{\lambda}{\sin\theta_1 + \sin\theta_2} \quad [3.134]$$

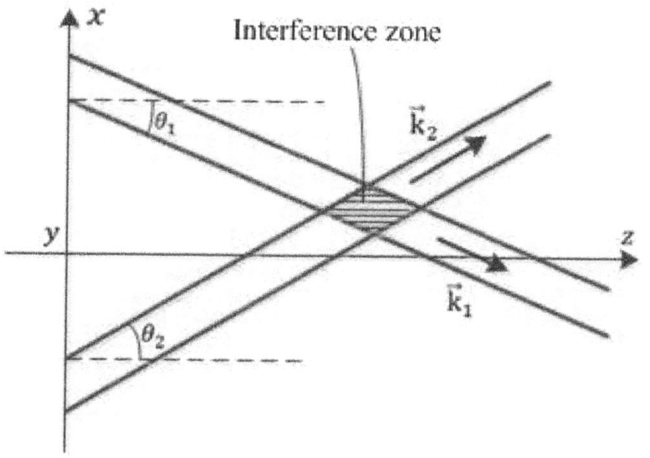

Figure 3.11. *Interference of two plane waves*

Similarly, in the Oz-direction, two consecutive bright fringes are separated by the distance:

$$\Delta z = \frac{\lambda}{\cos\theta_1 - \cos\theta_2} \quad [3.135]$$

In particular, when $\theta_1 = \theta_2 = \theta$, we get:

$$\Delta x = \frac{\lambda}{2\sin\theta}$$
$$\Delta z \longrightarrow \infty \qquad [3.136]$$

The fringes are, therefore, parallel to the Oz axis.

9) Case of spherical waves: consider two parallel-polarized spherical waves propagating toward $+Oz$ in interference, as illustrated in Figure 3.12. In this case, we have: $\vec{\mathbf{E}}_1(\vec{\mathbf{r}}, t) = \frac{A_1}{r_1} e^{-i\omega t} e^{i(k_1 r_1 - \varphi_1)} \vec{\mathbf{j}}$ and $\vec{\mathbf{E}}_2(\vec{\mathbf{r}}, t) = \frac{A_2}{r_2} e^{-i\omega t} e^{i(k_2 r_2 - \varphi_2)} \vec{\mathbf{j}}$. Let (x_1, y_1, z_1) and (x_2, y_2, z_2) be the coordinates of the source points of the two spherical waves. The distances r_1 and r_2 separating the sources from the observation point are written as $r_1 = \sqrt{(x-x_1)^2 + (y-y_1)^2 + (z-z_1)^2}$ and $r_2 = \sqrt{(x-x_2)^2 + (y-y_2)^2 + (z-z_2)^2}$. Thus, the interference phase is given by $\varphi = \frac{2\pi}{\lambda}(r_1 - r_2) + \varphi_2 - \varphi_1$. Taking these results into account, the interference signal is written as:

$$I(x,y,z) = \frac{A_1^2}{r_1^2} + \frac{A_2^2}{r_2^2} + 2\frac{A_1}{r_1}\frac{A_2}{r_2}\cos\left[\frac{2\pi}{\lambda}\left(\sqrt{(x-x_1)^2 + (y-y_1)^2 + (z-z_1)^2}\right.\right.$$
$$\left.\left. - \sqrt{(x-x_2)^2 + (y-y_2)^2 + (z-z_2)^2}\right) + \varphi_2 - \varphi_1\right] \qquad [3.137]$$

A bright fringe is therefore described by the equation $\frac{2\pi}{\lambda}(r_1 - r_2) + \varphi_2 - \varphi_1 = 2\pi k$ or alternatively:

$$\sqrt{(x-x_1)^2 + (y-y_1)^2 + (z-z_1)^2} -$$
$$\sqrt{(x-x_2)^2 + (y-y_2)^2 + (z-z_2)^2} = \alpha_k \lambda \qquad [3.138]$$

with $\alpha_k = k + \frac{\varphi_1 - \varphi_2}{2\pi}$. The fringes are curved, which was not the case for the interferences obtained with plane waves. In the plane $z = d$, the fringes are arcs of hyperbolas. A simplified form of equation [3.138] can be obtained within the framework of the Fresnel approximation. Let $d_1 = z - z_1$ and $d_2 = z - z_2$. Equation [3.138] can then be approximately rewritten as:

$$d_1 - d_2 + \frac{(x-x_1)^2 + (y-y_1)^2}{2d_1} - \frac{(x-x_2)^2 + (y-y_2)^2}{2d_2} = \alpha_k \lambda \qquad [3.139]$$

which leads to:

$$\frac{x^2}{2}\left(\frac{1}{d_1}-\frac{1}{d_2}\right)+\frac{y^2}{2}\left(\frac{1}{d_1}-\frac{1}{d_2}\right)-x\left(\frac{x_1}{d_1}-\frac{x_2}{d_2}\right)-y\left(\frac{y_1}{d_1}-\frac{y_2}{d_2}\right)=\alpha_k'\lambda \quad [3.140]$$

with:

$$\alpha_k' = k + \frac{\varphi_1-\varphi_2}{2\pi} - \frac{d_1-d_2}{\lambda} - \frac{x_1^2+y_1^2}{2\lambda d_1} + \frac{x_2^2+y_2^2}{2\lambda d_2} \quad [3.141]$$

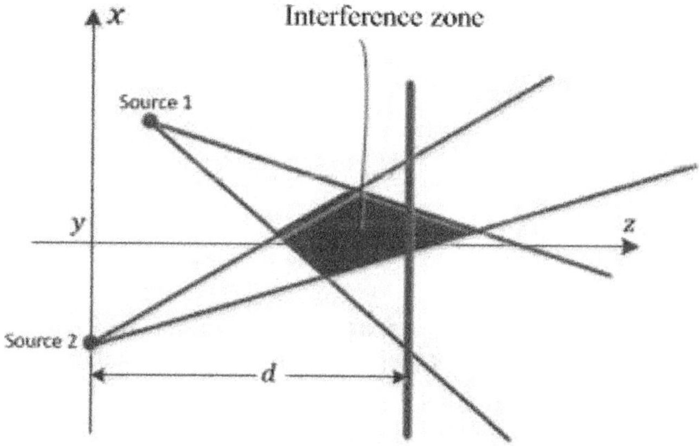

Figure 3.12. *Interference of two spherical waves. For a color version of this figure, see www.iste.co.uk/abdesselam/electrodynamics2.zip*

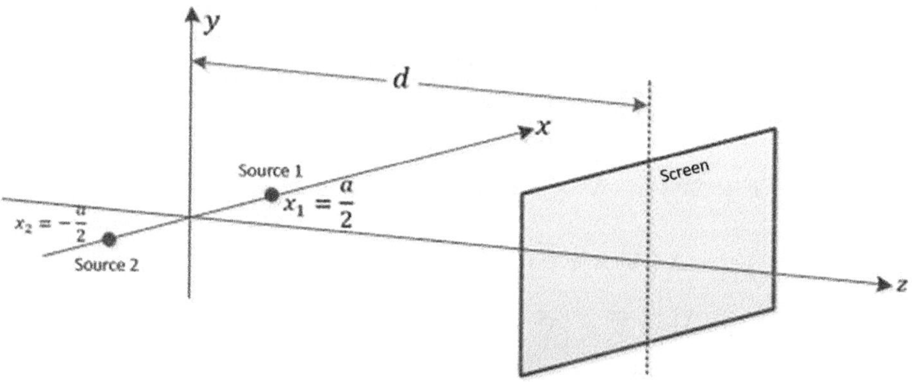

Figure 3.13. *Interference of two spherical waves in the Fresnel approximation*

In particular, suppose that the sources are symmetrically located on the Ox axis on either side of the origin at $x_1 = \frac{a}{2}$ and $x_2 = -\frac{a}{2}$ with $y_1 = y_2 = 0$ and $d_1 = d_2 = d$. We will assume $d \gg a$ to satisfy the Fresnel approximation, as illustrated in Figure 3.13. In this case, expression [3.140] reduces to:

$$\frac{ax}{d} = \alpha_k \lambda \qquad [3.142]$$

In this approximation, the fringes are straight and parallel to the Oy axis. The bright fringe of order k is located at $x_k = \frac{d\alpha_k \lambda}{a}$, and the next bright fringe, of order $k+1$, is located at $x_{k+1} = \frac{d(\alpha_{k+1})\lambda}{a}$. The fringe separation is therefore:

$$\Delta x = x_{k+1} - x_k = \frac{\lambda d}{a} \qquad [3.143]$$

3.6. Transverse electric (TE) and transverse magnetic (TM) spherical waves

In this section, we will prove an important theorem that will allow us to deduce the vectorial solutions of the d'Alembert equation from the scalar solution. In other words, if ψ is a scalar solution of the d'Alembert equation, the vectorial solution can be generated from the field ψ. In fact, the vector fields $\vec{\nabla}\psi$ and $\vec{\mathbf{r}} \times \vec{\nabla}\psi$ both satisfy the vectorial version of the d'Alembert equation. To see this, let us use the identity $\Delta\vec{\mathbf{A}} = \vec{\nabla}\left(\vec{\nabla}\cdot\vec{\mathbf{A}}\right) - \vec{\nabla}\times\left(\vec{\nabla}\times\vec{\mathbf{A}}\right)$. By choosing $\vec{\mathbf{A}} = \vec{\nabla}\psi$, we get:

$$\begin{aligned}\Delta\left(\vec{\nabla}\psi\right) &= \vec{\nabla}\left(\vec{\nabla}\cdot\vec{\nabla}\psi\right) - \vec{\nabla}\times\left(\vec{\nabla}\times\vec{\nabla}\psi\right) = \vec{\nabla}\left(\Delta\psi\right) \\ &= \vec{\nabla}\left(\frac{1}{c^2}\frac{\partial^2\psi}{\partial t^2}\right) = \frac{1}{c^2}\frac{\partial^2\left(\vec{\nabla}\psi\right)}{\partial t^2}\end{aligned} \qquad [3.144]$$

Similarly, $\vec{\mathbf{r}} \times \vec{\nabla}\psi = -\vec{\nabla} \times \vec{\mathbf{r}}\psi$ is also a solution. Indeed, we can write:

$$\begin{aligned}&\Delta\left(\vec{\mathbf{r}}\times\vec{\nabla}\psi\right) = -\Delta\left(\vec{\nabla}\times\vec{\mathbf{r}}\psi\right) = \vec{\nabla}\times\left[\vec{\nabla}\times\left(\vec{\nabla}\times\vec{\mathbf{r}}\psi\right)\right] \\ &\vec{\nabla}\left[\vec{\nabla}\cdot\left(\vec{\nabla}\times\vec{\mathbf{r}}\psi\right)\right] = \vec{\nabla}\times\left[\vec{\nabla}\times\left(\vec{\nabla}\times\vec{\mathbf{r}}\psi\right)\right] = \vec{\nabla}\times\left[\vec{\nabla}\times\left(\vec{\nabla}\psi\times\vec{\mathbf{r}}\right)\right] \\ &= \vec{\nabla}\times\left[\left(\vec{\mathbf{r}}\cdot\vec{\nabla}\right)\left(\vec{\nabla}\psi\right) - \vec{\mathbf{r}}\Delta\psi + 2\vec{\nabla}\psi\right] = -\vec{\nabla}\times\left[\vec{\mathbf{r}}\left(\Delta\psi\right)\right] \\ &= -\vec{\nabla}\times\left[\vec{\mathbf{r}}\left(\frac{1}{c^2}\frac{\partial^2\psi}{\partial t^2}\right)\right] = -\frac{1}{c^2}\frac{\partial^2}{\partial t^2}\left(\vec{\nabla}\times\vec{\mathbf{r}}\psi\right) = \frac{1}{c^2}\frac{\partial^2}{\partial t^2}\left(\vec{\mathbf{r}}\times\vec{\nabla}\psi\right)\end{aligned} \qquad [3.145]$$

Therefore, the vector fields $\vec{\nabla}\psi$ and $\vec{\mathbf{r}} \times \vec{\nabla}\psi$ are valid solutions. However, only the second solution, $\vec{\nabla} \times \vec{\mathbf{r}}\psi$, is of physical interest because it has zero divergence – an essential requirement for $\vec{\mathbf{B}}$ and also for $\vec{\mathbf{E}}$ in a source-free region of space. If we set $\vec{\mathbf{E}} \sim \vec{\nabla} \times \vec{\mathbf{r}}\psi$, we obtain $\vec{\mathbf{B}} \sim -\frac{1}{\mathrm{i}\omega}\vec{\nabla} \times \left(\vec{\nabla} \times \vec{\mathbf{r}}\psi\right)$. This solution is known as a *TE electromagnetic wave or type M*. Inversely, if $\vec{\mathbf{B}} \sim \vec{\nabla} \times \vec{\mathbf{r}}\psi$, we find $\vec{\mathbf{E}} \sim \frac{c^2}{\mathrm{i}\omega}\vec{\nabla} \times \left(\vec{\nabla} \times \vec{\mathbf{r}}\psi\right)$. This solution is known as a *TM electromagnetic wave or type E*. The function ψ, from which the fields can be derived, is known as the *Debye potential*. A superposition of the two types of waves is generally possible. By introducing a scaling factor, we obtain:

$$\begin{cases} \vec{\mathbf{E}}_{\mathrm{TE}} = \mathrm{i}\mathrm{k}\vec{\nabla} \times \vec{\mathbf{r}}\psi \\ \vec{\mathbf{B}}_{\mathrm{TE}} = \dfrac{1}{c}\vec{\nabla} \times \left(\vec{\nabla} \times \vec{\mathbf{r}}\psi\right) \end{cases} \qquad [3.146]$$

$$\begin{cases} \vec{\mathbf{E}}_{\mathrm{TM}} = \vec{\nabla} \times \left(\vec{\nabla} \times \vec{\mathbf{r}}\psi\right) \\ \vec{\mathbf{B}}_{\mathrm{TM}} = -\dfrac{\mathrm{i}\mathrm{k}}{c}\vec{\nabla} \times \vec{\mathbf{r}}\psi \end{cases} \qquad [3.147]$$

The scale factors were introduced to give the corresponding TE and TM fields the same dimensions and, consequently, the same energy flux.

In the following, we will briefly examine the solutions of the scalar d'Alembert equation [3.9] in the spherical coordinate system and then derive the vector solutions of [3.5] and [3.6] based on the previously mentioned theorem. Here, we explicitly present the vector solutions. In general, solving the scalar d'Alembert equation [3.9] is done using the method of *separation of variables*. This approach first separates the spatial and temporal components in the expression $\psi\left(\vec{\mathbf{r}}, t\right)$, that is, by setting $\psi\left(\vec{\mathbf{r}}, t\right) = \Psi\left(\vec{\mathbf{r}}\right)T\left(t\right)$. Substituting this expression into [3.9] gives $\frac{\Delta\Psi}{\Psi} = \frac{1}{c^2 T}\frac{d^2T}{dt^2}$, where the right-hand side depends only on time t, while the left-hand side depends only on $\vec{\mathbf{r}}$. This equality can hold only if $\frac{1}{c^2 T}\frac{d^2T}{dt^2} = \frac{\Delta\Psi}{\Psi} = -\mathrm{k}^2$, where k is a constant. This leads to:

$$\frac{d^2T}{dt^2} + \omega^2 T = 0 \qquad [3.148]$$

$$\left(\Delta + \mathrm{k}^2\right)\Psi = 0 \qquad [3.149]$$

with $\omega = kc$. The first equation admits the solution $Ae^{-\mathrm{i}\omega t}$ or $Ae^{\mathrm{i}\omega t}$. We will adopt the following temporal convention: $T\left(t\right) \sim e^{-\mathrm{i}\omega t}$, that is, $\psi\left(\vec{\mathbf{r}}, t\right) = \Psi\left(\vec{\mathbf{r}}\right) e^{-\mathrm{i}\omega t}$. The second equation [3.149] is called the *Helmholtz equation*. In spherical coordinates, it is written as:

$$\frac{1}{r^2}\frac{\partial}{\partial r}\left(r^2\frac{\partial\Psi}{\partial r}\right) + \frac{1}{r^2\sin\theta}\frac{\partial}{\partial\theta}\left(\sin\theta\frac{\partial\Psi}{\partial\theta}\right) + \frac{1}{r^2\sin^2\theta}\frac{\partial^2\Psi}{\partial\varphi^2} + \mathrm{k}^2\Psi = 0 \qquad [3.150]$$

Let us look for a solution of the form $\Psi(r,\theta,\varphi) = R(r)\,Y(\theta,\varphi)$. Consequently, we obtain $\frac{1}{R}\frac{d}{dr}\left(r^2\frac{dR}{dr}\right) + \mathsf{k}^2 r^2 = -\left[\frac{1}{\sin\theta Y}\frac{d}{d\theta}\left(\sin\theta\frac{dY}{d\theta}\right) + \frac{1}{\sin^2\theta Y}\frac{d^2Y}{d\varphi^2}\right] = K$. This last equation can be split into two differential equations:

$$\frac{1}{\sin\theta Y}\frac{d}{d\theta}\left(\sin\theta\frac{dY}{d\theta}\right) + \frac{1}{\sin^2\theta Y}\frac{d^2Y}{d\varphi^2} = -K \qquad [3.151]$$

$$\frac{1}{r^2}\frac{d}{dr}\left(r^2\frac{dR}{dr}\right) + \left(\mathsf{k}^2 - \frac{K}{r^2}\right)R = 0 \qquad [3.152]$$

We will see later that when $K = l(l+1)$, where l is a positive integer or zero, equation [3.151] admits the solution $Y_l^m(\theta,\varphi)$, the *spherical harmonic function.*

By performing the change of variable $x = \mathsf{k}r$ in [3.152] and taking $K = l(l+1)$, we deduce that:

$$\frac{d^2R}{dx^2} + \frac{2}{x}\frac{dR}{dx} + \left(1 - \frac{l(l+1)}{x^2}\right)R = 0 \qquad [3.153]$$

This is the *spherical Bessel equation,* whose solution is $A_l j_l(\mathsf{k}r) + B_l n_l(\mathsf{k}r)$, where $j_l(\mathsf{k}r)$ and $n_l(\mathsf{k}r)$ are the *spherical Bessel and Neumann functions,* respectively.

By virtue of additivity, the general solution to the d'Alembert equation [3.3] is:

$$\psi(r,\theta,\varphi,t) = \sum_{l=0}^{\infty}\sum_{m=-l}^{l}\int [A_l j_l(\mathsf{k}r) + B_l n_l(\mathsf{k}r)]\,Y_l^m(\theta,\varphi)\,e^{-i\omega t}d\mathsf{k} \qquad [3.154]$$

Setting $f_l(\mathsf{k}r) = A_l j_l(\mathsf{k}r) + B_l n_l(\mathsf{k}r)$ and then using [3.146] and [3.147], we obtain:

1) $(\mathrm{TE})_{lm}$ *wave*:

$$\begin{cases} E_r = 0 \\ E_\theta = -\dfrac{\mathsf{k}m}{\sin\theta} f_l(\mathsf{k}r)\,Y_l^m(\theta,\varphi) \\ E_\varphi = -i\mathsf{k} f_l(\mathsf{k}r)\dfrac{\partial Y_l^m(\theta,\varphi)}{\partial\theta} \end{cases} \quad \begin{cases} B_r = \dfrac{1}{c}\dfrac{l(l+1)}{r} f_l(\mathsf{k}r)\,Y_l^m(\theta,\varphi) \\ B_\theta = \dfrac{1}{cr}\dfrac{d(rf_l(\mathsf{k}r))}{dr}\dfrac{\partial Y_l^m(\theta,\varphi)}{\partial\theta} \\ B_\varphi = \dfrac{im}{cr\sin\theta}\dfrac{d(rf_l(\mathsf{k}r))}{dr}Y_l^m(\theta,\varphi) \end{cases} \qquad [3.155]$$

The electric field is indeed transverse to the direction of propagation. By expressing $f_l(\mathsf{k}r)$ in terms of B_r and then substituting the result into the other components, we find:

$$\begin{cases} E_r = 0 \\ E_\theta = -\dfrac{\mathrm{k}mc}{l\,(l+1)}\dfrac{rB_r}{\sin\theta} \\ E_\varphi = -\dfrac{\mathrm{ik}crB_r}{l\,(l+1)}\dfrac{1}{Y_l^m\,(\theta,\varphi)}\dfrac{\partial Y_l^m\,(\theta,\varphi)}{\partial\theta} \\ B_\theta = \dfrac{1}{l\,(l+1)\,r}\dfrac{d\left(r^2B_r\right)}{dr}\dfrac{1}{Y_l^m\,(\theta,\varphi)}\dfrac{\partial Y_l^m\,(\theta,\varphi)}{\partial\theta} \\ B_\varphi = \dfrac{im}{l\,(l+1)\,r\sin\theta}\dfrac{d\left(r^2B_r\right)}{dr} \end{cases} \qquad [3.156]$$

Knowing B_r, we see that all the other components can be determined. These results will be used in Volume 3.

2) $(\mathrm{TM})_{lm}$ *wave*:

$$\begin{cases} E_r = \dfrac{l\,(l+1)}{r}f_l\,(\mathrm{k}r)\,Y_l^m\,(\theta,\varphi) \\ E_\theta = \dfrac{1}{r}\dfrac{d\,(rf_l\,(\mathrm{k}r))}{dr}\dfrac{\partial Y_l^m\,(\theta,\varphi)}{\partial\theta} \\ E_\varphi = \dfrac{im}{r\sin\theta}\dfrac{d\,(rf_l\,(\mathrm{k}r))}{dr}Y_l^m\,(\theta,\varphi) \end{cases} \quad \begin{cases} B_r = 0 \\ B_\theta = \dfrac{\mathrm{k}m}{c\sin\theta}f_l\,(\mathrm{k}r)\,Y_l^m\,(\theta,\varphi) \\ B_\varphi = \dfrac{\mathrm{ik}}{c}f_l\,(\mathrm{k}r)\,\dfrac{\partial Y_l^m\,(\theta,\varphi)}{\partial\theta} \end{cases} \qquad [3.157]$$

In terms of E_r, we find:

$$\begin{cases} E_\theta = \dfrac{1}{l\,(l+1)\,r}\dfrac{d\left(r^2E_r\right)}{dr}\dfrac{1}{Y_l^m\,(\theta,\varphi)}\dfrac{\partial\ln Y_l^m\,(\theta,\varphi)}{\partial\theta} \\ E_\varphi = \dfrac{im}{l\,(l+1)\,r\sin\theta}\dfrac{d\left(r^2E_r\right)}{dr} \\ B_r = 0 \\ B_\theta = \dfrac{\mathrm{k}mrE_r}{l\,(l+1)\,c\sin\theta} \\ B_\varphi = \dfrac{\mathrm{ik}rE_r}{l\,(l+1)\,c}\dfrac{1}{Y_l^m\,(\theta,\varphi)}\dfrac{\partial Y_l^m\,(\theta,\varphi)}{\partial\theta} \end{cases} \qquad [3.158]$$

Therefore, it is sufficient to know E_r to calculate all the other components.

The fields described above are multipolar radiation fields of the order l. As we will see in the third chapter of Volume 3, TE waves are emitted by oscillating magnetic dipoles, while TM waves are emitted by oscillating electric dipoles. Note that unlike plane waves, where the fields $\vec{\mathbf{E}}$ and $\vec{\mathbf{B}}$ are both perpendicular to $\vec{\mathrm{k}}$ (the direction of propagation), for spherical waves, only one of the two can be perpendicular to $\vec{\mathrm{k}}$.

Chapter 4

Electromagnetic Energies

4.1. Energy stored in a charged distribution

Up until now, we have considered fixed or moving charge distributions without discussing how these distributions were assembled. In this section, we will calculate the work required to assemble a given distribution. This work is stored as an *electrostatic potential energy* in the electric field, because if the charges are allowed to move, this work can be recovered as kinetic energy. We have already seen that the electrostatic potential energy of a point charge placed in an electrostatic field, denoted here as W_e, is equal to the work required to bring this charge quasi-statically from infinity to its current position. Consider a charge q placed in a field $\overrightarrow{\mathbf{E}}$. To move it from infinity to a point M, an external force must be applied at all times to counteract the Coulomb force, that is, $\overrightarrow{\mathbf{F}}_{\text{ext}} = -\overrightarrow{\mathbf{F}} = -q\overrightarrow{\mathbf{E}}$. If this displacement is carried out slowly enough, the charge will not gain any kinetic energy. The work supplied to the charge will therefore be $W_e(M) = \int_\infty^M \overrightarrow{\mathbf{F}}_{\text{ext}} \cdot d\overrightarrow{\mathbf{r}} = -\int_\infty^M q\overrightarrow{\mathbf{E}} \cdot d\overrightarrow{\mathbf{r}} = q[V(M) - V(\infty)]$. This energy is independent of the path taken. Since the potential is zero at infinity, we deduce that:

$$W_e(M) = qV(M) \qquad [4.1]$$

This is the *expression of the electrostatic energy of a point charge located at M.*

From expression [4.1], we see that the (static) electric potential at a point M is simply the electrostatic energy per unit of charge at that point. Formula [4.1] can be derived by reasoning in the following way: (1) applying the kinetic energy theorem to the charge, we write $W_c(\infty) - W_c(M) = \int_M^\infty q\overrightarrow{\mathbf{E}} \cdot d\overrightarrow{\mathbf{r}}$; (2) the conservation of mechanical energy theorem also shows that $W_c(M) + W_e(M) = W_c(\infty) + W_e(\infty)$, which gives $W_e(M) - W_e(\infty) = W_c(\infty) - W_c(M) = \int_M^\infty q\overrightarrow{\mathbf{E}} \cdot d\overrightarrow{\mathbf{r}}$; (3) since

$W_e(\infty) = 0$ (zero potential at infinity), we obtain $W_e(M) = \int_M^\infty q\vec{\mathbf{E}} \cdot d\vec{\mathbf{r}} = qV(M)$, since $V(\infty) = 0$. Note that we are considering a single charge placed in an external $\vec{\mathbf{E}}$ field and nothing has been said about the field created by the charge itself. When dealing with a system of several point-like charges, each charge creates an electrostatic field that acts on the others, thus introducing an electrostatic interaction energy. We now aim to calculate this energy.

4.1.1. *Electrostatic potential energy of a distribution of point charges*

Consider a system of N point charges $q_1, q_2, \ldots, q_N$, located at the points $\vec{\mathbf{r}}_1 \vec{\mathbf{r}}_2, \ldots \vec{\mathbf{r}}_N$, respectively, in equilibrium under the action of mutual electrostatic forces, as illustrated in Figure 4.1. Let $r_{12} = \left|\vec{\mathbf{r}}_1 - \vec{\mathbf{r}}_2\right|$ be the distance between charges q_1 and q_2, $r_{13} = \left|\vec{\mathbf{r}}_1 - \vec{\mathbf{r}}_3\right|$ the distance between q_1 and q_3, etc. Let us assume that all charges are initially at infinity and then attempt to calculate the electrostatic energy required to assemble them. First, let q_1 move from infinity to its position $\vec{\mathbf{r}}_1$ extremely slowly so that it does not acquire any kinetic energy. Since the external electric field is zero, the required work is also zero. Now, let q_2 move from infinity to its position $\vec{\mathbf{r}}_2$, so that the first particle always remains in equilibrium. The required work is equal to the product of charge q_2 and the electrostatic potential created by charge q_1 at position $\vec{\mathbf{r}}_2$:

$$W_2 = q_2 V_1\left(\vec{\mathbf{r}}_2\right) = q_1 V_2\left(\vec{\mathbf{r}}_1\right) = W_1 \qquad [4.2]$$

We see that this work is identical to that required to bring q_1 from infinity to $\vec{\mathbf{r}}_1$ in the presence of q_2 at position $\vec{\mathbf{r}}_2$. Consequently, a system consisting of two point charges has an electrostatic energy:

$$W_e = \frac{q_1 q_2}{4\pi\varepsilon_0 r_{12}} = W_1 = W_2 = \frac{1}{2}\left(W_1 + W_2\right) \qquad [4.3]$$

If we now bring the charge q_3 from infinity to $\vec{\mathbf{r}}_3$, additional work must be done:

$$W_3 = q_3\left[V_1\left(\vec{\mathbf{r}}_3\right) + V_2\left(\vec{\mathbf{r}}_3\right)\right] = \frac{1}{4\pi\varepsilon_0}\left(\frac{q_1 q_3}{r_{13}} + \frac{q_2 q_3}{r_{23}}\right) \qquad [4.4]$$

Thus, the electrostatic energy associated with a system of three charges is:

$$W_e = \frac{1}{4\pi\varepsilon_0}\left(\frac{q_1 q_2}{r_{12}} + \frac{q_1 q_3}{r_{13}} + \frac{q_2 q_3}{r_{23}}\right) \qquad [4.5]$$

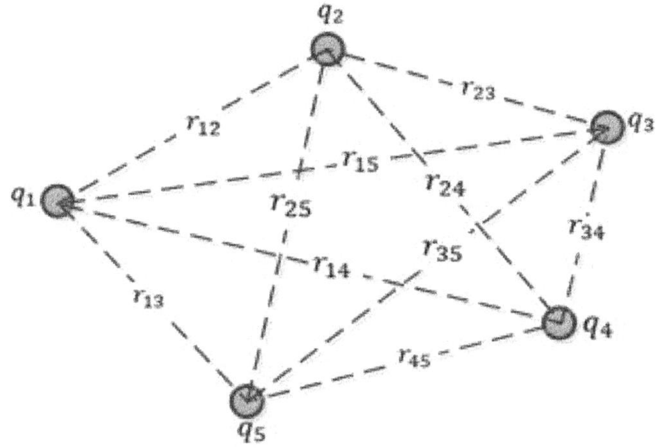

Figure 4.1. *To each pair $q_i q_j$ is associated an electrostatic energy $\frac{q_i q_j}{4\pi r_{ij}}$ (there are a total of 10 terms here)*

We therefore conclude that each pair $q_i q_j$ is associated with an electrostatic interaction potential energy. Continuing with the other charges, the series includes more and more terms. For the Nth charge, the work that must be done is:

$$\begin{aligned} W_N &= q_N\left[V_1\left(\vec{\mathbf{r}}_N\right)+V_2\left(\vec{\mathbf{r}}_N\right)+\cdots+V_{N-1}\left(\vec{\mathbf{r}}_N\right)\right] \\ &= \frac{1}{4\pi\varepsilon_0}\left(\frac{q_1 q_N}{r_{1N}}+\frac{q_2 q_N}{r_{2N}}+\cdots+\frac{q_{N-1}q_N}{r_{N-1,N}}\right) \end{aligned} \qquad [4.6]$$

The total electrostatic energy of the distribution (N charges) is therefore equal to:

$$\begin{aligned} W_e = W_2 + W_3 + \cdots + W_{N-1} &= \frac{1}{4\pi\varepsilon_0}\left[\frac{q_1 q_2}{r_{12}} + \left(\frac{q_1 q_3}{r_{13}}+\frac{q_2 q_3}{r_{23}}\right)\right. \\ &+ \cdots + \left(\frac{q_1 q_{N-1}}{r_{1,N-1}}+\frac{q_2 q_{N-1}}{r_{2,N-1}}+\cdots+\frac{q_{N-2}q_{N-1}}{r_{N-2,N-1}}\right) \\ &\left.+\left(\frac{q_1 q_N}{r_{1N}}+\frac{q_2 q_N}{r_{2N}}+\cdots+\frac{q_{N-1}q_N}{r_{N-1,N}}\right)\right] = \frac{1}{4\pi\varepsilon_0}\sum_{\text{pairs}}\frac{q_i q_j}{r_{ij}} \end{aligned} \qquad [4.7]$$

This formula contains $\frac{1}{2}N(N-1)$ terms. Equation [4.7] is quite convenient for systems composed of a few point charges. There is an equivalent form of equation [4.7] that proves to be more useful when generalizing electrostatic energy to a continuous distribution. Indeed, we have:

$$\begin{aligned}
W_2 &= \frac{1}{8\pi\varepsilon_0}\left(\frac{q_1q_2+q_2q_1}{r_{12}}\right)\\
W_3 &= \frac{1}{8\pi\varepsilon_0}\left(\frac{q_1q_3+q_3q_1}{r_{13}}+\frac{q_2q_3+q_3q_2}{r_{23}}\right)\\
W_4 &= \frac{1}{8\pi\varepsilon_0}\left(\frac{q_1q_4+q_4q_1}{r_{14}}+\frac{q_2q_4+q_4q_2}{r_{24}}+\frac{q_3q_4+q_4q_3}{r_{34}}\right)\\
&\vdots\\
W_{N-1} &= \frac{1}{8\pi\varepsilon_0}\left(\frac{q_1q_{N-1}+q_{N-1}q_1}{r_{1,N-1}}+\cdots+\frac{q_{N-2}q_{N-1}+q_{N-1}q_{N-2}}{r_{N-2,N-1}}\right)\\
W_N &= \frac{1}{8\pi\varepsilon_0}\left(\frac{q_1q_N+q_Nq_1}{r_{1N}}+\cdots+\frac{q_{N-1}q_N+q_Nq_{N-1}}{r_{N-1,N}}\right)
\end{aligned} \qquad [4.8]$$

(there are a total of $N\,(N-1)$ terms) and, consequently:

$$\begin{aligned}
W_e = \frac{1}{8\pi\varepsilon_0}&\left[q_1\left(\frac{q_2}{r_{12}}+\frac{q_3}{r_{13}}+\cdots+\frac{q_N}{r_{1N}}\right)+q_2\left(\frac{q_1}{r_{21}}+\frac{q_3}{r_{23}}+\cdots+\right.\right.\\
&\left.\frac{q_N}{r_{2N}}\right)+\cdots+q_{N-1}\left(\frac{q_1}{r_{N-1,1}}+\frac{q_2}{r_{N-1,2}}+\cdots+\frac{q_N}{r_{N-1,N}}\right)+\\
&\left.q_N\left(\frac{q_1}{r_{N1}}+\frac{q_2}{r_{N2}}+\cdots+\frac{q_{N-1}}{r_{N,N-1}}\right)\right]=\frac{1}{8\pi\varepsilon_0}\sum_{i=1}^{N}q_i\sum_{j=1,j\neq i}^{N}\frac{q_j}{r_{ij}}
\end{aligned} \qquad [4.9]$$

The electrostatic energy of a set of N point charges can alternatively (but equivalently to [4.7]) be written as:

$$W_e = \frac{1}{2}\sum_{i=1}^{N} q_i V_i\left(\vec{\mathbf{r}}_i\right) \qquad [4.10]$$

where $V_i\left(\vec{\mathbf{r}}_i\right) = \sum_{j=1,j\neq i}^{n}\frac{q_j}{4\pi\varepsilon_0 r_{ij}}$ is the potential created at $\vec{\mathbf{r}}_i$ by all the other charges.

EXAMPLE 4.1.– Consider an electric dipole (a system of two charges, $-q$ at the point N and $+q$ at the point P, where $\overrightarrow{\mathbf{NP}} = \vec{\mathbf{d}}$) placed in an external electric field $\vec{\mathbf{E}}_{\text{ext}}$. We assume that the electric field varies very little over distances on the order of the dipole size. We are interested in the total electrostatic energy of the system, composed here of an external electrostatic energy and an internal electrostatic energy, that is:

$$W_e^{\text{tot}} = W_e^{\text{ext}} + W_e^{\text{int}} \qquad [4.11]$$

where:

$$W_e^{\text{ext}} = \int_{\infty}^{P} \left[-(+q)\,\vec{\mathbf{E}}_{\text{ext}}\right] \cdot d\vec{\mathbf{r}} + \int_{\infty}^{N} \left[-(-q)\,\vec{\mathbf{E}}_{\text{ext}}\right] \cdot d\vec{\mathbf{r}}$$
$$= -q \int_{N}^{P} \vec{\mathbf{E}}_{\text{ext}} \cdot d\vec{\mathbf{r}} \simeq -q\vec{\mathbf{E}}_{\text{ext}} \cdot \overrightarrow{\mathbf{NP}} = -\vec{\mathbf{E}}_{\text{ext}} \cdot \vec{\mathbf{p}} \qquad [4.12]$$

$$W_e^{\text{int}} = \frac{1}{2}\left[qV_{\text{int}}(q) + (-q)\,V_{\text{int}}(-q)\right] = \frac{1}{2}\left[\frac{q\times(-q)}{4\pi\varepsilon_0 d} + \frac{(-q)\times q}{4\pi\varepsilon_0 d}\right]$$
$$= -\frac{q^2}{4\pi\varepsilon_0 d} = \vec{\mathbf{E}}_{\text{int}} \cdot \vec{\mathbf{p}} \qquad [4.13]$$

$\vec{\mathbf{E}}_{\text{int}}$ is the field created by $-q$ at the point P, and $\vec{\mathbf{p}} = q\overrightarrow{\mathbf{NP}}$ represents the *dipole moment*. Thus, we obtain:

$$W_e^{\text{tot}} = -\left[\vec{\mathbf{E}}_{\text{ext}} - \vec{\mathbf{E}}_{\text{int}}\right] \cdot \vec{\mathbf{p}} \qquad [4.14]$$

When $\left|\vec{\mathbf{E}}_{\text{ext}}\right| \gg \left|\vec{\mathbf{E}}_{\text{int}}\right|$, the dipole is significantly altered by the field or even broken: the interaction energy exceeds the internal binding energy. However, since the distance d is very small, this does not happen and the dipole remains bound.

EXAMPLE 4.2. (Electrostatic energy of an ionic crystal).– One of the main applications of [4.10] is the calculation of the binding energy of an ionic crystal.

1) To simplify, consider a 1D crystal consisting of an infinitely long chain of positive and negative ions with charges $\pm q$ arranged alternately, at distances a, as shown in Figure 4.2. The work required to bring a positive ion (or a negative ion) from infinity to its position on the line is:

$$W_e = \frac{1}{2}\frac{2q^2}{4\pi\varepsilon_0 a}\left[-1 + \frac{1}{2} - \frac{1}{3} + \frac{1}{4} - \frac{1}{5} + \cdots\right] = -\frac{q^2}{4\pi\varepsilon_0 a}\sum_{m=1}^{\infty}\frac{(-1)^{m+1}}{m} \qquad [4.15]$$

The additional factor 2 in the numerator is necessary because the chain extends infinitely on both sides of the ion. In square brackets, we recognize the Taylor series expansion of the logarithmic function: $\ln(1+x) \simeq x - \frac{x^2}{2} + \frac{x^3}{3} - \frac{x^4}{4} + \frac{x^5}{5} - \cdots$ for $x = 1$. Consequently, we obtain:

$$W_e = -\frac{q^2 \ln 2}{4\pi\varepsilon_0 a} \equiv \frac{1}{2}\frac{q^2}{4\pi\varepsilon_0 a}\sum_{\substack{m\in\mathbb{Z}\\ m\neq 0}}^{\infty}\frac{(-1)^m}{|m|} \qquad [4.16]$$

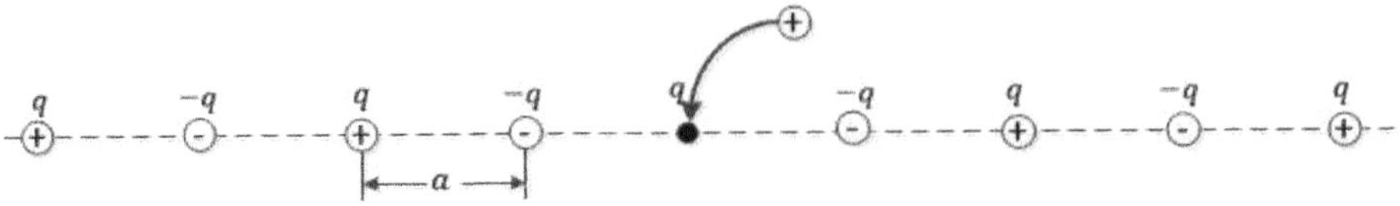

Figure 4.2. *1D crystal with alternating charge ions $\pm q$. A positive amount of work is required to remove the ion*

This work is negative because the crystal pulls on the ion as it arrives from infinity. In other words, a positive amount of work would be required to remove the ion (it is bound to the crystal).

2) If we now imagine a 2D crystal, as shown in Figure 4.3, and consider the adjacent line of ions placed just above, the nearest ion is negative and at a distance a. There are two positive ions at a distance $\sqrt{2}a$. The next pair is at a distance $\sqrt{5}a$, the following one at $\sqrt{10}a$ and so on. Thus, for the entire line, we obtain:

$$\frac{1}{2}\frac{2q^2}{4\pi\varepsilon_0 a}\left[-\frac{1}{1}+\frac{2}{\sqrt{2}}-\frac{2}{\sqrt{5}}+\frac{2}{\sqrt{10}}-\cdots\right] \qquad [4.17]$$

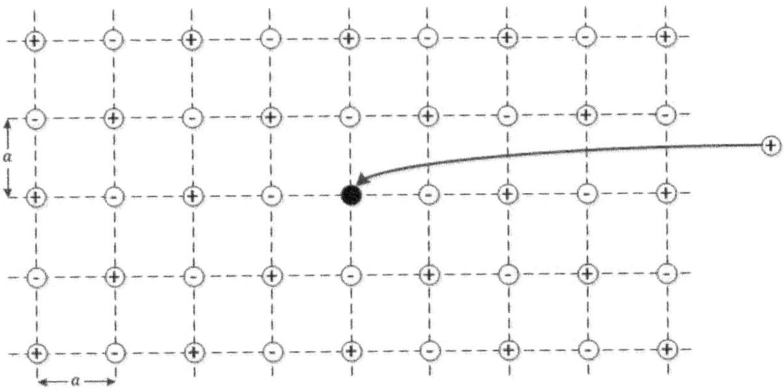

Figure 4.3. *2D crystal with alternating charge ions q arranged along the Ox and Oy directions*

The factor 2 appears in the numerator because there are two such lines: one above and one below. If we move a bit higher, there are also two more lines, one at the top and one at the bottom. For these two lines, we get:

$$\frac{1}{2}\frac{2q^2}{4\pi\varepsilon_0 a}\left[\frac{1}{2}-\frac{2}{\sqrt{5}}+\frac{2}{\sqrt{8}}-\frac{2}{\sqrt{13}}+\cdots\right] \qquad [4.18]$$

and so on. By summing these series, we obtain the binding energy of the ion:

$$W_e = \frac{1}{2}\frac{q^2}{4\pi\varepsilon_0 a}\sum_{\substack{m,n\in\mathbb{Z}\\(m,n)\neq(0,0)}}^{\infty}\frac{(-1)^{m+n}}{\sqrt{m^2+n^2}} \qquad [4.19]$$

3) For a 3D crystal, there are four lines: above, below, in front and behind. Then there are the four closest diagonal lines and so on. If we carry out this process for all lines and then sum them up, we find that the total binding energy is:

$$W_e = \frac{1}{2}\frac{q^2}{4\pi\varepsilon_0 a}\sum_{\substack{m,n,p\in\mathbb{Z}\\(m,n,p)\neq(0,0,0)}}^{\infty}\frac{(-1)^{m+n+p}}{\sqrt{m^2+n^2+p^2}} \qquad [4.20]$$

This is the *Madelung equation.*

4.1.2. *Electrostatic energy of a continuous charge distribution*

For a distribution confined within a volume τ with a continuous density $\rho\left(\vec{\mathbf{r}}\right)$, we simply replace, in [4.10], q_i with $\rho\left(\vec{\mathbf{r}}\right)d\tau$ and the sum $\sum$ with an integral, that is:

$$W_e = \frac{1}{2}\int_\tau V\left(\vec{\mathbf{r}}\right)\rho\left(\vec{\mathbf{r}}\right)d\tau \qquad [4.21]$$

where $V\left(\vec{\mathbf{r}}\right)$ is now the total potential created at the point $M\left(x,y,z\right)$ by the entire charge distribution, including the charge $\rho\left(\vec{\mathbf{r}}\right)d\tau$. This potential is given by the integral $V\left(\vec{\mathbf{r}}\right) = \frac{1}{4\pi\varepsilon_0}\int_{\tau'}\frac{\rho(\vec{\mathbf{r}}')d\tau'}{\left|\vec{\mathbf{r}}-\vec{\mathbf{r}}'\right|}$. This does not change the result. In fact, for an infinitesimal volume $d\tau$ with a characteristic size a, that is, $d\tau\sim a^3$, the potential in the center of this volume created by the charges inside $d\tau$ is proportional to a^{-1}. As a result, the region bounded by this volume contributes on the order of a^2 in the integral [4.21], which vanishes in the infinitesimal limit $a\longrightarrow 0$. W_e represents the formation energy of the charge system, that is, its self-energy. Clearly, when the system has not yet formed, the charge density is zero, and its energy must also be zero. When the system is fully formed, the charge density at M is $\rho\left(\vec{\mathbf{r}}\right)$. The following remarks should be noted:

1) The addition of a constant V_0 to the potential will change the energy unless the total charge of the system is zero. Indeed, we verify that:

$$\begin{aligned} W'_e &= \frac{1}{2}\int_\tau\left[V\left(\vec{\mathbf{r}}\right)+V_0\right]\rho\left(\vec{\mathbf{r}}\right)d\tau = \frac{1}{2}\int_\tau V\left(\vec{\mathbf{r}}\right)\rho\left(\vec{\mathbf{r}}\right)d\tau + \\ &\frac{1}{2}V_0\int_\tau\rho\left(\vec{\mathbf{r}}\right)d\tau = W_e+\frac{1}{2}V_0 Q \end{aligned} \qquad [4.22]$$

If $Q = 0$, the energy remains unchanged.

2) The integral [4.21] can be extended to the entire space (even outside τ):

$$W_e = \frac{1}{2}\int_{\text{Entirespace}} V\left(\vec{\mathbf{r}}\right)\rho\left(\vec{\mathbf{r}}\right)d\tau \qquad [4.23]$$

since $\rho\left(\vec{\mathbf{r}}\right)$ is zero outside τ. In practice, the integral is only performed over the regions containing the charge distribution.

3) In the case of a surface-charge distribution, equation [4.21] must be replaced by:

$$W_e = \frac{1}{2}\int_S V\left(\vec{\mathbf{r}}\right)\sigma\left(\vec{\mathbf{r}}\right)dS \qquad [4.24]$$

where S is the surface containing the charge. This formula cannot be extended to a linear distribution (a wire, for example), because the potential V becomes infinite in the immediate vicinity of the distribution. To solve this problem, we must assume that the wire is not infinitely thin and instead consider a volume distribution (e.g. the wire can be modeled as an infinite cylinder of radius R) or a surface distribution. Then, at the end of the calculations, we take the limit ($R \longrightarrow 0$).

4) The potential $V\left(\vec{\mathbf{r}}\right)$ in expression [4.21] is the one created by the entire charge distribution. By substituting the integral expression of $V\left(\vec{\mathbf{r}}\right)$ into [4.21], we obtain:

$$W_e = \frac{1}{8\pi\varepsilon_0}\int_\tau\int_{\tau'}\frac{\rho\left(\vec{\mathbf{r}}\right)\rho\left(\vec{\mathbf{r}}'\right)}{\left|\vec{\mathbf{r}} - \vec{\mathbf{r}}'\right|}d\tau d\tau' \qquad [4.25]$$

This formula is symmetric under the transformation $\vec{\mathbf{r}} \longleftrightarrow \vec{\mathbf{r}}'$.

EXAMPLE 4.3. (Electrostatic energy of a sphere of radius R).– Let us calculate the electrostatic energy of a sphere of radius R carrying a total charge Q. Two approaches are possible.

1) *Q is distributed on the surface of the sphere*: the surface charge density is $\sigma(R,\theta,\varphi) = \frac{Q}{4\pi R^2}$. The potential induced by the sphere is (a) inside the sphere: $V_{\text{int}}(r \leq R) = \frac{Q}{4\pi\varepsilon_0 R}\left(\vec{\mathbf{E}}_{\text{int}}(r \leq R) = 0\right)$; (b) outside the sphere: $V_{\text{ext}}(r > R) = \frac{Q}{4\pi\varepsilon_0 r}\left(\vec{\mathbf{E}}_{\text{ext}}(r > R) = \frac{Q}{4\pi\varepsilon_0}\frac{\vec{\mathbf{r}}}{r^3}\right)$. Consequently, the electrostatic (binding) energy of the sphere is equal to:

$$\begin{aligned} W_e &= \frac{1}{2}\int_{\theta=0}^{\pi}\int_{\varphi=0}^{2\pi} V_{\text{int}}(R)\,\sigma(R,\theta,\varphi)\,R^2\sin\theta d\theta d\varphi \\ &= \frac{Q}{8\pi}\int_{\theta=0}^{\pi}\int_{\varphi=0}^{2\pi} V_{\text{int}}(R)\sin\theta d\theta d\varphi = \frac{1}{2}QV_{\text{int}}(R) = \frac{Q}{8\pi\varepsilon_0 R} \end{aligned} \qquad [4.26]$$

2) *Q is uniformly distributed throughout the volume of the sphere*: the charge density is given by $\rho(r,\theta,\varphi) = \frac{3Q}{4\pi R^3}$ for $r \leq R$ and 0 for $r > R$. The induced potential is (a) inside $V_{\text{int}}(r \leq R) = \frac{Q}{8\pi\varepsilon_0 R}\left[3 - \left(\frac{r}{R}\right)^2\right]$ $\left(\vec{\mathbf{E}}_{\text{int}}(r \leq R) = \frac{Q}{4\pi\varepsilon_0 R^3}\vec{\mathbf{r}}\right)$ and (b) outside $V_{\text{ext}}(r > R) = \frac{Q}{4\pi\varepsilon_0 r}$ $\left(\vec{\mathbf{E}}_{\text{ext}}(r > R) = \frac{Q}{4\pi\varepsilon_0}\frac{\vec{\mathbf{r}}}{r^3}\right)$. It follows that the binding energy of the sphere is given by:

$$\begin{aligned} W_e &= \frac{1}{2}\int_{r=0}^{R}\int_{\theta=0}^{\pi}\int_{\varphi=0}^{2\pi} V_{\text{int}}(r \leq R)\,\frac{3Q}{4\pi R^3}r^2 dr \sin\theta d\theta d\varphi \\ &= \frac{3Q^2}{16\pi\varepsilon_0 R^4}\int_0^R \left[3r^2 - \frac{r^4}{R^2}\right] dr = \frac{3Q^2}{20\pi\varepsilon_0 R} \end{aligned} \quad [4.27]$$

This formula can be derived in a simpler way: imagine that we build this spherical charge distribution by successively stacking infinitely thin spherical shells. Each time, we take a small amount of charge and place it in a thin shell between r and $r + dr$. This process continues until we reach the final radius R, as shown in Figure 4.4.

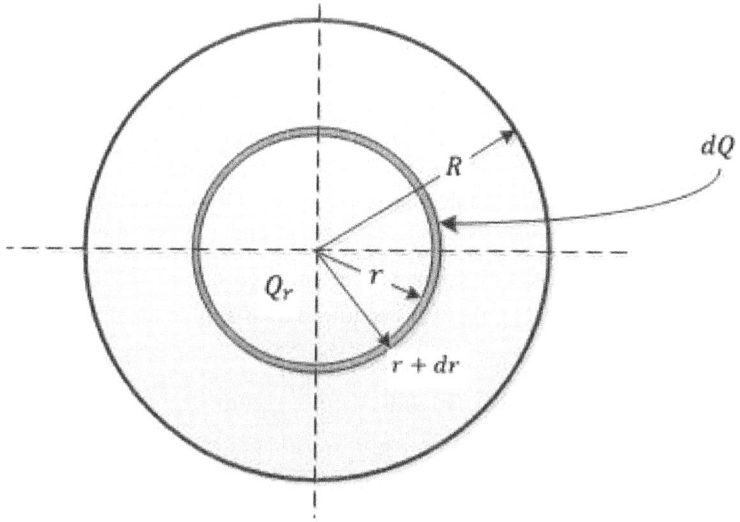

Figure 4.4. *Electrostatic energy of a uniformly charged sphere of radius R*

If Q_r is the charge of the sphere of radius r, the work required to bring an additional charge dQ is given by $dW_e = \frac{Q_r dQ}{4\pi\varepsilon_0 r}$. Since $Q_r = \rho\frac{4}{3}\pi r^3 = \frac{Qr^3}{R^3}$ and $dQ = \rho 4\pi r^2 dr = \frac{3Qr^2 dr}{R^3}$, we obtain $dW_e = \frac{3Q^2 r^4 dr}{4\pi\varepsilon_0 R^6}$. Therefore, the energy required to assemble the sphere is the integral of dW_e from $r = 0$ to $r = R$, that is:

$$W_e = \frac{3Q^2}{4\pi\varepsilon_0 R^6}\int_0^R r^4 dr = \frac{3Q^2}{20\pi\varepsilon_0 R} \qquad [4.28]$$

Note that (i) for a finite charge with zero radius, the energy W_e is infinite; (ii) Einstein's theory of relativity teaches that the work required to assemble the charge is stored as energy, which is related to the mass of the charge by $W_e = mc^2 = \frac{3Q^2}{20\pi\varepsilon_0 R}$ and, therefore:

$$R = \frac{3Q^2}{20\pi\varepsilon_0 mc^2} \qquad [4.29]$$

If we consider an electron (viewed here as a charged sphere in volume), we see that the radius r_e, the charge e and its mass m_e are related by the formula:

$$r_e = \frac{3e^2}{20\pi\varepsilon_0 m_e c^2} \qquad [4.30]$$

An electron can also be viewed as a surface-charged sphere. In this case, the expression for radius r_e is slightly different from [4.30].

EXAMPLE 4.4. (Atomic model).– An atom is modeled as a point-like nucleus with a fixed positive charge $+Ze$ located in the center of an imaginary sphere of radius R, surrounded by a spherical electron cloud that is uniformly distributed over the volume of the sphere with a total charge of $-Ze$. The induced potentials are as follows: (1) for the nucleus: $V_+(r) = \frac{Ze}{4\pi\varepsilon_0 r}$, and (2) for the electron cloud: $V_-(r) = -\frac{3Ze}{8\pi\varepsilon_0 R^3}\left(R^2 - \frac{r^2}{3}\right)$ for $r \leq R$ and $V_-(r) = -\frac{Ze}{4\pi\varepsilon_0 r}$ for $r > R$. According to [4.10] and [4.21], the work required to assemble an atom is:

$$\begin{aligned} W_e &= \frac{1}{2}ZeV_-(0) - \frac{1}{2}\int_{r=0}^{R}\left[V_+(r) + V_-(r)\right]\frac{3Zer^2}{R^3}dr \\ &= -\frac{3Z^2e^2}{16\pi\varepsilon_0 R} - \frac{3Z^2e^2}{8\pi\varepsilon_0 R^3}\int_0^R\left[r - \frac{3r^2}{2R} + \frac{r^4}{2R^3}\right]dr = -\frac{9Z^2e^2}{40\pi\varepsilon_0 R} \end{aligned} \qquad [4.31]$$

This is the binding energy of the atom. We thus conclude that to disassemble an atom by overcoming the Coulomb attractive forces between the nucleus and the electron cloud, a total energy equal to $\frac{9Z^2e^2}{40\pi\varepsilon_0 R}$ must be supplied. Note that formula [4.31] can be derived by superposition, that is, by first considering the uniformly charged negative sphere with a self-energy given by [4.27] and then adding the energy of the point-like positive nucleus:

$$W_e = \frac{3Z^2e^2}{20\pi\varepsilon_0 R} + ZeV_-(r=0) = -\frac{9Z^2e^2}{40\pi\varepsilon_0 R} \qquad [4.32]$$

4.1.3. *Energy stored in the electric field*

Equations [4.10] and [4.21] indicate that a charge distribution can store electrostatic energy. These expressions involve a summation or integration over the domain τ that contains the distribution, suggesting that the energy is confined within it. But where is this energy actually stored? In what follows, we will see that this energy is stored in the electric field itself, that is, it is delocalized throughout all of space. In fact, we can relate the electrostatic potential energy W_e to the electric field created by the charges as follows: using the Poisson equation, we write $\rho = -\varepsilon_0 \Delta V$ and, consequently:

$$W_e = -\frac{\varepsilon_0}{2} \int_{\text{Space}} V \Delta V d\tau \qquad [4.33]$$

From the identity $V\Delta V = \vec{\nabla} \cdot \left(V\vec{\nabla}V\right) - \left|\vec{\nabla}V\right|^2$, the latter can also be written as:

$$W_e = -\frac{\varepsilon_0}{2} \left[\int_{\text{Space}} \vec{\nabla} \cdot \left(V\vec{\nabla}V\right) d\tau - \int_{\text{Space}} \left|\vec{\nabla}V\right|^2 d\tau\right] \qquad [4.34]$$

Using the divergence theorem, we can also transform the first term on the right-hand side into a surface integral over S:

$$W_e = -\frac{\varepsilon_0}{2} \left[\oint_{S \longrightarrow \infty} \left(V\vec{\nabla}V\right) \cdot d\vec{\mathbf{S}} - \int_{\text{Space}} \left|\vec{\nabla}V\right|^2 d\tau\right] \qquad [4.35]$$

where S is a surface located at infinity. Since $V \sim \frac{1}{r}$, $\left|\vec{\nabla}V\right| \sim \frac{1}{r^2}$ and $S \sim r^2$, this integral tends to zero as S is sufficiently far away. Thus, the electrostatic potential energy of the charge distribution simplifies to:

$$W_e = \frac{\varepsilon_0}{2} \int_{\text{Space}} \left|\vec{\mathbf{E}}\right|^2 d\tau \qquad [4.36]$$

because $\vec{\mathbf{E}} = -\vec{\nabla}V$. Equation [4.36] reveals a *volumetric electrostatic energy density* equal to $w_e = \frac{\varepsilon_0}{2}\left|\vec{\mathbf{E}}\right|^2$. This energy density is present at any point where the field is non-zero, including regions outside the charge distribution. This demonstrates that electrostatic energy is distributed throughout space, not just confined within the region containing the charges. Note that this density has the same physical dimension as pressure and is therefore also referred to as *electrostatic pressure*. When a charge is placed in the field created by the distribution, this pressure acts to push it in the direction of the field or in the opposite direction, depending on its sign.

Equation [4.36] appears to lead to a paradox. Indeed, a simple calculation of the electrostatic energy of two charges of opposite signs shows that this energy is negative. However, equation [4.36] can never give a value less than zero, since $\left|\vec{\mathbf{E}}\right|^2 > 0$. The problem arises in the reasoning used to generalize [4.10]–[4.36]. In [4.10], the contribution of charge q_i to the potential $V\left(\vec{\mathbf{r}}_i\right)$ is not taken into account, but in [4.36], this contribution is included. In fact, this excluded part of the energy represents the energy required to assemble the charge q_i, which is infinite for a point charge. Recall that the electric field created by a point charge is given by $\left|\vec{\mathbf{E}}\right| = \frac{q}{4\pi\varepsilon_0 r^2}$ and, consequently, the energy density at a distance r from the charge is $\frac{\varepsilon_0}{2}\left|\vec{\mathbf{E}}\right|^2 = \frac{q^2}{32\pi^2\varepsilon_0 r^4}$. We can take as a volume element a spherical shell of thickness dr and surface area $4\pi r^2$. The total energy is given as:

$$W_e = \int_{r=0}^{\infty} \frac{q^2 dr}{8\pi\varepsilon_0 r^2} = \lim_{r\longrightarrow 0} \frac{q^2}{8\pi\varepsilon_0 r} = +\infty \qquad [4.37]$$

EXAMPLE 4.5. (Electrostatic energy of a spherical charge).– To test equation [4.36], let us calculate the electrostatic potential energy required to assemble a spherical charge distribution of radius R and total charge Q. Similarly, we consider two cases.

1) *Q is distributed on the surface of the sphere*:

$$\begin{aligned} W_e &= \frac{\varepsilon_0}{2}\int_0^R \vec{\mathbf{E}}_{\text{int}}^2(4\pi r^2 dr) + \frac{\varepsilon_0}{2}\int_R^{\infty} \vec{\mathbf{E}}_{\text{ext}}^2(4\pi r^2 dr) \\ &= \frac{\varepsilon_0}{2}\int_0^R 0\times(4\pi r^2 dr) + \frac{\varepsilon_0}{2}\int_R^{\infty} \frac{Q^2}{16\pi^2\varepsilon_0^2 r^4}(4\pi r^2 dr) = \frac{Q^2}{8\pi\varepsilon_0 R} \end{aligned} \qquad [4.38]$$

2) *Q is uniformly distributed throughout the volume of the sphere*:

$$\begin{aligned} W_e &= \frac{\varepsilon_0}{2}\int_0^R \vec{\mathbf{E}}_{\text{int}}^2(4\pi r^2 dr) + \frac{\varepsilon_0}{2}\int_R^{\infty} \vec{\mathbf{E}}_{\text{ext}}^2(4\pi r^2 dr) \\ &= \frac{\varepsilon_0}{2}\int_0^R \frac{Q^2 r^4 dr}{4\pi\varepsilon_0^2 R^2} + \frac{\varepsilon_0}{2}\int_R^{\infty} \frac{Q^2 dr}{16\pi\varepsilon_0^2 r^2} = \frac{3Q^2}{20\pi\varepsilon_0 R} \end{aligned} \qquad [4.39]$$

EXAMPLE 4.6. (Electrostatic energy in a conductor at equilibrium).– Let us consider an isolated conductor with charge Q distributed on its surface. Since the conductor is equipotential, its electrostatic energy is:

$$W_e = \frac{1}{2}\int dqV\left(\vec{\mathbf{r}}\right) = \frac{V}{2}\int dq = \frac{VQ}{2} = \frac{1}{2}CV^2 = \frac{1}{2}\frac{Q^2}{C} \qquad [4.40]$$

This is the energy required to bring a conductor with capacitance C to a potential V. It is always positive, regardless of V (and thus its charge). For a system of N charged conductors located in a region τ, at equilibrium, they have charges Q_i and potentials V_i. Outside the volume occupied by each conductor, there are no charges. Therefore, the total electrostatic energy is:

$$W_e = \frac{1}{2}\int dqV\left(\vec{\mathbf{r}}\right) = \frac{1}{2}\sum_{i=1}^{N}\int_{S_i} V_i\sigma\left(\vec{\mathbf{r}}\right)dS_i = \frac{1}{2}\sum_{i=1}^{N} Q_iV_i \qquad [4.41]$$

EXAMPLE 4.7. (Energy stored in a capacitor).– In a capacitor, all the charge is localized on the electrodes. Consider two electrodes (1) and (2) at potentials V_1 and V_2, respectively, as shown in Figure 4.5. The charges measured on (1) and (2) are $-Q$ and Q, respectively. The energy stored inside the capacitor is calculated using the following formula: $W_e = \frac{1}{2}\left[\int_{S_1} V_1\sigma_1 dS_1 + \int_{S_2} V_2\sigma_2 dS_2\right]$. Since each electrode is an equipotential, the potentials V_1 and V_2 can be factored out of the integrals, reducing them to the total charges $\pm Q$ on the electrodes:

$$W_e = \frac{1}{2}\left[V_1\int_{S_1}\sigma_1 dS_1 + V_2\int_{S_2}\sigma_2 dS_2\right] = \frac{1}{2}\left[V_1 \times (-Q) + V_2 \times (+Q)\right] = \frac{1}{2}\left(V_2 - V_1\right)Q \qquad [4.42]$$

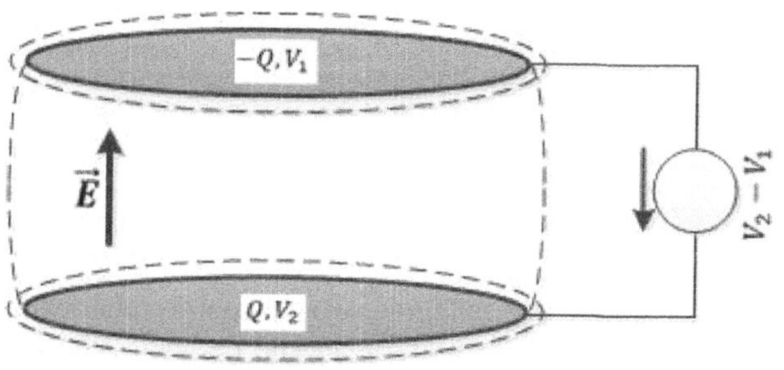

Figure 4.5. *Energy stored in a capacitor*

Here are two remarks:

1) When the capacitor is additionally immersed in an external electric field $\vec{\mathbf{E}}_{\text{ext}}$, the previous formula must be replaced by:

$$W_e = QV_{\text{ext}}(P) - QV_{\text{ext}}(N) + \frac{1}{2}\left(V_2 - V_1\right)Q \simeq -Q\vec{\mathbf{E}}_{\text{ext}}\cdot\overrightarrow{\mathbf{NP}} + \frac{1}{2}\left(V_2 - V_1\right)Q \simeq \frac{1}{2}Q\left(V_2 - V_1 - 2\vec{\mathbf{E}}_{\text{ext}}\cdot\vec{\mathbf{d}}\right) \equiv \frac{1}{2}QV' \qquad [4.43]$$

where P and N are the positions of the centers of the positive and negative plates, respectively.

2) For a capacitor, the charge and voltage are linearly related by the capacitance $Q = C(V_2 - V_1)$. Consequently, the energy stored in the capacitor can also be written as: $W_e = \frac{1}{2}Q(V_2 - V_1) = \frac{1}{2}C(V_2 - V_1)^2 = \frac{1}{2}\frac{Q^2}{C}$. This energy is equivalent to [4.36] and thus provides an alternative method of capacitance calculation when the electric field repartition is known. In fact, we have $\frac{1}{2}\frac{Q^2}{C} = \frac{\varepsilon_0}{2}\int_{\text{space}} \vec{\mathbf{E}}^2 d\tau$. This gives:

$$C = \frac{Q^2}{\varepsilon_0 \displaystyle\int_{\text{space}} \vec{\mathbf{E}}^2 d\tau} \qquad [4.44]$$

To verify this formula, let us apply it to the following cases.

a) *Parallel plate capacitor*: recall that a parallel plate capacitor consists of two plates with a surface area S separated by a very small distance d compared to the dimensions of the plates. The two plates are in near-total electrostatic influence. Inside the space between the plates, there is an electric field $\frac{\sigma}{\varepsilon_0}$ (σ is the charge density of the positive plate), directed from the positive plate to the negative plate, and zero outside, as shown in Figure 4.5. Consequently, the capacitance of the capacitor is given by:

$$C = \frac{Q^2}{\varepsilon_0 \displaystyle\int_{\text{space between plates}} \vec{\mathbf{E}}^2 d\tau} = \frac{(\sigma S)^2}{\varepsilon_0 \times \left(\frac{\sigma}{\varepsilon_0}\right)^2 \times Sd} = \frac{\varepsilon_0 S}{d} \qquad [4.45]$$

b) *Capacitance of a spherical capacitor*: two concentric spherical electrodes, with radii R_1 and $R_2 > R_1$, carry charges $Q_1 = Q$ and $Q_2 = -Q$, respectively. Let us determine the capacitance of this capacitor. Equation [4.44] gives:

$$C = \frac{Q^2}{\varepsilon_0 \displaystyle\int_{R_1}^{R_2} \frac{Q^2}{16\pi^2\varepsilon_0^2 r^4} 4\pi r^2 dr} = \frac{4\pi\varepsilon_0}{\frac{1}{R_1} - \frac{1}{R_2}} \qquad [4.46]$$

4.1.4. *Electrostatic interaction energy between two fields*

Formula [4.36] also allows us to describe electrostatic interaction in terms of fields. To see this, let us consider two distributions $\mathcal{D}_1$ and $\mathcal{D}_1$, which generate at any point

M the fields $\overrightarrow{\mathbf{E}}_1(M)$ and $\overrightarrow{\mathbf{E}}_2(M)$, respectively. The global distribution $\mathcal{D}$, the union of $\mathcal{D}_1$ and $\mathcal{D}_2$, creates at the point M the total field $\overrightarrow{\mathbf{E}}(M) = \overrightarrow{\mathbf{E}}_1(M) + \overrightarrow{\mathbf{E}}_2(M)$ and has a self-energy given by:

$$W_e = \frac{1}{2}\varepsilon_0 \int_{\text{space}} \overrightarrow{\mathbf{E}}^2 d\tau = \frac{1}{2}\varepsilon_0 \int_{\text{space}} \left(\overrightarrow{\mathbf{E}}_1^2 + \overrightarrow{\mathbf{E}}_2^2 + 2\overrightarrow{\mathbf{E}}_1 \cdot \overrightarrow{\mathbf{E}}_2\right) d\tau \qquad [4.47]$$

We see that this energy appears as the sum of the self-electrostatic energies $W_e^{(1)} = \frac{1}{2}\varepsilon_0 \int_{\text{space}} \overrightarrow{\mathbf{E}}_1^2 d\tau$ and $W_e^{(2)} = \frac{1}{2}\varepsilon_0 \int_{\text{space}} \overrightarrow{\mathbf{E}}_2^2 d\tau$ of the distributions $\mathcal{D}_1$ and $\mathcal{D}_2$, respectively, along with an additional term:

$$W_e^{(12)} = \varepsilon_0 \int_{\text{space}} \overrightarrow{\mathbf{E}}_1 \cdot \overrightarrow{\mathbf{E}}_2 d\tau \qquad [4.48]$$

which represents the electrostatic interaction energy between the fields of the two systems. The corresponding volumetric energy density is:

$$w_e^{(12)} = \varepsilon_0 \overrightarrow{\mathbf{E}}_1 \cdot \overrightarrow{\mathbf{E}}_2 \qquad [4.49]$$

EXAMPLE 4.8. (Interaction of two charged spheres).– Consider two identical spheres (1) and (2) of radius R, with centers O_1 and O_2 located at a distance $D > 2R$ from each other and uniformly carrying charges Q_1 and Q_2, respectively. The field created by the sphere (1) at a point M is zero inside (1) and is given by $\overrightarrow{\mathbf{E}}_1(M) = \frac{Q_1}{4\pi\varepsilon_0} \frac{\overrightarrow{r}_1}{r_1^3}$ outside (1). Similarly, the field created at M by (2) is zero inside (2) and $\overrightarrow{\mathbf{E}}_1(M) = \frac{Q_1}{4\pi\varepsilon_0} \frac{\overrightarrow{r}_1}{r_1^3}$ outside (2). Let us define a Cartesian coordinate system with origin at O_1. We assume that O_2 is located on the Oz-axis. In this coordinate system, the position of the point M is described by its spherical coordinates (r_1, θ, φ), as shown in Figure 4.6. Outside the two spheres, we have $w_e^{(12)} = \varepsilon_0 \overrightarrow{\mathbf{E}}_1 \cdot \overrightarrow{\mathbf{E}}_2 = \frac{Q_1 Q_2}{16\pi^2\varepsilon_0}\left(\frac{r_1 - D\cos\theta}{r_1^2 r_2^3}\right)$, and consequently:

$$W_e^{(12)} = \frac{Q_1 Q_2}{16\pi^2\varepsilon_0} \int_{r_1 > R, r_2 > R} \left(\frac{r_1 - D\cos\theta}{r_1^2 r_2^3}\right) r_1^2 dr_1 \sin\theta d\theta d\varphi \qquad [4.50]$$

where $r_2 = \sqrt{r_1^2 + D^2 - 2Dr_1\cos\theta} \equiv r_2(r_1, \theta)$. By integrating over the variable φ, we obtain:

$$W_e^{(12)} = \frac{Q_1 Q_2}{8\pi\varepsilon_0} \int_{r_1 > R, r_2 > R} \left(\frac{r_1 - D\cos\theta}{r_2^3}\right) dr_1 \sin\theta d\theta \qquad [4.51]$$

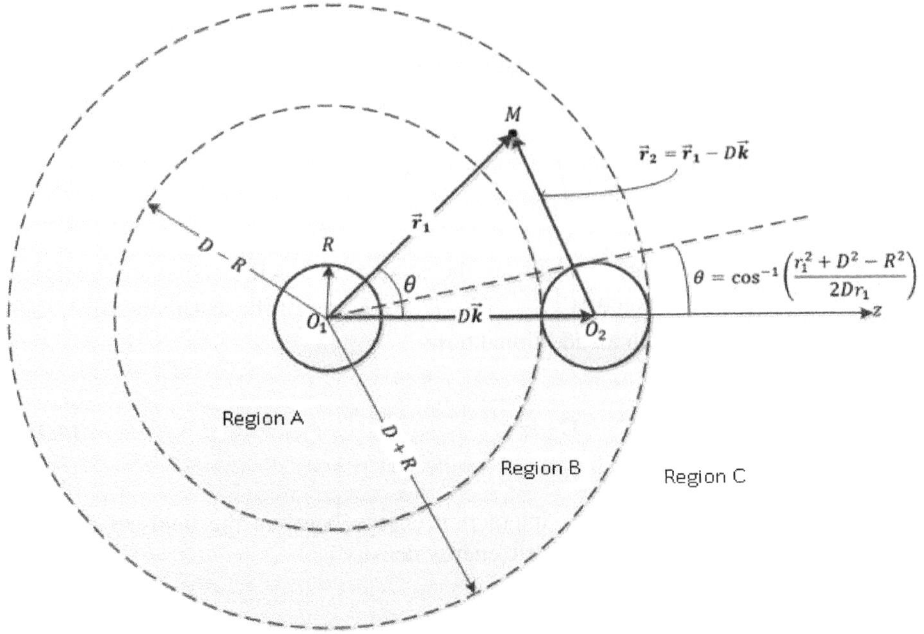

Figure 4.6. *Electrostatic interaction energy between two identical spheres*

To satisfy conditions $r_1 > R$ and $r_2 > R$ of the integral [4.51], let us divide the space into three regions (A), (B) and (C), as shown in Figure 4.6. For region (B), the interior of sphere (2) is excluded. Under these conditions, the integral [4.51] is written as:

$$W_e^{(12)} = \frac{Q_1 Q_2}{8\pi\varepsilon_0} \left(I_A + I_B + I_C\right) \tag{4.52}$$

where the integrals I_A, I_B and I_C are given by:

$$\begin{aligned}
I_A &= \int_R^{D-R} \int_0^{\pi} \left(\frac{r_1 - D\cos\theta}{r_2^3}\right) dr_1 \sin\theta d\theta = -\int_R^{D-R} \int_0^{\pi} \frac{\partial}{\partial r_1}\left(\frac{1}{r_2}\right) dr_1 \sin\theta d\theta \\
&= -\int_R^{D-R} \frac{\partial}{\partial r_1}\left[\int_0^{\pi} \frac{\sin\theta d\theta}{r_2}\right] dr_1 = -\int_R^{D-R} \frac{\partial}{\partial r_1}\left[\frac{r_2\left(r_1,\pi\right) - r_2\left(r_1,0\right)}{r_1 D}\right] dr_1 \\
&= -\int_R^{D-R} \frac{\partial}{\partial r_1}\left[\frac{\left(D + r_1\right) - \left(D - r_1\right)}{r_1 D}\right] dr_1 = -\int_R^{D-R} \frac{\partial}{\partial r_1}\left[\frac{2}{D}\right] dr_1 = 0
\end{aligned}$$

$$
\begin{aligned}
I_B &= \int_{D-R}^{D+R} \int_{\cos^{-1}\left(\frac{r_1^2+D^2-R^2}{2Dr_1}\right)}^{\pi} \left(\frac{r_1 - D\cos\theta}{r_2^3}\right) dr_1 \sin\theta d\theta \\
&= -\int_{D-R}^{D+R} \frac{\partial}{\partial r_1} \left[\frac{r_2(r_1,\pi) - r_2\left(r_1, \cos^{-1}\left(\frac{r_1^2+D^2-R^2}{2Dr_1}\right)\right)}{r_1 D}\right] dr_1 \\
&= -\int_{D-R}^{D+R} \frac{\partial}{\partial r_1} \left[\frac{(D+r_1)-R}{r_1 D}\right] dr_1 = \frac{2}{D} - \frac{2}{D+R} \\
I_C &= \int_{D+R}^{\infty} \int_0^{\pi} \left(\frac{r_1 - D\cos\theta}{r_2^3}\right) dr_1 \sin\theta d\theta = -\int_{D+R}^{\infty} \frac{\partial}{\partial r_1} \left[\frac{r_2(r_1,\pi) - r_2(r_1,0)}{r_1 D}\right] dr_1 \\
&= -\int_{D+R}^{\infty} \frac{\partial}{\partial r_1} \left[\frac{(D+r_1)-(r_1-D)}{r_1 D}\right] dr_1 = -\int_{D+R}^{\infty} \frac{\partial}{\partial r_1} \left[\frac{2}{r_1}\right] dr_1 = \frac{2}{D+R}
\end{aligned}
\quad [4.53]
$$

It thus follows that:

$$
W_e^{(12)} = \frac{Q_1 Q_2}{4\pi\varepsilon_0 D} \quad [4.54]
$$

Of course, we retrieve the interaction energy of two charges. The self-energies of the two spheres are given, respectively, by $W_e^{(1)} = \frac{Q_1^2}{8\pi\varepsilon_0 R}$ and $W_e^{(2)} = \frac{Q_2^2}{8\pi\varepsilon_0 R}$.

EXAMPLE 4.9. (Stability of an electron).– Considering an electron as a sphere of radius r_e with charge distributed on its surface and equating its electrostatic self-energy to its rest energy given by the Einstein equation $E_0 = mc^2$, where m is the mass of the electron and c is the speed of light, we can estimate the value of r_e. We write $mc^2 = \frac{e^2}{8\pi\varepsilon_0 r_e}$, which gives $r_e = \frac{e^2}{8\pi\varepsilon_0 mc^2}$. Let us now examine the stability of such a charged sphere. We know that a pressure is exerted on the surface of this sphere, which tends to expel the charges from the sphere and cause it to burst. This pressure is given by:

$$
P = \frac{\sigma^2}{2\varepsilon_0} = \frac{e^2}{32\pi^2\varepsilon_0 r_e^4} = \frac{1}{4\pi}\frac{mc^2}{r_e^3} \quad [4.55]
$$

This pressure is enormous, indicating that modeling a particle as a solid sphere of matter is not very realistic and goes against common sense.

4.2. Electrostatic actions on a conductor in equilibrium

4.2.1. *Concepts of rigid body mechanics*

A conductor is a solid. Therefore, the laws of rigid body mechanics must be applied to it. Some basic concepts are presented below.

1) *Direct calculation of action (force and moment of a force)*: consider a rigid body. By choosing a reference point O and an appropriate coordinate system that respects the symmetries of the body, the force and moment of this force (or torque) applied to the body with respect to O are given as:

$$\overrightarrow{\mathbf{F}} = \int_{\text{solid}} d\overrightarrow{\mathbf{F}} \tag{4.56}$$

$$\overrightarrow{\mathbf{N}} = \int_{\text{solid}} d\overrightarrow{\mathbf{N}} = \int_{\text{solid}} \overrightarrow{\mathbf{r}} \times d\overrightarrow{\mathbf{F}} \tag{4.57}$$

where $d\overrightarrow{\mathbf{F}}$ is the force acting on an infinitesimal element of the solid, located at position $\overrightarrow{\mathbf{r}}$. The integral is taken over all the points of the solid. Rigid body mechanics shows that the total force $\overrightarrow{\mathbf{F}}$ acts in the center of mass of the solid. If $\overrightarrow{\mathbf{r}}'$ represents the position of the infinitesimal element relative to the center of mass, that is, $\overrightarrow{\mathbf{r}} = \overrightarrow{\mathbf{R}}_G + \overrightarrow{\mathbf{r}}'$, equation [4.57] simplifies to:

$$\overrightarrow{\mathbf{N}} = \overrightarrow{\mathbf{R}}_G \times \overrightarrow{\mathbf{F}} + \int_{\text{solid}} \overrightarrow{\mathbf{r}}' \times d\overrightarrow{\mathbf{F}} = \overrightarrow{\mathbf{R}}_G \times \overrightarrow{\mathbf{F}} + \overrightarrow{\mathbf{N}}_G \tag{4.58}$$

2) *Relationship between work and force or moment of a force*: during a pure infinitesimal translation of a rigid body, each point M undergoes the same translation ds in a given direction $\overrightarrow{\mathbf{u}}$ (where $\overrightarrow{\mathbf{u}}$ is the unit vector along the direction of translation), meaning that $d\overrightarrow{\mathbf{r}} = \overrightarrow{\mathbf{r}}(t+dt) - \overrightarrow{\mathbf{r}}(t) = ds\overrightarrow{\mathbf{u}}$. The total force responsible for this translation must perform the work:

$$dW = \int_{\text{solid}} d\overrightarrow{\mathbf{F}} \cdot d\overrightarrow{\mathbf{r}} = \left(\int_{\text{solid}} d\overrightarrow{\mathbf{F}}\right) \cdot \overrightarrow{\mathbf{u}} ds = \left(\overrightarrow{\mathbf{F}} \cdot \overrightarrow{\mathbf{u}}\right) ds = F_s ds \tag{4.59}$$

where F_s represents the projection of the total force $\overrightarrow{\mathbf{F}}$ along $\overrightarrow{\mathbf{u}}$. Thus, we obtain:

$$F_s = \frac{dW}{ds} \tag{4.60}$$

If, instead of translation, the solid undergoes a pure infinitesimal rotation around an axis $\overrightarrow{\mathbf{n}}$ by an angle $d\alpha$, each point M of the solid moves by the quantity $d\overrightarrow{\mathbf{r}} = d\alpha\overrightarrow{\mathbf{n}} \times \overrightarrow{\mathbf{r}}$. During this rotation, the work done by the force is:

$$\begin{aligned} W &= \int_{\text{solid}} d\overrightarrow{\mathbf{F}} \cdot d\overrightarrow{\mathbf{r}} = \int_{\text{solid}} d\overrightarrow{\mathbf{F}} \cdot \left(d\alpha\overrightarrow{\mathbf{n}} \times \overrightarrow{\mathbf{r}}\right) \\ &= d\alpha\overrightarrow{\mathbf{n}} \cdot \left(\int_{\text{solid}} \overrightarrow{\mathbf{r}} \times d\overrightarrow{\mathbf{F}}\right) = d\alpha\overrightarrow{\mathbf{n}} \cdot \overrightarrow{\mathbf{N}} = d\alpha N_n \end{aligned} \tag{4.61}$$

and, consequently:

$$N_n = \frac{dW}{d\alpha} \tag{4.62}$$

If the motion is more complex, consisting of both translation and rotation, each component will contribute separately to the expression of W.

3) *Calculation of action from potential energy*: the total mechanical energy of a solid is given by $E_{\text{tot}} = W_{\text{kin}} + W_p$, where W_{kin} is its kinetic energy and W_p its potential energy. If the solid is isolated, its total energy remains constant over time, that is, $dE_{\text{tot}} = 0$, leading to:

$$dW_{\text{kin}} = dW = -dW_p \tag{4.63}$$

We therefore conclude that if the expression for the potential energy W_p is known, the force (or its moment) can be expressed in terms of W_p as follows:

$$F_s = -\frac{dW_p}{ds} \qquad \text{(for a pure translation)} \tag{4.64}$$

$$N_n = -\frac{dW_p}{d\alpha} \qquad \text{(for a pure rotation)} \tag{4.65}$$

If, over time, the solid is not isolated and gains or loses energy, then $dE_{\text{tot}} \neq 0$, and consequently, $dW_{\text{kin}} = dW = dE_{\text{tot}} - dW_p$. Thus, in this case, the relationship between the force (or its moment) and the potential energy is no longer direct. To establish this relationship, we must subtract from the work the portion corresponding to the gain or loss of mechanical energy.

4.2.2. *Electrostatic actions from electrostatic energy*

Consider a charged conductor placed in an external electrostatic field $\vec{\mathbf{E}}_{\text{ext}}$. After reaching equilibrium, a force (due to electrostatic pressure) acts on each electric charge distributed on the surface S of the conductor. This force is given by $d\vec{\mathbf{F}} = \frac{\sigma^2}{2\varepsilon_0}\vec{\mathbf{n}}\,dS$. Thus, the total force and its moment are given by:

$$\begin{aligned} \vec{\mathbf{F}} &= \int_{\mathcal{T}} d\vec{\mathbf{F}} = \int_{\mathcal{T}} \frac{\sigma^2}{2\varepsilon_0}\vec{\mathbf{n}}\,dS \\ \vec{\mathbf{N}} &= \int_{\mathcal{T}} \vec{\mathbf{r}} \times d\vec{\mathbf{F}} = \int_{\mathcal{T}} \frac{\sigma^2}{2\varepsilon_0}\left(\vec{\mathbf{r}} \times \vec{\mathbf{n}}\right) dS \end{aligned} \tag{4.66}$$

These expressions are applicable only if σ is known. When σ is not known, the energy method must be used. If the electrostatic energy of the conductor can be expressed in terms of physical parameters, then it becomes relatively easy to determine the forces and/or the moments of the forces.

– *Isolated system*: consider an isolated system composed of multiple charged conductors. Each conductor in equilibrium is characterized by a potential V_i and a charge Q_i. To calculate the electrostatic force acting on a given conductor, we imagine (or induce) a small modification in its physical parameters (position and/or orientation) and examine the resulting change in electrostatic energy. This allows us to determine the forces and/or torques associated with these modifications. This method is called the *virtual work method*. Suppose that the conductor undergoes a displacement ds in the direction $\vec{u}$. The force $\vec{\mathbf{F}}$ acting on the conductor performs mechanical work $\vec{\mathbf{F}} \cdot d\vec{s}$, at the expense of the electrostatic energy W_e stored in the system. The conservation of energy in the isolated system implies that:

$$F_s d\zeta = -dW_e \qquad [4.67]$$

The isolation of the system means that no charge enters or leaves it. The projection F_s of the force along the direction of the conductor's translation is given by:

$$F_s = -\left.\frac{dW_e}{ds}\right|_Q \qquad [4.68]$$

This formula is entirely consistent with the usual form $\vec{\mathbf{F}} = -\vec{\nabla} W$. If, instead of a translation, the conductor undergoes a pure rotation around an axis defined by the unit vector $\vec{n}$ by an angle $d\alpha$, we write the following instead of [4.67]:

$$dW = N_n d\alpha = -dW_e = -\frac{dW_e}{d\alpha} d\alpha \qquad [4.69]$$

where N_n is the projection of the moment $\vec{\mathbf{N}}$ along the axis of rotation. Thus, we obtain:

$$N_n = -\left.\frac{dW_e}{d\alpha}\right|_Q \qquad [4.70]$$

The use of formulas [4.68] and [4.70] requires knowing the expression of W_e as a function of the position and/or orientation of the conductor or conductors. The negative sign indicates that electrostatic forces always tend to return the conductor to a position of minimal energy.

– *Non-isolated system*: when the system is not isolated, meaning that charges are allowed to enter or leave its components, voltage generators must be added to maintain a constant potential for each component. These generators ensure that the potentials remain unchanged. The work done on the moving component can derive its energy either from a decrease in the stored electrostatic energy W_e or from the energy supplied by the generators $E_{(\text{gen})}$, allowing us to write:

$$F_s ds = -\left. dW_e\right|_V - \left. dE_{(\text{gen})}\right|_V \qquad [4.71]$$

The energy supplied by the generators to add charges dQ_i to each component (i) of the system, while keeping its potential constant at V_i, is:

$$dE_{(\text{gen})}\Big|_V = \sum_i V_i dQ_i \tag{4.72}$$

The electrostatic energy of the system is also given by $W_e = \frac{1}{2}\sum_i V_i Q_i$, which implies:

$$dW_e|_V = \frac{1}{2}\sum_i V_i dQ_i \tag{4.73}$$

Consequently, we deduce that:

$$F_s ds = -\frac{1}{2}\sum_i V_i dQ_i + \sum_i V_i dQ_i = +\frac{1}{2}\sum_i V_i dQ_i = +\, dW_e|_V \tag{4.74}$$

and therefore

$$F_s = +\left.\frac{dW_e}{ds}\right|_V \tag{4.75}$$

If the motion involves a rotation around an axis instead of a translation, the previous formula must be replaced by:

$$N_n = +\left.\frac{dW_e}{d\alpha}\right|_V \tag{4.76}$$

EXAMPLE 4.10. (Electric dipole).– Let us consider an electrostatic dipole with dipole moment $\overrightarrow{\mathbf{p}}$ placed in an external field $\overrightarrow{\mathbf{E}}_{\text{ext}}$. We aim to calculate the electrostatic force exerted by this field on the dipole. Since the charges of the dipole remain fixed, we use the expression derived for an isolated system. The electrostatic energy is given by $W_e = -\overrightarrow{\mathbf{p}} \cdot \overrightarrow{\mathbf{E}}_{\text{ext}}$. It follows that $F_s = -\left.\frac{dW_e}{ds}\right|_Q = \frac{d}{ds}\left(\overrightarrow{\mathbf{p}} \cdot \overrightarrow{\mathbf{E}}_{\text{ext}}\right)$ and consequently:

$$\overrightarrow{\mathbf{F}} = \overrightarrow{\nabla}\left(\overrightarrow{\mathbf{p}} \cdot \overrightarrow{\mathbf{E}}_{\text{ext}}\right) \tag{4.77}$$

We see that the dipole always moves toward the regions where the electrostatic field is strongest. To determine the torque of this force, we use the formula:

$$N_n = -\left.\frac{dW_e}{d\alpha}\right|_Q = \frac{d\left(\overrightarrow{\mathbf{p}} \cdot \overrightarrow{\mathbf{E}}_{\text{ext}}\right)}{d\alpha} = \frac{d\overrightarrow{\mathbf{p}}}{d\alpha} \cdot \overrightarrow{\mathbf{E}}_{\text{ext}} \tag{4.78}$$

During a rotation, the dipole moment varies as follows: $d\overrightarrow{\mathbf{p}} = d\alpha\overrightarrow{\mathbf{n}} \times \overrightarrow{\mathbf{p}}$; consequently, $\frac{d\overrightarrow{\mathbf{p}}}{d\alpha} = \overrightarrow{\mathbf{n}} \times \overrightarrow{\mathbf{p}}$. Thus, we obtain $N_n = \left(\overrightarrow{\mathbf{n}} \times \overrightarrow{\mathbf{p}}\right) \cdot \overrightarrow{\mathbf{E}}_{\text{ext}} = \overrightarrow{\mathbf{n}} \cdot \left(\overrightarrow{\mathbf{p}} \times \overrightarrow{\mathbf{E}}_{\text{ext}}\right)$, which leads to:

$$\overrightarrow{\mathbf{N}} = \overrightarrow{\mathbf{p}} \times \overrightarrow{\mathbf{E}}_{\text{ext}} \qquad [4.79]$$

The torque thus tends to align the dipole with the direction of the external field.

EXAMPLE 4.11. (Capacitor).– Let us recall that the electrostatic potential energy of a capacitor is given by $W_e = \frac{1}{2}QV = \frac{1}{2}CV^2 = \frac{1}{2}\frac{Q^2}{C}$. When the capacitor is isolated (i.e. not connected to a generator), the electrostatic force acting between the two plates is:

$$F_s = -\left.\frac{dW_e}{ds}\right|_Q = \frac{Q^2}{2C^2}\frac{dC}{ds} = \frac{V^2}{2}\frac{dC}{ds} \qquad [4.80]$$

On the other hand, when it is connected to a generator, the expression becomes:

$$F_s = \left.\frac{dW_e}{ds}\right|_V = \frac{V^2}{2}\frac{dC}{ds} \qquad [4.81]$$

We therefore conclude that the electrostatic force exerted mutually between the two plates is, in both cases:

$$\overrightarrow{\mathbf{F}} = \frac{V^2}{2}\overrightarrow{\nabla} C \qquad [4.82]$$

Similarly, we find that the projection of the electrostatic force moment along the axis of rotation is given, in both cases, by the formula:

$$N_n = \frac{V^2}{2}\frac{dC}{d\alpha} \qquad [4.83]$$

Note that (1) electrostatic forces always tend to increase the capacitance of the capacitor; (2) the force obtained using the virtual work method must be the same as that given by the expression $\overrightarrow{\mathbf{F}} = \int \frac{\sigma^2}{2\varepsilon_0}\overrightarrow{\mathbf{n}}\, dS$, which means that the charge distribution σ must modify itself to satisfy this condition. For a parallel plate capacitor with capacitance $C = \frac{\varepsilon_0 S}{z_2 - z_1}$, where S is the common surface area of the plates located at z_1 and z_2, respectively, the force exerted by plate (1) on plate (2) is given by:

$$\overrightarrow{\mathbf{F}}_{1\to 2} = \frac{V^2}{2}\overrightarrow{\nabla}_2 C = \frac{V^2}{2}\frac{dC}{dz_2}\overrightarrow{\mathbf{k}} = -\frac{V^2}{2}\frac{\varepsilon_0 S}{\left(z_2 - z_1\right)^2}\overrightarrow{\mathbf{k}} = -\overrightarrow{\mathbf{F}}_{2\to 1} \qquad [4.84]$$

We can clearly see that this force is indeed attractive (meaning that it tends to bring the plates closer together).

4.3. Magnetic actions

4.3.1. *Laplace force – Moment of Laplace force*

Consider an electrically neutral conductor that moves in an electromagnetic field relative to a Galilean reference frame $Oxyz$. Suppose that the conductor contains α different species of charged particles, each with a density n_α, charge q_α and velocity $\vec{\mathbf{v}}_\alpha = \vec{\mathbf{v}}'_\alpha + \vec{\mathbf{v}}_e$, where $\vec{\mathbf{v}}'_\alpha$ is the velocity of the charge carriers of species α relative to the conductor (i.e. relative to a reference frame $O'x'y'z'$ attached to the conductor), and $\vec{\mathbf{v}}_e$ is the driving velocity of the conductor (the frame $O'x'y'z'$ undergoes, in addition to translation, a rotational motion relative to $Oxyz$). These charge carriers generate the local current density:

$$\vec{\mathbf{J}} = \sum_\alpha n_\alpha q_\alpha \vec{\mathbf{v}}'_\alpha \qquad [4.85]$$

and, consequently, each volume element $d\tau$ will experience the following force:

$$\begin{aligned} d\vec{\mathbf{F}} &= \sum_\alpha n_\alpha q_\alpha \left[\vec{\mathbf{E}} + (\vec{\mathbf{v}}'_\alpha + \vec{\mathbf{v}}_e) \times \vec{\mathbf{B}}\right] d\tau \\ &= \sum_\alpha n_\alpha q_\alpha \vec{\mathbf{v}}'_\alpha \times \vec{\mathbf{B}} d\tau = \vec{\mathbf{J}} \times \vec{\mathbf{B}} d\tau \equiv d\vec{\mathbf{F}}_{\text{Lap}} \end{aligned} \qquad [4.86]$$

Since the conductor is electrically neutral ($\sum_\alpha n_\alpha q_\alpha = 0$). The conductor will thus experience the total force:

$$\vec{\mathbf{F}}_{\text{Lap}} = \int_\tau \vec{\mathbf{J}} \times \vec{\mathbf{B}} d\tau \qquad [4.87]$$

This is the *Laplace force*. For a filamentary conductor, formula [4.87] should be replaced by:

$$d\vec{\mathbf{F}}_{\text{Lap}} = \left(\vec{\mathbf{I}} \times \vec{\mathbf{B}}\right) dl = I \left(d\vec{\mathbf{l}} \times \vec{\mathbf{B}}\right) \qquad [4.88]$$

Thus, the total force applied to a closed circuit carrying a current I is:

$$\vec{\mathbf{F}}_{\text{Lap}} = \oint_{\text{circuit}} d\vec{\mathbf{F}}_{\text{Lap}} = I \oint_{\text{circuit}} d\vec{\mathbf{l}} \times \vec{\mathbf{B}} \qquad [4.89]$$

If it is a conductor carrying a surface current density $\vec{\mathbf{K}}$, the total Laplace force acting on it is:

$$\vec{\mathbf{F}}_{\text{Lap}} = \int_S \vec{\mathbf{K}} \times \vec{\mathbf{B}} dS \qquad [4.90]$$

Consider a volume element $d\tau$ located at $\vec{\mathbf{r}}'$ relative to the point O'. The total moment of the forces applied to all the carriers within $d\tau$ is given by:

$$
\begin{aligned}
&d\vec{\mathbf{N}}' = \sum_{\alpha} q_\alpha n_\alpha \vec{\mathbf{r}}' \times \left[\vec{\mathbf{E}} + (\vec{\mathbf{v}}'_\alpha + \vec{\mathbf{v}}_e) \times \vec{\mathbf{B}}\right] d\tau = \\
&\sum_{\alpha} q_\alpha n_\alpha \vec{\mathbf{r}}' \times \left(\vec{\mathbf{v}}'_\alpha \times \vec{\mathbf{B}}\right) d\tau = \vec{\mathbf{r}}' \times \left(\vec{\mathbf{J}} \times \vec{\mathbf{B}}\right) d\tau = \vec{\mathbf{r}}' \times d\vec{\mathbf{F}}_{\text{Lap}}
\end{aligned}
\qquad [4.91]
$$

which implies that the total moment exerted on the entire conductor is:

$$
\vec{\mathbf{N}}' = \int_{\mathcal{T}} \vec{\mathbf{r}}' \times d\vec{\mathbf{F}}_{\text{Lap}} \qquad [4.92]
$$

If we now want to obtain the moment of the forces relative to O, it is enough to add the term $\vec{\mathbf{R}} \times \vec{\mathbf{F}}_{\text{Lap}}$. Indeed, it can be verified that:

$$
\vec{\mathbf{N}} = \int_{\mathcal{T}} \vec{\mathbf{r}} \times d\vec{\mathbf{F}}_{\text{Lap}} = \vec{\mathbf{R}} \times \vec{\mathbf{F}}_{\text{Lap}} + \vec{\mathbf{N}}' \qquad [4.93]
$$

For a filamentary conductor in the shape of a loop, each element $d\vec{\mathbf{l}}$ experiences the Laplace force $d\vec{\mathbf{F}}_{\text{Lap}} = I d\vec{\mathbf{l}} \times \vec{\mathbf{B}}$, and consequently, the total moment, relative to O', experienced by the entire circuit is given by:

$$
\vec{\mathbf{N}}' = \oint_{\text{circuit}} \vec{\mathbf{r}}' \times d\vec{\mathbf{F}}_{\text{Lap}} = I \oint_{\text{circuit}} \vec{\mathbf{r}}' \times \left(d\vec{\mathbf{l}} \times \vec{\mathbf{B}}\right) \qquad [4.94]
$$

4.3.2. *Magnetic moment of a closed circuit in a constant magnetic field*

Consider a filamentary conductor in the shape of a loop (closed circuit, denoted here as Γ), fixed in space, carrying a current I, and placed in a constant external magnetic field $\vec{\mathbf{B}}$. The Laplace force acting on the circuit is:

$$
\vec{\mathbf{F}}_{\text{Lap}} = I \oint_{\Gamma} d\vec{\mathbf{l}} \times \vec{\mathbf{B}} = \left(I \oint_{\Gamma} d\vec{\mathbf{l}}\right) \times \vec{\mathbf{B}} = \vec{0} \times \vec{\mathbf{B}} = \vec{0} \qquad [4.95]
$$

We conclude that the total Laplace force applied to the entire circuit is zero. However, the moment of this force with respect to O' is not zero. Since the circuit is fixed, the two reference frames, $Oxyz$ and $O'x'y'z'$, coincide. For this reason, we will omit the prime. Indeed, we find that:

$$
\begin{aligned}
\vec{\mathbf{N}} &= \oint_{\Gamma} \vec{\mathbf{r}} \times d\vec{\mathbf{F}}_{\text{Lap}} = I \oint_{\Gamma} \vec{\mathbf{r}} \times \left(d\vec{\mathbf{r}} \times \vec{\mathbf{B}}\right) \\
&= I \oint_{\Gamma} d\vec{\mathbf{r}} \left(\vec{\mathbf{r}} \cdot \vec{\mathbf{B}}\right) - I\vec{\mathbf{B}} \oint_{\Gamma} \vec{\mathbf{r}} \cdot d\vec{\mathbf{r}}
\end{aligned}
\qquad [4.96]
$$

The second integral is identically zero. If B_x, B_y and B_z are the components of the magnetic field, we get:

$$\begin{aligned}\vec{\mathbf{N}} &= I\oint_\Gamma \left(dx\,\vec{\mathbf{i}} + dy\,\vec{\mathbf{j}} + dz\,\vec{\mathbf{k}}\right)(xB_x + yB_y + zB_z) \\ &= I\left(B_x\vec{\mathbf{j}} - B_y\vec{\mathbf{i}}\right)\oint_\Gamma xdy + I\left(B_y\vec{\mathbf{k}} - B_z\vec{\mathbf{j}}\right)\oint_\Gamma ydz \\ &\quad + I\left(B_z\vec{\mathbf{i}} - B_x\vec{\mathbf{k}}\right)\oint_\Gamma zdx = I\left[(B_zS_y - B_yS_z)\,\vec{\mathbf{i}}\right. \\ &\quad \left. + (B_xS_z - B_zS_x)\,\vec{\mathbf{j}} + (B_yS_x - B_xS_y)\,\vec{\mathbf{k}}\right] = IS\vec{\mathbf{n}} \times \vec{\mathbf{B}}\end{aligned} \quad [4.97]$$

where $\vec{\mathbf{n}}$ is the normal to an arbitrary surface S that is bounded by the circuit. To establish [4.97], the results from the *Green theorem* were used, meaning the fact that $\oint_\Gamma ydx = -\oint_\Gamma xdy$ and $\oint_\Gamma xdy = \int_{S_z} dxdy = S_z$, etc. Setting $\vec{\mathbf{m}} = IS\vec{\mathbf{n}}$ (where $\vec{\mathbf{m}}$ is the *magnetic dipole moment of the circuit*), we get:

$$\vec{\mathbf{N}} = \vec{\mathbf{m}} \times \vec{\mathbf{B}} \quad [4.98]$$

Although the total Laplace force $\vec{\mathbf{F}}_{\text{Lap}}$ is zero, the magnetic field exerts a torque that tends to rotate the circuit, causing its magnetic moment $\vec{\mathbf{m}}$ to align with the direction of $\vec{\mathbf{B}}$.

4.3.3. *Expressions of actions from magnetic energy*

Consider an element $d\vec{\mathbf{l}}$ of a filamentary circuit Γ carrying a current I. This element experiences a force $d\vec{\mathbf{F}}_{\text{Lap}}$. Recall that:

a) To move the circuit by $d\vec{\mathbf{r}}$, the work done is:

$$dW_{\text{Lap}} = \oint_\Gamma d\vec{\mathbf{F}}_{\text{Lap}} \cdot d\vec{\mathbf{r}} = I\oint_\Gamma \left(d\vec{\mathbf{r}} \times d\vec{\mathbf{l}}\right) \cdot \vec{\mathbf{B}} = Id\vec{\mathbf{S}}_c \cdot \vec{\mathbf{B}} \quad [4.99]$$

where $d\vec{\mathbf{S}}_c$ is the *surface cut* by the circuit during the displacement. The quantity $d\vec{\mathbf{S}}_c \cdot \vec{\mathbf{B}}$ represents the flux through this cut surface, that is, *the cut flux*. Therefore, we obtain:

$$dW_{\text{Lap}} = Id\Phi_c \quad [4.100]$$

This is the expression of the *Maxwell theorem*: *the work done by magnetic forces is equal to the product of the current flowing through the circuit and the flux cut by it during its displacement.*

b) The equation of conservation of magnetic flux $\vec{\nabla} \cdot \vec{\mathbf{B}} = 0$ links the cut flux to the variation of the flux passing through the circuit itself, that is, $d\Phi_c = \Phi(t + dt) - \Phi(t) = d\Phi$, which gives:

$$\Phi_c = \Phi_2 - \Phi_1 \quad [4.101]$$

(The circuit, at any instant t, defines a surface $S(t)$ bounded by $\Gamma(t)$, and $\Phi(t)$ will thus represent the flux passing through this surface $S(t)$). Therefore, we obtain $dW_{\text{Lap}} = Id\Phi$, and consequently:

$$W_{\text{Lap}} = I(\Phi_2 - \Phi_1) \qquad [4.102]$$

c) During its movement, the circuit will acquire kinetic energy, which can be calculated using the equation $dW_{\text{kin}} = dW_{\text{Lap}} = Id\Phi$. Since the total mechanical energy of the system is conserved, that is, $E = W_{\text{kin}} + W_m$, we deduce that $dW_m = -dW_{\text{Lap}}$, meaning that the circuit will deplete its energy into the *magnetic potential energy* W_m stored in the circuit. Therefore, we conclude that any closed circuit immersed in a magnetic field possesses magnetic potential energy, which can be calculated using the expression:

$$W_m = -I\Phi \qquad [4.103]$$

When the Laplace force $\vec{\mathbf{F}}_{\text{Lap}}$ and its moment are not easily calculated directly from equations [4.89] and [4.94], they can always be obtained from the magnetic potential energy W_m (if this can be evaluated as a function of the problem parameters). Suppose that this force should produce a translation ds in the direction $\vec{\mathbf{u}}$. It will therefore do the work $\vec{\mathbf{F}}_{\text{Lap}} \cdot d\vec{\mathbf{s}} = F_{\text{Lap},s}ds$, which is converted into the magnetic potential energy stored in the circuit, that is, $\vec{\mathbf{F}}_{\text{Lap}} \cdot d\vec{\mathbf{r}} = F_{\text{Lap},s}ds = -dW_m$. Therefore, we obtain:

$$F_{\text{Lap},s} = -\frac{dW_m}{ds} \qquad [4.104]$$

This result is consistent with the vectorial expression:

$$\vec{\mathbf{F}}_{\text{Lap}} = -\vec{\nabla}W_m = I\vec{\nabla}\Phi \qquad [4.105]$$

valid only when the current is steady. We can thus conclude that the total Laplace force acting on the circuit tends to push the circuit toward regions where the flux is maximized. In other words, *a circuit Γ, moving in an external static magnetic field, moves in such a way that it aligns with its constraints until the magnetic flux through the circuit is maximized.* This is known as the *maximum flux rule*.

If, instead of translation, there is pure rotation around an axis Δ, represented by the unit vector $\vec{\mathbf{n}}$ and passing through the point O, we get:

$$\begin{aligned} dW_{\text{Lap}} &= \oint_\Gamma d\vec{\mathbf{F}}_{\text{Lap}} \cdot d\vec{\mathbf{r}} = \oint_\Gamma d\vec{\mathbf{F}}_{\text{Lap}} \cdot (d\alpha\,\vec{\mathbf{n}} \times \vec{\mathbf{r}}) \\ &= d\alpha\,\vec{\mathbf{n}} \cdot \left(\oint_\Gamma \vec{\mathbf{r}} \times d\vec{\mathbf{F}}_{\text{Lap}}\right) = d\alpha\,\vec{\mathbf{n}} \cdot \vec{\mathbf{N}} \\ &= d\alpha N_n = -dW_m = Id\Phi \end{aligned} \qquad [4.106]$$

and, consequently:

$$N_n = I\frac{d\Phi}{d\alpha} \qquad [4.107]$$

We conclude that the moment of $d\overrightarrow{\mathbf{F}}_{\text{Lap}}$, relative to an axis Δ passing through O, depends on the variation of the flux during the rotation of the circuit around this axis.

EXAMPLE 4.12. (Magnetic moment).– Assuming that the external magnetic field is constant on the scale of the magnetic moment $\overrightarrow{\mathbf{m}} = IS\overrightarrow{\mathbf{n}}$, we obtain a flux $\Phi = S\overrightarrow{\mathbf{n}} \cdot \overrightarrow{\mathbf{B}}$.The total Laplace force is written as:

$$\overrightarrow{\mathbf{F}}_{\text{Lap}} = I\overrightarrow{\nabla}\Phi = \overrightarrow{\nabla}\left(\overrightarrow{\mathbf{m}} \cdot \overrightarrow{\mathbf{B}}\right) = \left(\overrightarrow{\mathbf{m}} \cdot \overrightarrow{\nabla}\right)\overrightarrow{\mathbf{B}} \qquad [4.108]$$

Similarly, the moment of the Laplace force is given by $N_n = I\frac{d\Phi}{d\alpha} = I\frac{d\left(\overrightarrow{\mathbf{B}} \cdot S\overrightarrow{\mathbf{n}}\right)}{d\alpha} = \overrightarrow{\mathbf{B}} \cdot \frac{d\left(IS\overrightarrow{\mathbf{n}}\right)}{d\alpha} = \overrightarrow{\mathbf{B}} \cdot \frac{d\overrightarrow{\mathbf{m}}}{d\alpha}$. However, the magnetic moment changes during a rotation as follows: $\frac{d\overrightarrow{\mathbf{m}}}{d\alpha} = \overrightarrow{\mathbf{n}} \times \overrightarrow{\mathbf{m}}$, and we obtain $N_n = \overrightarrow{\mathbf{B}} \cdot \left(\overrightarrow{\mathbf{n}} \times \overrightarrow{\mathbf{m}}\right) = \overrightarrow{\mathbf{n}} \cdot \left(\overrightarrow{\mathbf{m}} \times \overrightarrow{\mathbf{B}}\right)$. Therefore, $\overrightarrow{\mathbf{N}} = \overrightarrow{\mathbf{m}} \times \overrightarrow{\mathbf{B}}$. It is important to note that this calculation is much easier than a direct calculation.

4.4. Action of a magnetic field on a conductor: magnetoresistance

When a conductor is exposed to a magnetic field, its electrical resistance changes. This phenomenon is known as *magnetoresistance* and was discovered by *Thomson*. Similarly to the Hall effect, this phenomenon results from the action of the Lorentz force on charge carriers, which tends to curve their trajectories and thus increase their travel time within the conductor. To study this phenomenon, let us examine the behavior of a single type of charge carrier in a conducting material under the influence of an electric field $\overrightarrow{\mathbf{E}}$ and a magnetic field $\overrightarrow{\mathbf{B}}$. Let m be the mass of the carrier, q its charge and n its numerical density (number per unit volume). The motion of the carriers is described by equation:

$$m\frac{d\overrightarrow{\mathbf{v}}}{dt} = q\left(\overrightarrow{\mathbf{E}} + \overrightarrow{\mathbf{v}} \times \overrightarrow{\mathbf{B}}\right) - f\overrightarrow{\mathbf{v}} \qquad [4.109]$$

The term $-f\overrightarrow{\mathbf{v}}$ is due to the collisions of charge carriers within the conductor. This is a first-order differential equation. It has a transient solution (without the second term) that quickly tends to zero and a particular solution that describes the steady-state regime of the carriers. In steady-state conditions, the average velocity of the carriers must satisfy the equation $f\overrightarrow{\mathbf{v}} = q\left(\overrightarrow{\mathbf{E}} + \overrightarrow{\mathbf{v}} \times \overrightarrow{\mathbf{B}}\right)$. By introducing the mobility $\mu = \frac{q}{f}$, this equation becomes $\overrightarrow{\mathbf{v}} = \mu\left(\overrightarrow{\mathbf{E}} + \overrightarrow{\mathbf{v}} \times \overrightarrow{\mathbf{B}}\right)$. Assuming that $\overrightarrow{\mathbf{B}}$ is oriented along the Oz axis, that is, $\overrightarrow{\mathbf{B}} = B\overrightarrow{\mathbf{k}}$ with $B > 0$, and projecting onto the three coordinate axes, we obtain:

$$\begin{cases} v_x = \mu\left(E_x + v_y B\right) \\ v_y = \mu\left(E_y - v_x B\right) \\ v_z = \mu E_z \end{cases} \Longrightarrow \begin{cases} v_x = \dfrac{\mu\left(E_x + \mu E_y B\right)}{1+\mu^2 B^2} \\ v_y = \dfrac{\mu\left(E_y - \mu E_x B\right)}{1+\mu^2 B^2} \\ v_z = \mu E_z \end{cases} \qquad [4.110]$$

These last equations can be rewritten in vector notation as follows:

$$\overrightarrow{\mathbf{v}} = \frac{\mu}{1+\mu^2 B^2}\left[\overrightarrow{\mathbf{E}} + \mu\overrightarrow{\mathbf{E}} \times \overrightarrow{\mathbf{B}} + \mu^2\left(\overrightarrow{\mathbf{E}} \cdot \overrightarrow{\mathbf{B}}\right)\overrightarrow{\mathbf{B}}\right] \qquad [4.111]$$

In the following, we assume that the electric field is always perpendicular to the magnetic field, that is, $E_z = 0$, and consequently, $v_z = 0$.

To describe the Hall effect, we must apply an electric field in the Oy direction, that is, $\overrightarrow{\mathbf{E}}_0 = E_y\overrightarrow{\mathbf{j}}$, to ensure the movement of charge carriers (the flow of an electric current) in this direction. The Lorentz force deflects the charges along Ox, causing positive charges to accumulate on one side and negative charges on the other, where they settle on the lateral faces (perpendicular to Ox). As a result, an electric field develops along Ox, leading to Hall voltage between these lateral faces of the conductor. This electric field gradually counteracts the charge deflection until equilibrium is reached, that is, until $v_x = 0$. From this, we deduce that:

$$E_x = -\mu B E_y \qquad [4.112]$$

$$v_y = \mu E_y \qquad [4.113]$$

Equation [4.113] shows that the relationship between current density $\overrightarrow{\mathbf{J}}$ and applied electric field $\overrightarrow{\mathbf{E}}_0$ remains unchanged, that is, $\overrightarrow{\mathbf{J}} = \gamma\overrightarrow{\mathbf{E}}$, where $\gamma = nq\mu$ is the *conductivity* of the material. We thus conclude that the resistance of the material is not affected by the magnetic field. The Hall effect is characterized by a constant R_H, called the *Hall constant*, defined by:

$$E_x = E_H = R_H J B \qquad [4.114]$$

where J is the volume current density. The model used here shows that if conduction occurs through a single type of carrier, the Hall constant is given by:

$$R_H = \frac{1}{nq} \qquad [4.115]$$

The Hall voltage that appears between the lateral faces of the conductor is given by:

$$V_H = E_H l = R_H J B l = R_H B l \frac{I}{S} = R_H \frac{IBl}{S} \qquad [4.116]$$

where l, S and I are, respectively, the distance between the two lateral faces perpendicular to Ox, the surface area of the lateral faces perpendicular to Oy, and the current intensity flowing in the Oy direction.

When the conductor contains multiple types of charge carriers, let q_i, m_i, $\vec{\mathbf{v}}_i$, μ_i and n_i denote, respectively, the charge, mass, average velocity, mobility and density of the carrier type i. We will assume here that there are two types, that is, $i = 1, 2$. Defining $\gamma_i = n_i q_i \mu_i$, the components of the current density are given by:

$$\begin{aligned} J_{ix} &= n_i q_i v_{ix} = \frac{\gamma_i}{1+\mu_i^2 B^2}\left(E_x + \mu_i E_y B\right) \\ J_{iy} &= n_i q_i v_{iy} = \frac{\gamma_i}{1+\mu_i^2 B^2}\left(E_y - \mu_i E_x B\right) \end{aligned} \qquad [4.117]$$

The total current density $\vec{\mathbf{J}}$ is the vector sum of these current densities and has the following components:

$$J_x = \left(\frac{\gamma_1}{1+\mu_1^2 B^2} + \frac{\gamma_2}{1+\mu_2^2 B^2}\right) E_x + \left(\frac{\mu_1\gamma_1}{1+\mu_1^2 B^2} + \frac{\mu_2\gamma_2}{1+\mu_2^2 B^2}\right) E_y B \qquad [4.118]$$

$$J_y = \left(\frac{\gamma_1}{1+\mu_1^2 B^2} + \frac{\gamma_2}{1+\mu_2^2 B^2}\right) E_y - \left(\frac{\mu_1\gamma_1}{1+\mu_1^2 B^2} + \frac{\mu_2\gamma_2}{1+\mu_2^2 B^2}\right) E_x B \qquad [4.119]$$

The steady-state regime is reached when $J_x = 0$, meaning that E_x is given by:

$$E_x = E_H = -\frac{\mu_1 K_1 + \mu_2 K_2}{K_1 + K_2} E_y B \qquad [4.120]$$

where $K_i = \frac{\gamma_i}{1+\mu_i^2 B^2}$. By substituting [4.120] into the expression for J_y, we obtain:

$$\gamma = \frac{J_y}{E_y} = -\frac{\left(K_1+K_2\right)^2 + B^2\left(\mu_1 K_1 + \mu_2 K_2\right)^2}{K_1+K_2} \qquad [4.121]$$

and consequently, the Hall constant is:

$$R_H = \frac{E_H}{J_y B} = -\frac{\mu_1 K_1 + \mu_2 K_2}{\left(K_1+K_2\right)^2 + B^2\left(\mu_1 K_1 + \mu_2 K_2\right)^2} \qquad [4.122]$$

We see that the conductivity now depends on the field B, indicating a *magneto-resistive effect*. In particular, when both species have the same mobility, the conductivity γ reduces to its expression obtained in the absence of a magnetic field. If $|\mu_i B| \ll 1$, a series expansion leads to the following expression for the conductivity:

$$\gamma = \gamma_1 + \gamma_2 - B^2 \frac{\gamma_1\gamma_2}{\gamma_1+\gamma_2}\left(\mu_1 - \mu_2\right)^2 \qquad [4.123]$$

At zero order, we recover the expression $\gamma_1+\gamma_2$ for the conductivity in the absence of a magnetic field. The correction term is of B^2 order and is negative, indicating that the conductivity decreases in the presence of a magnetic field. At zero order, the Hall constant is given by:

$$R_H = \frac{\mu_1\gamma_1 + \mu_2\gamma_2}{(\gamma_1+\gamma_2)^2} \qquad [4.124]$$

In the case of a semiconductor where conduction occurs through electrons ($q=-e$, μ_e, n_e) and holes ($q=+e$, μ_p, n_p), we find:

$$R_H = \frac{1}{e}\frac{n_p\mu_p^2 - n_e\mu_e^2}{(n_p\mu_p - n_e\mu_e)^2} \qquad [4.125]$$

4.5. Magnetic inductances

4.5.1. *Neumann formula*

Let us consider two filamentary circuits, Γ_i and Γ_j, which are either closed or quasi-closed (as in the case of coils) and are, respectively, carrying steady currents I_i and I_j. The circuit Γ_j generates a magnetic field $\vec{\mathbf{B}}_j$, whose flux through a surface S_i bounded by Γ_i is given by:

$$\begin{aligned}\Phi_{i,j} &= \int_{S_i} \vec{\mathbf{B}}_j(\vec{\mathbf{r}}_i)\cdot d\vec{\mathbf{S}}_i = \oint_{\Gamma_i} \vec{\mathbf{A}}_j(\vec{\mathbf{r}}_i)\cdot d\vec{\mathbf{l}}_i \\ &= \oint_{\Gamma_i}\left[\frac{\mu_0 I_j}{4\pi}\oint_{\Gamma_j}\frac{d\vec{\mathbf{l}}_j}{\left|\vec{\mathbf{r}}_i - \vec{\mathbf{r}}_j\right|}\right]\cdot d\vec{\mathbf{l}}_i = M_{i,j}I_j\end{aligned} \qquad [4.126]$$

where:

$$M_{i,j} = \frac{\mu_0}{4\pi}\oint_{\Gamma_i}\oint_{\Gamma_j}\frac{d\vec{\mathbf{l}}_i\cdot d\vec{\mathbf{l}}_j}{\left|\vec{\mathbf{r}}_i - \vec{\mathbf{r}}_j\right|} \qquad [4.127]$$

Note that the two circuits Γ_i and Γ_j play symmetrical roles in [4.127], that is, $M_{i,j} = M_{j,i}$. The coefficient $M_{i,j}$ is called the *mutual inductance coefficient* between the two circuits and depends only on the geometry of the system. Equation [4.127] is known as the *Neumann formula*. Its generalization to volumetric current distributions is given by:

$$M_{i,j} = \frac{\mu_0}{4\pi I_i I_j}\int_{\tau_i}\int_{\tau_j}\frac{\vec{\mathbf{J}}_i(\vec{\mathbf{r}}_i)\cdot\vec{\mathbf{J}}_j(\vec{\mathbf{r}}_j)}{\left|\vec{\mathbf{r}}_i - \vec{\mathbf{r}}_j\right|}d\tau_i d\tau_j \qquad [4.128]$$

where $\vec{\mathbf{J}}_i(\vec{\mathbf{r}}_i)$ and $\vec{\mathbf{J}}_j(\vec{\mathbf{r}}_j)$ are the current density vectors in τ_i and τ_j, respectively.

A circuit Γ_i carrying a current I_i generates its own magnetic field, which induces not only a flux through any surface S_j bounded by the circuit Γ_j, but also a flux through itself, known as *self-flux* or *self-inductance flux*. However, the Neumann formula diverges when setting $\Gamma_i = \Gamma_j$, that is, when $\overrightarrow{\mathbf{r}}_i$ and $\overrightarrow{\mathbf{r}}_j$ coincide within the integral [4.127], due to the term $\left|\overrightarrow{\mathbf{r}}_i - \overrightarrow{\mathbf{r}}_j\right|$ in the denominator. Therefore, this formula must be modified in the case where $\Gamma_i = \Gamma_j$. To overcome this problem, we must consider the circuit Γ_i as a volume τ_i (i.e. take into account the thickness of the circuit) rather than as a thin wire when expressing the field or potential. Thus, we write:

$$\Phi_{i,i} = \int_{S_i} \overrightarrow{\mathbf{B}}_i\left(\overrightarrow{\mathbf{r}}_i\right)\cdot d\overrightarrow{\mathbf{S}}_i = \oint_{\Gamma_i} \overrightarrow{\mathbf{A}}_i\left(\overrightarrow{\mathbf{r}}_i\right)\cdot d\overrightarrow{\mathbf{l}}_i = \oint_{\Gamma_i} \left(\frac{\mu_0}{4\pi}\int_{\tau_i} \frac{\overrightarrow{\mathbf{J}}_i\left(\overrightarrow{\mathbf{r}}'_i\right) d\tau'_i}{\left|\overrightarrow{\mathbf{r}}_i - \overrightarrow{\mathbf{r}}'_i\right|}\right)\cdot d\overrightarrow{\mathbf{l}}_i \quad [4.129]$$

where $\overrightarrow{\mathbf{r}}_i$ and $\overrightarrow{\mathbf{r}}'_i$ denote the positions of $d\overrightarrow{\mathbf{l}}_i$ and $d\tau'_i$, respectively. Permuting the two integrals, we obtain:

$$\begin{aligned}\Phi_{i,i} &= \int_{\tau_i} \overrightarrow{\mathbf{J}}_i\left(\overrightarrow{\mathbf{r}}'_i\right)\cdot\left[\frac{\mu_0}{4\pi}\oint_{\Gamma_i}\frac{d\overrightarrow{\mathbf{l}}_i}{\left|\overrightarrow{\mathbf{r}}_i - \overrightarrow{\mathbf{r}}'_i\right|}\right] d\tau'_i \\ &= \frac{1}{I_i}\int_{\tau_i} \overrightarrow{\mathbf{J}}_i\left(\overrightarrow{\mathbf{r}}'_i\right)\cdot\left[\frac{\mu_0 I_i}{4\pi}\oint_{\Gamma_i}\frac{d\overrightarrow{\mathbf{l}}_i}{\left|\overrightarrow{\mathbf{r}}'_i - \overrightarrow{\mathbf{r}}_i\right|}\right] d\tau'_i \\ &= \frac{1}{I_i}\int_{\tau_i} \overrightarrow{\mathbf{J}}_i\left(\overrightarrow{\mathbf{r}}'_i\right)\cdot \overrightarrow{\mathbf{A}}_i\left(\overrightarrow{\mathbf{r}}'_i\right) d\tau'_i\end{aligned} \quad [4.130]$$

As the current density $\overrightarrow{\mathbf{J}}_i$ in a thin-wire circuit and the potential $\overrightarrow{\mathbf{A}}_i$ that it produces are both proportional to the current I_i flowing through the circuit, it follows that:

$$\Phi_{i,i} = M_{i,i} I_i \quad [4.131]$$

where:

$$M_{i,i} = \frac{1}{I_i^2}\int_{\tau_i} \overrightarrow{\mathbf{J}}_i\left(\overrightarrow{\mathbf{r}}_i\right)\cdot \overrightarrow{\mathbf{A}}_i\left(\overrightarrow{\mathbf{r}}_i\right) d\tau_i \equiv L_i \quad [4.132]$$

is called the *self-inductance coefficient* (or *self-inductance*) of Γ_i. This coefficient can also be obtained by directly calculating the self-flux and dividing it by the current intensity or by using the equation:

$$M_{i,i} = \frac{\mu_0}{4\pi I_i^2}\int_{\tau_i}\int_{\tau_i} \frac{\overrightarrow{\mathbf{J}}_i\left(\overrightarrow{\mathbf{r}}\right)\cdot \overrightarrow{\mathbf{J}}_i\left(\overrightarrow{\mathbf{r}}'\right)}{\left|\overrightarrow{\mathbf{r}} - \overrightarrow{\mathbf{r}}'\right|} d\tau d\tau' \quad [4.133]$$

(derived from [4.128]), where the integrals are independently evaluated within the same volume τ_i of the circuit. The terms $\overrightarrow{\mathbf{J}}_i\left(\overrightarrow{\mathbf{r}}\right) d\tau$ and $\overrightarrow{\mathbf{J}}_i\left(\overrightarrow{\mathbf{r}}'\right) d\tau'$ represent two current elements located at a distance $\left|\overrightarrow{\mathbf{r}} - \overrightarrow{\mathbf{r}}'\right|$ from each other within this volume.

EXAMPLE 4.13. (Self-induction of a loop).– Consider a loop of radius R carrying a current I. The magnetic field created in the center of the loop is parallel to its axis and has a magnitude of $B = \frac{\mu_0 I}{2R}$. We assume that the field is uniform inside the loop, since R is small. To determine the self-inductance of the loop, we first need to calculate the flux of $\overrightarrow{\mathbf{B}}$ through the loop and then divide it by I, that is, $\Phi = BS = \frac{1}{2}\mu_0 IR$, which corresponds to:

$$L = \frac{\Phi}{I} = \frac{1}{2}\mu_0 R \quad [4.134]$$

EXAMPLE 4.14. (Self-inductance of a solenoid).– A current I flows through the turns of an infinite solenoid. The magnetic field created by this current is uniform inside the solenoid, parallel to its axis, and has a magnitude of $B = \mu_0 nI$, where n is the number of turns per unit length. Outside the solenoid, the field is zero. To determine the self-inductance of a segment of length l of the solenoid, we need to find the flux through the $nl \equiv N$ turns contained within it: $\Phi = B \cdot nlS = \mu_0 n^2 lSI = \frac{\mu_0 N^2 SI}{l}$, where S is the cross-sectional area of the solenoid. Consequently, we obtain:

$$L = \mu_0 n^2 lS = \frac{\mu_0 N^2 S}{l} \quad [4.135]$$

We can clearly see that the coefficient L is proportional to N^2, which emphasizes the advantage of having a large N.

EXAMPLE 4.15. (Self-inductance of a toroidal coil).– Consider a toroidal coil with a square or circular cross section, an inner radius R, an outer radius $R + a$ and a height a. A very thin conducting wire is wound tightly and uniformly around the coil. The number of turns N is assumed to be large enough so that the winding can be modeled as a series of adjacent closed loops. The current flowing through the winding is indicated by I, as shown in Figure 4.7. It is clear that any plane containing Oz is a plane of symmetry for this current distribution, which shows that the magnetic field it creates is orthoradial. Moreover, the rotational symmetry of the distribution around Oz implies that the orthoradial component of the field cannot depend on the azimuthal angle. Thus, the field at a point M, identified by cylindrical coordinates r, φ and z, is of the form: $\overrightarrow{\mathbf{B}}(M) = B(r, z)\,\overrightarrow{\mathbf{e}}_\varphi$ and its field lines are circles centered on the Oz axis. Along these field lines, defined by r = constant and z = constant, the circulation of the field is given by $2\pi r B(r, z)$. Applying Ampére's theorem shows that this quantity is not zero only if the considered field line passes inside the toroid and is worth $\mu_0 NI$. It follows that the field is zero outside the toroid, whereas inside the toroid it is given by $\overrightarrow{\mathbf{B}}(M) = \frac{\mu_0 NI}{2\pi r}\overrightarrow{\mathbf{e}}_\varphi$. Now, let us compute the self-flux of the coil. We first determine the flux ϕ_1 of the field through the surface of a single turn. For a square turn, we obtain:

$$\phi_1 = \int_{-a/2}^{a/2} dz \int_R^{R+a} dr \frac{\mu_0 NI}{2\pi r} = \frac{\mu_0 NIa}{2\pi} \ln\left(\frac{R+a}{R}\right) \quad [4.136]$$

This flux is independent of the position of the loop on the toroid. Consequently, the total self-flux is $\phi = N\phi_1 = \frac{\mu_0 N^2 I a}{2\pi} \ln\left(\frac{R+a}{R}\right)$. From this, we deduce the self-inductance coefficient:

$$L = \frac{\phi}{I} = \frac{\mu_0 N^2 a}{2\pi} \ln\left(\frac{R+a}{R}\right) \quad [4.137]$$

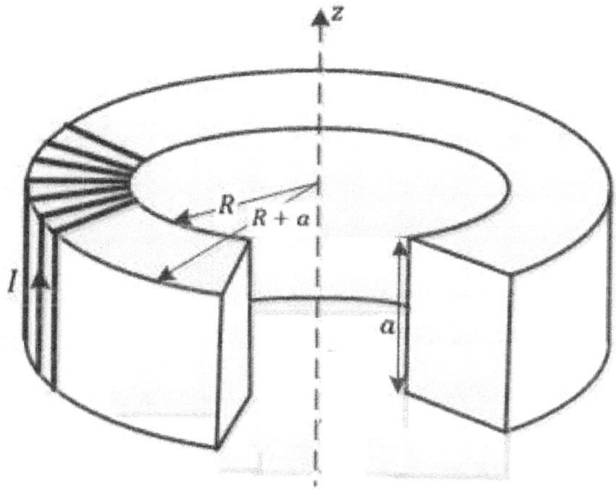

Figure 4.7. *Toroidal coil with a square cross-section carrying a current I. For a color version of this figure, see www.iste.co.uk/abdesselam/electrodynamics2.zip*

If the cross-section is circular, the situation becomes more complex. Let C be the center of the loop. The position of a point M on the surface bounded by the loop is defined by the distance $\rho = CM$ (varying between 0 and $\frac{a}{2}$) and the angle ψ between $\overrightarrow{\mathrm{CM}}$ and $\overrightarrow{\mathrm{OC}}$ (varying between 0 and 2π), as shown in Figure 4.8. A surface element inside the loop is given by $\rho d\rho d\psi$. Noting that $r = R + \frac{a}{2} + \rho\cos\psi$, we obtain:

$$\phi_1 = \frac{\mu_0 NI}{2\pi} \int_0^{a/2} \rho d\rho \int_0^{2\pi} dr \frac{d\psi}{R + \frac{a}{2} + \rho\cos\psi} \quad [4.138]$$

Using the result $\int_0^{2\pi} dr \frac{d\psi}{R+\frac{a}{2}+\rho\cos\psi} = \frac{2\pi}{\sqrt{\left(R+\frac{a}{2}\right)^2 - \rho^2}}$, it follows that:

$$\phi_1 = \mu_0 NI \int_0^{a/2} \frac{\rho d\rho}{\sqrt{\left(R + \frac{a}{2}\right)^2 - \rho^2}} = \mu_0 NI \left[R + \frac{a}{2} - \sqrt{R^2 + aR}\right] \quad [4.139]$$

and consequently:

$$L = \frac{\phi}{I} = \frac{N\phi_1}{I} = \mu_0 N^2 \left[R + \frac{a}{2} - \sqrt{R^2 + aR}\right] \quad [4.140]$$

When $R \gg a$, both formulas [4.137] and [4.140] approximately reduce to $L = \frac{\mu_0 N^2 S}{2\pi R}$, where S is the cross-sectional area of the toroid.

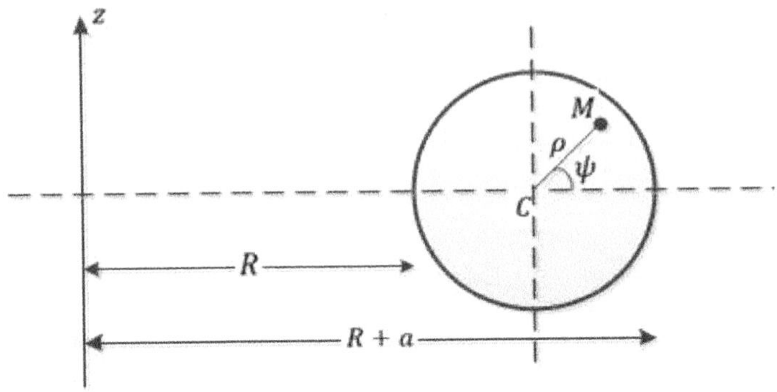

Figure 4.8. *Toroidal coil with a circular cross-section. Each point of the cross-section is identified by the distance ρ from the center of the section and the angle ψ with the horizontal*

4.5.2. *Inductance matrix*

Let us assume that we have a system composed of n closed or quasi-closed circuits Γ_i, each carrying a current I_i. The flux ϕ_i through the ith circuit is given by:

$$\Phi_i = \int_{S_i} \vec{\mathbf{B}} \cdot d\vec{\mathbf{S}}_i = \int_{S_i} \left[\sum_{j=1}^{n} \vec{\mathbf{B}}_j \left(\vec{\mathbf{r}}_i \right) \right] \cdot d\vec{\mathbf{S}}_i = \sum_{j=1}^{n} \Phi_{i,j} = \sum_{j=1}^{n} M_{i,j} I_j \quad [4.141]$$

These equations can also be expressed in the form:

$$\begin{pmatrix} \Phi_1 \\ \Phi_2 \\ \vdots \\ \Phi_n \end{pmatrix} = \begin{pmatrix} M_{1,1} & M_{1,2} & \cdots & M_{1,n} \\ M_{2,1} & M_{2,2} & \cdots & M_{2,n} \\ \vdots & \vdots & \ddots & \vdots \\ M_{n,1} & M_{n,2} & \cdots & M_{n,n} \end{pmatrix} \cdot \begin{pmatrix} I_1 \\ I_2 \\ \vdots \\ I_n \end{pmatrix} \quad [4.142]$$

The matrix $(M_{i,j})$ is the *inductance matrix*. Note that the inductance coefficients $M_{i,j}$ depend only on the geometry of the system, that is, the shapes of the circuits and their relative positions in space. Therefore, they remain constant, unless the geometry of the system changes over time. These expressions are analogous to those previously encountered in electrostatics, which relate the charges Q_i to the potentials V_i of different conductors: $Q_i = \sum_{j=1}^{n} C_{ij} V_j$, where C_{ij} are the *coefficients of mutual electrostatic influence*.

Using calculations similar to those performed to obtain the coefficients $M_{i,i}$, it can also be shown that:

$$\begin{aligned} M_{i,j} &= \frac{1}{I_j} \oint_{\Gamma_i} \vec{\mathbf{A}}_j(\vec{\mathbf{r}}_i) \cdot \vec{\mathbf{l}}_i = \frac{1}{I_j} \oint_{\Gamma_i} \left[\int_{\tau'_j} \frac{\vec{\mathbf{J}}_j(\vec{\mathbf{r}}'_j)\, d\tau'_j}{|\vec{\mathbf{r}}_i - \vec{\mathbf{r}}'_j|} \right] \cdot \vec{\mathbf{l}}_i \\ &= \frac{1}{I_i I_j} \int_{\tau'_j} \vec{\mathbf{J}}_j(\vec{\mathbf{r}}'_j) \cdot \left[\frac{\mu_0 I_i}{4\pi} \oint_{\Gamma_i} \frac{\vec{\mathbf{l}}_i}{|\vec{\mathbf{r}}'_j - \vec{\mathbf{r}}_i|} \right] d\tau'_j \\ &= \frac{1}{I_i I_j} \int_{\tau_j} \vec{\mathbf{J}}_j(\vec{\mathbf{r}}_j) \cdot \vec{\mathbf{A}}_i(\vec{\mathbf{r}}_j)\, d\tau_j = \frac{1}{I_i I_j} \int_{\tau_i} \vec{\mathbf{J}}_i(\vec{\mathbf{r}}_i) \cdot \vec{\mathbf{A}}_j(\vec{\mathbf{r}}_i)\, d\tau_i \end{aligned} \quad [4.143]$$

Since the density $\vec{\mathbf{J}}_i$ is zero outside the volume τ_i, the previous integral can be extended over the entire space without altering the result:

$$M_{i,j} = \frac{1}{I_i I_j} \int_{\text{Space}} \vec{\mathbf{J}}_i(\vec{\mathbf{r}}_i) \cdot \vec{\mathbf{A}}_j(\vec{\mathbf{r}}_i)\, d\tau_i \quad [4.144]$$

By replacing $\vec{\mathbf{J}}_i(\vec{\mathbf{r}}_i)$ with $\frac{\vec{\nabla}_i \times \vec{\mathbf{B}}_i(\vec{\mathbf{r}}_i)}{\mu_0}$ and using the identity $\vec{\nabla} \cdot \left(\vec{\mathbf{A}} \times \vec{\mathbf{B}}\right) = \vec{\mathbf{B}} \cdot \left(\vec{\nabla} \times \vec{\mathbf{A}}\right) - \vec{\mathbf{A}} \cdot \left(\vec{\nabla} \times \vec{\mathbf{B}}\right)$, we obtain:

$$\begin{aligned} M_{i,j} &= \frac{1}{\mu_0 I_i I_j} \int_{\text{Space}} \left[\vec{\nabla}_i \times \vec{\mathbf{B}}_i(\vec{\mathbf{r}}_i)\right] \cdot \vec{\mathbf{A}}_j(\vec{\mathbf{r}}_i)\, d\tau_i \\ &= \frac{1}{\mu_0 I_i I_j} \int_{\text{Space}} \left[\vec{\mathbf{B}}_i(\rightarrow_i) \cdot \left[\vec{\nabla}_i \times \vec{\mathbf{A}}_j(\vec{\mathbf{r}}_i)\right] - \right. \\ &\quad \left. \vec{\nabla}_i \cdot \left(\vec{\mathbf{A}}_j(\vec{\mathbf{r}}_i) \times \vec{\mathbf{B}}_i(\vec{\mathbf{r}}_i)\right)\right] d\tau_i \end{aligned} \quad [4.145]$$

Apply the divergence theorem to the second term on the right-hand side:

$$\int_{\text{Space}} \vec{\nabla}_i \cdot \left(\vec{\mathbf{A}}_j(\vec{\mathbf{r}}_i) \times \vec{\mathbf{B}}_i(\vec{\mathbf{r}}_i)\right) d\tau_i = \lim_{S_i \longrightarrow \infty} \oint_{S_i} \left(\vec{\mathbf{A}}_j(\vec{\mathbf{r}}_i) \times \vec{\mathbf{B}}_i(\vec{\mathbf{r}}_i)\right) \cdot d\vec{\mathbf{S}}_i \quad [4.146]$$

However, $\vec{\mathbf{A}}_j \sim \frac{1}{r_i}$, $\vec{\mathbf{B}}_i \sim \frac{1}{r_i^2}$ and $S_j \sim r_i^2$. Thus:

$$\begin{aligned} &\lim_{S_i \longrightarrow \infty} \oint_{S_i} \left(\vec{\mathbf{A}}_j(\vec{\mathbf{r}}_i) \times \vec{\mathbf{B}}_i(\vec{\mathbf{r}}_i)\right) \cdot d\vec{\mathbf{S}}_i = \\ &\qquad \lim_{r_i \longrightarrow \infty} \oint_{S_i} \left(\vec{\mathbf{A}}_j(\vec{\mathbf{r}}_i) \times \vec{\mathbf{B}}_i(\vec{\mathbf{r}}_i)\right) \cdot d\vec{\mathbf{S}}_i = 0 \end{aligned} \quad [4.147]$$

We conclude that this integral is zero, and consequently:

$$\begin{aligned} M_{i,j} &= \frac{1}{\mu_0 I_i I_j} \int_{\text{Space}} \vec{\mathbf{B}}_i(\vec{\mathbf{r}}_i) \cdot \left[\vec{\nabla}_i \times \vec{\mathbf{A}}_j(\vec{\mathbf{r}}_i)\right] d\tau_i \\ &= \frac{1}{\mu_0 I_i I_j} \int_{\text{Space}} \vec{\mathbf{B}}_i(\vec{\mathbf{r}}_i) \cdot \vec{\mathbf{B}}_j(\vec{\mathbf{r}}_i)\, d\tau_i \end{aligned} \quad [4.148]$$

In particular, for $i = j$, we obtain:

$$L_i = M_{i,i} = \frac{1}{\mu_0 I_i^2} \int_{\text{Space}} \left| \overrightarrow{\mathbf{B}}_i \left(\overrightarrow{\mathbf{r}}_i \right) \right|^2 d\tau_i \qquad [4.149]$$

Equations [4.148] and [4.149] provide an alternative method of calculation for L_i and M_{ij} if the distributions of the magnetic fields are known. The self-inductance coefficients L_i are obviously positive. Equation [4.148] has the properties of a scalar product. Indeed, for $i \neq j$, we can write:

$$\begin{aligned} M_{i,j}^2 &= \left(\frac{1}{\mu_0 I_i I_j} \right)^2 \left(\int_{\text{Space}} \overrightarrow{\mathbf{B}}_i \left(\overrightarrow{\mathbf{r}}_i \right) \cdot \overrightarrow{\mathbf{B}}_j \left(\overrightarrow{\mathbf{r}}_i \right) d\tau_i \right) \\ &\times \left(\int_{\text{Space}} \overrightarrow{\mathbf{B}}_j \left(\overrightarrow{\mathbf{r}}_j \right) \cdot \overrightarrow{\mathbf{B}}_i \left(\overrightarrow{\mathbf{r}}_j \right) d\tau_j \right) \leq \left(\frac{1}{\mu_0 I_i I_j} \right)^2 \times \\ &\int_{\text{Space}} \int_{\text{Space}} \left| \overrightarrow{\mathbf{B}}_i \left(\overrightarrow{\mathbf{r}}_i \right) \right|^2 \left| \overrightarrow{\mathbf{B}}_j \left(\overrightarrow{\mathbf{r}}_j \right) \right|^2 d\tau_i d\tau_j \leq M_{i,i} M_{j,j} \leq L_i L_j \end{aligned} \qquad [4.150]$$

where we have used the Cauchy–Schwarz inequality $\left| \int f \overline{g} d\tau \right|^2 \leq \left(\int |f|^2 d\tau \right) \times \left(\int |g|^2 d\tau \right)$ (for measurable complex functions). It follows that:

$$|M_{i,j}| \leq \sqrt{L_i L_j} \qquad [4.151]$$

We say that the *coupling is weak*, that is, that the influence between circuits i and j is negligible if $|M_{i,j}| \simeq 0$. In this case, we have $\Phi_i \simeq L_i I_i$. Similarly, the coupling is said to be *strong or maximal* if $|M_{i,j}| \simeq \sqrt{L_i L_j}$.

EXAMPLE 4.16. (Self-inductance of a coaxial cable).– A coaxial cable consists of two infinite coaxial cylindrical shells with radii R_1 and R_2 ($R_1 < R_2$). A current I flows through the inner cylinder in the positive Oz direction and returns through the outer cylinder, as shown in Figure 4.9.

The magnetic field $\overrightarrow{\mathbf{B}}$ at a point M is perpendicular to the plane of the sheet, and its magnitude depends only on the distance r from the axis to M . Applying the Ampére theorem, we can determine the field $\overrightarrow{\mathbf{B}}$. In the region between the two cylinders, that is, for $R_1 < r < R_2$, the field $\overrightarrow{\mathbf{B}}$ is given by $B = \frac{\mu_0 I}{2\pi r}$. Outside this region, that is, for $r < R_1$ and $r > R_2$, the magnetic field is zero. We can then calculate the flux of $\overrightarrow{\mathbf{B}}$ through any cross-section dS of the conductor over a length l, as shown in Figure 4.9:

$$\Phi = \int_{R_1}^{R_2} B dS = \int_{R_1}^{R_2} \frac{\mu_0 I l}{2\pi} \frac{dr}{r} = \frac{\mu_0 I l}{2\pi} \ln \frac{R_2}{R_1} \qquad [4.152]$$

The self-inductance of a length l of the coaxial cable is therefore:

$$L = \frac{\mu_0 l}{2\pi} \ln \frac{R_2}{R_1} \qquad [4.153]$$

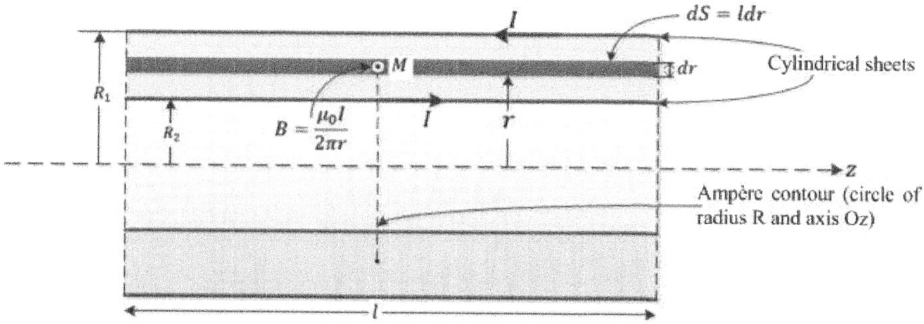

Figure 4.9. *Self-inductance of a coaxial cable (I forms a closed circuit by flowing through the inner cylinder and returning through the outer cylinder)*

This formula can be derived using [4.149]. Since the magnetic field is non zero only in the region between the two cylinders, we get:

$$L = \frac{1}{\mu_0 I^2}\int_{R_1}^{R_2}\left|\overrightarrow{\mathbf{B}}\right|^2 2\pi r dr l = \frac{1}{\mu_0 I^2}\int_{R_1}^{R_2}\frac{\mu_0^2 I^2}{4\pi^2 r^2}2\pi r dr l = \frac{\mu_0 l}{2\pi}\ln\frac{R_2}{R_1} \quad [4.154]$$

EXAMPLE 4.17. (Mutual inductance between two solenoids).– We consider a solenoid with a circular cross-section of radius R_1, length l_1 and composed of N_1 turns. Inside this solenoid, we place a second solenoid with radius R_2, length l_2 and composed of N_2 turns, as shown in Figure 4.10. Let us calculate the mutual inductances between the two solenoids assuming that they are infinitely long. Under this approximation, the respective magnetic fields are given by: $\overrightarrow{\mathbf{B}}_1 = \mu_0 n_1 I_1 \overrightarrow{\mathbf{k}} = \mu_0 \frac{N_1}{l_1} I_1 \overrightarrow{\mathbf{k}}$ and $\overrightarrow{\mathbf{B}}_2 = \mu_0 n_2 I_2 \overrightarrow{\mathbf{k}} = \mu_0 \frac{N_2}{l_2} I_2 \overrightarrow{\mathbf{k}}$. We now express the flux ϕ_1 through circuit 1. The self-inductance of the circuit and the mutual inductance relate the currents in the circuits to the flux ϕ_1 as follows: $\phi_1 = L_1 I_1 + M_{12} I_2$. Similarly, the flux in circuit 2 is given by: $\phi_2 = M_{21} I_1 + L_2 I_2$. To determine the self-inductances, we first consider each solenoid independently. We begin with solenoid 1 alone, setting $I_2 = 0$. Using the expression for $\overrightarrow{\mathbf{B}}_1$, the magnetic flux through each of the N_1 turns of this solenoid is: $\phi_{1,\text{turns}} = \int_{S_1} \overrightarrow{\mathbf{B}}_1 \cdot \overrightarrow{\mathbf{k}}\, dS_1 = \mu_0 \frac{N_1}{l_1} \pi R_1^2 I_1$. Thus, the total flux through the solenoid is $\phi_1 = \mu_0 \frac{N_1^2}{l_1} \pi R_1^2 I_1$. Comparing this with the expression for ϕ_1 (with $I_2 = 0$), we deduce the expression for the self-inductance of solenoid 1:

$$L_1 = \mu_0 \pi R_1^2 \frac{N_1^2}{l_1} \quad [4.155]$$

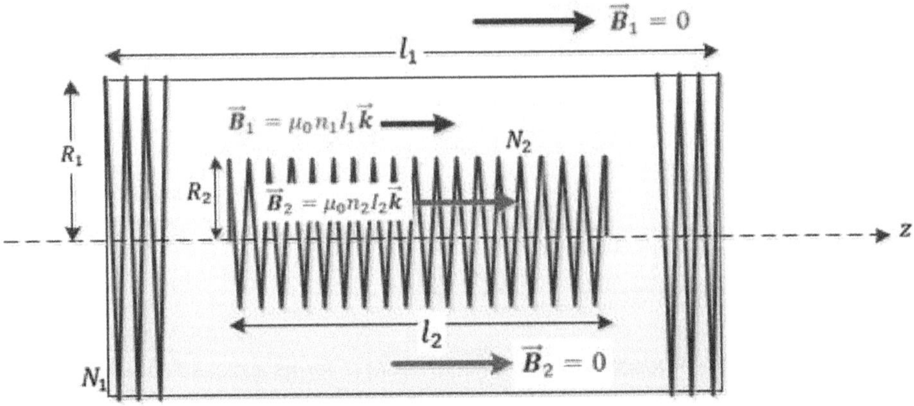

Figure 4.10. *Mutual inductances of two solenoids. For a color version of this figure, see www.iste.co.uk/abdesselam/electrodynamics2.zip*

By following the same reasoning with $I_1 = 0$, we obtain the self-inductance of solenoid 2:

$$L_2 = \mu_0 \pi R_2^2 \frac{N_2^2}{l_2} \qquad [4.156]$$

For mutual inductances, we calculate the flux of the field $\vec{\mathbf{B}}_1$ through solenoid 2. Since solenoid 2 is fully immersed in $\vec{\mathbf{B}}_1$, we get $\phi_{21} = \left|\vec{\mathbf{B}}_1\right| \pi R_2^2 N_2 = \mu_0 \pi R_2^2 \frac{N_1 N_2}{l_1} I_1 \equiv M_{21} I_1$. We can therefore conclude that the mutual inductance M_{21} is:

$$M_{21} = \mu_0 \pi R_2^2 \frac{N_1 N_2}{l_1} \qquad [4.157]$$

To determine the mutual inductance M_{12}, we note that the magnetic field produced by solenoid 2 partially immerses solenoid 1. Consequently, the flux of the field $\vec{\mathbf{B}}_2$ through solenoid 1 is $\phi_{12} = \left|\vec{\mathbf{B}}_2\right| \pi R_2^2 n_1 l_2 = \mu_0 \frac{N_2}{l_2} I_2 \pi R_2^2 \frac{N_1}{l_1} l_2 = \mu_0 \pi R_2^2 \frac{N_1 N_2}{l_1} I_2 \equiv M_{12} I_2$, which gives:

$$M_{12} = \mu_0 \pi R_2^2 \frac{N_1 N_2}{l_1} \qquad [4.158]$$

We can naturally verify that the mutual inductances are equal, $M_{21} = M_{12} = M$, and that $M^2 \leq L_1 L_2$, since $l_1 > l_2$ and $R_1 > R_2$. It is more convenient to define a magnetic coupling factor k between the two circuits as follows:

$$k = \frac{M}{\sqrt{L_1 L_2}} = \frac{R_2}{R_1} \sqrt{\frac{l_2}{l_1}} \qquad [4.159]$$

We observe that $k \leq 1$. This coefficient approaches 1 in the limit where the dimensions of solenoid 2 approach those of solenoid 1, that is, when $l_2 \longrightarrow l_1$ and $R_2 \longrightarrow R_1$, leading to an almost perfect coupling, $M \simeq \sqrt{L_1 L_2}$.

4.5.3. *Mutual inductance coupling*

1) Consider a closed or quasi-closed Γ thin-wire circuit. When a time-varying external flux Φ_{ext} passes through the circuit, the induced ohmic current i generates an electric field $E = \frac{J}{\gamma} = \frac{i}{S\gamma}$, where S and γ are the cross-sectional area and the conductivity of the wire that forms the circuit, respectively. Applying Faraday's law, we obtain an EMF equal to the ohmic voltage drop $\oint_\Gamma \overrightarrow{\mathbf{E}} \cdot d\overrightarrow{\mathbf{l}} = \frac{il}{S\gamma} = Ri \simeq -\frac{d\Phi_{\text{ext}}}{dt}$ and, consequently:

$$i \simeq -\frac{1}{R}\frac{d\Phi_{\text{ext}}}{dt} \qquad [4.160]$$

where R is the total resistance of Γ (where l is the length of the circuit). Here, the effects of mutual inductance have not been taken into account. Note that the direction of the induced current i is opposite to the direction of variation of Φ_{ext}. This current i, in turn, generates a magnetic field whose flux Φ_{auto} through the circuit is given by:

$$\Phi_{\text{auto}} = Li \simeq -\frac{L}{R}\frac{d\Phi_{\text{ext}}}{dt} \qquad [4.161]$$

where L is the self-inductance of Γ. We thus conclude that *the induced current produces effects that oppose the causes that gave rise to it. In other words, Φ_{auto} has a sign opposite to the variation of Φ_{ext}*. This is *Lenz's law*.

2) Consider two circuits Γ_1 and Γ_2, carrying the currents I_1 and I_2, respectively. If I_1 varies over time, then ϕ_2 also changes. Consequently, an induced EMF $\mathcal{E}_2^{\text{ind}}$ appears in the circuit Γ_2: $\mathcal{E}_2^{\text{ind}} = -M\frac{dI_1}{dt}$. Thus, the current I_2 flowing in Γ_2 varies, as does the magnetic field $\overrightarrow{\mathbf{B}}_2$ (created by Γ_2) in Γ_1. An induced EMF also appears in Γ_1: $\mathcal{E}_1^{\text{ind}} = -M\frac{dI_2}{dt}$. The circuits are said to be *coupled by mutual induction*. These induced electromotive forces will be added to the self-induced electromotive forces of the circuits, resulting in:

$$\begin{aligned} \mathcal{E}_1 &= -L_1\frac{dI_1}{dt} - M\frac{dI_2}{dt} \\ \mathcal{E}_2 &= -M\frac{dI_1}{dt} - L_2\frac{dI_2}{dt} \end{aligned} \qquad [4.162]$$

If the coupling between the two circuits is strong, that is, $M = \sqrt{L_1 L_2}$, we can verify that:

$$\frac{\mathcal{E}_1}{\mathcal{E}_2} = \sqrt{\frac{L_1}{L_2}} \qquad [4.163]$$

These equations are the basis of the transformer operating principle.

3) Consider multiple circuits $\Gamma_1, \ldots, \Gamma_n$ coupled by mutual induction. The induced EMF in circuit Γ_i is given by:

$$\mathcal{E}_i = -\frac{d\Phi_i}{dt} = -\frac{d}{dt}\left(\sum_{j=1}^{n} M_{i,j} I_j\right) = -\sum_{j=1}^{n}\left(M_{i,j}\frac{dI_j}{dt} + I_j\frac{dM_{i,j}}{dt}\right) \quad [4.164]$$

$\mathcal{E}_i$ arises from both the movement and/or deformation of the circuits, and from the variation of the currents. In particular, when the circuits are fixed and rigid, $\mathcal{E}_i$ is given by:

$$\mathcal{E}_i = -\sum_{j=1}^{n} M_{i,j}\frac{dI_j}{dt} \quad [4.165]$$

If the couplings between the circuits are weak (independent circuits), the previous equations reduce to:

$$\mathcal{E}_i = -L_i\frac{dI_i}{dt} \quad [4.166]$$

The minus sign indicates that the EMF opposes the change in current.

4) Since every coil has self-inductance, which opposes changes in current, the current in the coil exhibits a form of inertia. In fact, to vary the current in a coil, we must overcome this inertia by connecting it to an external voltage source. In such a circuit, the current I is related to the voltage E by the equation $E = L\frac{dI}{dt}$. This equation resembles Newton's law governing the one-dimensional motion of a particle. By analogy, if we associate the applied external voltage E with the external force F, and the current I in the coil with the velocity of the particle, then the inductance of the coil corresponds to the mass m of the particle. As a result, we can establish the correspondence shown in Table 4.1. Clearly, the analogy is not only mathematical, but also physical.

Particle	**Coil**
F (force)	E (voltage)
v (velocity)	I (current)
x (displacement)	q (charge)
$F = m\frac{dv}{dt}$	$E = L\frac{dI}{dt}$
mv (momentum)	LI
$\frac{1}{2}mv^2$ (kinetic energy)	$\frac{1}{2}LI^2$ (magnetic energy)
Fv (power)	EI (power)

Table 4.1. *Correspondence diagram*

EXAMPLE 4.18. (Rectangular loop near an infinite wire carrying a variable current).– Let us consider a rectangular loop consisting of a wire of length $l = 2(D+d)$, a cross-sectional area S and conductivity γ. The loop is located close to an infinite wire through which a current $I(t)$ flows at a distance r, as shown in Figure 4.11. The magnetic flux through the loop due to $I(t)$ is:

$$\begin{aligned}\Phi_m &= \int_{-D/2}^{D/2} dz \int_r^{d+d} B_\varphi(r')\, dr' \\ &= \frac{\mu_0 I D}{2\pi}\int_r^{d+d}\frac{dr'}{r'} = \frac{\mu_0 I D}{2\pi}\ln\left(\frac{r+d}{r}\right)\end{aligned} \qquad [4.167]$$

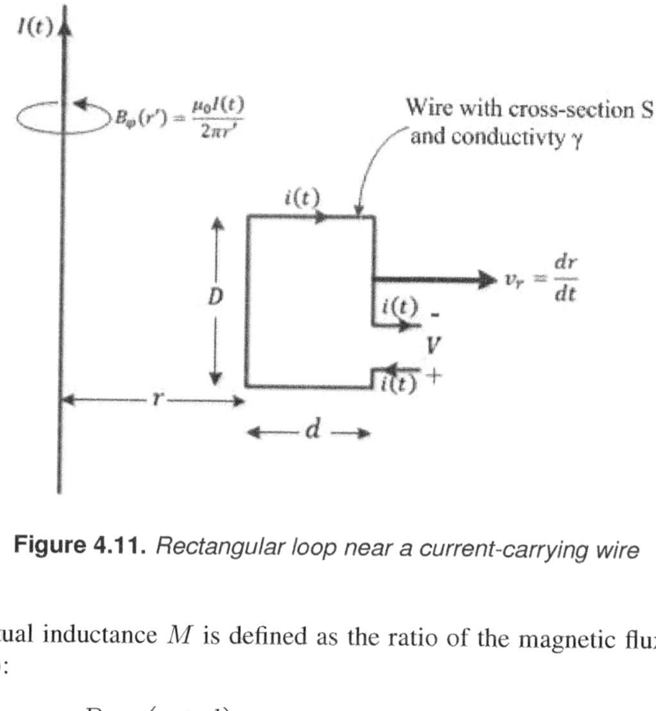

Figure 4.11. *Rectangular loop near a current-carrying wire*

The mutual inductance M is defined as the ratio of the magnetic flux Φ_m to the current $I(t)$:

$$M(r) = \frac{\mu_0 D}{2\pi}\ln\left(\frac{r+d}{r}\right) \qquad [4.168]$$

When the loop is short-circuited ($V = 0$), the induced ohmic current appears as:

$$i = -\frac{1}{R}\frac{d\Phi}{dt} \qquad [4.169]$$

where R is the resistance of the loop. The flux in [4.169] is due both to the imposed current, represented by the term $M(r)\,I(t)$, and to the induced current, which arises from the self-inductance of the loop:

$$\Phi = M(r) I + Li \quad [4.170]$$

If the loop also moves radially outward with the radial velocity $v_r = \frac{dr}{dt}$, then equation [4.169] becomes:

$$\begin{aligned} -Ri = \frac{d\Phi}{dt} &= M(r)\frac{dI}{dt} + I\frac{dM(r)}{dt} + L\frac{di}{dt} \\ &= M(r)\frac{dI}{dt} + I\frac{dM(r)}{dr}\frac{dr}{dt} + L\frac{di}{dt} \end{aligned} \quad [4.171]$$

where L does not depend on r. If the loop is stationary, only the first and third terms on the right-hand side contribute, and they are not zero only when the currents vary over time. The second term arises due to motion and contributes even for steady currents. Some simple cases are presented below.

– *Case 1, excitation by a constant current* I_0: if the loop is stationary at $r = r_0$, equation [4.171] simplifies to $L\frac{di}{dt} + Ri = -M(r_0)\frac{dI}{dt}$, or equivalently:

$$\frac{di}{dt} + \frac{i}{\tau} = -\frac{M(r_0)}{L}\frac{dI}{dt} \quad [4.172]$$

where $\tau = \frac{L}{R}$. If the applied current $I(t)$ is a DC turned on at $t = 0$, as in Figure 4.12, the solution of [4.172] is:

$$\lim_{\epsilon \longrightarrow 0+} \int_{-\epsilon}^{t} \frac{d}{dt'}\left(e^{t'/\tau} i(t')\right) dt' = -\frac{M(r_0)}{L} \lim_{\epsilon \longrightarrow 0+} \int_{-\epsilon}^{t} \frac{dI}{dt} e^{t'/\tau} dt' \quad [4.173]$$

which gives $e^{t/\tau} i(t) = -\frac{M(r_0)I_0}{R} \lim_{\Delta t \longrightarrow 0} \frac{e^{\Delta t/\tau} - 1}{\Delta t} = -\frac{M(r_0)I_0}{L}$ and, consequently:

$$i(t) = -\frac{M(r_0) I_0}{L} e^{-t/\tau} \quad t > 0 \quad [4.174]$$

where the impulse term on the right-hand side of [4.174] imposes the initial condition $i(0) = -\frac{M(r_0)I_0}{L}$. The current is negative because Lenz's law requires the self-induced flux to oppose the applied flux.

If, after a long time T, the current I is instantly cut off, as illustrated in Figure 4.13, the solution is given by $e^{t/\tau} i(t) = \frac{M(r_0)I_0}{R} \lim_{\Delta t \longrightarrow 0} \frac{e^{(T+\Delta t)/\tau} - e^{T/\tau}}{\Delta t} = \frac{M(r_0)I_0}{L} e^{T/\tau}$, which results in:

$$i(t) = \frac{M(r_0) I_0}{L} e^{-(t-T)/\tau} \quad t > T \quad [4.175]$$

We observe that the instantaneous suppression of the current (I) reverses the direction of the initial current.

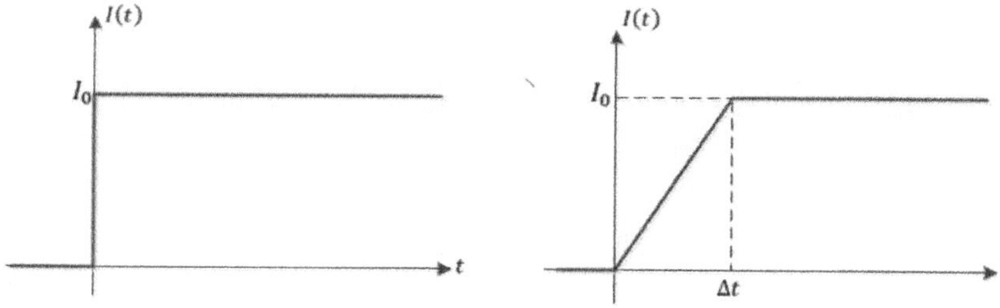

Figure 4.12. *The Heaviside step is the function $H(t)$ equal to zero until $t = 0$ and 1 from $t = 0$. $I(t)$ is written as $I(t) = I_0 H(t)$. The function $H(t)$ is modeled by $I(t) = \lim_{\Delta t \to 0} \begin{cases} 0 & \text{for } t < 0 \\ \dfrac{I_0 t}{\Delta t} & \text{for } 0 \leq t \leq \Delta t \\ I_0 & \text{for } t > \Delta t \end{cases}$. For a color version of this figure, see www.iste.co.uk/abdesselam/electrodynamics2.zip*

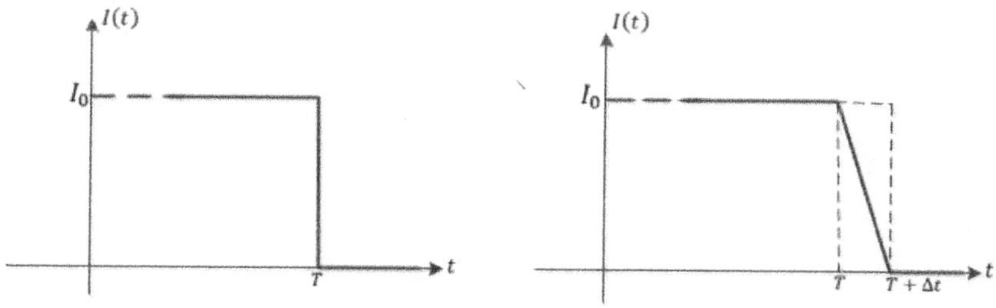

Figure 4.13. *When the current $I(t)$ is instantaneously cut off, it can be written as $I(t) = I_0 H(T - t)$. For a color version of this figure, see www.iste.co.uk/abdesselam/electrodynamics2.zip*

– *Case 2, motion with a direct current (DC)*: with a DC, the first term on the right-hand side of [4.171] is zero, which implies that:

$$L\frac{di}{dt} + Ri = \frac{\mu_0 I D d}{2\pi r(r+d)}\frac{dr}{dt} \qquad [4.176]$$

To solve this equation, we need to specify the motion of the loop, that is, determine the expression of r as a function of time. Let us consider the simplest case, where the loop has no resistance ($R = 0$). In this case, the previous equation can be directly integrated as follows:

$$Li = -\frac{\mu_0 I D}{2\pi}\ln\frac{1+\frac{d}{r}}{1+\frac{d}{r_0}} \qquad [4.177]$$

where the current is zero when $r = r_0$. We can clearly see that the current is positive when $r > r_0$ because the imposed flux decreases. In contrast, the current is negative when $r < r_0$ because the imposed flux increases. In all cases, the total force exerted on the loop results from the forces applied to the branches oriented along Oz at r and $r + dr$:

$$f_r = \frac{\mu_0 DiI}{2\pi}\left(\frac{1}{r+d} - \frac{1}{r}\right) = -\frac{\mu_0 DiId}{2\pi r\,(r+d)} \qquad [4.178]$$

This force is attractive for $iI > 0$ and repulsive for $iI < 0$.

– *Case 3, open-circuit loop*: if the loop is open-circuit, no current can flow, and therefore the electric field in the wire is zero ($J = \gamma E = 0$). Applying Faraday's law to the loop gives: $\oint_L \vec{\mathbf{E}} \cdot d\,\vec{\mathbf{l}} = \int_{\text{negative terminal}}^{\text{positive terminal}} \vec{\mathbf{E}} \cdot d\,\vec{\mathbf{l}} = -V = -\dfrac{d\Phi}{dt}$ and thus:

$$V = \frac{d\Phi}{dt} \qquad [4.179]$$

The flux in [4.179] is now solely due to the mutual induction flux given by [4.167], since $i = 0$ (the self-induced flux is zero). Therefore, the voltage observed across the moving (open-circuited) loop is:

$$V = M\,(r)\,\frac{dI}{dt} + I\frac{dM\,(r)}{dr}\frac{dr}{dt} \qquad [4.180]$$

EXAMPLE 4.19. (Principle of transformers).– Consider two solenoids, 1 and 2, which are strongly coupled. To ensure this coupling, the solenoids can be wound face-to-face, having the same cross-sectional area S and the same length, on a laminated core (referred to as a *magnetic circuit*) to facilitate magnetic field transfer from one solenoid to the other, as shown in Figure 4.14. The magnetic flux through solenoid 2 due to the magnetic field $\vec{\mathbf{B}}_1$ generated by solenoid 1 is given by: $\Phi_{21} = B_1 \cdot n_2 lS = \mu_0 n_1 n_2 lSI_1$, which leads to $M_{2,1} = \mu_0 n_1 n_2 lS$. Similarly, the flux through solenoid 1 due to the magnetic field $\vec{\mathbf{B}}_2$ generated by solenoid 2 is $\Phi_{12} = B_2 \cdot n_1 lS = \mu_0 n_1 n_2 lSI_2$, which implies $M_{1,2} = \mu_0 n_1 n_2 lS$. It is clear that the coupling between the two solenoids is maximal, that is, $M_{1,2} = M_{2,1} = M = \sqrt{L_1 L_2}$, where L_1 and L_2 are the self-inductances of solenoids 1 and 2, respectively. Using the generator convention and neglecting their resistances, the voltages across their terminals are given by:

$$\begin{aligned} V_1 &= -\mathcal{E}_1 = L_1\frac{dI_1}{dt} + M\frac{dI_2}{dt} \\ V_2 &= -\mathcal{E}_2 = M\frac{dI_1}{dt} + L_2\frac{dI_2}{dt} \end{aligned} \qquad [4.181]$$

We therefore deduce that:

$$V_2 = \frac{M_{12}}{L_1}V_1 + \frac{L_1 L_2 - M^2}{L_1}\frac{dI_2}{dt} = \frac{M}{L_1}V_1 = \sqrt{\frac{L_2}{L_1}}V_1 = \frac{N_2}{N_1}V_1 \qquad [4.182]$$

where N_1 (respectively, N_2) is the number of turns in circuit 1 (respectively, 2). Adjusting the ratio $\frac{N_2}{N_1}$, we can transform V_1 into another voltage V_2, either higher or lower. This is the *principle of a transformer*.

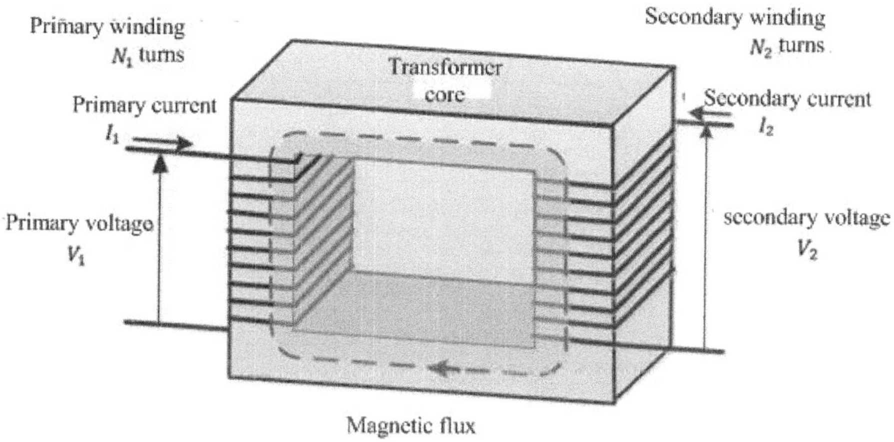

Figure 4.14. *Transformer principle. For a color version of this figure, see www.iste.co.uk/abdesselam/electrodynamics2.zip*

EXAMPLE 4.20. (RL Circuit).– Let us consider the circuit shown in Figure 4.15(a). At time $t = 0$, the switch is closed in position A:

1) *the current starts flowing through the coil*;

2) *the flux of the magnetic field* $\vec{\mathbf{B}}$ *created by the coil through itself starts to increase*;

3) *appearance of an induced EMF, which acts to oppose the causes that generated it:* $\mathcal{E} = -\frac{d\phi}{dt} = -L\frac{dI}{dt}$;

4) *since the circuit is conductive, an induced current appears, whose direction is such that the magnetic induction field it creates opposes the increase in flux.*

This is a transient phenomenon that affects only the establishment of the steady-state regime for a short period of time. To determine the evolution of the current in the circuit, we must first apply Ohm's law, $E + \mathcal{E} = E - L\frac{dI}{dt} = RI(t)$, or equivalently:

$$\frac{dI}{dt} + \frac{I}{\tau} = \frac{E}{L} \qquad [4.183]$$

where $\tau = \frac{L}{R}$. This is a first-order differential equation with a non-homogeneous term. The general solution is the sum of a particular solution of the equation with the non-homogeneous term and the solution of the corresponding homogeneous equation:

– *particular solution*: $I(t) = \frac{E}{R} = I_0$;

– *solution of the equation without the second member*: $\frac{dI}{dt} = -\frac{I}{\tau}$, which gives $\frac{dI}{I} = -\frac{dt}{\tau} \Longrightarrow I(t) = Ke^{-t/\tau}$.

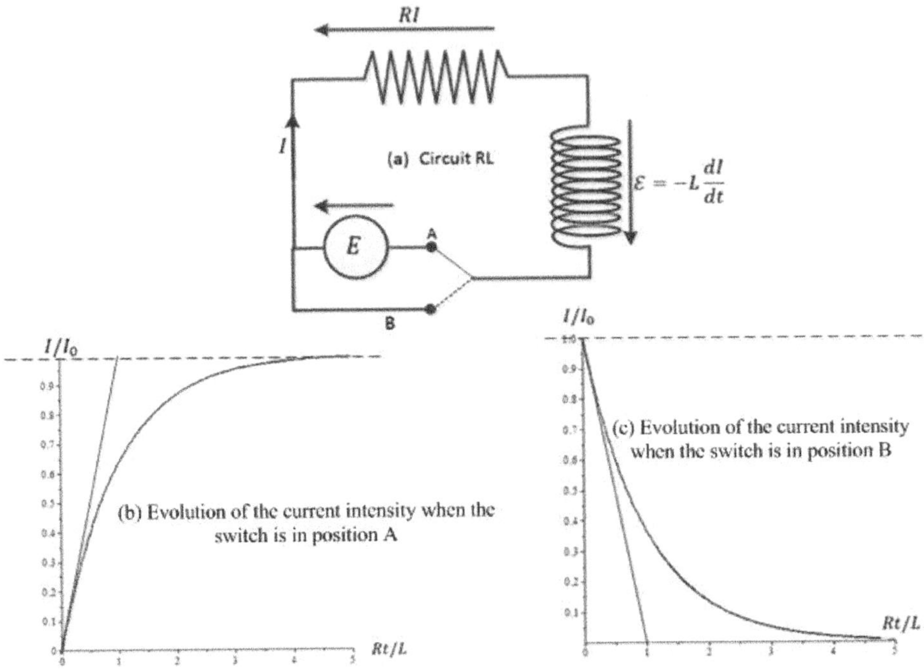

Figure 4.15. *Evolution of the current intensity in an RL circuit*

Hence, the general solution is given by $I(t) = I_0 + Ke^{-t/\tau}$. The constant K is determined by the initial conditions: at $t = 0$, we have $I(0) = 0$. This gives $K = -I_0$, and consequently:

$$I(t) = I_0 \left[1 - e^{-t/\tau}\right] \qquad [4.184]$$

The evolution of the current is illustrated in Figure 4.15(b). Multiplying equation [4.183] by I, we obtain $EI = RI^2 + \frac{d}{dt}\left(\frac{1}{2}LI^2\right)$. This confirms the expression for the energy stored in the coil in the form of magnetic energy, $W_m = \frac{1}{2}LI^2$. Here, EI and RI^2 represent the power supplied by the voltage source E and the power dissipated due to Joule effect, respectively. Similarly, if the switch is flipped to position B, the current will not drop instantaneously because the coil will respond by generating a current that opposes the decrease in the magnetic flux $\vec{\mathbf{B}}$ through itself, as shown in Figure 4.15(c). This current is given by:

$$I(t) = I_0 e^{-t/\tau} \qquad [4.185]$$

EXAMPLE 4.21. (Mutual inductance between two circuits).– We consider the two RLC circuits coupled by mutual induction, as illustrated in Figure 4.16. The circuit on the left is excited by a sinusoidal voltage $V(t)$. Let us analyze the current in each circuit in the steady-state regime. At any instant t, the equations are:

$$\left\{\begin{array}{l} L\dfrac{dI_1}{dt}+RI_1+\dfrac{Q_1}{C_1}+M\dfrac{dI_2}{dt}=V(t) \\ L\dfrac{dI_2}{dt}+RI_2+\dfrac{Q_2}{C_2}+M\dfrac{dI_1}{dt}=0 \end{array}\right.$$
$$\Longrightarrow \left\{\begin{array}{l} L\dfrac{d^2I_1}{dt^2}+R\dfrac{dI_1}{dt}+\dfrac{I_1}{C_1}+M\dfrac{d^2I_2}{dt^2}=\dfrac{dV(t)}{dt} \\ L\dfrac{d^2I_2}{dt^2}+R\dfrac{dI_2}{dt}+\dfrac{I_2}{C_2}+M\dfrac{d^2I_1}{dt^2}=0 \end{array}\right. \quad [4.186]$$

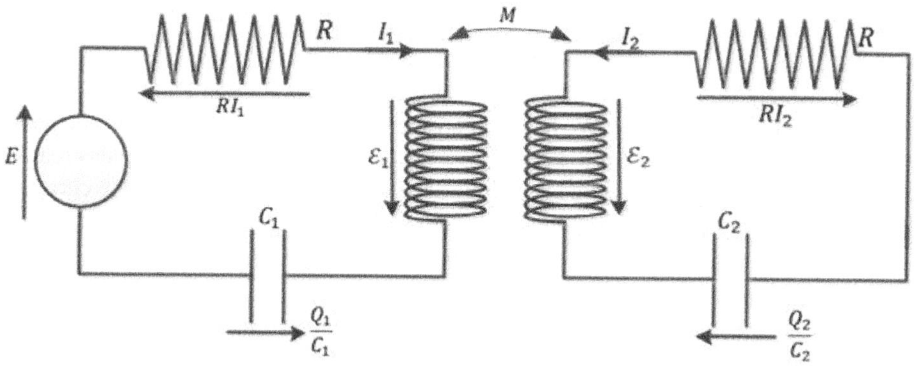

Figure 4.16. *Two circuits with mutual inductance*

In complex notation (i.e. taking $V(t)=\hat{V}e^{i\omega t}$, $I_1(t)=\hat{I}_1e^{i\omega t}$ and $I_2(t)=\hat{I}_2e^{i\omega t}$), these equations simplify to $-L\omega^2\hat{I}_1+iR\omega\hat{I}_1+\frac{\hat{I}_1}{C_1}-M\omega^2\hat{I}_2=i\omega\hat{V}$ and $-L\omega^2\hat{I}_2+iR\omega\hat{I}_2+\frac{\hat{I}_2}{C_2}-M\omega^2\hat{I}_1=0$, which leads to:

$$\begin{array}{l} (R+iZ_1)\,\hat{I}_1+i\omega M\hat{I}_2=\hat{V} \\ (R+iZ_2)\,\hat{I}_2+i\omega M\hat{I}_1=0 \end{array} \quad [4.187]$$

where $Z_1=L\omega-\frac{1}{C_1\omega}$ and $Z_2=L\omega-\frac{1}{C_2\omega}$. The solutions are:

$$\begin{array}{l} \hat{I}_1=\dfrac{Z_2\hat{V}}{Z_1Z_2+M^2\omega^2} \\ \hat{I}_2=\dfrac{-i\omega M\hat{V}}{Z_1Z_2+M^2\omega^2} \end{array} \quad [4.188]$$

By multiplying the first set of equations in [4.186], respectively, by I_1 and I_2, and then summing them term by term, we obtain:

$$\begin{aligned}\underbrace{V(t)\,I_1}_{\text{Power supplied}} &= \underbrace{RI_1^2 + RI_2^2}_{\text{Dissipated power}} + \frac{d}{dt}\underbrace{\left(\frac{1}{2}\frac{Q_1^2}{C_1} + \frac{1}{2}\frac{Q_2^2}{C_2}\right)}_{\text{Electrostatic energy}} \\ &+ \frac{d}{dt}\underbrace{\left(\frac{1}{2}LI_1^2 + \frac{1}{2}LI_2^2 + MI_1I_2\right)}_{\text{Electromagnetic energy}}\end{aligned} \qquad [4.189]$$

4.6. Electromagnetic energy

4.6.1. *Magnetic energy of a set of n thin-wire circuits*

Let a system be composed of n closed or quasi-closed circuits $\Gamma_1, \ldots, \Gamma_n$. Each circuit Γ_i carries a current I_i. We have already seen that the magnetic flux Φ_i through the circuit Γ_i must take the form $\Phi_i = \sum_{j=1}^{n} M_{i,j} I_j$, since the magnetic fields created by the other circuits Γ_j (including the circuit Γ_i itself) are proportional to the currents I_j flowing through Γ_j. The flux Φ_i induces an EMF around circuit Γ_i given by $\mathcal{E}_i = -\frac{d\Phi_i}{dt} = -\sum_{j=1}^{n} M_{i,j}\frac{dI_j}{dt}$. This EMF supplies, during the motion of the circuit Γ_i over the time interval dt, a work that is stored as magnetic potential energy W_m^i within the circuit Γ_i, given by $dW^i = \mathcal{E}_i I_i dt = -dW_m^i$. The total variation of the magnetic potential energy W_m stored in the system over the time dt is obtained by summing over all circuits:

$$\begin{aligned} dW_m &= -\sum_{i=1}^{n} \mathcal{E}_i I_i dt = \sum_{i,j=1}^{n} I_i M_{i,j} dI_j \\ &= d\left[\frac{1}{2}\sum_{i,j=1}^{n} M_{i,j} I_i I_j\right] \end{aligned} \qquad [4.190]$$

since $M_{i,j} = M_{j,i}$. It follows that:

$$W_m = \frac{1}{2}\sum_{i,j=1}^{n} M_{i,j} I_i I_j = \frac{1}{2}\sum_{i=1}^{n} \Phi_i I_i \qquad [4.191]$$

4.6.2. *Magnetic energy of a volume current distribution*

To express W_m as a function of the field $\vec{\mathbf{B}}$, we first replace Φ_i with the expression, $\Phi_i = \int_{S_i} \vec{\mathbf{B}} \cdot d\vec{\mathbf{S}}_i = \oint_{\Gamma_i} \vec{\mathbf{A}} \cdot d\vec{\mathbf{l}}_i$, where $\vec{\mathbf{B}}$ represents the total magnetic field generated by all the circuits:

$$W_m = \frac{1}{2}\sum_{i=1}^{n}\oint_{\Gamma_i} \vec{\mathbf{A}} \cdot I_i d\vec{\mathbf{l}}_i \qquad [4.192]$$

This equation is easily generalized for a volumetric current distribution by substituting $I_i d\vec{\mathbf{l}}_i$ with $\vec{\mathbf{J}}\, d\tau$:

$$W_m = \frac{1}{2}\int_{\tau} \vec{\mathbf{A}} \cdot \vec{\mathbf{J}}\, d\tau \qquad [4.193]$$

where $\vec{\mathbf{J}}$ is the current density of the system. This integral can be extended to the entire space without altering the result, as $\vec{\mathbf{J}}$ is zero outside the current distribution. Thus, we can write:

$$W_m = \frac{1}{2}\int_{\text{Space}} \vec{\mathbf{A}} \cdot \vec{\mathbf{J}}\, d\tau \qquad [4.194]$$

Let us substitute $\vec{\mathbf{J}}$ with $\frac{\vec{\nabla}\times\vec{\mathbf{B}}}{\mu_0}$, and then use the identity, $\left(\vec{\nabla}\times\vec{\mathbf{B}}\right)\cdot\vec{\mathbf{A}} = \vec{\mathbf{B}}\cdot\left(\vec{\nabla}\times\vec{\mathbf{A}}\right) - \vec{\nabla}\cdot\left(\vec{\mathbf{A}}\times\vec{\mathbf{B}}\right) = \left|\vec{\mathbf{B}}\right|^2 - \vec{\nabla}\cdot\left(\vec{\mathbf{A}}\times\vec{\mathbf{B}}\right)$:

$$\begin{aligned} W_m &= \frac{1}{2\mu_0}\int_{\text{Space}} \left(\vec{\nabla}\times\vec{\mathbf{B}}\right)\cdot\vec{\mathbf{A}} d\tau = \frac{1}{2\mu_0}\int_{\text{Space}} \left|\vec{\mathbf{B}}\right|^2 d\tau \\ &- \frac{1}{2\mu_0}\int_{\text{Space}} \vec{\nabla}\cdot\left(\vec{\mathbf{A}}\times\vec{\mathbf{B}}\right)\cdot d\tau = \frac{1}{2\mu_0}\int_{\text{Space}} \left|\vec{\mathbf{B}}\right|^2 d\tau \\ &- \frac{1}{2\mu_0}\lim_{S\longrightarrow\infty}\oint_S \left(\vec{\mathbf{A}}\times\vec{\mathbf{B}}\right)\cdot d\vec{\mathbf{S}} \end{aligned} \qquad [4.195]$$

Since $\vec{\mathbf{A}}\times\vec{\mathbf{B}} \sim \frac{1}{r^3}$ and $\vec{\mathbf{S}} \sim r^2$, it follows that $\left(\vec{\mathbf{A}}\times\vec{\mathbf{B}}\right)\cdot d\vec{\mathbf{S}} \sim \frac{1}{r}$. We therefore conclude that the second integral vanishes as $r \longrightarrow \infty$. Consequently, equation [4.195] simplifies to:

$$W_m = \frac{1}{2\mu_0}\int_{\text{Space}} \left|\vec{\mathbf{B}}\right|^2 d\tau \qquad [4.196]$$

demonstrating that it can be expressed in terms of a *magnetic energy density*:

$$w_m = \frac{\left|\vec{\mathbf{B}}\right|^2}{2\mu_0} \qquad [4.197]$$

The magnetic energy is therefore stored in the system's magnetic field. By replacing $\vec{\mathbf{B}}$ with $\sum_{i=1}^{n}\vec{\mathbf{B}}_i$, we get:

$$W_m = \frac{1}{2\mu_0}\sum_{i,j=1}^{n}\int_{\text{Space}} \vec{\mathbf{B}}_i\cdot\vec{\mathbf{B}}_j d\tau = \frac{1}{2}\sum_{i,j=1}^{n} M_{i,j} I_i I_j \qquad [4.198]$$

The quantities $W_m^i = \frac{1}{2\mu_0}\int_{\text{Space}}\left|\overrightarrow{\mathbf{B}}_i\right|^2 d\tau$ and $W_m^{ij} = \frac{1}{\mu_0}\int_{\text{Space}}\overrightarrow{\mathbf{B}}_i \cdot \overrightarrow{\mathbf{B}}_j d\tau$ represent, respectively, the self-energies and the interaction energies.

EXAMPLE 4.22. (Force between solenoids).– Consider two solenoids (1) and (2) sharing the same axis Oz, with the following characteristics: (1) lengths l_1 and l_2; (2) radii R_1 and R_2 with $R_1 > R_2$; (3) n_1 and n_2 turns of wire per unit length; (4) current intensities I_1 and I_2. Assume that solenoid (2) is partially inside solenoid (1) over a distance $z \ll l_1, l_2$, and let us compute their interaction energy. Since the magnetic fields $\overrightarrow{\mathbf{B}}_1$ and $\overrightarrow{\mathbf{B}}_2$ are, respectively, zero outside solenoids (1) and (2), the only region that contributes to the integral W_m^{12} is the portion of the volume of (2) that is inside (1), as illustrated in Figure 4.17. From this, we immediately deduce:

$$W_m^{12} = \mu_0 I_1 I_2 n_1 n_2 \pi R_2^2 z \quad [4.199]$$

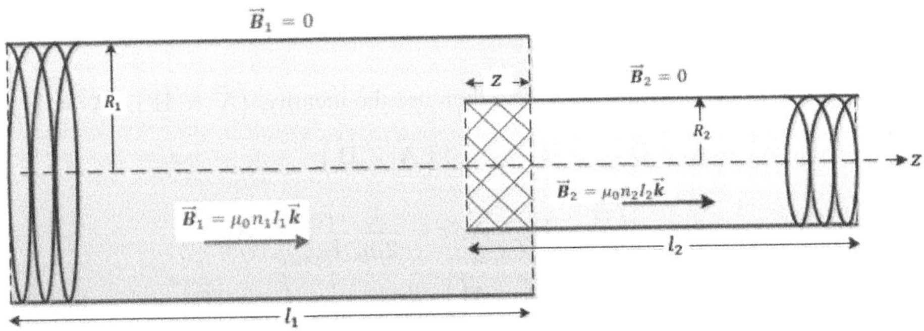

Figure 4.17. *When $I_1 I_2 > 0$, solenoid (2) tends to move further inside solenoid (1), whereas when $I_1 I_2 < 0$, solenoid (2) tends to be pushed out of solenoid (1). For a color version of this figure, see www.iste.co.uk/abdesselam/ electrodynamics2.zip*

The force exerted between the two solenoids can be directly determined from the expression:

$$F_{12} = \frac{\partial W_m^{12}}{\partial z} = \mu_0 I_1 I_2 n_1 n_2 \pi R_2^2 \quad [4.200]$$

It is oriented in the positive direction of Oz if $I_1 I_2 > 0$ and in the opposite direction if $I_1 I_2 < 0$. This result aligns with the maximum flux rule: when $I_1 I_2 > 0$, the flux of $\overrightarrow{\mathbf{B}}_1$ through the turns of (2) is entirely positive. The force acting on (2) will therefore tend to pull the maximum number of turns of (2) inside (1) to increase the flux of $\overrightarrow{\mathbf{B}}_1$.

EXAMPLE 4.23. (Compression effect of turns in a solenoid).– The magnetic energy of a solenoid with length l and containing N turns is given by

$W_m = \frac{1}{2}LI^2 = \frac{1}{2}\mu_0\pi R^2\frac{N^2}{l}I^2$. The resulting force acting on the turns will be calculated in the same way:

$$F = \frac{\partial W_m}{\partial l} = \frac{1}{2}I^2\frac{\partial L}{\partial l} = -\mu_0 I^2\pi R^2\frac{N^2}{l^2} \qquad [4.201]$$

Thus, to keep the length of a solenoid constant, a tensile force equal to $\mu_0 I^2\pi R^2\frac{N^2}{l^2}$ must be applied.

EXAMPLE 4.24. (Magnetic energy of two parallel wires).– Let us calculate the magnetic energy per unit length of a system consisting of two identical parallel wires, each with radius R, separated by a distance d, and carrying equal currents in opposite directions, as shown in Figure 4.18. To do this, we first need to determine the vector potential $\overrightarrow{\mathbf{A}}$ both inside and outside a wire that carries a current I along the Oz direction by solving equation $\Delta\overrightarrow{\mathbf{A}} = -\mu_0\overrightarrow{\mathbf{J}}$. The current density $\overrightarrow{\mathbf{J}}(r)$ is given by $\frac{I\overrightarrow{\mathbf{k}}}{\pi R^2}$ for $r \leq R$ (inside the wire) and 0 for $r > R$ (outside the wire). Since $\overrightarrow{\mathbf{A}}$ is parallel to $\overrightarrow{\mathbf{J}}$ in the Coulomb gauge, we only need to compute A_z. In cylindrical coordinates, the equation $\Delta A_z = -\mu_0 J_z$ can be written as:

$$\frac{1}{r}\frac{d}{dr}\left(r\frac{dA_z}{dr}\right) = \begin{cases} -\frac{\mu_0 I}{\pi R^2} & r \leq R \\ 0 & r > R \end{cases} \qquad [4.202]$$

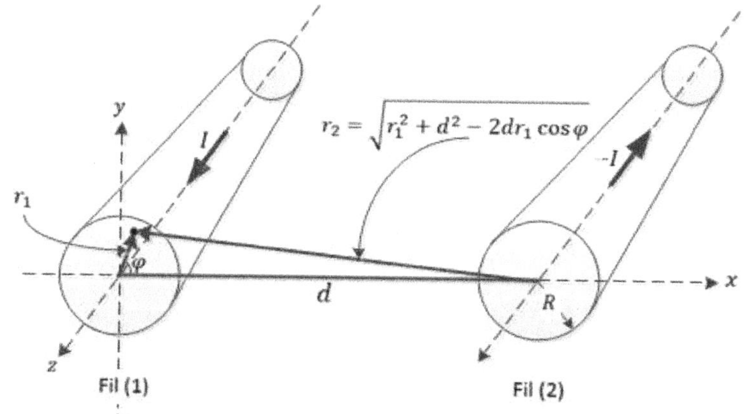

Figure 4.18. *System of two identical parallel wires carrying equal currents in opposite directions*

A_z does not depend on the other variables φ and z (due to symmetry). After integration, we obtain:

$$A_z = \begin{cases} \dfrac{\mu_0 I}{4\pi}\left(D - \dfrac{r^2}{R^2}\right) + C\ln\dfrac{r}{R} & r \leq R \\ E\ln\dfrac{r}{R} + F & r > R \end{cases} \qquad [4.203]$$

where C, D, E and F are constants. To determine them, it suffices to observe that:

– since A_z is finite when $r \longrightarrow 0$, C must be zero;

– the constant E is determined by calculating the magnetic field outside the wire from $\overrightarrow{\mathbf{A}}$ and then comparing the obtained result with the expression $\frac{\mu_0 I}{2\pi r}$, i.e. $B_\varphi = \left(\overrightarrow{\nabla} \times \overrightarrow{\mathbf{A}}\right)_\varphi = -\frac{\partial A_z}{\partial r} = -\frac{E}{r} = \frac{\mu_0 I}{2\pi r}$, we obtain $E = -\frac{\mu_0 I}{2\pi}$;

– the potential A_z is continuous at the boundary $r = R$, that is, $\frac{\mu_0 I}{4\pi}(D-1) = F$, which gives, $F = \frac{\mu_0 I}{4\pi}(D-1)$.

Therefore, equation [4.203] reduces to:

$$A_z = \begin{cases} \dfrac{\mu_0 I}{4\pi}\left(D - \dfrac{r^2}{R^2}\right) & r \leq R \\ \dfrac{\mu_0 I}{4\pi}\left(D - 1 - \ln\dfrac{r^2}{R^2}\right) & r > R \end{cases} \qquad [4.204]$$

At point M, located inside the wire (1), the vector potential is the superposition of its own potential $A_z^{(1)}(r_1 \leq R, I)$ and the external potential $A_z^{(2)}(r_2 > R, -I)$ created by wire (2), which carries the same current but in the opposite direction. Adding $A_z^{(2)}(r_2 > R, -I)$ to $A_z^{(1)}(r_1 \leq R, I)$, we obtain:

$$\begin{aligned} A_z(M) &= \frac{\mu_0 I}{4\pi}\left(D - \frac{r_1^2}{R^2}\right) - \frac{\mu_0 I}{4\pi}\left(D - 1 - \ln\frac{r_2^2}{R^2}\right) \\ &= \frac{\mu_0 I}{4\pi}\left(1 - \frac{r_1^2}{R^2} + \ln\frac{r_2^2}{R^2}\right) \end{aligned} \qquad [4.205]$$

where $r_2^2 = r_1^2 + d^2 - 2r_1 d\cos\varphi$. Substitute this result into $\overrightarrow{\mathbf{J}} \cdot \overrightarrow{\mathbf{A}}$ and then integrate over a segment of length l of the wire (1):

$$\begin{aligned} &\int_{\text{wire(1)}} \overrightarrow{\mathbf{J}} \cdot \overrightarrow{\mathbf{A}}\, d\tau \\ &= \frac{\mu_0 I^2}{4\pi^2 R^2}\int_0^R\int_0^{2\pi}\int_0^l \left(1 - \frac{r_1^2}{R^2} + \ln\frac{r_1^2 + d^2 - 2r_1 d\cos\varphi}{R^2}\right) r_1 dr_1 d\varphi dz \\ &= \frac{\mu_0 I^2 l}{4\pi^2 R^2}\int_0^R\int_0^{2\pi} \left(1 - \frac{r_1^2}{R^2} + \ln\frac{r_1^2 + d^2 - 2r_1 d\cos\varphi}{R^2}\right) r_1 dr_1 d\varphi \end{aligned} \qquad [4.206]$$

To compute the second integral on the right-hand side, we observe that:

$$\begin{aligned}
&\int_0^{2\pi} \ln\left[\frac{r_1^2+d^2}{R^2} - 2\frac{r_1 d}{R^2}\cos\varphi\right] d\varphi = \int_0^{\pi} \ln\left[\left(\frac{r_1^2+d^2}{R^2}\right)^2 - 4\frac{r_1^2 d^2}{R^4}\cos^2\varphi\right] d\varphi \\
&= \frac{1}{2}\int_0^{2\pi} \ln\left[\frac{r_1^4+d^4}{R^4} - 2\frac{r_1^2 d^2}{R^4}\cos\varphi\right] d\varphi \\
&= \frac{1}{2^2}\int_0^{2\pi} \ln\left[\frac{r_1^8+d^8}{R^8} - 2\frac{r_1^4 d^4}{R^8}\cos\varphi\right] d\varphi \\
&\vdots \\
&= \frac{1}{2^{n-1}}\int_0^{2\pi} \ln\left[\frac{r_1^{2^n}+d^{2^n}}{R^{2^n}} - 2\frac{r_1^{2^{n-1}} d^{2^{n-1}}}{R^{2^n}}\cos\varphi\right] d\varphi
\end{aligned} \qquad [4.207]$$

By taking $n \longrightarrow \infty$, the previous integral simplifies to:

$$\int_0^{2\pi} \ln\left[\frac{r_1^2+d^2}{R^2} - 2\frac{r_1 d}{R^2}\cos\varphi\right] d\varphi = 2\int_0^{2\pi} \ln\frac{d}{R} d\varphi = 4\pi \ln\frac{d}{R} \qquad [4.208]$$

since $r_1 \leq R$. It follows that:

$$\begin{aligned}
\int_{\text{wire(1)}} \vec{\mathbf{J}} \cdot \vec{\mathbf{A}} d\tau &= \frac{\mu_0 I^2 l}{4\pi^2 R^2} \int_0^R \left[\left(1 - \frac{r_1^2}{R^2}\right) 2\pi + 4\pi \ln\frac{d}{R}\right] r_1 dr_1 \\
&= \frac{\mu_0 I^2 l}{4\pi}\left(\frac{1}{2} + 2\ln\frac{d}{R}\right)
\end{aligned} \qquad [4.209]$$

To ultimately determine the total magnetic energy of the system, we compute the sum:

$$W_m = \frac{1}{2}\int_{\text{wire(1)}} \vec{\mathbf{J}} \cdot \vec{\mathbf{A}} d\tau + \frac{1}{2}\int_{\text{wire(2)}} \vec{\mathbf{J}} \cdot \vec{\mathbf{A}} d\tau = \frac{\mu_0 I^2 l}{4\pi}\left(\frac{1}{2} + 2\ln\frac{d}{R}\right) \qquad [4.210]$$

By writing $W_m = \frac{1}{2}LI^2$, we obtain the following expression for the system's self-inductance:

$$L = \frac{\mu_0 l}{4\pi}\left(1 + 4\ln\frac{d}{R}\right) \qquad [4.211]$$

Between the two wires, a repulsive force is exerted and given by:

$$F = \frac{\partial W_m}{\partial d} = \frac{\mu_0 l I^2}{2\pi d} \qquad [4.212]$$

4.6.3. *Poynting's theorem*

Consider a charge distribution $\mathcal{D}$ with a density ρ immersed in an electromagnetic field. The work done by the Lorentz force on the charge element dq contained in an infinitesimal volume $d\tau$, moving through the field with velocity $\vec{\mathbf{v}}$, during a displacement $d\vec{\mathbf{l}}$, is given by:

$$d^2W = dq\left(\vec{\mathbf{E}} + \vec{\mathbf{v}} \times \vec{\mathbf{B}}\right)\cdot d\vec{\mathbf{l}} = dq\left(\vec{\mathbf{E}} + \vec{\mathbf{v}} \times \vec{\mathbf{B}}\right)\cdot \vec{\mathbf{v}}dt = \rho\vec{\mathbf{E}}\cdot\vec{\mathbf{v}}d\tau dt \quad [4.213]$$

Thus, the total *electromagnetic power* received by the distribution is:

$$\mathsf{P}_{\mathcal{D}} = \frac{dW}{dt} = \int_{\mathcal{D}} \rho\vec{\mathbf{E}}\cdot\vec{\mathbf{v}}\,d\tau = \int_{\mathcal{D}} \vec{\mathbf{E}}\cdot\vec{\mathbf{J}}\,d\tau \quad [4.214]$$

Two possibilities can be envisioned:

– Integral [4.214] can be extended to the entire space, including regions where $\vec{\mathbf{J}}$ is zero:

$$\mathsf{P}_{\mathcal{D}} = \int_{\text{Space}} \vec{\mathbf{E}}\cdot\vec{\mathbf{J}}\,d\tau \quad [4.215]$$

To express [4.215] in terms of $\vec{\mathbf{E}}$ and $\vec{\mathbf{B}}$, we use the Maxwell–Ampére equation, $\vec{\mathbf{J}} = \frac{\vec{\nabla}\times\vec{\mathbf{B}}}{\mu_0} - \varepsilon_0\frac{\partial\vec{\mathbf{E}}}{\partial t}$:

$$\mathsf{P}_{\mathcal{D}} = \int_{\text{Space}} \vec{\mathbf{E}}\cdot\left[\frac{\vec{\nabla}\times\vec{\mathbf{B}}}{\mu_0} - \varepsilon_0\frac{\partial\vec{\mathbf{E}}}{\partial t}\right]d\tau = \int_{\text{Space}}\left[-\vec{\nabla}\cdot\left(\frac{\vec{\mathbf{E}}\times\vec{\mathbf{B}}}{\mu_0}\right)\right. \quad [4.216]$$

$$\left. + \frac{1}{\mu_0}\vec{\mathbf{B}}\cdot\left(\vec{\nabla}\times\vec{\mathbf{E}}\right) - \varepsilon_0\vec{\mathbf{E}}\cdot\frac{\partial\vec{\mathbf{E}}}{\partial t}\right]d\tau = -\int_{\text{Space}}\frac{\partial}{\partial t}\left(\frac{\varepsilon_0\left|\vec{\mathbf{E}}\right|^2}{2} + \frac{\left|\vec{\mathbf{B}}\right|^2}{2\mu_0}\right)d\tau$$

$$-\int_{\text{Space}}\vec{\nabla}\cdot\left(\frac{\vec{\mathbf{E}}\times\vec{\mathbf{B}}}{\mu_0}\right)d\tau = -\int_{\text{Space}}\frac{\partial w_{em}}{\partial t}d\tau - \lim_{S\longrightarrow\infty}\oint_S\frac{\vec{\mathbf{E}}\times\vec{\mathbf{B}}}{\mu_0}\cdot d\vec{\mathbf{S}}$$

where:

$$w_{em} = \frac{1}{2}\varepsilon_0\left|\vec{\mathbf{E}}\right|^2 + \frac{1}{2\mu_0}\left|\vec{\mathbf{B}}\right|^2 \quad [4.217]$$

The last integral in [4.216] is zero because $\vec{\mathbf{E}}\times\vec{\mathbf{B}} \sim \frac{1}{r^4}$ and $\vec{\mathbf{S}} \sim r^2$, which implies that $\left(\vec{\mathbf{E}}\times\vec{\mathbf{B}}\right)\cdot d\vec{\mathbf{S}} \sim \frac{1}{r^2} \to 0$ as $r \to 0$. Finally, we obtain:

$$\mathsf{P}_{\mathcal{D}} = -\int_{\text{Space}}\frac{\partial w_{em}}{\partial t}d\tau = -\frac{d}{dt}\int_{\text{Space}} w_{em}d\tau \quad [4.218]$$

– By restricting ourselves to the distribution and repeating the same procedure as in [4.216], we obtain:

$$\mathsf{P}_{\mathcal{D}} = -\frac{d}{dt}\int_{\mathcal{D}} w_{em} d\tau - \oint_{S_{\mathcal{D}}} \frac{\vec{\mathbf{E}} \times \vec{\mathbf{B}}}{\mu_0} \cdot d\vec{\mathbf{S}} \qquad [4.219]$$

where $S_{\mathcal{D}}$ represents the surface surrounding the distribution.

4.6.3.1. *Integral form of Poynting's theorem*

By comparing the results of [4.218] and [4.219], we deduce that:

$$\frac{d}{dt}\left[\int_{\text{Space}} w_{em} d\tau - \int_{\mathcal{D}} w_{em} d\tau\right] = \oint_{S_{\mathcal{D}}} \frac{\vec{\mathbf{E}} \times \vec{\mathbf{B}}}{\mu_0} \cdot d\vec{\mathbf{S}} \qquad [4.220]$$

This is the integral form of *Poynting's theorem*. The quantity w_{em} can be interpreted as the *electromagnetic energy density* associated with the field. Its integral over the distribution (or, over the whole space) represents the electromagnetic energy $W_{em}^{\mathcal{D}} = W_e^{\mathcal{D}} + W_m^{\mathcal{D}}$ ((or, $W_{em}^{\text{Space}} = W_e^{\text{Space}} + W_m^{\text{Space}}$) contained within $\mathcal{D}$ (or within the entire space). The vector quantity:

$$\vec{\mathbf{\Pi}} = \frac{\vec{\mathbf{E}} \times \vec{\mathbf{B}}}{\mu_0} \qquad [4.221]$$

is the *Poynting vector*. The quantity $\vec{\mathbf{g}} = \frac{\vec{\mathbf{\Pi}}}{c^2} = \frac{\vec{\mathbf{E}} \times \vec{\mathbf{B}}}{c^2 \mu_0} = \varepsilon_0 \vec{\mathbf{E}} \times \vec{\mathbf{B}}$ is also interpreted as the *momentum density* or *impulse density* associated with the electromagnetic field. It corresponds to the flux or radiation of the electromagnetic energy and represents the *electromagnetic energy current density vector*. Its flux across a surface, whether closed or not, represents the electromagnetic power passing through that surface. Poynting's theorem expresses the conservation of energy: *the variation of electromagnetic energy within a volume is due to either the work done by electromagnetic forces on the charges in that volume or (inclusively) to an exchange of energy with the exterior through the surface of the volume*. In other words, the transfer of electromagnetic energy from the exterior to the distribution occurs through the Poynting vector.

4.6.3.2. *Differential form of Poynting's theorem*

To obtain the differential form of Poynting's theorem, it suffices to write:

$$-\int_{\mathcal{D}} \frac{\partial w_{em}}{\partial t} d\tau - \int_{\mathcal{D}} \vec{\nabla} \cdot \vec{\mathbf{\Pi}}\tau = \int_{\mathcal{D}} \vec{\mathbf{E}} \cdot \vec{\mathbf{J}} d\tau \qquad [4.222]$$

Since the distribution $\mathcal{D}$ is arbitrary, it follows that the following expression holds at every point in space:

$$\frac{\partial w_{em}}{\partial t} + \vec{\nabla} \cdot \vec{\mathbf{\Pi}} = -\vec{\mathbf{E}} \cdot \vec{\mathbf{J}} \qquad [4.223]$$

This is the *differential form of Poynting's theorem.* On the right-hand side, we recognize the work done per unit time and per unit volume by the electric field on moving charges, which causes the dissipation of electromagnetic energy. To illustrate this, here are a few examples.

EXAMPLE 4.25. (Cylindrical conductor).– Consider an infinite cylindrical conductor of radius R, carrying an electric current I along its length in the Oz direction. We focus on the section of the conductor located between the planes $z = 0$ and $z = l$. Let V denote the potential difference between these two planes. The electric field inside the conductor is uniform and given by $\overrightarrow{\mathbf{E}} = \frac{V}{l}\overrightarrow{\mathbf{k}}$. In cylindrical coordinates, the magnetic field created by the current on the surface of the conductor is $\overrightarrow{\mathbf{B}} = \frac{\mu_0 I}{2\pi R}\overrightarrow{\mathbf{e}}_{\varphi}$. Therefore, the Poynting vector is given by:

$$\overrightarrow{\Pi} = \frac{\overrightarrow{\mathbf{E}} \times \overrightarrow{\mathbf{B}}}{\mu_0} = \frac{VI}{S}(-\overrightarrow{\mathbf{e}}_r) \qquad [4.224]$$

with $S = 2\pi Rl$. It is directed toward the interior of the conductor, as shown in Figure 4.19(a). Its flux across the surface of the conductor segment is given as:

$$\overrightarrow{\Pi} \cdot (-\overrightarrow{\mathbf{e}}_r S) = VI \qquad [4.225]$$

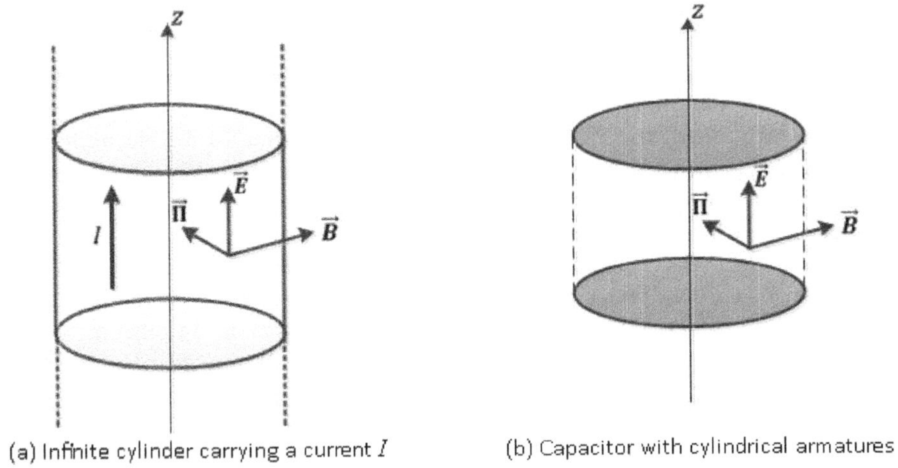

Figure 4.19. *Interpretation of the Poynting vector: (a) infinite cylinder; (b) cylindrical capacitor*

It is equal to the electrical power supplied between the two planes $z = 0$ and $z = l$ by the electric field that drives the current flow. Thus, we conclude that the Poynting vector transports the power necessary for the movement of charges within the conductor through its surface.

EXAMPLE 4.26.– Consider a capacitor with disk-shaped electrodes sharing the same Oz axis, each with a radius R, where the first electrode is at $z = 0$ and the second is at $z = l$. The space between the electrodes is vacuum. We assume $l \ll R$ to neglect edge effects. At $t = 0$, the capacitor is connected to a voltage source E. Let $V(t)$ be the potential difference between the two electrodes at time t. At that moment, the electric field between the electrodes is given by $\overrightarrow{\mathbf{E}} = \frac{V(t)}{l}\overrightarrow{\mathbf{k}}$. Since this field varies over time, the Maxwell–Ampére equation predicts the emergence of a magnetic field, which can be determined using $\overrightarrow{\nabla} \times \overrightarrow{\mathbf{B}} = \varepsilon_0\mu_0\frac{\partial \overrightarrow{\mathbf{E}}}{\partial t} = \mu_0\frac{\varepsilon_0}{l}\frac{dV(t)}{dt}\overrightarrow{\mathbf{k}}$. For symmetry reasons, this magnetic field must take the form $\overrightarrow{\mathbf{B}} = B\overrightarrow{\mathbf{k}}$, leading to $\overrightarrow{\nabla} \times \overrightarrow{\mathbf{B}} = \frac{1}{r}\frac{\partial(rB)}{\partial r}\overrightarrow{\mathbf{e}}_\varphi = \mu_0\frac{\varepsilon_0}{l}\frac{dV(t)}{dt}\overrightarrow{\mathbf{e}}_\varphi$, which gives $\overrightarrow{\mathbf{B}} = \mu_0\frac{\varepsilon_0}{l}\frac{dV(t)}{dt}\frac{r}{2}\overrightarrow{\mathbf{e}}_\varphi$. This magnetic field also varies over time and should, in principle, modify the previously described electric field. For simplicity, we assume that these modifications are negligible. Under these conditions, at the edge of the capacitor (i.e. for $r = R$), the Poynting vector appears as:

$$\overrightarrow{\Pi} = \frac{\overrightarrow{\mathbf{E}} \times \overrightarrow{\mathbf{B}}}{\mu_0} = \frac{\varepsilon_0 R}{2l^2}V\frac{dV}{dt}\left(-\overrightarrow{\mathbf{e}}_r\right) \qquad [4.226]$$

oriented toward the interior of the capacitor, as shown in Figure 4.19(b). Its inward flux through the cylindrical lateral surface enclosing the space between the electrodes is given as:

$$\overrightarrow{\Pi} \cdot \left(-2\pi Rl\overrightarrow{\mathbf{e}}_r\right) = \frac{\varepsilon_0\pi R^2}{l}V\frac{dV}{dt} = CV\frac{dV}{dt} \qquad [4.227]$$

where $C = \frac{\varepsilon_0\pi R^2}{l}$ is the capacitance of the capacitor. Equation [4.227] represents the electromagnetic power that accumulates between the electrodes during the capacitor charging process. Thus, the total energy stored in the capacitor is obtained by integrating this power and is given by:

$$W_e = \int_0^\infty \overrightarrow{\Pi} \cdot \left(-2\pi Rl\overrightarrow{\mathbf{e}}_r\right)dt = \frac{1}{2}CV^2 = \frac{1}{2}\varepsilon_0 E^2\pi R^2 l \qquad [4.228]$$

which correctly corresponds to the expected result for the final potential energy of the charged capacitor. A similar calculation can be performed to determine the magnetic energy stored in a solenoid once the current has been established.

EXAMPLE 4.27. (Energy transported by an electromagnetic wave).– Let us calculate the energy flux of a progressive plane electromagnetic wave. For a progressive plane wave propagating along $+\overrightarrow{\mathbf{n}}$, the fields $\overrightarrow{\mathbf{E}}$ and $\overrightarrow{\mathbf{B}}$ are related by equations $\overrightarrow{\mathbf{E}} = -c\overrightarrow{\mathbf{n}} \times \overrightarrow{\mathbf{B}}$ and $\overrightarrow{\mathbf{B}} = \frac{1}{c}\overrightarrow{\mathbf{n}} \times \overrightarrow{\mathbf{E}}$. The vectors $\overrightarrow{\mathbf{n}}$, $\overrightarrow{\mathbf{E}}$ and $\overrightarrow{\mathbf{B}}$ form a right-handed triad, and we have $\left|\overrightarrow{\mathbf{E}}\right| = c\left|\overrightarrow{\mathbf{B}}\right|$. For the progressive wave, we verify that:

$$\vec{\mathbf{\Pi}} = \frac{\vec{\mathbf{E}} \times \vec{\mathbf{B}}}{\mu_0} = \frac{\left|\vec{\mathbf{E}}\right|^2}{\mu_0 c}\vec{\mathbf{n}}$$
$$w_{\mathrm{EB}} = \frac{\varepsilon_0 \left|\vec{\mathbf{E}}\right|^2}{2} + \frac{\left|\vec{\mathbf{B}}\right|^2}{2\mu_0} = \varepsilon_0 \left|\vec{\mathbf{E}}\right|^2 = \frac{\left|\vec{\mathbf{B}}\right|^2}{\mu_0} \qquad [4.229]$$

We verify that, during a time interval dt, the energy $\vec{\mathbf{\Pi}} \cdot d\vec{\mathbf{S}}\,dt$ crossing a surface element dS is equal to the energy contained in the volume element $d\tau$, given by $d\tau$ $d\tau = d\vec{\mathbf{S}} \cdot cdt\,\vec{\mathbf{n}}$, which means $w_{\mathrm{EB}} d\tau$. This confirms that the energy propagates at the wave velocity $c\vec{\mathbf{n}}$ and crosses the surface $d\vec{\mathbf{S}}$ during dt. The intensity of a plane wave is defined as the time-averaged magnitude of the electromagnetic energy flux density, that is, the magnitude of the Poynting vector:

$$I_{\mathrm{Lum.}} = \left\langle \left|\vec{\mathbf{\Pi}}\right| \right\rangle \qquad [4.230]$$

4.7. Maxwell tensor

4.7.1. *Maxwell stress tensor for the electric field*

Consider a charge distribution with density ρ immersed in an electrostatic field $\vec{\mathbf{E}}$. The total electrostatic force acting on the entire distribution is given by $\vec{\mathbf{F}} = \int_\tau \rho \vec{\mathbf{E}}\, d\tau$, where τ represents the volume occupied by the distribution. Using the Maxwell–Gauss equation, this force can be expressed exclusively in terms of $\vec{\mathbf{E}}$ as follows:

$$\vec{\mathbf{F}} = \varepsilon_0 \int_\tau \left(\vec{\nabla} \cdot \vec{\mathbf{E}}\right) \vec{\mathbf{E}}\, d\tau \qquad [4.231]$$

which, in tensor notation, also takes the form:

$$F^i = \varepsilon_0 \int_\tau \left(\partial_j E^j\right) E^i d\tau = \varepsilon_0 \int_\tau \left[\partial_j \left(E^i E^j\right) - E^j \partial_j E^i\right] d\tau \qquad [4.232]$$

To apply the divergence theorem, we need to transform the second term into an appropriate form using the Maxwell–Faraday equation, $\vec{\nabla} \times \vec{\mathbf{E}} = 0$. In tensor notation, this equation is written as $\epsilon^{ijk}\partial_j E_k = 0$, which implies that $\partial_j E_k = \partial_k E_j$. Multiplying this equation by g^{ik} (the metric tensor), we obtain: $\partial_j E^i = g^{ik}\partial_j E_k = g^{ik}\partial_k E_j = \partial^i E_j$, since the metric is independent of the position. From this, we deduce that: $E^j \partial_j E^i = E^j \partial^i E_j = \frac{1}{2}\partial^i \left(E^j E_j\right) = \frac{1}{2}\delta^{ij}\partial_j \left(E^2\right)$. By substituting this result into [4.232], we get:

$$F^i = -\varepsilon_0 \int_\tau \partial_j \left(-E^i E^j + \frac{1}{2}\delta^{ij} E^2 \right) d\tau = -\oint_S T_E^{ij} dS_j \qquad [4.233]$$

$$T_E^{ij} = \varepsilon_0 \left(-E^i E_j + \frac{1}{2}\delta^{ij} E^2 \right) \qquad [4.234]$$

T_E^{ij} is the *Maxwell stress tensor for the electric field.* In tensor notation, it is expressed as:

$$\overrightarrow{\mathbf{F}} = -\oint_S \overleftrightarrow{\mathbf{T}}_{E} \overline{\otimes} \overrightarrow{\mathbf{n}} dS \qquad [4.235]$$

where $\overrightarrow{\mathbf{n}}$ is the unit vector normal to the surface element dS, and:

$$\overleftrightarrow{\mathbf{T}}_E = \varepsilon_0 \left(-\overrightarrow{\mathbf{E}} \otimes \overrightarrow{\mathbf{E}} + \frac{1}{2} E^2 \mathbf{I} \right) = \varepsilon_0 \begin{pmatrix} \frac{1}{2}E^2 - E_x^2 & -E_x E_y & -E_x E_z \\ -E_x E_y & \frac{1}{2}E^2 - E_y^2 & -E_y E_z \\ -E_x E_z & -E_y E_z & \frac{1}{2}E^2 - E_z^2 \end{pmatrix} \qquad [4.236]$$

The tensor [4.236] is a symmetric second-order tensor and has eigenvalues $\lambda_1 = -\frac{1}{2}\varepsilon_0 E^2$, $\lambda_2 = \lambda_3 = \frac{1}{2}\varepsilon_0 E^2$. The eigenvector associated with λ_1 is parallel to $\overrightarrow{\mathbf{E}}$: $\overrightarrow{\mathbf{v}}_{-\frac{1}{2}\varepsilon_0 E^2} = a\overrightarrow{\mathbf{E}}$. The other two eigenvectors associated with λ_2 and λ_3, respectively, are perpendicular to $\overrightarrow{\mathbf{E}}$: $\overrightarrow{\mathbf{v}}_{\frac{1}{2}\varepsilon_0 E^2} = b\overrightarrow{\mathbf{m}}$ and $\overrightarrow{\mathbf{v}}_{\frac{1}{2}\varepsilon_0 E^2} = c\overrightarrow{\mathbf{m}} \times \overrightarrow{\mathbf{E}}$, where $\overrightarrow{\mathbf{m}}$ is an arbitrary vector perpendicular to $\overrightarrow{\mathbf{E}}$.

The Maxwell electric stress is given by:

$$\overleftrightarrow{\mathbf{T}}_{E} \overline{\otimes} \overrightarrow{\mathbf{n}} = \varepsilon_0 \begin{pmatrix} \frac{1}{2}E^2 - E_x^2 & -E_x E_y & -E_x E_z \\ -E_x E_y & \frac{1}{2}E^2 - E_y^2 & -E_y E_z \\ -E_x E_z & -E_y E_z & \frac{1}{2}E^2 - E_z^2 \end{pmatrix} \begin{pmatrix} n_x \\ n_y \\ n_z \end{pmatrix} \qquad [4.237]$$

$$= \varepsilon_0 \begin{pmatrix} \frac{1}{2}E^2 n_x - E_x (E_x n_x + E_y n_y + E_z n_z) \\ \frac{1}{2}E^2 n_y - E_y (E_x n_x + E_y n_y + E_z n_z) \\ \frac{1}{2}E^2 n_z - E_z (E_x n_x + E_y n_y + E_z n_z) \end{pmatrix} \equiv \varepsilon_0 \left[\frac{1}{2} E^2 \overrightarrow{\mathbf{n}} - \left(\overrightarrow{\mathbf{n}} \cdot \overrightarrow{\mathbf{E}} \right) \overrightarrow{\mathbf{E}} \right]$$

We therefore have three possibilities:

1) $\overrightarrow{\mathbf{n}}$ is parallel to $\overrightarrow{\mathbf{E}}$, that is, $\overrightarrow{\mathbf{n}} = \frac{\overrightarrow{\mathbf{E}}}{E}$: $\overleftrightarrow{\mathbf{T}}_{E} \overline{\otimes} \overrightarrow{\mathbf{n}} = -\frac{\varepsilon_0}{2} E^2 \overrightarrow{\mathbf{n}}$. Thus, $d\overrightarrow{\mathbf{F}} = \frac{\varepsilon_0}{2} E^2 \overrightarrow{\mathbf{n}} dS$. The force is normal to the surface and directed outward.

2) $\overrightarrow{\mathbf{n}}$ is perpendicular to $\overrightarrow{\mathbf{E}}$, that is, $\overrightarrow{\mathbf{n}} \cdot \overrightarrow{\mathbf{E}} = 0$: $\overleftrightarrow{\mathbf{T}}_{E} \overline{\otimes} \overrightarrow{\mathbf{n}} = \frac{\varepsilon_0}{2} E^2 \overrightarrow{\mathbf{n}}$. Thus, $d\overrightarrow{\mathbf{F}} = -\frac{\varepsilon_0}{2} E^2 \overrightarrow{\mathbf{n}} dS$. The force is perpendicular to the surface and directed inward.

3) $\vec{\mathbf{E}}$ is inclined at an angle θ relative to $\vec{\mathbf{n}}$, so that $\vec{\mathbf{n}} \cdot \vec{\mathbf{E}} = E\cos\theta$: $\overleftrightarrow{\mathbf{T}}_{E}\overline{\otimes}\vec{\mathbf{n}} = \frac{\varepsilon_0}{2}E^2\vec{\mathbf{n}} - \varepsilon_0 E\cos\theta\vec{\mathbf{E}}$. Projecting $\vec{\mathbf{E}}$ onto $\vec{\mathbf{n}}$ and $\vec{\mathbf{t}}$ ($\vec{\mathbf{t}}$ being the unit tangent vector to dS, lying in the plane defined by $\vec{\mathbf{n}}$ and $\vec{\mathbf{E}}$): $\vec{\mathbf{E}} = E\cos\theta\vec{\mathbf{n}} + E\sin\theta\vec{\mathbf{t}}$. Thus, $\overleftrightarrow{\mathbf{T}}_{E}\overline{\otimes}\vec{\mathbf{n}} = -\frac{\varepsilon_0}{2}E^2\cos 2\theta\vec{\mathbf{n}} - \frac{\varepsilon_0}{2}E^2\sin 2\theta\vec{\mathbf{t}}$. Therefore, $d\vec{\mathbf{F}} = \frac{\varepsilon_0}{2}E^2\cos 2\theta\vec{\mathbf{n}}dS + \frac{\varepsilon_0}{2}E^2\sin 2\theta\vec{\mathbf{t}}dS$. The stresses transmitted through an infinitesimal cylinder aligned with the field consist of tension on the bases and compression on the lateral surfaces. Consequently, the field lines experience tension, while the field tubes are compressed, as illustrated in Figure 4.20.

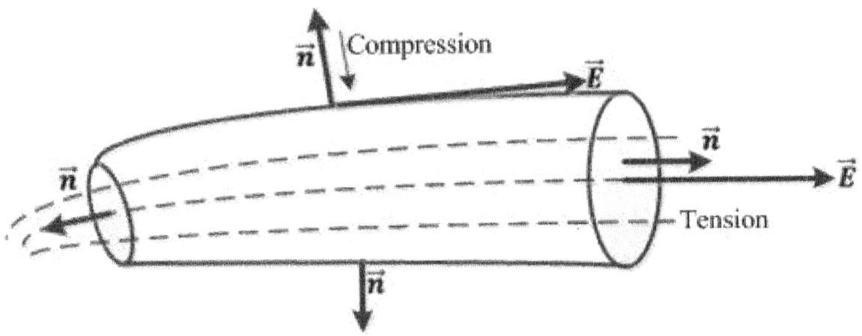

Figure 4.20. *Stresses on electric field line tubes*

EXAMPLE 4.28.– Let us calculate the force exerted between two electric charges using Maxwell's electric stress tensor. We place the positive charge on the right at a distance $2d$ from the negative charge along the Oz axis, as shown in Figure 4.21. We enclose the charge $+q$ within an imaginary hemisphere of radius R. The total force exerted on the hemisphere is:

$$\vec{\mathbf{F}} = \frac{\varepsilon_0}{2}\oint_{\text{hemisphere}} E^2\left(\cos 2\theta\vec{\mathbf{n}} + \sin 2\theta\vec{\mathbf{t}}\right)dS = \frac{\varepsilon_0}{2}\int_{S_1} E^2 \quad [4.238]$$

$$\left(\cos 2\theta_1\vec{\mathbf{n}}_1 + \sin 2\theta_1\vec{\mathbf{t}}_1\right)dS_1 + \frac{\varepsilon_0}{2}\int_{S_2} E^2\left(\cos 2\theta_2\vec{\mathbf{n}}_2 + \sin 2\theta_2\vec{\mathbf{t}}_2\right)dS_2$$

When the radius of the sphere is very large, the field on the shell S_2 approximates that of a dipole. This field, on the order of $\frac{1}{R^3}$, is parallel to $\vec{\mathbf{n}}_2$, that is, $\theta = 0$. This shows that the integral over S_2 is proportional to $\frac{1}{R^4}$ and vanishes in the limit as $R \longrightarrow \infty$. It follows that:

$$\vec{\mathbf{F}} = \frac{\varepsilon_0}{2}\int_{Oxy} planeE^2\left(\cos 2\theta_1\vec{\mathbf{n}}_1 + \sin 2\theta_1\vec{\mathbf{t}}_1\right)dS_1 \quad [4.239]$$

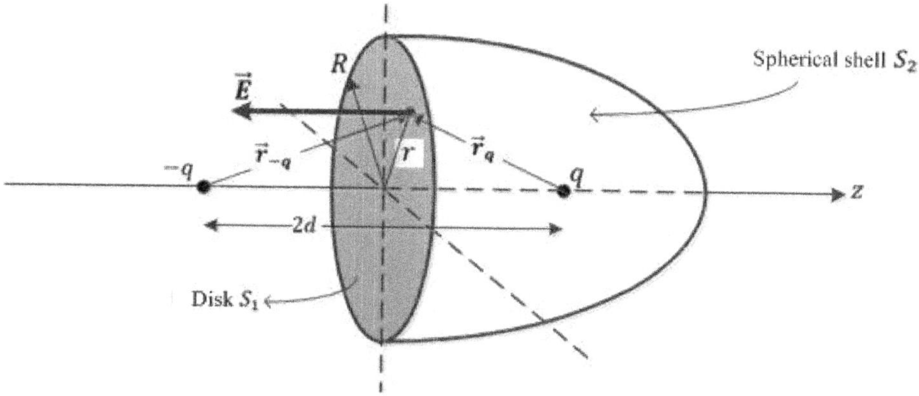

Figure 4.21. *Calculation of the force exerted between two electric charges*

In the plane Oxy, we have: $\overrightarrow{\mathbf{E}} = -\frac{2qd}{4\pi\varepsilon_0(d^2+r^2)^{3/2}}\overrightarrow{\mathbf{k}}$. Consequently, the total force applied at the median plane is (noting that $\theta_1 = 0$):

$$\overrightarrow{\mathbf{F}} = \frac{\varepsilon_0}{2}\int_0^\infty E^2\left(-\overrightarrow{\mathbf{k}}2\pi r dr\right) = -\frac{q^2d^2}{4\pi\varepsilon_0}\int_0^\infty \frac{rdr\,\overrightarrow{\mathbf{k}}}{(d^2+r^2)^3} = -\frac{q^2\overrightarrow{\mathbf{k}}}{4\pi\varepsilon_0\,(2d)^2} \quad [4.240]$$

4.7.2. *Maxwell stress tensor for the magnetic field*

We will now construct the Maxwell magnetic stress tensor $\overleftrightarrow{\mathbf{T}}_B$ from the equation describing the Laplace force, following the same procedure. The equations $\overrightarrow{\nabla}\times\overrightarrow{\mathbf{B}} = \mu_0\overrightarrow{\mathbf{J}}$ ($\epsilon_{ijk}\partial^j B^k = \mu_0 J_i$ in tensor notation) and $\overrightarrow{\nabla}\cdot\overrightarrow{\mathbf{B}} = 0$ ($\partial^l B_l = 0$) allow us to express the components of the Laplace force solely in terms of the magnetic field:

$$F^j_{\text{Lap}} = \int_\tau d\tau \epsilon^{jim} J_i B_m = \frac{1}{\mu_0}\int_\tau d\tau\epsilon^{jim}\epsilon_{ilk}B_m\partial^l B^k = \frac{1}{\mu_0}\int_\tau d\tau\left(\delta^m_l\delta^j_k - \right.$$
$$\left.\delta^m_k\delta^j_l\right)B_m\partial^l B^k = \frac{1}{\mu_0}\int_\tau d\tau\left(B_l\partial^l B^j - B_k\partial^j B^k\right) = \frac{1}{\mu_0}\int_\tau d\tau\left[\partial^l\left(B_lB^j\right) - \right.$$
$$\left.B^j\partial^l B_l - B_k\partial^j B^k\right] = \frac{1}{\mu_0}\int_\tau d\tau\left[\partial^l\left(B_lB^j\right) - B_k\partial^j B^k\right] = \frac{1}{\mu_0}\int_\tau d\tau\partial_i\left[B^iB^j\right.$$
$$\left.-\frac{1}{2}\delta^{ij}\left(B_kB^k\right)\right] = -\int_\tau d\tau\partial_i T^{ij}_B = -\oint_S T^{ij}_B n_j dS \quad [4.241]$$

with:

$$T^{ij}_B = \frac{1}{\mu_0}\left(-B^iB^j + \frac{1}{2}\delta^{ij}B_kB^k\right) \quad [4.242]$$

This is the *Maxwell magnetic stress tensor*. In tensor form, it is written as:

$$\overleftrightarrow{\mathbf{T}}_B = \frac{1}{\mu_0}\left(-\overrightarrow{\mathbf{B}} \otimes \overrightarrow{\mathbf{B}} + \frac{1}{2}B^2\mathbf{I}\right) = \frac{1}{\mu_0}\begin{pmatrix} \frac{1}{2}B^2 - B_x^2 & -B_xB_y & -B_xB_z \\ -B_xB_y & \frac{1}{2}B^2 - B_y^2 & -B_yB_z \\ -B_xB_z & -B_yB_z & \frac{1}{2}B^2 - B_z^2 \end{pmatrix} \quad [4.243]$$

and:

$$\overrightarrow{\mathbf{F}}_{\text{Lap}} = -\oint_S \overleftrightarrow{\mathbf{T}}_B \overline{\otimes} \overrightarrow{\mathbf{n}} dS \quad [4.244]$$

The tensor [4.243] has the eigenvalues $\lambda_1 = -\frac{B^2}{2\mu_0}$ and $\lambda_2 = \lambda_3 = \frac{B^2}{2\mu_0}$. The eigenvector associated with λ_1 is parallel to $\overrightarrow{\mathbf{B}}$, while the two eigenvectors associated with λ_2 and λ_3 are perpendicular to $\overrightarrow{\mathbf{B}}$.

Let us use, as before, the local basis $\left(\overrightarrow{\mathbf{n}}, \overrightarrow{\mathbf{t}}\right)$ at each point on a surface S so that $\overrightarrow{\mathbf{B}} = B\cos\theta\overrightarrow{\mathbf{n}} + B\sin\theta\overrightarrow{\mathbf{t}}$, where θ is the angle measured between $\overrightarrow{\mathbf{B}}$ and $\overrightarrow{\mathbf{n}}$:

$$\begin{aligned} \overleftrightarrow{\mathbf{T}}_B \overline{\otimes} \overrightarrow{\mathbf{n}} &= -\frac{B^2}{2\mu_0}\left(\cos 2\theta\overrightarrow{\mathbf{n}} + \sin 2\theta\overrightarrow{\mathbf{t}}\right) \\ d\overrightarrow{\mathbf{F}}_{\text{Lap}} &= \frac{B^2}{2\mu_0}\left(\cos 2\theta\overrightarrow{\mathbf{n}} + \sin 2\theta\overrightarrow{\mathbf{t}}\right) dS \end{aligned} \quad [4.245]$$

The stresses transmitted through a cylinder aligned with the field are therefore a traction/tension on the bases and a compression on the lateral surfaces: (1) field line: $\overrightarrow{\mathbf{B}} \parallel \overrightarrow{\mathbf{n}}$ (i.e. $\theta = 0$) $\Longrightarrow d\overrightarrow{\mathbf{F}}_{\text{Lap}} = \frac{B^2}{2\mu_0}\overrightarrow{\mathbf{n}}$ (tension); (2) field tube: $\overrightarrow{\mathbf{B}} \perp \overrightarrow{\mathbf{n}}$ (i.e. $\theta = \frac{\pi}{2}$) $\Longrightarrow d\overrightarrow{\mathbf{F}}_{\text{Lap}} = -\frac{B^2}{2\mu_0}\overrightarrow{\mathbf{n}}$ (compression).

In the case of the magnetic field, it is more insightful to consider a field line rather than a flux tube and to rewrite the Laplace force as the sum of two terms, interpreted, respectively, as the *pressure term* and the *tension term*:

$$\overrightarrow{\mathbf{J}} \times \overrightarrow{\mathbf{B}} = \frac{1}{\mu_0}\overrightarrow{\nabla}\overline{\otimes}\left(\overrightarrow{\mathbf{B}} \otimes \overrightarrow{\mathbf{B}} - \frac{1}{2}B^2\mathbf{I}\right) = \frac{1}{\mu_0}\left(\overrightarrow{\nabla} \cdot \overrightarrow{\mathbf{B}}\right)\overrightarrow{\mathbf{B}} - \overrightarrow{\nabla}\left(\frac{B^2}{2\mu_0}\right) \quad [4.246]$$

Thus, the Laplace force is the sum of the gradient of the magnetic pressure $\frac{B^2}{2\mu_0}$ and a magnetic tension term $\frac{1}{\mu_0}\left(\overrightarrow{\nabla} \cdot \overrightarrow{\mathbf{B}}\right)\overrightarrow{\mathbf{B}}$.

4.7.3. *Maxwell stress tensor for the electromagnetic field*

In the presence of an electromagnetic field, the Lorentz force applied to an infinitesimal volume element $d\tau$ is given by: $\vec{\mathbf{f}} = \rho\vec{\mathbf{E}} + \vec{\mathbf{J}} \times \vec{\mathbf{B}} \equiv \frac{\partial \vec{\mathbf{P}}}{\partial t}$, where $\vec{\mathbf{P}}$ is the *momentum density* imparted to the system. In tensor notation, it is expressed as:

$$\frac{\partial P^i}{\partial t} = \rho E^i + \epsilon^{ijk} J_j B_k = \varepsilon_0 \left(\partial_j E^j\right) E^i + \epsilon^{ijk} \left[\frac{1}{\mu_0}\left(\vec{\nabla} \times \vec{\mathbf{B}}\right)_j - \varepsilon_0 \frac{\partial E_j}{\partial t}\right] B_k$$
$$= \varepsilon_0 \left(\partial_j E^j\right) E^i + \frac{1}{\mu_0}\epsilon^{ijk}\epsilon_{jlm} B_k \partial^l B^m - \varepsilon_0 \epsilon^{ijk} \frac{\partial E_j}{\partial t} B_k = \varepsilon_0 \left(\partial_j E^j\right) E^i$$
$$+ \frac{1}{\mu_0}\left(\delta^k_l \delta^i_m - \delta^k_m \delta^i_l\right) B_k \partial^l B^m - \varepsilon_0 \epsilon^{ijk} \left[\frac{\partial}{\partial t}\left(E_j B_k\right) - E_j \frac{\partial B_k}{\partial t}\right] = \varepsilon_0 \left(\partial_j E^j\right) E^i$$
$$+ \frac{1}{\mu_0}\left(B_l \partial^l B^i - B_m \partial^i B^m\right) - \varepsilon_0 \frac{\partial \left(\vec{\mathbf{E}} \times \vec{\mathbf{B}}\right)^i}{\partial t} - \varepsilon_0 \epsilon^{ijk} E_j \left(\vec{\nabla} \times \vec{\mathbf{E}}\right)_k \qquad [4.247]$$

and, consequently:

$$\frac{\partial P^i}{\partial t} + \varepsilon_0 \frac{\partial \left(\vec{\mathbf{E}} \times \vec{\mathbf{B}}\right)^i}{\partial t} = \varepsilon_0 \left[E^i \partial_j E^j - \epsilon^{ijk} \epsilon_{klm} E_j \partial^l E^m\right] + \frac{1}{\mu_0}\left[B_l \partial^l B^i\right.$$
$$\left.- B_m \partial^i B^m\right] = \varepsilon_0 \left[E^i \partial_j E^j - \left(\delta^i_l \delta^j_m - \delta^i_m \delta^j_l\right) E_j \partial^l E^m\right] + \frac{1}{\mu_0}\left[B_l \partial^l B^i\right.$$
$$\left.- B_m \partial^i B^m\right] = \varepsilon_0 \left[E^i \partial_j E^j - E_m \partial^i E^m + E^j \partial_j E^i\right] + \frac{1}{\mu_0}\left[B_l \partial^l B^i -\right.$$
$$\left.B_m \partial^i B^m\right] = \varepsilon_0 \left[\partial_j \left(E^i E^j\right) - E_m \partial^i E^m\right] + \frac{1}{\mu_0}\left[\partial_l \left(B^i B^l\right) - B^i \partial_l B^l -\right.$$
$$\left.B_m \partial^i B^m\right] = \partial_j \left[\varepsilon_0 \left(E^i E^j - \frac{1}{2}\delta^{ij} E^2\right) + \frac{1}{\mu_0}\left(B^i B^j - \frac{1}{2}\delta^{ij} B^2\right)\right] \qquad [4.248]$$

since $\partial_l B^l = 0$. Here, we recognize (in the second term on the left) the expression of the Poynting vector $\frac{\vec{\mathbf{E}} \times \vec{\mathbf{B}}}{\mu_0}$. Let us define:

$$T^{ij} = -\left[\varepsilon_0 \left(E^i E^j - \frac{1}{2}\delta^{ij} E^2\right) + \frac{1}{\mu_0}\left(B^i B^j - \frac{1}{2}\delta^{ij} B^2\right)\right]$$
$$\overleftrightarrow{\mathbf{T}} = -\left[\varepsilon_0 \vec{\mathbf{E}} \otimes \vec{\mathbf{E}} + \frac{1}{\mu_0}\vec{\mathbf{B}} \otimes \vec{\mathbf{B}} - \frac{1}{2}\left(\varepsilon_0 E^2 + \frac{B^2}{\mu_0}\right)\mathbf{I}\right] \qquad [4.249]$$

This is the *Maxwell stress tensor for an electromagnetic field*. In terms of T^{ij}, the equation of motion [4.248] takes the form:

$$\begin{aligned} \frac{\partial P^i}{\partial t} + \varepsilon_0\mu_0\frac{\partial \Pi^i}{\partial t} &= -\partial_j T^{ij} \\ \frac{\partial \vec{\mathbf{P}}}{\partial t} + \varepsilon_0\mu_0\frac{\partial \vec{\mathbf{\Pi}}}{\partial t} &= -\vec{\nabla}\overline{\otimes}\overleftrightarrow{\mathbf{T}} \end{aligned} \qquad [4.250]$$

The second term on the left-hand side can be interpreted as the time derivative of the momentum density of the electromagnetic field. We use the notation: $\vec{\mathbf{P}}_{\mathrm{EM}} = \varepsilon_0\vec{\mathbf{E}}\times\vec{\mathbf{B}}$. The first term is, of course, the time derivative of the mechanical momentum density (associated with massive particles). We write: $\vec{\mathbf{P}}_{\mathrm{mec}} = \vec{\mathbf{P}}$. By integrating [4.250] over the entire volume, we obtain:

$$\frac{d}{dt}\int_\tau \left[P^i + \varepsilon_0\left(\vec{\mathbf{E}}\times\vec{\mathbf{B}}\right)\right]d\tau = -\int_\tau \partial_j T^{ij}d\tau = -\oint_S T^{ij}n_j dS \qquad [4.251]$$

where S is the surface enclosing the volume τ. In particular, if the integration volume is chosen such that the stress tensor T^{ij} vanishes at the boundary, then $\frac{d}{dt}\int_\tau \left[P^i + \varepsilon_0\left(\vec{\mathbf{E}}\times\vec{\mathbf{B}}\right)\right]d\tau = 0$. We thus conclude that even when the system is no longer subjected to any external force, there is still an exchange of momentum, meaning a transfer of energy between the mechanical system and the electromagnetic field. Equation [4.251] can also be interpreted differently: the momentum within the volume decreases at the rate at which T^{ij} crosses the surface and exits the volume. It is therefore evident that T^{ij} represents the flux of electromagnetic momentum leaving the volume.

A similar result can be obtained for the angular momentum of the fields. By writing $\vec{\mathbf{L}}_{\mathrm{EM}} = \vec{\mathbf{r}}\times\vec{\mathbf{P}}_{\mathrm{EM}}$ and $\vec{\mathbf{L}}_{\mathrm{mec}} = \vec{\mathbf{r}}\times\vec{\mathbf{P}}_{\mathrm{mec}}$, we deduce that:

$$\frac{d}{dt}\left(\vec{\mathbf{L}}_{\mathrm{EM}} + \vec{\mathbf{L}}_{\mathrm{mec}}\right) + \vec{\nabla}\overline{\otimes}\overleftrightarrow{\mathbf{M}} = 0 \qquad [4.252]$$

where $\overleftrightarrow{\mathbf{M}} = \vec{\mathbf{r}}\times\overleftrightarrow{\mathbf{T}}\ \left(M_i^{\ j} = \epsilon_{ikl}x^k T^{jl}\right)$.

4.7.4. *Eigenvalues of the electromagnetic stress tensor*

Before calculating the eigenvalues of the matrix $\mathbf{T}$, let us recall some results from linear algebra. If $\mathbf{A}$ is an invertible matrix, then:

$$\left(\mathbf{A}+\mathbf{u}\mathbf{v}^T\right)^{-1} = \mathbf{A}^{-1} - \frac{\mathbf{A}^{-1}\mathbf{u}\mathbf{v}^T\mathbf{A}^{-1}}{1+\mathbf{v}^T\mathbf{A}^{-1}\mathbf{u}} \qquad [4.253]$$

where $\mathbf{u}$ and $\mathbf{v}$ are two column vectors. It is also shown that:

$$\det\left(\mathbf{A} + \mathbf{u}\mathbf{v}^T\right)^{-1} = \left(1 + \mathbf{v}^T\mathbf{A}^{-1}\mathbf{u}\right)\det\left(\mathbf{A}\right) \qquad [4.254]$$

These are the *Sherman–Morrison formulas*. The stress tensor [4.249] has the eigenvalues: $\lambda_1 = \frac{1}{2}\left(\varepsilon_0 E^2 + \frac{B^2}{\mu_0}\right)$ and $\lambda_2 = \sqrt{\left(\frac{\varepsilon_0 E^2 - B^2/\mu_0}{2}\right)^2 + \frac{\varepsilon_0}{\mu_0}\vec{\mathbf{E}}\cdot\vec{\mathbf{B}}} = -\lambda_3$. In fact, these values are obtained from the characteristic equation $\mathbf{T} - \lambda\mathbf{I} = (w_{em} - \lambda)\mathbf{I} - \varepsilon_0\mathbf{E}\mathbf{E}^T - \frac{1}{\mu_0}\mathbf{B}\mathbf{B}^T$, where $w_{em} = \frac{\varepsilon_0 E^2}{2} + \frac{B^2}{2\mu_0}$. Setting $U = (\lambda - w_{em})\mathbf{I} + \varepsilon_0\mathbf{E}\mathbf{E}^T$, we can write that: $\mathbf{T} - \lambda\mathbf{I} = -\mathbf{U} - \frac{1}{\mu_0}\mathbf{B}\mathbf{B}^T$. By applying the *Sherman--Morrison formulas* consecutively, we obtain:

$$\begin{aligned}\det\left(\mathbf{T} - \lambda\mathbf{I}\right) &= -\det\left(U + \frac{1}{\mu_0}\mathbf{B}\mathbf{B}^T\right) = -\left(1 + \mu_0^{-1}\mathbf{B}^T\mathbf{U}^{-1}\mathbf{B}\right)\det\left(\mathbf{U}\right) \\ &= -\left(1 + \mu_0^{-1}\mathbf{B}^T\mathbf{U}^{-1}\mathbf{B}\right)\left(1 + \frac{\varepsilon_0\mathbf{E}^T\mathbf{E}}{\lambda - w_{em}}\right)\left(\lambda - w_{em}\right)^3 \qquad [4.255]\end{aligned}$$

To find the inverse of the matrix $\mathbf{U}$, it is sufficient to use equation [4.253], that is, $\mathbf{U}^{-1} = \dfrac{\mathbf{I}}{\lambda - w_{em}} - \dfrac{1}{\lambda - w_{em}}\dfrac{\varepsilon_0\mathbf{E}\mathbf{E}^T}{\lambda - w_{em} + \varepsilon_0 E^2}$. It follows that:

$$\begin{aligned}&\det\left(\mathbf{T} - \lambda\mathbf{I}\right) = -\left(1 + \frac{\mu_0^{-1}}{\lambda - w_{em}}\mathbf{B}^T\left[\mathbf{I} - \frac{\varepsilon_0\mathbf{E}\mathbf{E}^T}{\lambda - w_{em} + \varepsilon_0 E^2}\right]\mathbf{B}\right) \\ &\times\left(1 + \frac{\varepsilon_0 E^2}{\lambda - w_{em}}\right)\left(\lambda - w_{em}\right)^3 = -\left(1 + \frac{\mu_0^{-1}B^2}{\lambda - w_{em}} - \right. \\ &\left.\frac{1}{\lambda - w_{em}}\frac{\varepsilon_0\mu_0^{-1}\left(\vec{\mathbf{E}}\cdot\vec{\mathbf{B}}\right)^2}{\lambda - w_{em} + \varepsilon_0 E^2}\right)\left(1 + \frac{\varepsilon_0 E^2}{\lambda - w_{em}}\right)\left(\lambda - w_{em}\right)^3 \\ &= -\left[\lambda^2 - \left(\frac{\varepsilon_0 E^2}{2} - \frac{B^2}{2\mu_0}\right)^2 - \frac{\varepsilon_0}{\mu_0}\left(\vec{\mathbf{E}}\cdot\vec{\mathbf{B}}\right)^2\right]\left(\lambda - w_{em}\right) \qquad [4.256]\end{aligned}$$

We can conclude that λ_1, λ_2 and λ_3 are indeed the solutions of the characteristic equation.

4.8. Magnetic monopoles and dual transformations

Although, to date, there has been no confirmed experimental observation of magnetic monopoles, a number of theoretical arguments in favor of their existence, notably put forward by Dirac, have led to a continuous search for primitive magnetic monopoles. In this section, we will demonstrate that a magnetic monopole placed near an electric charge would produce a field whose angular momentum is not zero. It is also remarkable that this angular momentum is independent of the distance between the two monopoles, electric and magnetic. To this end, let us calculate the angular momentum $\overrightarrow{\mathbf{L}}$ of the fields produced by a magnetic monopole q_m located at the origin and an electric charge located at the point $(0, 0, a)$:

$$\overrightarrow{\mathbf{L}} = \int \overrightarrow{\mathbf{r}} \times \varepsilon_0 \left(\overrightarrow{\mathbf{E}} \times \overrightarrow{\mathbf{B}}\right) d\tau = \varepsilon_0 \int \left[\left(\overrightarrow{\mathbf{r}} \cdot \overrightarrow{\mathbf{B}}\right) \overrightarrow{\mathbf{E}} - \left(\overrightarrow{\mathbf{r}} \cdot \overrightarrow{\mathbf{E}}\right) \overrightarrow{\mathbf{B}}\right] d\tau \qquad [4.257]$$

$\overrightarrow{\mathbf{E}}$ and $\overrightarrow{\mathbf{B}}$ are the electric and magnetic fields created, respectively, by the electric charge and the magnetic monopole. The magnetic field is given by:

$$\overrightarrow{\mathbf{B}} = \frac{\mu_0 q_m \overrightarrow{\mathbf{r}}}{4\pi r^3} \qquad [4.258]$$

By substituting [4.258] into [4.257], we obtain:

$$\overrightarrow{\mathbf{L}} = \frac{\mu_0 \varepsilon_0 q_m}{4\pi} \int \left(\frac{\overrightarrow{\mathbf{E}}}{r} - \frac{\overrightarrow{\mathbf{r}} \left(\overrightarrow{\mathbf{r}} \cdot \overrightarrow{\mathbf{E}}\right)}{r^3} \right) d\tau \qquad [4.259]$$

which, in tensor notation, is also written as:

$$\begin{aligned} L^i &= \frac{q_m}{4\pi c^2} \int \left(\frac{E^i}{r} - \frac{x^i \left(E^j x_j\right)}{r^3} \right) d\tau = \frac{q_m}{4\pi c^2} \int E^j \partial_j \left(\frac{x^i}{r}\right) d\tau \\ &= \frac{q_m}{4\pi c^2} \int \partial_j \left(\frac{E^j x^i}{r}\right) d\tau - \frac{q_m}{4\pi c^2} \int \frac{x^i}{r} \left(\partial_j E^j\right) d\tau \\ &= \frac{q_m}{4\pi c^2} \oint_S \frac{E^j x^i}{r} dS_j - \frac{q_m}{4\pi c^2} \int \frac{x^i}{r} \frac{\rho}{\varepsilon_0} d\tau \end{aligned} \qquad [4.260]$$

Since $E^j dS_j = \overrightarrow{\mathbf{E}} \cdot d\overrightarrow{\mathbf{S}} = \left(\frac{e\overrightarrow{\mathbf{r}}}{4\pi\varepsilon_0 r^3}\right) \cdot \left(r^2 d\Omega \frac{\overrightarrow{\mathbf{r}}}{r}\right) = \frac{e d\Omega}{4\pi\varepsilon_0}$, we deduce that:

$$L^i = \frac{q_m}{4\pi c^2} \int_{\theta=0}^{\pi} \int_{\varphi=0}^{2\pi} \frac{e x^i d\Omega}{4\pi\varepsilon_0 r} - \frac{q_m}{4\pi c^2} \int \frac{x^i}{r} \frac{\rho}{\varepsilon_0} d\tau \qquad [4.261]$$

The first integral is obviously zero. It is easily verified that $\int \frac{x^i d\Omega}{r} = 0$, for all i. Since the charge is localized at the position $a\overrightarrow{\mathbf{k}}$, we take $\rho = e\delta\left(\overrightarrow{\mathbf{r}} - a\overrightarrow{\mathbf{k}}\right)$. The second integral simplifies to:

$$L^i = -\frac{q_m e}{4\pi c^2 \varepsilon_0} \int \frac{x^k}{r} \delta\left(\overrightarrow{\mathbf{r}} - a\overrightarrow{\mathbf{k}}\right) d\tau \qquad [4.262]$$

and obviously gives 0 for L_x and L_y. For the component along the Oz-axis, we find:

$$L_z = -\frac{q_m e}{4\pi c^2 \varepsilon_0} \quad [4.263]$$

It is clear that the projection L_z is independent of the distance a that separates the two monopoles. Even if the magnetic and electric monopoles are separated by very large distances, the angular momentum of their fields remains constant. In other words, their interaction is always the same, even if they are light years apart.

According to quantum mechanics, the smallest allowed angular momentum is $\frac{\hbar}{2}$, which implies that the smallest charge of a magnetic monopole is:

$$q_m = \frac{2\pi\varepsilon_0 c^2 \hbar}{e} \quad [4.264]$$

To account for the existence of magnetic monopoles, the Maxwell equations must be modified. It is therefore clear that equation $\vec{\nabla} \cdot \vec{\mathbf{B}} = 0$ should be replaced by $\vec{\nabla} \cdot \vec{\mathbf{B}} = \mu_0 \rho_m$, by analogy with Gauss' law. These monopoles must also produce a magnetic current density $\vec{\mathbf{J}}_m$ when in motion, which requires us to add a term to the modified Maxwell–Faraday equation, that is, to modify it as follows: $\vec{\nabla} \times \vec{\mathbf{E}} = -\frac{\partial \vec{\mathbf{B}}}{\partial t} - \mu_0 \vec{\mathbf{J}}_m$. Applying the operator $\vec{\nabla}$ to the modified Maxwell–Faraday equation, we obtain: $0 = \vec{\nabla} \cdot \left(\vec{\nabla} \times \vec{\mathbf{E}}\right) = -\frac{\partial}{\partial t}\left(\vec{\nabla} \cdot \vec{\mathbf{B}}\right) - \mu_0 \vec{\nabla} \cdot \vec{\mathbf{J}}_m$. Magnetic current (by analogy with electrical current) must obey the continuity equation (for magnetic monopoles):

$$\frac{\partial \rho_m}{\partial t} + \vec{\nabla} \cdot \vec{\mathbf{J}}_m = 0 \quad [4.265]$$

Thus, the Maxwell equations are generalized as follows:

$$\begin{array}{ll} \vec{\nabla} \cdot \vec{\mathbf{E}} = \frac{\rho_e}{\varepsilon_0} & \vec{\nabla} \cdot \vec{\mathbf{B}} = \mu_0 \rho_m \\ \vec{\nabla} \times \vec{\mathbf{E}} = -\frac{\partial \vec{\mathbf{B}}}{\partial t} - \mu_0 \vec{\mathbf{J}}_m & \vec{\nabla} \times \vec{\mathbf{B}} = \frac{1}{c^2}\frac{\partial \vec{\mathbf{E}}}{\partial t} + \mu_0 \vec{\mathbf{J}}_e \end{array} \quad [4.266]$$

They are invariant under *dual transformation*. Indeed, the global transformation:

$$\begin{array}{ll} \text{Field transformation:} & \left\{ \begin{array}{l} \vec{\mathbf{E}}' = \vec{\mathbf{E}} \cos\theta + c\vec{\mathbf{B}} \sin\theta \\ c\vec{\mathbf{B}}' = -\vec{\mathbf{E}} \sin\theta + c\vec{\mathbf{B}} \cos\theta \end{array} \right. \\ \text{Charge transformation:} & \left\{ \begin{array}{l} c\rho_e' = c\rho_e' \cos\theta + \rho_m \sin\theta \\ \rho_m' = -c\rho_e' \sin\theta + \rho_m \cos\theta \end{array} \right. \\ \text{Current transformation:} & \left\{ \begin{array}{l} c\vec{\mathbf{J}}'_e = c\vec{\mathbf{J}}_e \cos\theta + \vec{\mathbf{J}}_m \sin\theta \\ \vec{\mathbf{J}}'_m = -c\vec{\mathbf{J}}_e \sin\theta + \vec{\mathbf{J}}_m \cos\theta \end{array} \right. \end{array} \quad [4.267]$$

leaves the generalized Maxwell equations invariant and thus all the physics that follows. With the appropriate choice of *mixing angle* θ, magnetic monopoles can appear or disappear at will. Therefore, the existence or nonexistence of monopoles is merely a matter of convention.

Chapter 5

Polarization, Polarized Media (Dielectrics)

5.1. Polarization vector (field), polarization charge

A *perfect dielectric* is a medium without any free charges. Although electrically neutral on a macroscopic scale, this medium still has electrical properties that become apparent when exposed to an electric field. Consider the example of an atom: its positive charge, owing to protons, is concentrated in its nucleus, which can be considered as a point charge. The negative charge that compensates for the charge on the protons is due to the electrons. This charge is uniformly distributed (in an electron cloud) around the nucleus. In the absence of an electric field, the center of mass of the electron cloud coincides with that of the protons. In this case, the atom does not have an electric dipole The application of an electric field causes a displacement of the electrons in one direction and of the nucleus in the opposite direction. This results in a new distribution of the negative charges (which remain bound to the nucleus); and as a consequence, the atom acquires a non-zero dipole moment (or electric dipole). These induced electric dipoles in the material are capable of generating a field that will significantly alter the structure of the field at every point.

Experiments have shown that the electric dipole model is often sufficient to describe the behavior of the constituents of dielectric objects. For this reason, the contribution of the quadrupole, etc., will be neglected. Recall that material media or macroscopic liquids are composed of a very large number of atoms and/or molecules. Some constituents possess permanent electric dipoles. However, in the absence of an electric field, these dipoles are randomly oriented so that the average global dipole moment is zero. The presence of an electric field will cause these dipoles to align in its direction. Other constituents have no dipole moment of their own, but can acquire induced electric dipoles under the influence of an electric field (as mentioned above). These dipoles cease to exist when the field disappears. Some rare materials or substances, known as *ferroelectrics*, have permanent electric dipoles that tend to align with each other, even in the absence of an electric field. All of these materials are called *dielectrics*.

Consider an infinitesimal elementary volume $\Delta\tau = \Delta x \Delta y \Delta z$ inside the dielectric, as shown in Figure 5.1(a). Let us calculate the total net charge contained within the parallelepiped due to the dipoles. Note that the dipoles located inside the volume do not contribute to the net charge of the volume because they are electrically neutral. Only dipoles cut by the volume boundaries will contribute to the net charge of the elementary volume. If n represents the number of dipoles per unit volume, it is convenient to define the dipole density (or dipolar density) as the polarization vector:

$$\vec{\mathbf{P}} = n\vec{\mathbf{p}} = nq\vec{\mathbf{d}} \tag{5.1}$$

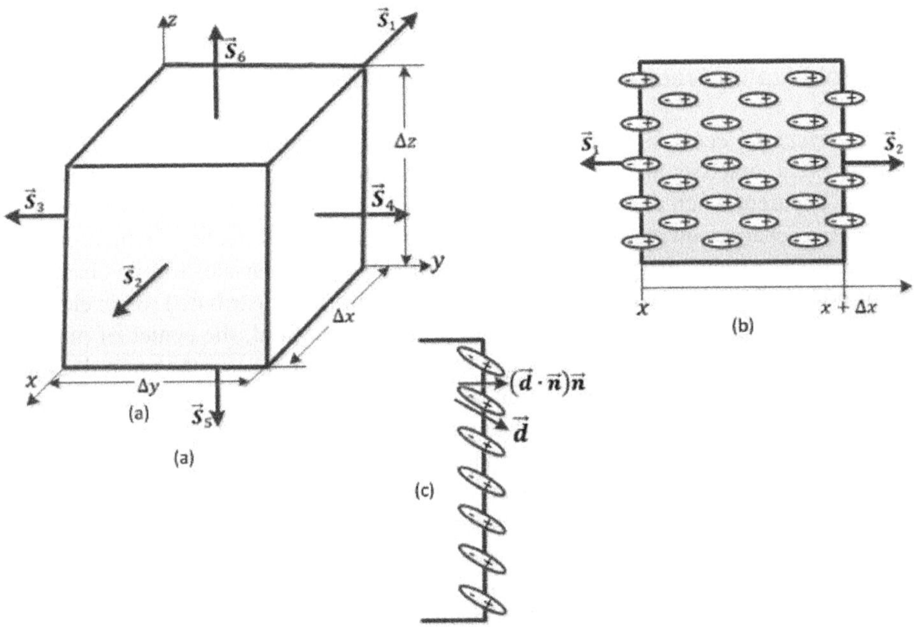

Figure 5.1. *Only the dipoles cut by the boundary surfaces contribute to the net charge of the parallelepiped*

The net charge measured near the surface S_1 is given as (Figure 5.1(b)):

$$dq_1 = \left.(nqd_x)\right|_x \Delta y \Delta z = P_x(x)\,\Delta y \Delta z \tag{5.2}$$

while the net charge near the opposite surface S_2 is given as:

$$dq_2 = -\left.(nqd_x)\right|_{x+\Delta x} \Delta y \Delta z = -P_x(x+\Delta x)\,\Delta y \Delta z \tag{5.3}$$

where it is assumed that the dimensions Δy and Δz are small enough for the polarization $\vec{\mathbf{P}}$ to be essentially constant at the surface. The polarization of the surface S_1 at coordinate x may differ from the polarization of surface S_2 at coordinate $x + \Delta x$ if the density n, charge q, or distance d (the characteristic distance of the dipole) is a function of x. The difference in sign between [5.2] and [5.3] arises because in the vicinity of the surface S_1, only the positive charge of the dipole is enclosed within the volume, while at the boundary of the surface S_2, it is the negative charge of the dipole that lies within the volume. Note also that only the normal component of $\vec{\mathbf{d}}$ to the surface contributes to the net charge of the volume, as illustrated in Figure 5.1(c). Similarly, near surfaces S_3, S_4, S_5 and S_6, the net charges contained within the volume are:

$$\begin{aligned} dq_3 &= (nqd_y)|_y \, \Delta x \Delta z = P_y(y) \, \Delta x \Delta z \\ dq_4 &= -\left(nqd_y\right)|_{y+\Delta y} \Delta x \Delta z = -P_y(y+\Delta y) \, \Delta x \Delta z \\ dq_5 &= (nqd_z)|_z \, \Delta x \Delta y = P_z(z) \, \Delta x \Delta y \\ dq_6 &= -\left(nqd_z\right)|_{z+\Delta z} \Delta x \Delta y = -P_z(z+\Delta z) \, \Delta x \Delta y \end{aligned} \quad [5.4]$$

Therefore, the total net charge contained within the elementary volume $\Delta\tau$ is the sum of terms [5.2], [5.3] and [5.4]: $dq_T = dq_1 + dq_2 + dq_3 + dq_4 + dq_5 + dq_6 = \left[\frac{P_x(x)-P_x(x+\Delta x)}{\Delta x} + \frac{P_y(y)-P_y(y+\Delta y)}{\Delta y} + \frac{P_z(z)-P_z(z+\Delta z)}{\Delta z}\right] \Delta x \Delta y \Delta z$. When Δx, Δy and Δz tend to zero (i.e. the elementary volume will contain only the point (x, y, z)), the terms in square brackets will reduce to partial derivatives so that the volume density of the polarization charges (also called the *polarization volume charge density*), noted here ρ_l, becomes:

$$\rho_l = \lim_{\Delta x, \Delta y, \Delta z \longrightarrow 0} \frac{dq_T}{\Delta x \Delta y \Delta z} = -\left[\frac{\partial P_x}{\partial x} + \frac{\partial P_y}{\partial y} + \frac{\partial P_z}{\partial z}\right] = -\vec{\nabla} \cdot \vec{\mathbf{P}} \quad [5.5]$$

The polarization charge density ρ_l is also a source of the electric field and must be included in the Maxwell–Gauss equation, that is:

$$\vec{\nabla} \cdot \left(\varepsilon_0 \vec{\mathbf{E}}\right) = \rho + \rho_l \quad [5.6]$$

where ρ here denotes the free charge volume density (also called the *free volume charge density*). The polarization charge in a region τ is obtained by integrating [5.5] over the volume:

$$q_p = \int_\tau \rho_l d\tau = -\int_\tau \vec{\nabla} \cdot \vec{\mathbf{P}} d\tau = -\oint_S \vec{\mathbf{P}} \cdot d\vec{\mathbf{S}} \quad [5.7]$$

where the divergence theorem has been used to relate the polarization charge to a surface integral of the polarization vector. In other words, the polarization charge accumulated within a volume τ is equal to the flux of the polarization vector through the surface S surrounding τ.

5.2. Electric field created by a dielectric

Consider a dielectric that contains charges, electric dipoles, quadrupoles, etc., distributed throughout the material. The free charge density is denoted by $\rho\left(\overrightarrow{\mathbf{r}}'\right)$, the dipole moment density, or more simply the polarization, by $\overrightarrow{\mathbf{P}}\left(\overrightarrow{\mathbf{r}}'\right)$ and the quadrupole moment density by $Q_{ij}\left(\overrightarrow{\mathbf{r}}'\right)$. The potential induced by these distributions at a point $M\left(x = x_1, y = x_2, z = x_3\right)$ within the dielectric is:

$$V\left(\overrightarrow{\mathbf{r}}\right) = \int_{\tau'} \frac{d\tau'}{4\pi\varepsilon_0}\left[\frac{\rho\left(\overrightarrow{\mathbf{r}}'\right)}{\left|\overrightarrow{\mathbf{r}} - \overrightarrow{\mathbf{r}}'\right|} + \frac{\left(\overrightarrow{\mathbf{r}} - \overrightarrow{\mathbf{r}}'\right)\cdot\overrightarrow{\mathbf{P}}\left(\overrightarrow{\mathbf{r}}'\right)}{\left|\overrightarrow{\mathbf{r}} - \overrightarrow{\mathbf{r}}'\right|^3}\right.$$
$$\left.+\sum_{ij}\frac{\left(x_i - x_i'\right)\left(x_j - x_j'\right)Q_{ij}}{2\left|\overrightarrow{\mathbf{r}} - \overrightarrow{\mathbf{r}}'\right|^5} + \cdots\right] = \frac{1}{4\pi\varepsilon_0}\int_{\tau'}\left[\frac{\rho\left(\overrightarrow{\mathbf{r}}'\right)}{\left|\overrightarrow{\mathbf{r}} - \overrightarrow{\mathbf{r}}'\right|}\right. \qquad [5.8]$$
$$+\sum_{i}\frac{\partial}{\partial x_i'}\left(\frac{1}{\left|\overrightarrow{\mathbf{r}} - \overrightarrow{\mathbf{r}}'\right|}\right)P_i\left(\overrightarrow{\mathbf{r}}'\right) + \frac{1}{6}\sum_{ij}\frac{\partial^2}{\partial x_i'\partial x_j'}\left(\frac{1}{\left|\overrightarrow{\mathbf{r}} - \overrightarrow{\mathbf{r}}'\right|}\right)Q_{ij}$$
$$\left.+\cdots\right]d\tau'$$

where τ' is the volume occupied by the dielectric.

If we neglect the effect of quadrupole moments and only consider the effect of dipole moments, the previous equation reduces to:

$$V\left(\overrightarrow{\mathbf{r}}\right) = \frac{1}{4\pi\varepsilon_0}\int_{\tau'}\left[\frac{\rho\left(\overrightarrow{\mathbf{r}}'\right)}{\left|\overrightarrow{\mathbf{r}} - \overrightarrow{\mathbf{r}}'\right|} + \overrightarrow{\mathbf{P}}\left(\overrightarrow{\mathbf{r}}'\right)\cdot\overrightarrow{\nabla}'\left(\frac{1}{\left|\overrightarrow{\mathbf{r}} - \overrightarrow{\mathbf{r}}'\right|}\right)\right]d\tau' \qquad [5.9]$$

In the following, we aim to examine the effect of matter on Maxwell's equations. Let us begin with its effect on $\overrightarrow{\nabla}\cdot\overrightarrow{\mathbf{E}}$, that is, to express the divergence of the electric field in terms of the free charge density $\rho\left(\overrightarrow{\mathbf{r}}\right)$ and the polarization $\overrightarrow{\mathbf{P}}\left(\overrightarrow{\mathbf{r}}\right)$, as we have already done in equation [5.6]. To do this, we calculate $\overrightarrow{\mathbf{E}} = -\overrightarrow{\nabla}V$, and then $\overrightarrow{\nabla}\cdot\overrightarrow{\mathbf{E}} = -\Delta V$. It follows that:

$$\overrightarrow{\nabla}\cdot\overrightarrow{\mathbf{E}} = -\frac{1}{4\pi\varepsilon_0}\Delta\int_{\tau'}\left[\frac{\rho\left(\overrightarrow{\mathbf{r}}'\right)}{\left|\overrightarrow{\mathbf{r}} - \overrightarrow{\mathbf{r}}'\right|} + \overrightarrow{\mathbf{P}}\left(\overrightarrow{\mathbf{r}}'\right)\cdot\overrightarrow{\nabla}'\left(\frac{1}{\left|\overrightarrow{\mathbf{r}} - \overrightarrow{\mathbf{r}}'\right|}\right)\right]d\tau'$$
$$= -\frac{1}{4\pi\varepsilon_0}\int_{\tau'}\left[\rho\left(\overrightarrow{\mathbf{r}}'\right)\Delta\left(\frac{1}{\left|\overrightarrow{\mathbf{r}} - \overrightarrow{\mathbf{r}}'\right|}\right) + \overrightarrow{\mathbf{P}}\left(\overrightarrow{\mathbf{r}}'\right)\cdot\overrightarrow{\nabla}'\left[\Delta\left(\frac{1}{\left|\overrightarrow{\mathbf{r}} - \overrightarrow{\mathbf{r}}'\right|}\right)\right]\right]d\tau'$$
$$= \frac{1}{\varepsilon_0}\int_{\tau'}\left[\rho\left(\overrightarrow{\mathbf{r}}'\right)\delta^3\left(\overrightarrow{\mathbf{r}} - \overrightarrow{\mathbf{r}}'\right) + \overrightarrow{\mathbf{P}}\left(\overrightarrow{\mathbf{r}}'\right)\cdot\overrightarrow{\nabla}'\left(\delta^3\left(\overrightarrow{\mathbf{r}} - \overrightarrow{\mathbf{r}}'\right)\right)\right]d\tau' \qquad [5.10]$$

where we have used the relation $\Delta\left(\frac{1}{\left|\overrightarrow{\mathbf{r}} - \overrightarrow{\mathbf{r}}'\right|}\right) = -4\pi\delta^3\left(\overrightarrow{\mathbf{r}} - \overrightarrow{\mathbf{r}}'\right)$. The first integral on the right-hand side gives $\rho\left(\overrightarrow{\mathbf{r}}\right)$, while the second integral is written as (using the notation $\overrightarrow{\mathbf{P}}' \equiv \overrightarrow{\mathbf{P}}\left(\overrightarrow{\mathbf{r}}'\right)$):

$$\int_{\tau'} \overrightarrow{\mathbf{P}}' \cdot \overrightarrow{\nabla}' \delta^3\left(\overrightarrow{\mathbf{r}} - \overrightarrow{\mathbf{r}}'\right) d\tau' = \int_{\tau'} [P'_x \delta'\left(x - x'\right) \delta\left(y - y'\right) \delta\left(z - z'\right)$$

$$+P'_y \delta\left(x - x'\right) \delta'\left(y - y'\right) \delta\left(z - z'\right) + P'_z \delta\left(x - x'\right) \delta\left(y - y'\right) \delta'\left(z - z'\right)] \, d\tau'$$

$$= \int P_x\left(x', y, z\right) \delta'\left(x - x'\right) dx' + \int P_y\left(x, y', z\right) \delta'\left(y - y'\right) dy'$$

$$+ \int P_z\left(x, y, z'\right) \delta'\left(z - z'\right) dz' = -\frac{\partial P_x(x, y, z)}{\partial x} - \frac{\partial P_y(x, y, z)}{\partial y}$$

$$-\frac{\partial P_z(x, y, z)}{\partial z} = -\overrightarrow{\nabla} \cdot \overrightarrow{\mathbf{P}}\left(\overrightarrow{\mathbf{r}}\right) \qquad [5.11]$$

where $\int f(x) \delta'(x - a) \, dx = -f'(a)$. Based on these results, we deduce that:

$$\overrightarrow{\nabla} \cdot \overrightarrow{\mathbf{E}} = \frac{\rho}{\varepsilon_0} - \frac{\overrightarrow{\nabla} \cdot \overrightarrow{\mathbf{P}}}{\varepsilon_0} \qquad [5.12]$$

which confirms the accuracy of formula [5.6]. It is clear that $-\overrightarrow{\nabla} \cdot \overrightarrow{\mathbf{P}}$ acts as a source of the electric field. This is called the *bound charge density or polarization*, $\rho_l = -\overrightarrow{\nabla} \cdot \overrightarrow{\mathbf{P}}$. If $\overrightarrow{\nabla} \cdot \overrightarrow{\mathbf{P}} > 0$ (respectively, $\overrightarrow{\nabla} \cdot \overrightarrow{\mathbf{P}} < 0$), then $\rho_l < 0$ (respectively, $\rho_l > 0$), that is, negative charges (respectively, positive charges) will accumulate in the center and act as a source of the electric field, as illustrated in Figure 5.2(a) (respectively, Figure 5.2(b)).

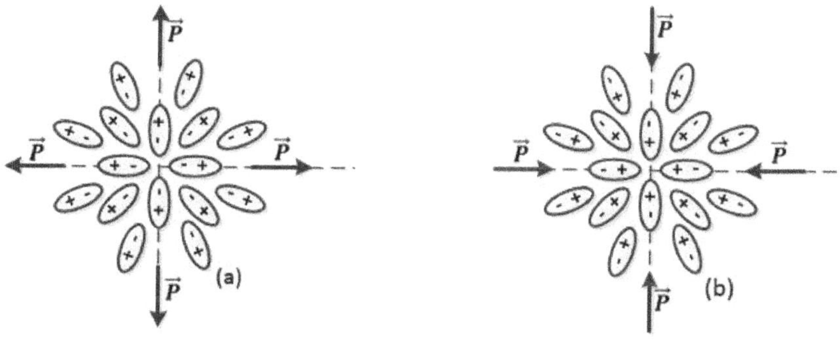

Figure 5.2. *Polarization charge density inside a dielectric*

The equation for the curl of $\overrightarrow{\mathbf{E}}$ is obviously not modified:

$$\overrightarrow{\nabla} \times \overrightarrow{\mathbf{E}} = 0 \qquad [5.13]$$

5.3. Boundary conditions for $\vec{\mathbf{E}}$ and bound surface charge density

At the surface of the dielectric, there is also an accumulation of bound surface charge, even for uniform polarizations, as illustrated in Figure 5.3. In fact, if the free charge ρ is zero, equation [5.12] reduces to:

$$\vec{\nabla} \cdot \vec{\mathbf{E}} = -\frac{\vec{\nabla} \cdot \vec{\mathbf{P}}}{\varepsilon_0} \quad [5.14]$$

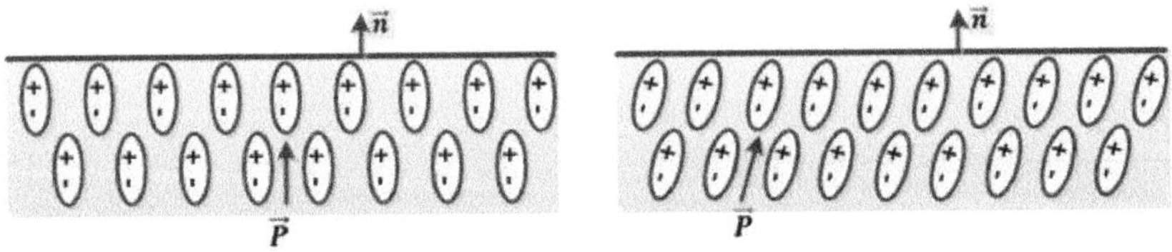

Figure 5.3. *On the different surfaces of the dielectric, the dipoles generate a bound surface charge equal to $\vec{\mathbf{P}} \cdot \vec{\mathbf{n}}$, where $\vec{\mathbf{n}}$ is the normal to the surface*

Let us apply Gauss's theorem to a fictitious, infinitesimal cylindrical surface crossing the surface of a dielectric, as shown in Figure 5.4. The lateral surface of the cylinder is denoted S_L, while the inner base S_i and the outer (top) base S_e are parallel to the surface of the dielectric. If τ is the volume of the cylinder, we can write $\int_\tau \vec{\nabla} \cdot \vec{\mathbf{E}} d\tau = -\int_\tau \frac{\vec{\nabla} \cdot \vec{\mathbf{P}}}{\varepsilon_0} d\tau$, and consequently, $\oint_S \vec{\mathbf{E}} \cdot d\vec{\mathbf{S}} = -\frac{1}{\varepsilon_0} \oint_S \vec{\mathbf{P}} \cdot d\vec{\mathbf{S}}$, which leads to:

$$\int_{S_i} \vec{\mathbf{E}} \cdot d\vec{\mathbf{S}}_i + \int_{S_e} \vec{\mathbf{E}} \cdot d\vec{\mathbf{S}}_e + \int_{S_L} \vec{\mathbf{E}} \cdot d\vec{\mathbf{S}}_L = -\frac{1}{\varepsilon_0} \int_{S_i} \vec{\mathbf{P}} \cdot d\vec{\mathbf{S}}_i - \frac{1}{\varepsilon_0} \int_{S_L} \vec{\mathbf{P}} \cdot d\vec{\mathbf{S}}_L \quad [5.15]$$

When $S_L \longrightarrow 0$ (that is, $S_i \longrightarrow S_e$), we get:

$$\left(\vec{\mathbf{E}}_e - \vec{\mathbf{E}}_i\right) \cdot \vec{\mathbf{n}} dS = -\frac{1}{\varepsilon_0} \vec{\mathbf{P}} \cdot (-\vec{\mathbf{n}})\, dS \quad [5.16]$$

Here, $\vec{\mathbf{n}}$ is the normal to the surface of the dielectric, oriented outward. Since dS is arbitrary, we finally obtain:

$$\left(\vec{\mathbf{E}}_e - \vec{\mathbf{E}}_i\right) \cdot \vec{\mathbf{n}} = \frac{1}{\varepsilon_0} \vec{\mathbf{P}} \cdot \vec{\mathbf{n}} \quad [5.17]$$

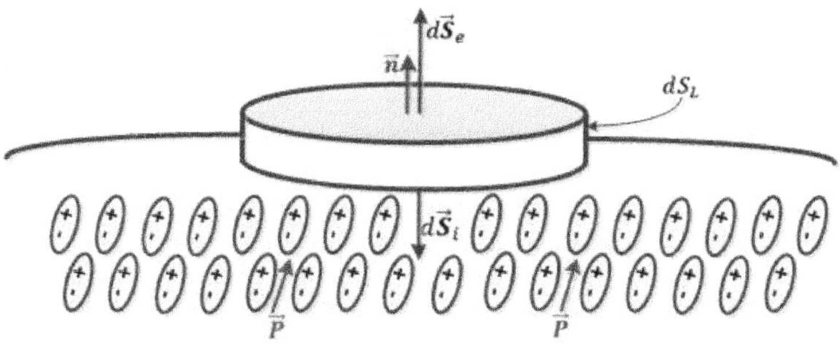

Figure 5.4. *Discontinuity of the electric field at the surface of a dielectric. For a color version of this figure, see www.iste.co.uk/abdesselam/electrodynamics2.zip*

This discontinuity in the field is exactly the effect that would be caused by a surface charge density:

$$\sigma_l = \vec{\mathbf{P}} \cdot \vec{\mathbf{n}} \qquad [5.18]$$

bound to the dielectric surface. This is the *polarization surface charge*. We therefore conclude that the electric field produced by a polarized dielectric medium is identical to that produced by a bound volume charge $\rho_l = -\vec{\nabla} \cdot \vec{\mathbf{P}}$ and a bound surface charge $\sigma_l = -\vec{\mathbf{P}} \cdot \vec{\mathbf{n}}$. The resulting electric field is written as:

$$\begin{aligned} \vec{\mathbf{E}}\left(\vec{\mathbf{r}}\right) = &\frac{1}{4\pi\varepsilon_0}\int_{\tau'} \frac{-\vec{\nabla}' \cdot \vec{\mathbf{P}}\left(\vec{\mathbf{r}} - \vec{\mathbf{r}}'\right)}{\left|\vec{\mathbf{r}} - \vec{\mathbf{r}}'\right|^3} d\tau' + \\ &\frac{1}{4\pi\varepsilon_0}\oint_{S'} \frac{\left(\vec{\mathbf{P}} \cdot \vec{\mathbf{n}}'\right)\left(\vec{\mathbf{r}} - \vec{\mathbf{r}}'\right)}{\left|\vec{\mathbf{r}} - \vec{\mathbf{r}}'\right|^3} dS' \end{aligned} \qquad [5.19]$$

Now, consider a rectangular contour of length L and width Δy crossing the interface of two media (1) and (2). Half of the contour is located in medium (1), while the other half is in medium (2). For simplicity, we choose the rectangle in the Oxy plane, as shown in Figure 5.5. By integrating the equation $\vec{\nabla} \times \vec{\mathbf{E}} = 0$ in the surface bounded by the contour, we obtain $0 = \int_S \left(\vec{\nabla} \times \vec{\mathbf{E}}\right) \cdot d\vec{\mathbf{S}} = \oint_{\text{rectangular}} \vec{\mathbf{E}} \cdot d\vec{\mathbf{l}} \simeq \int_{x_0}^{x_0+L} \left(E_x^{(2)} - E_x^{(1)}\right) dx + \Delta y \left[E_y\left(x_0 + L\right) - E_y\left(x_0\right)\right]$. Taking $\Delta y \longrightarrow 0$ (i.e. approaching the interface on both sides), we find that $\int_{x_0}^{x_0+L} \left(E_x^{(2)} - E_x^{(1)}\right) dx = 0$. This integral holds for any contour L, which leads to $E_x^{(2)} - E_x^{(1)} = 0$. By changing the orientation of the contour, we also verify that $E_z^{(2)} - E_z^{(1)} = 0$. In general, we write:

$$\left(\vec{\mathbf{E}}^{(2)} - \vec{\mathbf{E}}^{(1)}\right) \times \vec{\mathbf{n}} = 0 \qquad [5.20]$$

that is, the tangential component of $\vec{\mathbf{E}}$ remains continuous as it crosses the interface between two media, (1) and (2). We therefore conclude that $\left(\vec{\mathbf{E}}\right)_{\perp}$ is discontinuous due to σ_l and $\left(\vec{\mathbf{E}}\right)_{\parallel}$ is continuous.

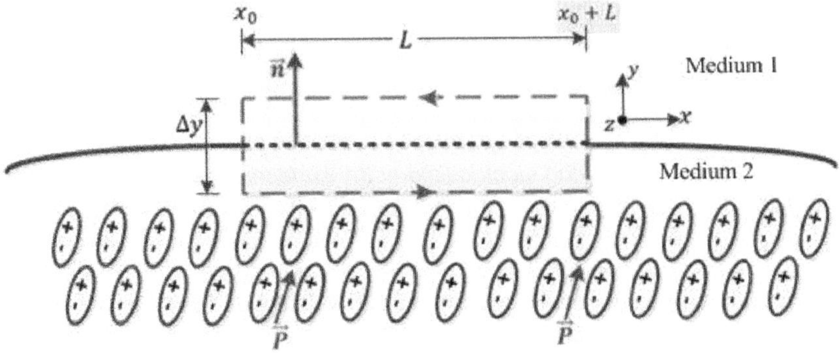

Figure 5.5. *Infinitesimal surface bounded by a rectangular contour L at the interface of two media (1) and (2). For a color version of this figure, see www.iste.co.uk/abdesselam/electrodynamics2.zip*

EXAMPLE 5.1. (Cylinder uniformly polarized along its axis).– Consider a dielectric in the shape of a cylinder, with length L and radius R, uniformly polarized along its axis, denoted here by Oz, as shown in Figure 5.6. The lower base is assumed to be in the Oxy plane, that is, at $z = 0$. Let us calculate the electric field induced by the cylinder along Oz. According to [5.19], the electric field is determined from the following formula:

$$\vec{\mathbf{E}}(z) = \frac{1}{4\pi\varepsilon_0} \oint_{S'} \frac{\left(\vec{\mathbf{P}} \cdot \vec{\mathbf{n}}'\right)\left(z\vec{\mathbf{k}} - \vec{\mathbf{r}}'\right)}{\left|z\vec{\mathbf{k}} - \vec{\mathbf{r}}'\right|^3} dS' \qquad [5.21]$$

Polarization $\vec{\mathbf{P}}$ is here perpendicular to the normal of the lateral surface. This does not contribute to the value of the integral [5.21]. What remains is:

$$\begin{aligned}\vec{\mathbf{E}} &= \frac{1}{4\pi\varepsilon_0} \int_0^R \int_0^{2\pi} \left[\frac{-P\left(z\vec{\mathbf{k}} - r'\vec{\mathbf{e}}'_r\right)}{\left(z^2 + r'^2\right)^{3/2}} + \frac{P\left(z\vec{\mathbf{k}} - r'\vec{\mathbf{e}}'_r - L\vec{\mathbf{k}}\right)}{\left((z-L)^2 + r'^2\right)^{3/2}} \right] r'dr'd\varphi' \\ &= \frac{P\vec{\mathbf{k}}}{2\varepsilon_0} \left[\frac{z}{\sqrt{z^2 + R^2}} - \frac{z - L}{\sqrt{(z-L)^2 + R^2}} - \frac{z}{|z|} + \frac{z - L}{|z - L|} \right] \qquad [5.22]\end{aligned}$$

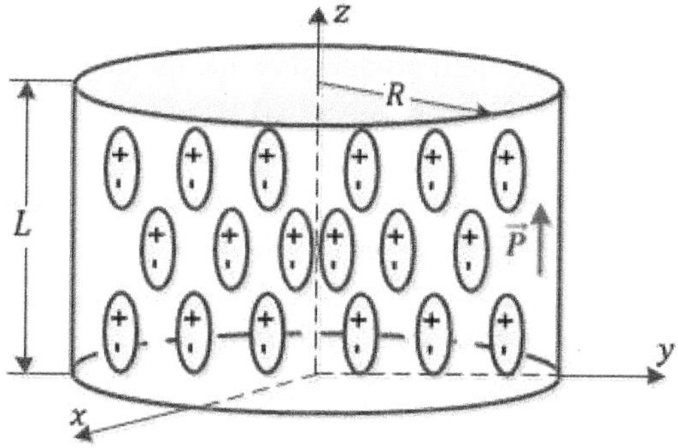

Figure 5.6. *Cylindrical dielectric uniformly polarized along the Oz-axis. For a color version of this figure, see www.iste.co.uk/abdesselam/electrodynamics2.zip*

Inside the cylinder (i.e. for $0 \leq z \leq L$), we get:

$$\vec{\mathbf{E}}_{\text{int}}(z) = \frac{P\vec{\mathbf{k}}}{2\varepsilon_0}\left[\frac{z}{\sqrt{z^2+R^2}} - \frac{z-L}{\sqrt{(z-L)^2+R^2}} - 2\right] \qquad [5.23]$$

while outside (i.e. for $z < 0$ and $z > L$), the electric field is:

$$\vec{\mathbf{E}}_{\text{ext}}(z) = \frac{P\vec{\mathbf{k}}}{2\varepsilon_0}\left[\frac{z}{\sqrt{z^2+R^2}} - \frac{z-L}{\sqrt{(z-L)^2+R^2}}\right] \qquad [5.24]$$

It is easy to check that at the cylinder boundaries (the lower and upper bases):

$$\begin{aligned} \vec{\mathbf{E}}_{\text{int}}(0) - \vec{\mathbf{E}}_{\text{ext}}(0) &= -\frac{P}{\varepsilon_0}\vec{\mathbf{k}} = \frac{\sigma_l}{\varepsilon_0}\left(-\vec{\mathbf{k}}\right) \\ \vec{\mathbf{E}}_{\text{ext}}(L) - \vec{\mathbf{E}}_{\text{int}}(L) &= \frac{P}{\varepsilon_0}\vec{\mathbf{k}} = \frac{\sigma_l}{\varepsilon_0}\vec{\mathbf{k}} \end{aligned} \qquad [5.25]$$

which is in agreement with the continuity equation [5.17]. Note that inside the cylinder, the first two terms in [5.23] are both positive and less than one, meaning that the electric field is directed in the negative direction of Oz. In particular, if $R \gg L$, the electric field inside becomes simply $\vec{\mathbf{E}}_{\text{int}} = -\frac{P\vec{\mathbf{k}}}{\varepsilon_0}$ (this is equivalent to the field between two parallel plates carrying the surface charge $\sigma = P$), while it becomes zero outside.

EXAMPLE 5.2. (Parallel plate capacitor with dielectric material).– Imagine a parallel plate capacitor with free surface charge σ, where the dielectric fills almost the entire space between the two plates, as shown in Figure 5.7. The electric field of the capacitor will therefore combine with the field produced by the dielectric:

$$\vec{\mathbf{E}}_{\text{int}}(z) = \frac{\sigma - P}{\varepsilon_0}\vec{\mathbf{k}} \qquad [5.26]$$

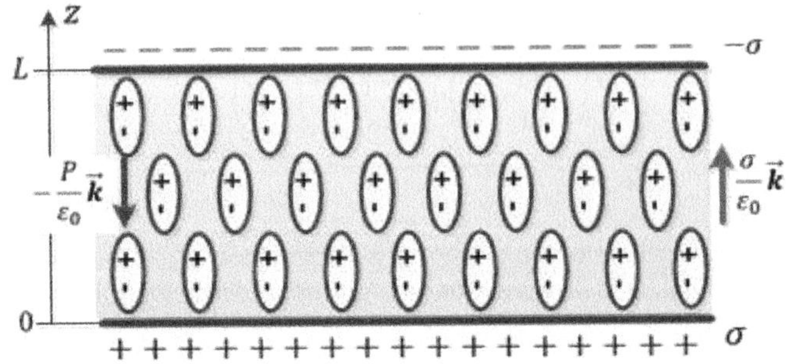

Figure 5.7. *Parallel plate capacitor with a dielectric material between the plates instead of a vacuum. For a color version of this figure, see www.iste.co.uk/abdesselam/electrodynamics2.zip*

This equation does not tell us what the electric field is, unless we know P. To solve this problem, we assume here that P depends on E – that it is, in fact, proportional to E. This proportionality is expressed by the equation:

$$\vec{\mathbf{P}} = \chi_e \varepsilon_0 \vec{\mathbf{E}} \qquad [5.27]$$

χ_e is called the *electric susceptibility of the dielectric*. Dielectric materials that satisfy the property [5.27] are called *linear dielectrics*. We will revisit the origin of this equation in detail. Thus, equation [5.26] becomes:

$$E = \frac{\sigma}{\varepsilon_0}\frac{1}{1+\chi_e} \qquad [5.28]$$

which gives the factor $\frac{1}{1+\chi_e}$ by which the field is reduced. The potential difference between the plates is the integral of the electric field. Since the field is uniform, this integral is the product of E and the distance d between the two plates. We thus obtain:

$$V = Ed = \frac{\sigma d}{\varepsilon_0(1+\chi_e)} \qquad [5.29]$$

The total capacitor charge is σS, where S is the common surface area of the plates. Therefore, the capacitance is:

$$C = \frac{\varepsilon_0 S(1+\chi_e)}{d} = \frac{\varepsilon_r \varepsilon_0 S}{d} \qquad [5.30]$$

We conclude that when a capacitor is filled with a dielectric, its capacitance is multiplied by the factor $\varepsilon_r = 1 + \chi_e$. ε_r is the *relative permittivity* of the material.

EXAMPLE 5.3. (Uniformly polarized sphere).– Consider a uniformly polarized sphere of radius R (uncharged). We will determine the potential at any point in space by solving Laplace's equation. Since the polarization is uniform $\vec{\mathbf{P}} = P\vec{\mathbf{k}}$, there is no bound charge density $\rho_l = -\vec{\nabla} \cdot \vec{\mathbf{P}} = 0$. In addition, there is no free charge density $\rho = 0$, which implies that $\rho + \rho_l = 0$. As a result, the Poisson equation $\Delta V = -\frac{\rho+\rho_l}{\varepsilon_0} = 0$ becomes the Laplace equation. The rotational symmetry of the problem around Oz (by an angle φ) indicates that the potential V does not depend on this coordinate. The Laplacian of V in spherical coordinates is:

$$\Delta V(r,\theta) = \frac{1}{r}\frac{\partial^2 (rV(r,\theta))}{\partial r^2} + \frac{1}{r^2 \sin\theta}\frac{\partial}{\partial\theta}\left(\sin\theta\frac{\partial V(r,\theta)}{\partial\theta}\right) \qquad [5.31]$$

Since the potentials V_{int} (inside the sphere) and V_{ext} (outside the sphere) must not diverge as $r \longrightarrow 0$ and $r \longrightarrow \infty$ (i.e. they must remain finite), the solutions are written as:

$$\begin{aligned} V_{\text{int}}(r \leq R, \theta) &= \sum_{l=0}^{\infty} A_l r^l P_l(\cos\theta) \\ V_{\text{ext}}(r > R, \theta) &= \sum_{l=0}^{\infty} \frac{B_l}{r^{l+1}} P_l(\cos\theta) \end{aligned} \qquad [5.32]$$

The continuity equation for the potential $V_{\text{int}}(R,\theta) = V_{\text{ext}}(R,\theta)$ and the discontinuity equation for the normal component of the vector $\vec{\mathbf{E}}$ at the dielectric–vacuum interface:

$$\left[\vec{\mathbf{E}}_{\text{ext}}(R,\theta) - \vec{\mathbf{E}}_{\text{int}}(R,\theta)\right] \cdot \vec{\mathbf{e}}_r = \frac{\vec{\mathbf{P}} \cdot \vec{\mathbf{e}}_r}{\varepsilon_0} = \frac{P\cos\theta}{\varepsilon_0} \qquad [5.33]$$

where the electric fields $\vec{\mathbf{E}}_{\text{ext}}$ and $\vec{\mathbf{E}}_{\text{int}}$ are:

$$\begin{aligned} \vec{\mathbf{E}}_{\text{int}} &= -\vec{\nabla} V_{\text{int}} = -\sum_{l=0}^{\infty} A_l r^{l-1}\left[l P_l(\cos\theta)\,\vec{\mathbf{e}}_r - P_l'(\cos\theta)\sin\theta\,\vec{\mathbf{e}}_\theta\right] \\ \vec{\mathbf{E}}_{\text{ext}} &= \vec{\nabla} V_{\text{ext}} = -\sum_{l=0}^{\infty} \frac{B_l}{r^{l+2}}\left[(l+1) P_l(\cos\theta)\,\vec{\mathbf{e}}_r + P_l'(\cos\theta)\sin\theta\,\vec{\mathbf{e}}_\theta\right] \end{aligned} \qquad [5.34]$$

This immediately leads to:

$$\begin{aligned} &\sum_{l=0}^{\infty} A_l R^l P_l(\cos\theta) = \sum_{l=0}^{\infty} \frac{B_l}{R^{l+1}} P_l(\cos\theta) \\ &\sum_{l=0}^{\infty} \left[\frac{(l+1) B_l}{R^{l+2}} + l A_l R^{l-1}\right] P_l(\cos\theta) = \frac{P\cos\theta}{\varepsilon_0} \end{aligned} \qquad [5.35]$$

which implies that $l = 1$, $A_1 R = \frac{B_1}{R^2}$, $\frac{2B_1}{R^3} + A_1 = \frac{P}{\varepsilon_0}$ and $A_l = B_l = 0$ for $l \neq 1$. By solving the previous constraints, we find $A_1 = \frac{P}{3\varepsilon_0}$ and $B_1 = \frac{PR^3}{3\varepsilon_0}$. Here are the main results:

$$\begin{aligned} V_{\text{ext}}(r > R, \theta) &= \frac{PR^3}{3\varepsilon_0}\frac{\cos\theta}{r^2} \\ V_{\text{int}}(r \leq R, \theta) &= \frac{P}{3\varepsilon_0} r\cos\theta \\ \vec{\mathbf{E}}_{\text{ext}}(r > R, \theta) &= -\vec{\nabla} V_{\text{ext.}} = \frac{PR^3}{3\varepsilon_0 r^3}\left(2\cos\theta\,\vec{\mathbf{e}}_r + \sin\theta\,\vec{\mathbf{e}}_\theta\right) \\ \vec{\mathbf{E}}_{\text{int}}(r \leq R, \theta) &= -\vec{\nabla} V_{\text{int.}} = -\frac{P}{3\varepsilon_0}\left(\cos\theta\,\vec{\mathbf{e}}_r - \sin\theta\,\vec{\mathbf{e}}_\theta\right) \end{aligned} \quad [5.36]$$

Now, we know that $2\cos\theta\,\vec{\mathbf{e}}_r + \sin\theta\,\vec{\mathbf{e}}_\theta = 3\cos\theta\,\vec{\mathbf{e}}_r - \vec{\mathbf{k}}$ and $\vec{\mathbf{P}} \cdot \vec{\mathbf{e}}_r = P\cos\theta$, because $\vec{\mathbf{k}} = \cos\theta\,\vec{\mathbf{e}}_r - \sin\theta\,\vec{\mathbf{e}}_\theta$. Thus, we deduce that:

$$\begin{aligned} \vec{\mathbf{E}}_{\text{ext}}(r, \theta) &= \frac{R^3}{3\varepsilon_0}\frac{3\left(\vec{\mathbf{P}} \cdot \vec{\mathbf{e}}_r\right)\vec{\mathbf{e}}_r - \vec{\mathbf{P}}}{r^3} \\ \vec{\mathbf{E}}_{\text{int}}(r, \theta) &= -\frac{\vec{\mathbf{P}}}{3\varepsilon_0} \end{aligned} \quad [5.37]$$

The external electric field is equivalent to the field created by a point electric dipole $\vec{\mathbf{p}}$ placed at the origin, with $\vec{\mathbf{p}} = \frac{4}{3}\pi R^3 \vec{\mathbf{P}}$: $\vec{\mathbf{E}} = \frac{1}{4\pi\varepsilon_0}\frac{3\left(\vec{\mathbf{p}} \cdot \vec{\mathbf{e}}_r\right)\vec{\mathbf{e}}_r - \vec{\mathbf{p}}}{r^3}$.

EXAMPLE 5.4. (Spherical cavity inside a dielectric).– Let us determine the electric field inside a spherical cavity carved in a uniformly polarized dielectric. The dielectric is placed in a uniform electric field E_0. For simplicity, we assume that the polarization $\vec{\mathbf{P}}$ and the field $\vec{\mathbf{E}}_0$ are aligned in the same direction (for example, Oz-axis). In spherical coordinates, the potential is:

$$V(r, \theta) = \begin{cases} \sum_{l=0}^{\infty} A_l r^l P_l(\cos\theta) & r < R \\ -E_0 r\cos\theta + \sum_{l=0}^{\infty} \frac{B_l}{r^{l+1}} P_l(\cos\theta) & r > R \end{cases} \quad [5.38]$$

As in the previous case, we use the continuity equation for the potential $V(R^-, \theta) = V(R^+, \theta)$ and the discontinuity equation for the electric field $\vec{\mathbf{E}}$, $\left[\vec{\mathbf{E}}(R^+, \theta) - \vec{\mathbf{E}}(R^-, \theta)\right] \cdot \vec{\mathbf{e}}_r = -\frac{\vec{\mathbf{P}} \cdot \vec{\mathbf{e}}_r}{\varepsilon_0} = -\frac{P\cos\theta}{\varepsilon_0}$ on the cavity interface. These two boundary conditions give:

$$\begin{aligned} &\sum_{l=1}^{\infty} A_l R^l P_l(\cos\theta) = -E_0 R\cos\theta + \sum_{l=0}^{\infty} \frac{B_l}{R^{l+1}} P_l(\cos\theta) \\ &E_0\cos\theta + \sum_{l=0}^{\infty} \frac{(l+1)B_l}{R^{l+2}} P_l(\cos\theta) + \sum_{l=0}^{\infty} l A_l R^{l-1} P_l(\cos\theta) = -\frac{P\cos\theta}{\varepsilon_0} \end{aligned} \quad [5.39]$$

From these equations, we deduce that $A_1 R = -E_0 R + \frac{B_1}{R^2}$, $E_0 + \frac{2B_1}{R^3} + A_1 = -\frac{P}{\varepsilon_0}$, $A_l R^l = \frac{B_l}{R^{l+1}}$ for $(l \neq 1)$ and $l A_l R^{l-1} + \frac{(l+1)B_l}{R^{l+2}} = 0$ for $(l \neq 1)$. Solving these constraints gives $A_1 = -E_0 - \frac{P}{3}$ and $B_1 = -\frac{R^3 P}{3\varepsilon_0}$. All other parameters are zero. It follows that:

$$V(r,\theta) = \begin{cases} -\left(E_0 + \frac{1}{3}\frac{P}{\varepsilon_0}\right) r\cos\theta = -\left(E_0 + \frac{1}{3}\frac{P}{\varepsilon_0}\right) z & r < R \\ -E_0 r\cos\theta - \frac{R^3 P}{3\varepsilon_0 r^2}\cos\theta & r > R \end{cases} \quad [5.40]$$

and, consequently:

$$\vec{\mathbf{E}} = \begin{cases} \vec{\mathbf{E}}_0 + \frac{1}{3}\frac{\vec{\mathbf{P}}}{\varepsilon_0} & r < R \\ \vec{\mathbf{E}}_0 - \frac{R^3 P}{3\varepsilon_0 r^3}\left(2\cos\theta\,\vec{\mathbf{e}}_r + \sin\theta\,\vec{\mathbf{e}}_\theta\right) & r > R \end{cases} \quad [5.41]$$

where $\vec{\mathbf{E}}_0 = E_0\vec{\mathbf{k}}$ and $\vec{\mathbf{P}} = P\vec{\mathbf{k}}$. We see that the electric field outside the cavity is the sum of $\vec{\mathbf{E}}_0$ and a field created by a point dipole $\vec{\mathbf{p}} = -\frac{4}{3}\pi R^3 \vec{\mathbf{P}}$ placed in the center of the cavity. Furthermore, the electric field induced inside the cavity is $\vec{\mathbf{E}} = \vec{\mathbf{E}}_0 + \frac{\vec{\mathbf{P}}}{3\varepsilon_0}$. When $R \longrightarrow 0$, the field everywhere becomes equal to $\vec{\mathbf{E}}_0$. We conclude that any object (for example, dipole) located within the cavity will experience the field $\vec{\mathbf{E}}_0 + \frac{\vec{\mathbf{P}}}{3\varepsilon_0}$. In dielectrics with low concentration (i.e. where the distances between molecules are large), there is little difference between the fields $\vec{\mathbf{E}}$ and $\vec{\mathbf{E}}_0$. On the other hand, in dense media, each dipole experiences a field different from $\vec{\mathbf{E}}_0$ due to the neighboring dipoles. The field $-\frac{\vec{\mathbf{P}}}{3\varepsilon_0}$ represents the average field created within the cavity by the dipoles removed when the cavity is carved out. Inside a vacuum cavity, this average field thus subtracts from the applied external field. The field experienced inside the cavity is equal to $\vec{\mathbf{E}}_0 - \left(-\frac{\vec{\mathbf{P}}}{3\varepsilon_0}\right) = \vec{\mathbf{E}}_0 + \frac{\vec{\mathbf{P}}}{3\varepsilon_0}$. To verify this, let us calculate the mean field inside a sphere filled with dipoles. We will assume that there are n dipoles per unit volume. We can estimate this field by first finding the average field due to a single point charge $+Q$ at a distance a along Oz from the center of a spherical volume of radius R, much larger than the radius of the electron cloud R_0 ($R \gg R_0$), as shown in Figure 5.8(a). The average field due to this charge on the spherical volume is:

$$\langle\vec{\mathbf{E}}\rangle = \frac{1}{\frac{4}{3}\pi R^3}\int_0^R\int_0^\pi\int_0^{2\pi}\frac{Q\left(r\,\vec{\mathbf{e}}_r - a\,\vec{\mathbf{k}}\right) r^2\sin\theta dr d\theta d\varphi}{4\pi\varepsilon_0\left(a^2 + r^2 - 2ra\cos\theta\right)^{3/2}} \quad [5.42]$$

where we used the relations $r_{QM}^2 = a^2 + r^2 - 2ra\cos\theta$ and $r\,\vec{\mathbf{r}}_{QM} = r\,\vec{\mathbf{e}}_r - a\,\vec{\mathbf{k}}$.

From the expression of $\vec{e}_r$, it is easy to verify that the components $\langle E_x \rangle$ and $\langle E_y \rangle$ of the field are zero. Only the component $\langle E_z \rangle$ is not zero:

$$\langle E_z \rangle = \frac{3Q}{8\pi\varepsilon_0 R^3} \int_0^R \int_0^\pi \frac{r^2 (r\cos\theta - a)\sin\theta dr d\theta}{(a^2 + r^2 - 2ra\cos\theta)^{3/2}} \qquad [5.43]$$

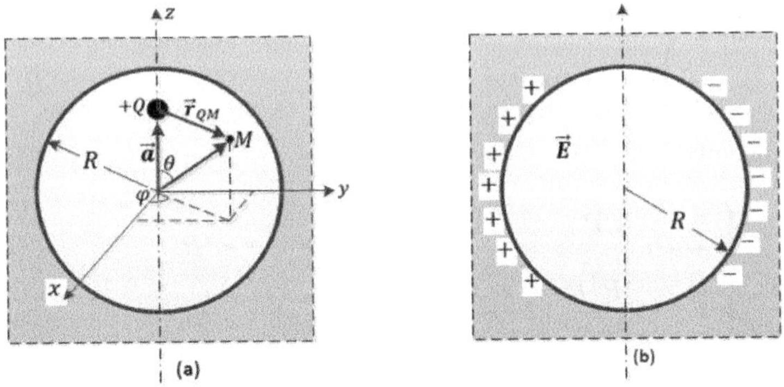

Figure 5.8. *Inside the cavity, the local electric field is the superposition of the applied external field observed by each dipole and the field $\frac{\vec{P}}{3\varepsilon_0}$ induced by the neighboring dipoles*

By performing the change of variables $u = r^2 + a^2 - 2ar\cos\theta$ and $du = 2ar\sin\theta d\theta$, the previous integral becomes:

$$\begin{aligned}\langle E_z \rangle &= \frac{3Q}{8\pi\varepsilon_0 R^3} \int_0^R \int_{(r-a)^2}^{(r+a)^2} \frac{r}{4a} \frac{r^2 - a^2 - u}{u^{3/2}} dr du \\ &= -\frac{3Q}{8\pi\varepsilon_0 R^3 a^2} \int_0^R \left(1 - \frac{(r-a)}{|r-a|}\right) r^2 dr\end{aligned} \qquad [5.44]$$

It is easy to see that this integral is zero for $r > a$. Therefore, the integral over r is reduced to the interval $[0, a]$:

$$\langle E_z \rangle = -\frac{3Q}{4\pi\varepsilon_0 R^3 a^2} \int_0^a r^2 dr = -\frac{Qa}{4\pi\varepsilon_0 R^3} \qquad [5.45]$$

To form a dipole, simply add a negative charge $-Q$ at a small distance d below the positive charge $+Q$. The average electric field due to the dipole is then the superposition of [5.45] for the two charges:

$$\langle E_z \rangle = -\frac{Q}{4\pi\varepsilon_0 R^3} [a - (a - d)] = -\frac{Qd}{4\pi\varepsilon_0 R^3} = -\frac{p}{4\pi\varepsilon_0 R^3} \qquad [5.46]$$

The total number of dipoles contained in the spherical volume is $n\frac{4}{3}\pi R^3$, and therefore, the average electric field due to all the dipoles is:

$$\langle \vec{\mathbf{E}} \rangle = -\frac{\frac{4}{3}\pi R^3 n \vec{\mathbf{p}}}{4\pi\varepsilon_0 R^3} = -\frac{\vec{\mathbf{P}}}{3\varepsilon_0} \quad [5.47]$$

To obtain the local field experienced inside the empty cavity, we must subtract the average field [5.47] from the applied field, that is:

$$\vec{\mathbf{E}}_{\text{loc}} = \vec{\mathbf{E}}_0 + \frac{\vec{\mathbf{P}}}{3\varepsilon_0} \quad [5.48]$$

$\vec{\mathbf{E}}_{\text{loc}}$ is called the *local electric field*. The term $\frac{\vec{\mathbf{P}}}{3\varepsilon_0}$ can also be interpreted as the field resulting from the dipoles placed on the cavity wall, with one side positively charged and the other negatively charged, as illustrated in Figure 5.8(b).

5.4. Electric displacement field

To simplify the Maxwell–Gauss equation [5.12], it is more convenient to introduce a new vector $\vec{\mathbf{D}}$, defined as a linear combination of $\vec{\mathbf{E}}$ and $\vec{\mathbf{P}}$:

$$\vec{\mathbf{D}} = \varepsilon_0 \vec{\mathbf{E}} + \vec{\mathbf{P}} \quad [5.49]$$

The vector $\vec{\mathbf{D}}$ is called the *electric displacement field*, because it differs from $\varepsilon_0 \vec{\mathbf{E}}$ due to the small charge displacements of the dipoles. Using [5.49], equation [5.12] can be rewritten as:

$$\vec{\nabla} \cdot \vec{\mathbf{D}} = \rho \quad [5.50]$$

where ρ represents only the free charge. Dipoles are not a source for $\vec{\mathbf{D}}$. Only free charges are sources of the field $\vec{\mathbf{D}}$. By integrating equation [5.50] on both sides over a volume τ enclosed by a closed surface S and applying the divergence theorem, we obtain the new integral form of the Gauss theorem:

$$\int_\tau \vec{\nabla} \cdot \vec{\mathbf{D}} d\tau = \oint_S \vec{\mathbf{D}} \cdot d\vec{\mathbf{S}} = \int_\tau \rho d\tau \quad [5.51]$$

When the polarization $\vec{\mathbf{P}}$ is zero (the vacuum case), the displacement field reduces to $\varepsilon_0 \vec{\mathbf{E}}$ and, consequently, [5.50] and [5.51] become the usual differential and integral Maxwell–Gauss equations.

In general, the polarization of a neutral medium is caused by an external field, but $\vec{\mathbf{P}}$ and $\vec{\mathbf{E}}$ are not necessarily parallel:

– if the medium is *isotropic*, $\vec{\mathbf{P}}$ is parallel to $\vec{\mathbf{E}}$;

– if $\left|\vec{\mathbf{P}}\right|$ is sufficiently small, we can generally assume that the relationship between $\vec{\mathbf{P}}$ and $\vec{\mathbf{E}}$ is *linear*;

– if the medium is *homogeneous*, the proportionality coefficient between $\left|\vec{\mathbf{P}}\right|$ and $\left|\vec{\mathbf{E}}\right|$ is a constant independent of the position.

Thus, in a linear isotropic homogeneous (LIH) medium, we assume the following:

$$\vec{\mathbf{P}} = \varepsilon_0 \chi_e \vec{\mathbf{E}} \qquad [5.52]$$

where χ_e is a dimensionless scalar. This is the *electrical susceptibility*. If $\vec{\mathbf{E}}$ is applied permanently, this is *a static susceptibility*. There is also *dynamic susceptibility*. We will discuss the microscopic origin of this relationship in the next section. Note that $\vec{\mathbf{E}}$ here represents the field within the material, typically called the *internal or local field*. The relation [5.52] is called the *constitutive relation* of the material.

In an LIH material, the displacement field is also parallel to $\vec{\mathbf{E}}$. Indeed, the relation [5.52] leads to:

$$\vec{\mathbf{D}} = \varepsilon_0 \vec{\mathbf{E}} + \chi_e \varepsilon_0 \vec{\mathbf{E}} = \varepsilon_0 (1 + \chi_e) \vec{\mathbf{E}} = \varepsilon_0 \varepsilon_r \vec{\mathbf{E}} = \varepsilon \vec{\mathbf{E}} \qquad [5.53]$$

where $\varepsilon_r = 1 + \chi_e$ and $\varepsilon = \varepsilon_0 \varepsilon_r$ are called the *relative dielectric permittivity* and *dielectric permittivity*, respectively. In vacuum, susceptibility is zero ($\chi_e = 0$) and therefore $\varepsilon_r = 1$. This implies that $\varepsilon = \varepsilon_0$ (where ε_0 is the permittivity of the vacuum).

5.5. Microscopic approach to linear dielectric media

It has been observed that the capacitance of a capacitor is multiplied by a factor of ε_r when the space between the plates is filled by a dielectric. Under constant charge, the difference in potential between the plates and the electric field derived from this potential is divided by the same factor ε_r, because there are bound charges in the dielectric that generate a depolarizing field opposite to that of the free charges. The electric field between the plates is thus the sum of the field $\vec{\mathbf{E}}$ created by the charges on the plates and the field created by the bound charges in the dielectric, denoted $\vec{\mathbf{E}}_p$ for the depolarizing field. In the first approximation, this field $\vec{\mathbf{E}}_p$ is created by the induced electric dipoles in the molecules that make up the dielectric material. These dipoles are formed from the positive and negative charges within a molecule (electrons and nuclei). In the absence of an electric field, the molecules can be classified as polar and non-polar. In the presence of an external electric field $\vec{\mathbf{E}}$, we observe:

– *Electronic polarization*, that is, a deformation of the atomic electron cloud under the effect of $\overrightarrow{\mathbf{E}}$. To the first order of approximation, this allows us to introduce a proportionality coefficient α between $\overrightarrow{\mathbf{p}}$ and $\overrightarrow{\mathbf{E}}$, $\overrightarrow{\mathbf{p}} = \varepsilon_0\alpha\overrightarrow{\mathbf{E}}$, here called the *polarizability of the material.*

– *Ionic polarization* that results in a deformation of the crystal lattice (deformation of the unit cells). The positive ions are displaced in the direction of the field, while the negative ions are displaced in the opposite direction, leading to the appearance of dipole moments. This ionic polarization adds to the electronic polarization, which occurs for weaker fields. Electronic and ionic polarizations are sometimes collectively referred to as *polarizations by deformation.*

– *Orientation polarization*: in the absence of an electric field, for a polar medium, the disorder induced by temperature results in a zero average dipole moment. However, under the influence of an electric field, the dipoles align in the direction of the field, which leads to the creation of a macroscopic dipole moment.

In this section, we will consider fields that may vary over time.

5.5.1. *Electronic polarization in steady-state*

Let us now try to find the relationship between the polarization $\overrightarrow{\mathbf{P}}$ and the applied electric field $\overrightarrow{\mathbf{E}}$. We will begin with the static case, where $\overrightarrow{\mathbf{E}}$ is independent of time. First, let us consider the case of non-polar molecules. When these molecules are placed in an electric field, the positive charges shift slightly in the direction of the field, whereas the negative charges move in the opposite direction, creating a polarization within the material. For polar molecules, which have permanent dipole moments that are initially oriented randomly, they tend to align with the electric field. This alignment leads to polarization. In both cases, the resulting polarization is a function of the *local electric field*. Therefore, we write:

$$P^i = P_0^i + \chi_j^i\varepsilon_0 E_{\text{loc}}^j + \chi_{jk}^i\varepsilon_0^2 E_{\text{loc}}^j E_{\text{loc}}^k + \chi_{jkl}^i\varepsilon_0^3 E_{\text{loc}}^j E_{\text{loc}}^k E_{\text{loc}}^l + \cdots \quad [5.54]$$

When the material is isotropic, only the diagonal terms of the so-called *electric susceptibility tensor* χ are not zero. Media or materials with strong spontaneous polarization are called *ferroelectrics*. A detailed discussion would require the use of quantum mechanics. However, many simplified classical models can be used here to qualitatively understand the most interesting case of a linear dielectric.

1) *Low-concentration dielectrics*: we will model an atom as a fixed positive nucleus with a uniform spherical negative electron cloud, as shown in Figure 5.9(a). In the absence of an electric field, the dipole moment of the atom is zero because the center of charge of the electron cloud coincides with that of the nucleus. To verify this, we simply use the formula [5.14] from Volume 1:

$$\vec{\mathbf{P}} = Q \times 0 - \int_0^{R_0} \int_0^{\pi} \int_0^{2\pi} \rho_0 r^3 \sin\theta dr d\theta d\varphi \vec{\mathbf{e}}_r \qquad [5.55]$$

where $\vec{\mathbf{e}}_r = \sin\theta\cos\varphi\,\vec{\mathbf{i}} + \sin\theta\sin\varphi\,\vec{\mathbf{j}} + \cos\theta\,\vec{\mathbf{k}}$, R_0 is the radius of the atom and $\rho_0 = -\frac{Q}{\frac{4}{3}\pi R_0^3}$ is the charge density of the electron cloud. Using the expression for $\vec{\mathbf{e}}_r$, it is easy to verify that the integral is zero. An applied electric field tends to push the negative charge in the opposite direction, causing a slight displacement d between the center of mass of the electron cloud and the center of mass of the positive nucleus, as shown in Figure 5.9(b). This movement will be opposed by the attractive Coulomb force. Considering the center of the electron cloud as our origin, the electric field inside the cloud is:

$$E_r = -\frac{Qr}{4\pi\varepsilon_0 R_0^3} \qquad [5.56]$$

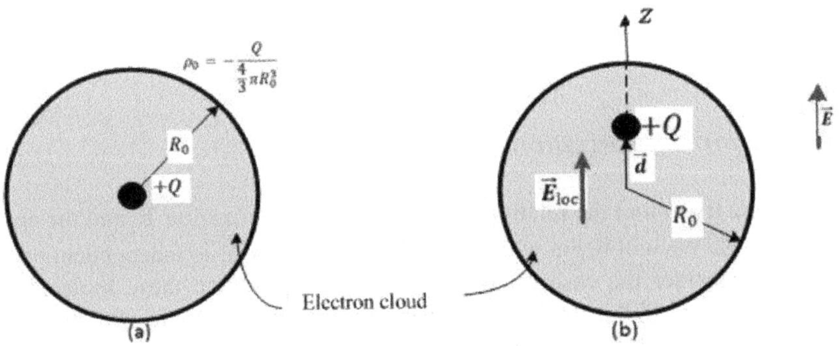

Figure 5.9. *Mechanism of creation of an electric dipole inside an atom under the effect of an external electric field. For a color version of this figure, see www.iste.co.uk/abdesselam/electrodynamics2.zip*

At equilibrium, the net force $\vec{\mathbf{F}}$ acting on the positive charge is zero:

$$\vec{\mathbf{F}} = Q\left(\vec{\mathbf{E}}_{loc} - \frac{Q\vec{\mathbf{d}}}{4\pi\varepsilon_0 R_0^3}\right) \qquad [5.57]$$

where the electric field is evaluated at $r = d$. $\vec{\mathbf{E}}_{loc}$ is the local electric field that acts on the dipole. From [5.57], we obtain:

$$\vec{\mathbf{d}} = \frac{4\pi\varepsilon_0 R_0^3}{Q}\vec{\mathbf{E}}_{loc} \qquad [5.58]$$

Thus, the dipole moment of the atom is given by:

$$\vec{\mathbf{p}} = Q\vec{\mathbf{d}} = 4\pi\varepsilon_0 R_0^3 \vec{\mathbf{E}}_{loc} \equiv \varepsilon_0 \alpha \vec{\mathbf{E}}_{loc} \qquad [5.59]$$

where $\alpha = 4\pi R_0^3$ is called the *polarizability*. It is approximately of the same order as the volume of an atom. Note that if n is the atomic concentration and the atoms are sufficiently dispersed so that the presence of induced dipoles does not significantly alter the value of the applied macroscopic field, that is, $\vec{\mathbf{E}} = \vec{\mathbf{E}}_{\text{loc}}$, then we can relate the macroscopic quantity χ_e to the microscopic quantity α by expressing the polarization $\vec{\mathbf{P}}$ in two different ways: $\vec{\mathbf{P}} = \varepsilon_0 \chi_e \vec{\mathbf{E}} = n\alpha\varepsilon_0 \vec{\mathbf{E}}$, which leads to:

$$\chi_e = n\alpha \qquad [5.60]$$

This is the *electrical susceptibility for low-concentration dielectrics.*

2) *Dense dielectrics*: if the dipole is isolated (i.e. sufficiently far from others), the local electric field is equal to the applied macroscopic field. However, it changes radically from point to point in a dense dielectric, due to the contributions of neighboring dipoles. For this type of dielectric, the polarization $\vec{\mathbf{P}}$ is related to the applied macroscopic field as follows:

$$\vec{\mathbf{P}} = n\vec{\mathbf{p}} = n\alpha\varepsilon_0 \vec{\mathbf{E}}_{\text{loc}} = n\alpha\varepsilon_0 \left(\vec{\mathbf{E}} + \frac{\vec{\mathbf{P}}}{3\varepsilon_0} \right) \qquad [5.61]$$

and, consequently:

$$\vec{\mathbf{P}} = \frac{n\alpha}{1 - (n\alpha)/3} \varepsilon_0 \vec{\mathbf{E}} \equiv \chi_e \varepsilon_0 \vec{\mathbf{E}} \qquad [5.62]$$

Where we introduce the *electrical susceptibility*:

$$\chi_e = \frac{n\alpha}{1 - (n\alpha)/3} \qquad [5.63]$$

as a constant of proportionality between $\vec{\mathbf{P}}$ and $\varepsilon_0 \vec{\mathbf{E}}$. Equation [5.63] is sometimes referred to as the *Clausius–Mossotti equation*. In a dense medium (solid or liquid), the product $n\alpha$ is of the order of unity (because the atoms are in contact). We conclude that for dense media: (a) the local field correction is significant; (b) the susceptibility is of the order of unity. On the other hand, for low-density media, especially gases, the product $n\alpha$ is much smaller than unity. The local field correction is negligible. The field exerted on the constituents is then the $\vec{\mathbf{E}}$ field alone. The electrical susceptibility is simply $\chi_e = n\alpha$.

In terms of ε_r, the so-called *electronic polarizability*, denoted α_{elc}, can be written as:

$$\frac{n\alpha_{\text{elc}}}{3} = \frac{\varepsilon_r - 1}{\varepsilon_r + 2} \qquad [5.64]$$

Since $n = \frac{N_A \rho_M}{M_A}$, where N_A is Avogadro's number, ρ_M is the mass density and M_A is the molar mass, the previous equation can also be expressed as:

$$\frac{N_A \alpha_{\text{elc}}}{3} = \frac{M_A}{\rho_M}\frac{\varepsilon_r - 1}{\varepsilon_r + 2} \equiv P_M \qquad [5.65]$$

P_M is known as the *molar polarizability*.

We can see that equation [5.63] diverges when $\frac{n\alpha}{3} = 1$, that is, $\chi_e \longrightarrow \infty$ for $\frac{n\alpha}{3} = 1$. Polarization can therefore not be zero even if the electric field is zero. In other words, the material will acquire a dipole moment even in the absence of any external electric field. This phenomenon is called the *ferroelectric effect*, and the material is *ferroelectric*. As the density n is a function of T, this can occur by lowering T below a critical temperature T_c, called the *Curie temperature*. Near the critical temperature, we write to the first order of approximation that:

$$n\alpha = 3 - \beta\left(T - T_c\right) \qquad [5.66]$$

where β is a very small constant. Substituting this relation into [5.63], we get:

$$\chi_e - 1 = \frac{3 - \beta\left(T - T_c\right)}{\beta\left(T - T_c\right)/3} \qquad [5.67]$$

Since we assumed that $\beta\left(T - T_c\right)$ is much smaller than unity, we can approximate this formula as:

$$\chi_e - 1 = \frac{9}{\beta\left(T - T_c\right)} \qquad [5.68]$$

5.5.2. *Electric polarization in a time-varying regime*

Now, let us try to define the meaning of polarizability α in the case of a variable regime, that is, when $\vec{\mathbf{E}}$ varies over time. For a non-polar atom or molecule, each electron inside the atom or molecule can be seen as a harmonic oscillator oscillating around an equilibrium position. This is the model of the elastically bound electron introduced by *Drude–Lorentz*. To demonstrate this, let us consider the propagation of a monochromatic plane wave with linear polarization in a sufficiently dilute dielectric medium (such as a gas). In this case, we assume that atoms consist of a fixed nucleus and a single electron (a situation that can be easily generalized to the case of multiple electrons per atom). The atoms are considered sufficiently far apart so that mutual influences can be neglected. Under the influence of the electric field:

$$\begin{aligned} E_x &= E_0 e^{-\mathrm{i}(\omega t - \mathrm{k}z)} \\ E_y &= 0 \\ E_z &= 0 \end{aligned} \qquad [5.69]$$

the electron is displaced from its equilibrium position. The force due to the magnetic field $f_{\overrightarrow{\mathbf{B}}}$ of the wave can be neglected compared to the electrostatic force $f_{\overrightarrow{\mathbf{E}}}$. In fact, for a plane wave, we have: $\frac{|\overrightarrow{\mathbf{B}}|}{|\overrightarrow{\mathbf{E}}|} = \frac{1}{c}$ and, consequently, $\frac{|f_{\overrightarrow{\mathbf{B}}}|}{|f_{\overrightarrow{\mathbf{E}}}|} \leq \frac{v}{c^2}$. Since the electron's velocity is much lower than the speed of light, we can therefore neglect the magnetic force. The equations of motion for an electron in 3D are thus:

$$\begin{aligned} m\frac{d^2x}{dt^2} &= -eE_x - m\omega_0^2 x - \lambda\frac{dx}{dt} \\ m\frac{d^2y}{dt^2} &= 0 \\ m\frac{d^2z}{dt^2} &= 0 \end{aligned} \quad [5.70]$$

where $-m\omega_0^2 x$ is the restoring force that pulls the electron back to its equilibrium position, and $-\lambda\frac{dx}{dt}$ is the frictional force accounting for energy dissipation. It follows that:

$$\frac{d^2x}{dt^2} + \gamma\frac{dx}{dt} + \omega_0^2 x = -\frac{e}{m}E_x \quad [5.71]$$

where $\gamma = \frac{\lambda}{m}$. The solutions take the form:

$$x = x_0 e^{-\mathrm{i}(\omega t - \mathrm{k}z)} \quad [5.72]$$

where x_0 must satisfy the constraint $-\omega^2 x_0 - \mathrm{i}\omega\gamma x_0 + \omega_0^2 x_0 = -\frac{e}{m}E_0$. This leads to:

$$x_0 = -\frac{\frac{e}{m}}{\omega_0^2 - \omega^2 - \mathrm{i}\omega\gamma} \quad [5.73]$$

If n is the number of electrons per unit volume, the components of the polarization are given by:

$$\begin{aligned} P_x &= -nex = P_0 e^{-\mathrm{i}(\omega t - \mathrm{k}z)} = \varepsilon_0 \chi_e(\omega) E_x \\ P_y &= 0 \\ P_z &= 0 \end{aligned} \quad [5.74]$$

where we have introduced the *complex electrical susceptibility* (depending on ω):

$$\chi_e(\omega) = \frac{\chi_e(0)\,\omega_0^2}{\omega_0^2 - \omega^2 - \mathrm{i}\omega\gamma} \quad [5.75]$$

with:

$$\chi_e(0) = \frac{ne^2}{m\varepsilon_0\omega_0^2} \quad [5.76]$$

where $\chi_e(0)$ is the *static electric susceptibility*. The complex nature of $\chi_e(\omega)$ reflects the phase shift between the medium response and the excitation (i.e. between $\overrightarrow{\mathbf{P}}$ and $\overrightarrow{\mathbf{E}}$). Three limiting cases are observed:

1) Low frequency ($\omega \ll \omega_0$) $\Longrightarrow \chi_e(\omega) \simeq \chi_e(0)$. $\overrightarrow{\mathbf{P}}$ is in phase with $\overrightarrow{\mathbf{E}}$. This corresponds to the stationary regimes previously examined, where χ_e is a real and positive constant.

2) Intermediate frequency regime ($\omega \simeq \omega_0$) $\Longrightarrow \chi_e(\omega) \simeq \frac{\mathrm{i}\omega_0\chi_e(0)}{\gamma} = \frac{\omega_0\chi_e(0)}{\gamma}e^{\mathrm{i}\frac{\pi}{2}}$. $\overrightarrow{\mathbf{P}}$ and $\overrightarrow{\mathbf{E}}$ are in quadrature, meaning they are phase-shifted by $\frac{\pi}{2}$.

3) High frequency ($\omega \gg \omega_0$) $\Longrightarrow \chi_e(\omega) \simeq -\frac{\chi_e(0)\omega_0^2}{\omega^2} = \frac{\chi_e(0)\omega_0^2}{\omega^2}e^{\mathrm{i}\pi}$. $\overrightarrow{\mathbf{P}}$ and $\overrightarrow{\mathbf{E}}$ are in opposite phase.

EXAMPLE 5.5. (Electronic polarizability of hydrogen).– Let us determine the dielectric constant of hydrogen at low frequency. It is already known that the energy required to ionize a hydrogen atom is approximately $E \simeq \frac{1}{2}\frac{me^4}{\hbar^2}$. To evaluate the natural frequency ω_0, we set this energy equal to $\hbar\omega_0$ (the energy of an atomic oscillator with natural frequency ω_0). Thus, we obtain:

$$\omega_0 \simeq \frac{1}{2}\frac{me^4}{\hbar^3} \qquad [5.77]$$

By substituting this value of ω_0 into equation [5.76], we find the electronic polarizability at low frequency:

$$\alpha_{\text{elec}} \simeq \frac{4}{\varepsilon_0}\left[\frac{\hbar^2}{me^2}\right]^3 \qquad [5.78]$$

■

In the general case, we decompose $\chi_e(\omega)$ as follows:

$$\chi_e(\omega) = \chi_e'(\omega) + \mathrm{i}\chi_e''(\omega) \qquad [5.79]$$

where:

$$\begin{aligned}\chi_e'(\omega) &= \frac{\chi_e(0)\left(\omega_0^2-\omega^2\right)\omega_0^2}{\left(\omega_0^2-\omega^2\right)^2+\omega^2\gamma^2}\\ \chi_e''(\omega) &= \frac{\chi_e(0)\,\omega\gamma\omega_0^2}{\left(\omega_0^2-\omega^2\right)^2+\omega^2\gamma^2}\end{aligned} \qquad [5.80]$$

The functions $\chi_e'(\omega)$ and $\chi_e''(\omega)$ are shown in Figure 5.10. (1) At very low frequencies $\omega \ll \omega_0$ (i.e. $\omega \longrightarrow 0$), we have $\chi_e'(\omega) \simeq \chi_e(0)$ and $\chi_e''(\omega) \approx 0$. The susceptibility $\chi_e(\omega)$ becomes real and positive. There is no phase shift between $\overrightarrow{\mathbf{P}}$ and $\overrightarrow{\mathbf{E}}$. (2) For $\omega = \omega_0$, we find $\chi_e'(\omega_0) = 0$ and $\chi_e''(\omega_0) = \frac{\chi_e(0)\omega_0}{\gamma}$. The vectors $\overrightarrow{\mathbf{P}}$ and $\overrightarrow{\mathbf{E}}$ are in quadrature. (3) At very high frequencies $\omega \gg \omega_0$ ($\omega \longrightarrow \infty$), both $\chi_e'(\omega)$ and $\chi_e''(\omega)$ tend toward zero. The susceptibility follows the relation $\chi_e(\omega) \approx -\chi_e(0)\frac{\omega_0^2}{\omega^2}$. The vectors $\overrightarrow{\mathbf{P}}$ and $\overrightarrow{\mathbf{E}}$ are in phase opposition. (4) Near ω_0, we use the approximation $\omega_0^2 - \omega^2 = (\omega_0 - \omega)(\omega_0 + \omega) \simeq 2\omega_0(\omega_0 - \omega)$, which leads to $\chi_e'(\omega) \simeq \frac{\omega_0 \chi_e(0)}{2} \frac{\omega_0 - \omega}{(\omega_0-\omega)^2 + \left(\frac{\gamma}{2}\right)^2}$ and $\chi_e''(\omega) \simeq \chi_e(\omega_0) \frac{\left(\frac{\gamma}{2}\right)^2}{(\omega_0-\omega)^2 + \left(\frac{\gamma}{2}\right)^2}$. In this region, the function $\chi_e'(\omega)$ is odd with respect to the line passing through ω_0. $\chi_e'(\omega)$ being positive on the left and negative on the right. The sign change is very sharp around ω_0. The derivative $\frac{d\chi_e'(\omega)}{d\omega}$ vanishes at $\omega_1 = \omega_0 - \frac{\gamma}{2}$ and $\omega_2 = \omega_0 + \frac{\gamma}{2}$, leading to $\Delta\omega = \gamma$. The function $\chi_e''(\omega)$ is even with respect to the line passing through ω_0 (bell-shaped curve). (5) In the case of weak damping, we have $\chi_e'(\omega) \simeq \frac{\chi_e(0)\omega_0^2}{\omega_0^2 - \omega^2}$ and $\chi_e''(\omega) \simeq \frac{\chi_e(0)\omega\gamma\omega_0^2}{\left(\omega_0^2-\omega^2\right)^2} \approx 0$. For $\omega \ll \omega_0$, we get $\chi_e'(\omega) \simeq \frac{\chi_e(0)\omega_0}{2(\omega_0 - \omega)}$. For $\omega \gg \omega_0$, we obtain $\chi_e'(\omega) \approx 0$.

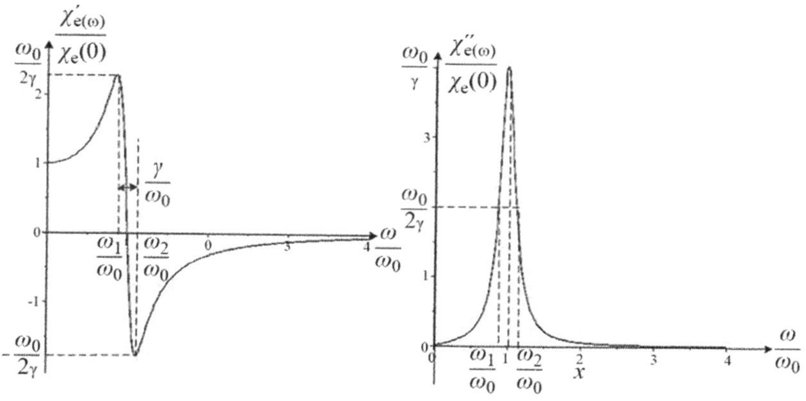

Figure 5.10. *Graphical representations of the functions $\chi_e'(\omega)$ and $\chi_e''(\omega)$*

It is the imaginary part $\chi_e''(\omega)$, which is at the origin of the term in the expression of the polarization, which is in quadrature with the electric field:

$$\begin{aligned} \mathrm{Re}(P_x) = & \frac{\varepsilon_0 \chi_e(0)\, \omega_0^2 \left(\omega_0^2 - \omega^2\right)}{\left(\omega_0^2 - \omega^2\right)^2 + \omega^2\gamma^2} E_0 \cos(\omega t - \mathrm{k}z) \\ & + \frac{\varepsilon_0 \chi_e(0)\, \omega_0^2 \omega\gamma}{\left(\omega_0^2 - \omega^2\right)^2 + \omega^2\gamma^2} E_0 \sin(\omega t - \mathrm{k}z) \end{aligned} \qquad [5.81]$$

In reality, various types of electrons coexist in a dielectric medium, identified by indices i. This is because electrons belong to different electronic layers of the atom; they are therefore pulled back to their equilibrium positions with restoring forces characterized by different ω_0 natural frequencies and γ (damping) coefficients. In this case, we write (instead of [5.74]):

$$\vec{\mathbf{P}} = \sum_i \frac{\frac{n_i e^2}{m\varepsilon_0}}{\omega_{0i}^2 - \omega^2 - \mathrm{i}\omega\gamma_i} \varepsilon_0 \vec{\mathbf{E}} \qquad [5.82]$$

which gives:

$$\chi_e(\omega) = \sum_i \frac{\frac{n_i e^2}{m\varepsilon_0}}{\omega_{0i}^2 - \omega^2 - \mathrm{i}\omega\gamma_i}$$
$$\alpha = \sum_i \frac{\frac{n_i e^2}{nm\varepsilon_0}}{\omega_{0i}^2 - \omega^2 - \mathrm{i}\omega\gamma_i} = \sum_i \frac{\frac{f_i e^2}{m\varepsilon_0}}{\omega_{0i}^2 - \omega^2 - \mathrm{i}\omega\gamma_i} \qquad [5.83]$$

where $f_i = \frac{n_i}{n}$ (n is the total number of electrons). Since $\overrightarrow{\mathbf{D}} = \varepsilon_0 \overrightarrow{\mathbf{E}} + \overrightarrow{\mathbf{P}} = \varepsilon \overrightarrow{\mathbf{E}}$, we deduce that the permittivity of the medium is complex:

$$\varepsilon(\omega) = \varepsilon_0 + \sum_k \frac{\frac{n_i e^2}{m}}{\omega_{0i}^2 - \omega^2 - \mathrm{i}\omega\gamma_k} \qquad [5.84]$$

At low frequencies $(\omega \longrightarrow \omega_0)$, equation [5.84] simplifies to $\varepsilon(\omega) = \varepsilon_0 + \sum_i \frac{n_i e^2}{m\omega_{0,k}^2}$ (static case).

The above considerations apply to atoms in a sufficiently dilute gas, meaning that the atoms are far apart from each other. For dense media (solids or liquids), we can no longer consider only the action of the external electric field $\overrightarrow{\mathbf{E}}$, but must take into account the additional field due to polarization (which adds to the external field $\overrightarrow{\mathbf{E}}$). In fact, in the case of a dilute gas, the equation of motion of an electron was written as, $\frac{d^2x}{dt^2} + \gamma\frac{dx}{dt} + \omega_0^2 x = -\frac{e}{m}E_x$. For a dense medium (where interaction forces are significant), E_x must be replaced by $E_x + \frac{P_x}{3\varepsilon_0} = E_x - \frac{nex}{3\varepsilon_0}$. Thus, the equation of motion becomes $\frac{d^2x}{dt^2} + \gamma\frac{dx}{dt} + \left(\omega_0^2 - \frac{ne^2}{3m\varepsilon_0}\right)x = -\frac{e}{m}E_x$. This equation has the solution $x = -\frac{\frac{e}{m}}{\omega_0^2 - \frac{ne^2}{3m\varepsilon_0} - \omega^2 - \mathrm{i}\omega\gamma} = -\frac{\frac{e}{m}}{\omega'^2_0 - \omega^2 - \mathrm{i}\omega\gamma}$, where $\omega'^2_0 = \omega_0^2 - \frac{ne^2}{3m\varepsilon_0}$. It follows that:

$$\overrightarrow{\mathbf{P}} = \frac{\frac{ne^2}{m\varepsilon_0}}{\omega'^2_0 - \omega^2 - \mathrm{i}\omega\gamma}\varepsilon_0 \overrightarrow{\mathbf{E}}$$
$$\chi_e(\omega) = \frac{n\alpha}{1 - \frac{n\alpha}{3}} = \frac{\frac{ne^2}{m\varepsilon_0}}{\omega'^2_0 - \omega^2 - \mathrm{i}\omega\gamma} \qquad [5.85]$$

These formulas can be easily generalized to multiple electrons. The expression for $\chi_e(\omega)$ remains the same. It is sufficient to replace ω_0^2 with $\omega'^2_0 = \omega_0^2 - \frac{ne^2}{3m\varepsilon_0}$. However, the expression for the polarizability becomes slightly more complex.

5.5.3. *Ionic polarization*

To simplify the discussion, let us consider the case of a one-dimensional crystal made up of an infinitely long chain of positive and negative ions with charge

$\pm Q\,(=\pm Ne)$, where N is the number of electrons removed (for the positive ion) or added (for the negative ion). These ions are alternately arranged at a distance a along an axis denoted as Ox, as illustrated in Figure 5.11. When an electric field is applied along the direction of Ox, it induces a local field $\overrightarrow{\mathbf{E}}_{\text{loc}}$ in the same direction. Under the influence of $\overrightarrow{\mathbf{E}}_{\text{loc}}$, the ions move by an amount Δx from their equilibrium positions (positive ions on one side and negative ions on the other). The displacement Δx results from a balance between the restoring forces (arising from valence interactions) and the force exerted by the local electric field. For small displacements, the restoring forces can be assumed to be proportional to Δx. Taking into account only the two nearest neighbors of the reference ion, the restoring force is given by $F = -2\gamma\Delta x$, where γ is determined from the expression of the binding energy $W(x)$. Taking into account the two nearest neighbors of the reference ion, the equilibrium condition is given by:

$$4\gamma\Delta x = QE_{\text{loc}} = NeE_{\text{loc}} \qquad [5.86]$$

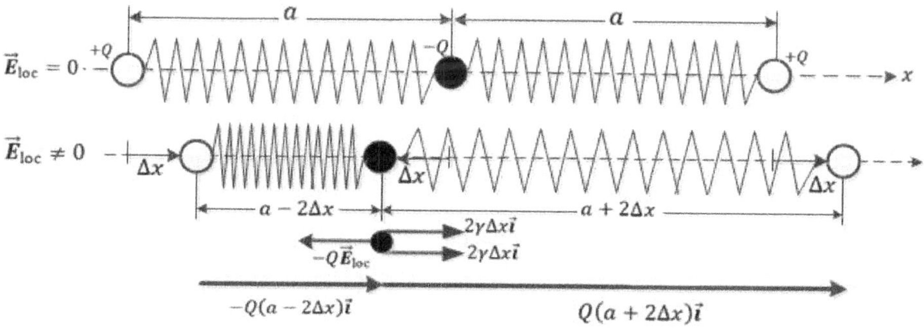

Figure 5.11. *Ionic polarization within a one-dimensional crystal. For a color version of this figure, see www.iste.co.uk/abdesselam/electrodynamics2.zip*

The dipole moment resulting from the displacements Δx is given by:

$$p = (Ne)\,(a + 2\Delta x) - Ne\,(a - 2\Delta x) = 4Ne\Delta x \qquad [5.87]$$

It follows that:

$$p = \frac{(Ne)^2}{\gamma} E_{\text{loc}} \qquad [5.88]$$

The *ionic polarizability* factor α_{ion} is therefore given by:

$$\alpha_{\text{ion}} = \frac{(Ne)^2}{\gamma} \qquad [5.89]$$

The binding energy $W(x)$ is independent of temperature, as is the ionic polarizability factor.

5.5.4. *Orientation polarization of polar molecules*

A polar molecule admits a permanent dipole moment. In a disordered medium composed of polar molecules, the average polarization is zero as a result of disorder, since the dipoles are randomly oriented (entropy must be maximal) in the absence of an electric field. However, when the local field $\vec{\mathbf{E}}_{\text{loc}}$ is not zero (due to the application of an external field), the dipole moment of the molecule $\vec{\mathbf{p}}$ tends to align in the direction of $\vec{\mathbf{E}}_{\text{loc}}$ to minimize the interaction potential energy $E_p = -\vec{\mathbf{p}} \cdot \vec{\mathbf{E}}_{\text{loc}}$ between the dipoles and the field. We now need to find the extent to which the dipoles align using statistical physics methods. We know that in a state of thermal equilibrium, the relative number of molecules with potential energy $E_p(x, y, z)$ is proportional to $\exp\left(-\frac{E_p}{k_B T}\right)$, where k_B is the *Boltzmann constant* and T is the *absolute temperature* of the system. In this case, considering the expression for the potential energy $E_p = -pE_{\text{loc}} \cos\theta$, we can state that the number of molecules in the direction θ per unit of solid angle is proportional to $\exp\left(\frac{pE_{\text{loc}} \cos\theta}{k_B T}\right)$. Let $dn(\theta)$ be the number of dipoles per unit volume in the solid angle $d\Omega$ defined around θ, then:

$$dn(\theta) = n_0 \exp\left(\frac{pE_{\text{loc}} \cos\theta}{k_B T}\right) d\Omega \qquad [5.90]$$

By symmetry, the polarization vector $\vec{\mathbf{P}}$, with magnitude P, is collinear with $\vec{\mathbf{E}}_{\text{loc}}$. Therefore, we write:

$$P = \int pn_0 \exp\left(\frac{pE_{\text{loc}} \cos\theta}{k_B T}\right) \cos\theta d\Omega \qquad [5.91]$$

Now, the solid angle element, when integrating over the angle φ, is given by $d\Omega = 2\pi \sin\theta d\theta$. It follows that:

$$P = 2\pi pn_0 \int_0^\pi \exp\left(\frac{pE_{\text{loc}} \cos\theta}{k_B T}\right) \cos\theta \sin\theta d\theta \qquad [5.92]$$

Moreover, as:

$$n = 2\pi n_0 \int_0^\pi \exp\left(\frac{pE_{\text{loc}} \cos\theta}{k_B T}\right) \sin\theta d\theta \qquad [5.93]$$

we can determine n_0 and, consequently, deduce that:

$$P = np \frac{\displaystyle\int_0^\pi \exp\left(\frac{pE_{\text{loc}} \cos\theta}{k_B T}\right) \cos\theta \sin\theta d\theta}{\displaystyle\int_0^\pi \exp\left(\frac{pE_{\text{loc}} \cos\theta}{k_B T}\right) \sin\theta d\theta} = np\mathcal{L}\left(\frac{pE_{\text{loc}}}{k_B T}\right) \qquad [5.94]$$

where $\mathcal{L}(a)$ is the *Langevin function*. Thus, the average polarization is equal to the product of the density per unit volume, the dipole moment of a molecule and the Langevin function, which is expressed as the quotient of the derivative of the function $D(a) = \int_0^{\pi} \exp(a\cos\theta)\sin\theta d\theta$ by this same function:

$$\mathcal{L}(a) = \frac{1}{D(a)}\frac{dD(a)}{da} = \frac{d}{da}(\ln D(a)) \qquad [5.95]$$

By performing the change of variable $u = \cos\theta$, we find $D(a) = \int_{-1}^{1} e^{au}du = \frac{(e^a - e^{-a})}{a} = \frac{2\sinh a}{a}$ and, consequently:

$$\mathcal{L}(a) = \frac{d}{da}(\ln 2 - \ln a + \ln\sinh a) = -\frac{1}{a} + \coth a \qquad [5.96]$$

By substituting [5.96] into [5.94], we finally obtain:

$$P = np\left(\coth\frac{pE_{\text{loc}}}{k_BT} - \frac{k_BT}{pE_{\text{loc}}}\right) \qquad [5.97]$$

This is the *Langevin formula*. Note that the Langevin function $\mathcal{L}(a)$ converges to 0 as $a \longrightarrow 0$ and to 1 as $a \longrightarrow \infty$, as illustrated in Figure 5.12. We therefore conclude that polarization vanishes at very high temperatures. In other words, at very high temperatures, thermal agitation causes the orientation polarization to average out to zero. At very low temperatures, we have $P = np$, that is, all dipoles are aligned with the field: *polarization saturation*. For ambient temperatures, we can equate $\mathcal{L}(a)$ with $\frac{a}{3}$ (even for a very intense E_{loc} field, a is very small) and, consequently:

$$\vec{\mathbf{P}} = \frac{np^2}{3k_BT}\vec{\mathbf{E}}_{\text{loc}} = \chi_e\varepsilon_0\vec{\mathbf{E}}_{\text{loc}} \qquad [5.98]$$

that is, the polarization is collinear with $\vec{\mathbf{E}}_{\text{loc}}$. This is the *Langevin–Debye equation*. The orientation polarizability is therefore:

$$\alpha_{\text{orient}} = \frac{p^2}{3\varepsilon_0 k_BT} \qquad [5.99]$$

and will be added to the other polarizabilities. Thus, we can modify the Clausius–Mossotti equation [5.65] as follows:

$$\frac{M_A}{\rho_M}\frac{\varepsilon_r - 1}{\varepsilon_r + 2} = \frac{N_A\alpha}{3} = \frac{N_A}{3}\left(\alpha_{\text{élec}} + \alpha_{\text{ion}} + \frac{p^2}{3\varepsilon_0 k_BT}\right) \qquad [5.100]$$

We have just established that dipoles (polar molecules) tend to orient themselves in the direction of the local electric field to minimize potential energy. This tendency is counteracted by thermal agitation. When the local field is instantly suppressed, depolarization is not instantaneous. Debye introduced a first-order differential equation to model the phenomenon or the *polarization relaxation effect* $\overrightarrow{\mathbf{P}}$:

$$\tau \frac{d\overrightarrow{\mathbf{P}}}{dt} + \overrightarrow{\mathbf{P}} = 0 \qquad [5.101]$$

where τ is the *relaxation time*. This differential equation has the solution:

$$\overrightarrow{\mathbf{P}} = \overrightarrow{\mathbf{P}}_0 e^{-t/\tau} \qquad [5.102]$$

where $\overrightarrow{\mathbf{P}}_0$ is the polarization at $t = 0$. In the presence of a local field $\overrightarrow{\mathbf{E}}_{\text{loc}}$ (assuming for simplicity that the local field $\overrightarrow{\mathbf{E}}_{\text{loc}}$ is approximately equal to the external field), equation [5.101] becomes:

$$\tau \frac{d\overrightarrow{\mathbf{P}}}{dt} + \overrightarrow{\mathbf{P}} = \varepsilon_0 n \alpha_0 \overrightarrow{\mathbf{E}} \qquad [5.103]$$

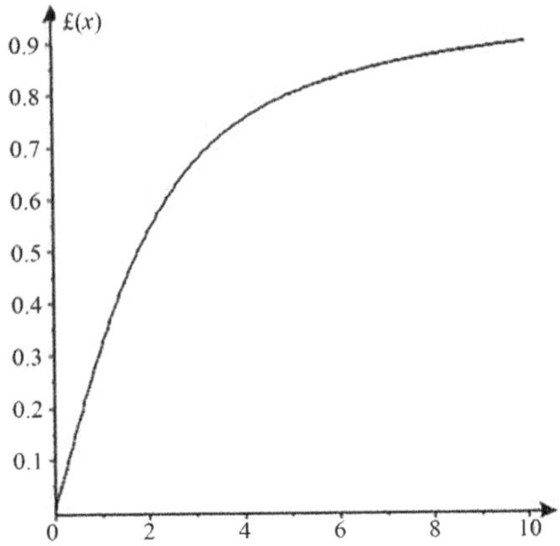

Figure 5.12. *Behavior of the Langevin function*

This new equation indicates that the creation of an average polarization in polar fluids under the influence of a field $\overrightarrow{\mathbf{E}}$ cannot be instantaneous either, as it involves the rotation of molecules, which is disrupted by intermolecular collisions. It is clear that, for long times in [5.103], the polarization tends toward the static value $\varepsilon_0 N \alpha_0 \overrightarrow{\mathbf{E}}$. In reality, [5.103] is equivalent to the equation of motion $\frac{d^2\overrightarrow{\mathbf{r}}}{dt^2} + \gamma \frac{d\overrightarrow{\mathbf{r}}}{dt} + \omega_0^2 \overrightarrow{\mathbf{r}} = -\frac{e}{m}\overrightarrow{\mathbf{E}}$ except that the acceleration term has been neglected in front of the friction term, as damping mechanisms are dominant. The Debye relaxation time thus appears as $\frac{\gamma}{\omega_0^2}$. Let us assume that the field $\overrightarrow{\mathbf{E}}$ is of the form $\overrightarrow{\mathbf{E}} = \overrightarrow{\mathbf{E}}_0 e^{-i\omega t}$. To determine the steady-state solution of [5.103] (i.e. for sufficiently long times), we can directly use the results already established in section 5.5.2. Neglecting the acceleration term in front of the damping term corresponds to assuming $\omega^2 \ll \omega\gamma$, which leads to the following expression for the orientational susceptibility:

$$\chi_{\text{orient}}(\omega) = \chi_{\text{orient}}(0)\frac{\omega_0^2}{\omega_0^2 - \mathrm{i}\omega\gamma} = \chi_{\text{orient}}(0)\frac{1}{1 - \mathrm{i}\omega\tau} \qquad [5.104]$$

where $\tau = \frac{\gamma}{\omega_0^2}$. The static susceptibility is the same as in the steady-state regime:

$$\chi_{\text{orient}}(0) = \frac{np^2}{3\varepsilon_0 k_B T} \qquad [5.105]$$

where p here denotes the permanent dipole moment of the molecule. The real and imaginary parts of $\chi_{\text{orient}}(\omega)$ constitute the *Debye equations*:

$$\chi'_{\text{orient}}(\omega) = \varepsilon'_r - 1 = \chi_{\text{orient}}(0)\frac{1}{1 + \omega^2\tau^2}$$

$$\chi''_{\text{orient}}(\omega) = \varepsilon''_r = \chi_{\text{orient}}(0)\frac{\omega\tau}{1 + \omega^2\tau^2} \qquad [5.106]$$

The Debye functions are shown in Figure 5.13. $\chi'_{\text{orient}}(\omega)$ is a decreasing function, approximately equal to $\chi'_{\text{orient}}(0)$ at low frequencies and tending to zero at high frequencies. $\chi''_{\text{orient}}(\omega)$ is zero at both low and high frequencies and reaches its maximum at $\omega = \frac{1}{\tau}$. Two distinct regions can be identified:

– $\omega\tau \ll 1$: $\chi_{\text{orient}}(\omega)$ is real and approximately equal to its low-frequency value, $\chi'_{\text{orient}}(\omega) \simeq \chi'_{\text{orient}}(0)$, while $\chi''_{\text{orient}}(\omega) \ll \chi''_{\text{orient}}(0)$.

– $\omega\tau \gg 1$: $\chi_{\text{orient}}(\omega)$ is negligible. This means that the dipole moments can no longer follow the oscillations, $\chi'_{\text{orient}}(\omega) \ll \chi'_{\text{orient}}(0)$ and $\chi''_{\text{orient}}(\omega) \ll \chi''_{\text{orient}}(0)$.

EXAMPLE 5.6. (Point charge inside a spherical dielectric).– Determine the fields and polarization charges generated by a point charge q placed in the center of a linear dielectric sphere with radius R and permittivity ε, as illustrated in Figure 5.14. Applying Gauss's theorem [5.51] to an imaginary sphere of arbitrary radius r centered in q, we obtain:

$$D_r = \frac{q}{4\pi r^2} \qquad [5.107]$$

The electric field is then discontinuous at $r = R$:

$$E_r = \begin{cases} \dfrac{D_r}{\varepsilon} = \dfrac{q}{4\pi\varepsilon r^2} & r \leq R^- \\[2ex] \dfrac{D_r}{\varepsilon_0} = \dfrac{q}{4\pi\varepsilon_0 r^2} & r \geq R^+ \end{cases} \qquad [5.108]$$

because of the change in permittivity at the boundary of the sphere. We observe that the electric field due to a point charge q inside a dielectric sphere is weaker than the electric field outside. The polarization field (or simply the polarization) is:

$$P_r = D_r - \varepsilon_0 E_r = \begin{cases} \dfrac{(\varepsilon - \varepsilon_0)\, q}{4\pi\varepsilon r^2} & r \leq R^- \\ 0 & r \geq R^+ \end{cases} \qquad [5.109]$$

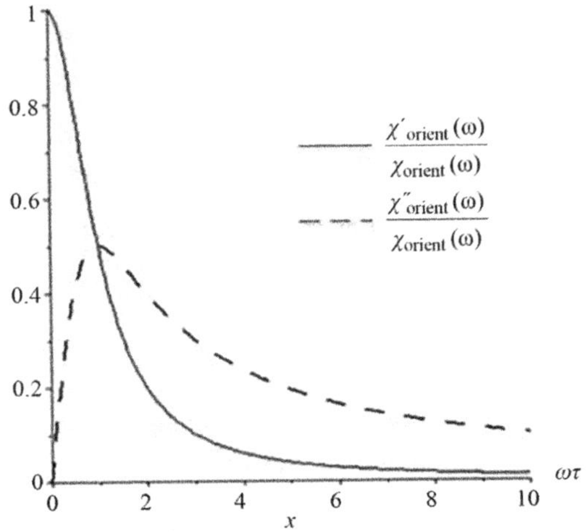

Figure 5.13. *Graphical representations of Debye functions. For a color version of this figure, see www.iste.co.uk/abdesselam/electrodynamics2.zip*

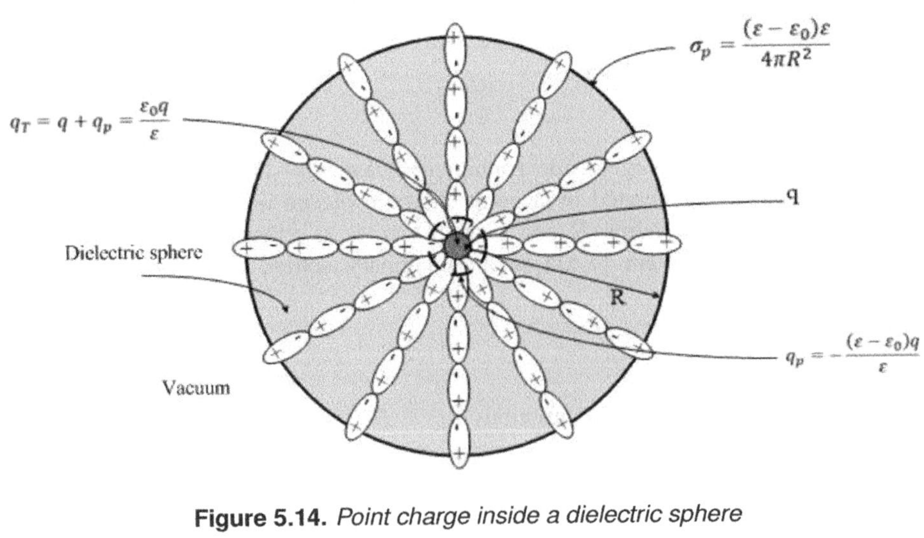

Figure 5.14. *Point charge inside a dielectric sphere*

The volume polarization charge density ρ_l is zero everywhere:

$$\rho_l = -\overrightarrow{\nabla} \cdot \overrightarrow{\mathbf{P}} = -\frac{1}{r^2}\frac{\partial}{\partial r}\left(r^2 P_r\right) = 0 \qquad [5.110]$$

except at $r = 0$, where a point polarization charge is present, and at $r = R^+$ (the outer surface of the dielectric), where a surface polarization charge exists. The total polarization charge contained within a concentric sphere of radius r, located either inside or outside the dielectric, is:

$$q_p = -\oint_S \overrightarrow{\mathbf{P}} \cdot d\overrightarrow{\mathbf{S}} = -P_r 4\pi r^2 = \begin{cases} -\dfrac{(\varepsilon - \varepsilon_0)\, q}{\varepsilon} & r \leq R^- \\ 0 & r \geq R^+ \end{cases} \qquad [5.111]$$

Now, we know that for $r \leq R^-$, this polarization charge must be concentrated at the origin, since the volume polarization charge is zero in the region $0 < r \leq R^-$. Thus, the total polarization charge at the origin is:

$$q_T = q_p + q = \frac{\varepsilon_0}{\varepsilon} q \qquad [5.112]$$

This reduction in net charge is explained by the fact that the ends of dipoles with opposite charges are attracted to the point charge, which leads to partial neutralization. This also explains why the electric field inside the sphere is weaker than outside. Since the total polarization charge contained within the sphere for $r > R$ is zero, a surface polarization charge of opposite polarity appears on the external surface of the material with a surface density:

$$\sigma_l = \overrightarrow{\mathbf{P}}\,(R) \cdot \overrightarrow{\mathbf{e}}_r = P_r\,(R) = \frac{(\varepsilon - \varepsilon_0)\, q}{4\pi\varepsilon R^2} \qquad [5.113]$$

because the total surface charge $\sigma_l 4\pi R^2 = \frac{(\varepsilon - \varepsilon_0) q}{\varepsilon}$ is equal to the opposite of the point polarization charge located at the origin. The total polarization charge always remains zero.

5.6. Boundary conditions for the field $\overrightarrow{\mathrm{D}}$

Consider a cylinder of height Δz located at the interface between two media (1) and (2). The upper half of the cylinder lies in medium (1), while the lower half is in medium (2). The cylinder axis is chosen along the Oz direction, perpendicular to the interface between the two media. The two bases of the cylinder (upper and lower) are parallel to the intersection surface between the cylinder and the interface, as illustrated in Figure 5.15. The free volume charge density (free charge density) inside the cylinder is given by the following equation:

$$\rho = \rho_v\,(x, y, z) + \sigma\,(x, y)\,\delta\,(z) \qquad [5.114]$$

where $\rho_v(x, y, z)$ represents the free volume charge density inside the cylinder but outside of the intersection surface (i.e. $\rho_v(x, y, 0) = 0$), and $\sigma(x, y)$ represents the free surface charge confined to the interface. Integrating the equation $\vec{\nabla} \cdot \vec{\mathbf{D}} = \rho$ over the volume of the cylinder gives:

$$\begin{aligned}\int_{\mathcal{T}} \vec{\nabla} \cdot \vec{\mathbf{D}} d\mathcal{T} &= \int_{\mathcal{T}} \rho d\mathcal{T} = \int_{\mathcal{T}} \rho_v(x, y, z)\, d\mathcal{T} + \int_S \sigma(x, y)\, dxdy \\ &= \langle \rho_v \rangle \mathcal{T} + \int_S \sigma(x, y)\, dxdy\end{aligned} \quad [5.115]$$

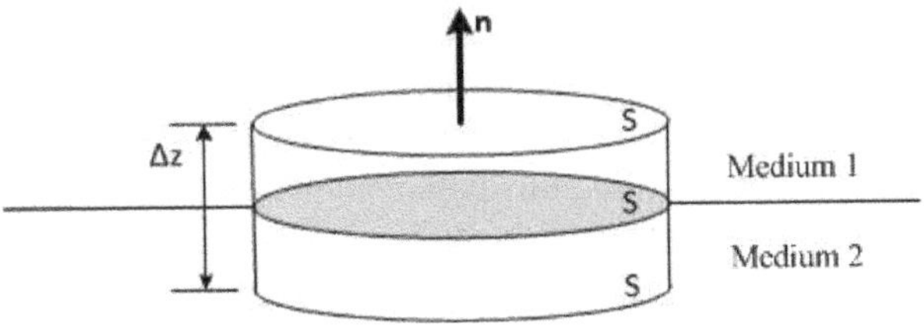

Figure 5.15. *Discontinuity of the normal component of the field $\vec{\mathbf{D}}$ between two dielectric media (1) and (2)*

According to the divergence theorem, the previous integral can also be written as (S' is the total surface area surrounding the cylinder):

$$\begin{aligned}\int_{\mathcal{T}} \vec{\nabla} \cdot \vec{\mathbf{D}} d\mathcal{T} = \int_{S'} \vec{\mathbf{D}} \cdot d\vec{\mathbf{S}}' = \int_S D_z^{(1)}\left(x, y, \frac{\Delta z}{2}\right) dxdy - \\ \int_S D_z^{(2)}\left(x, y, -\frac{\Delta z}{2}\right) dxdy + \int_{\text{surf.lat.}} \vec{\mathbf{D}} \cdot d\vec{\mathbf{S}}\end{aligned} \quad [5.116]$$

Taking the limit as $\Delta z \longrightarrow 0$ (i.e. at the interface), we find:

$$\int_S \left(D_z^{(1)} - D_z^{(2)}\right) dxdy = \int_S \sigma(x, y)\, dxdy \quad [5.117]$$

This equality must hold for any arbitrary surface S, which implies that $D_z^{(1)} - D_z^{(2)} = \sigma$. It can also be written in vector form as:

$$\left(\vec{\mathbf{D}}^{(1)} - \vec{\mathbf{D}}^{(2)}\right) \cdot \vec{\mathbf{n}}_{21} = \sigma \quad [5.118]$$

where $\vec{\mathbf{n}}_{21}$ is the unit normal vector to the interface, directed from medium (2) to medium (1). We therefore conclude that the perpendicular component of $\left(\vec{\mathbf{D}}\right)_{\perp}$ presents a discontinuity of σ at the interface between media (1) and (2).

Since the electric field $\vec{\mathbf{E}}$ satisfies the equation $\vec{\nabla} \times \vec{\mathbf{E}} = 0$, we can deduce that (i) the tangential component of $\vec{\mathbf{E}}$ is continuous at the interface between media (1) and

(2), that is, $\left(\vec{\mathbf{E}}\right)_{\parallel}^{(1)} = \left(\vec{\mathbf{E}}\right)_{\parallel}^{(2)}$; (ii) $\vec{\mathbf{E}}$ can be expressed as $\vec{\mathbf{E}} = -\vec{\nabla} V$; (iii) equation $\vec{\nabla} \cdot \vec{\mathbf{D}} = \rho$ (where ρ is the free charge density) gives, for a linear and isotropic dielectric: $\vec{\nabla} \cdot \left(\varepsilon \vec{\mathbf{E}}\right) = -\vec{\nabla} \cdot \left(\varepsilon \vec{\nabla} V\right) = \rho$. If ε is constant within a given region (homogeneous medium), then the potential V must satisfy the Poisson equation in each region: $\Delta V = -\frac{\rho}{\varepsilon}$. When $\rho = 0$, the equation simplifies to the Laplace equation, $\Delta V = 0$. To solve problems in dielectric media, we apply the known methods.

EXAMPLE 5.7. (Dielectric slab).– An infinite dielectric slab is placed in a uniform electric field, making an angle θ_v with the normal surface of the slab. Let us determine the angle θ_d that the electric field makes with the normal inside the slab, as shown in Figure 5.16. In fact, the components $D_\perp$ and $E_\parallel$ are continuous to the surface of the slab. In vacuum, the fields $D_\perp$ and $E_\parallel$ are, respectively, $(D_\perp)_v = \varepsilon_0 E_v \cos\theta_v$ and $\left(E_\parallel\right)_v = E_v \sin\theta_v$, while inside the dielectric slab, they are given by $(D_\perp)_d = \varepsilon_d E_d \cos\theta_d$ and $\left(E_\parallel\right)_d = E_d \sin\theta_d$. Thus, we obtain:

$$\begin{aligned} \varepsilon_0 E_v \cos\theta_v &= \varepsilon_d E_d \cos\theta_d \\ E_v \sin\theta_v &= E_d \sin\theta_d \end{aligned} \qquad [5.119]$$

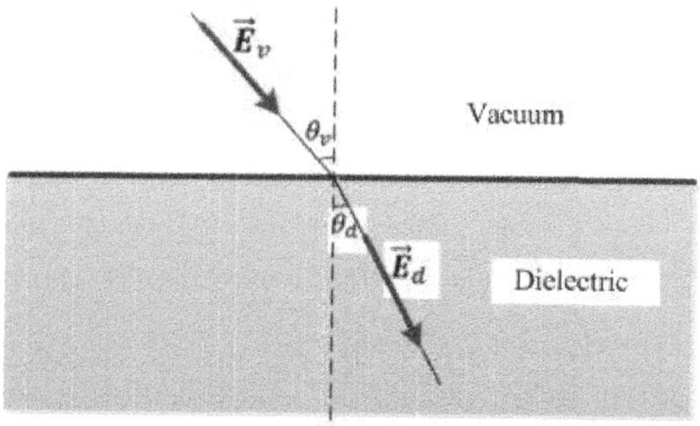

Figure 5.16. *Inclination of the electric field upon entering a dielectric slab. For a color version of this figure, see www.iste.co.uk/abdesselam/electrodynamics2.zip*

By eliminating E_v and E_d, it follows:

$$\tan\theta_v = \frac{\varepsilon_0}{\varepsilon_d} \tan\theta_d \qquad [5.120]$$

We therefore conclude that the electric field tilts slightly as it penetrates the plate. In particular, when $\theta_v = \frac{\pi}{2}$ (i.e. the external electric field is parallel to the dielectric surface), the electric field inside the dielectric is equal to the applied external field, $\vec{\mathbf{E}}_d = \vec{\mathbf{E}}_v$. The external field fully penetrates the dielectric (no depolarizing field) and leads to polarization $\vec{\mathbf{P}} = \varepsilon_0 (\varepsilon_r - 1) \vec{\mathbf{E}}_v$ parallel to the dielectric surface.

On the other hand, when $\theta_v = 0$ (i.e. the external electric field is perpendicular to the dielectric surface), we obtain $\vec{\mathbf{E}}_d - \vec{\mathbf{E}}_v = -\left(1 - \frac{1}{\varepsilon_r}\right)\vec{\mathbf{E}}_v$; this creates a depolarizing field that opposes the external field. The induced polarization inside the dielectric is given by $\vec{\mathbf{P}} = \left(1 - \frac{1}{\varepsilon_r}\right)\varepsilon_0\vec{\mathbf{E}}_v$.

EXAMPLE 5.8. (Spherical capacitor with a dielectric).–

Two concentric spherical electrodes with radii R_1 and R_2 $(R_2 > R_1)$ are maintained at potentials V_1 and V_2, respectively. The region between the electrodes, defined by $R_1 < r < R_2$, is filled with a dielectric material of permittivity ε. The only general solution to Laplace's equation, which possesses the spherical symmetry of the system, is of the form $V(r) = \frac{A}{r} + B$, where A and B are constants that must be determined to satisfy the boundary conditions $V(R_1) = V_1$ and $V(R_2) = V_2$. Consequently, the potential in the region $R_1 < r < R_2$ is given by:

$$V(r) = \frac{1}{R_2 - R_1}\left[(V_1 - V_2)\frac{R_1 R_2}{r} + V_2 R_2 - V_1 R_1\right] \qquad [5.121]$$

Consequently, the electrostatic and displacement fields can be expressed as:

$$\begin{aligned} \vec{\mathbf{E}} &= -\vec{\nabla}V = \frac{V_1 - V_2}{R_2 - R_1}\frac{R_1 R_2}{r^2}\vec{\mathbf{e}}_r \\ \vec{\mathbf{D}} &= \varepsilon\vec{\mathbf{E}} = \varepsilon\frac{V_1 - V_2}{R_2 - R_1}\frac{R_1 R_2}{r^2}\vec{\mathbf{e}}_r \end{aligned} \qquad [5.122]$$

The free surface charge densities on electrodes (1) and (2) are, respectively:

$$\begin{aligned} \sigma_1 &= \vec{\mathbf{D}}(R_1)\cdot\vec{\mathbf{e}}_r = \varepsilon\frac{V_1 - V_2}{R_2 - R_1}\frac{R_2}{R_1} \\ \sigma_2 &= -\vec{\mathbf{D}}(R_2)\cdot\vec{\mathbf{e}}_r = -\varepsilon\frac{V_1 - V_2}{R_2 - R_1}\frac{R_1}{R_2} \end{aligned} \qquad [5.123]$$

Thus, we deduce the charge of the capacitor as $Q = 4\pi R_1^2\sigma_1$, and subsequently, the capacitance:

$$C = \frac{Q}{V_1 - V_2} = 4\pi\varepsilon\frac{R_1 R_2}{R_2 - R_1} = \varepsilon_r C_0 \qquad [5.124]$$

where $C_0 = 4\pi\varepsilon_0\frac{R_1 R_2}{R_2 - R_1}$ represents the capacitance of a spherical capacitor with vacuum between the electrodes. The polarization vector inside the dielectric is:

$$\vec{\mathbf{P}} = \vec{\mathbf{D}} - \varepsilon_0\vec{\mathbf{E}} = (\varepsilon - \varepsilon_0)\frac{V_1 - V_2}{R_2 - R_1}\frac{R_1 R_2}{r^2}\vec{\mathbf{e}}_r \qquad [5.125]$$

and, consequently, the polarization charges densities accumulated at the electrode–dielectric interfaces $r = R_1$ and $r = R_2$ are, respectively:

$$\begin{aligned} \sigma_{p,1} &= -\vec{\mathbf{P}}(R_1)\cdot\vec{\mathbf{e}}_r = -(\varepsilon - \varepsilon_0)\frac{V_1 - V_2}{R_2 - R_1}\frac{R_2}{R_1} \\ \sigma_{p,2} &= -\vec{\mathbf{P}}(R_2)\cdot\left(-\vec{\mathbf{e}}_r\right) = (\varepsilon - \varepsilon_0)\frac{V_1 - V_2}{R_2 - R_1}\frac{R_1}{R_2} \end{aligned} \qquad [5.126]$$

EXAMPLE 5.9. (Spherical capacitor with two dielectrics).– Let us consider again the spherical capacitor from example 5.8, but this time, the space between the electrodes contains two dielectric materials. The region defined by $R_1 < r < R_2$ and $z > 0$ contains a dielectric with permittivity ε_1, while the region defined by $R_1 < r < R_2$ and $z < 0$ contains a dielectric with permittivity ε_2. The potential and the electrostatic field have the same form as before. However, the displacement vector differs. In the inter-electrode region, we now have:

$$\vec{\mathbf{D}} = \begin{cases} \varepsilon_1 \vec{\mathbf{E}} = \varepsilon_1 \dfrac{V_1 - V_2}{R_2 - R_1} \dfrac{R_1 R_2}{r^2} \vec{\mathbf{e}}_r & z > 0 \\ \varepsilon_2 \vec{\mathbf{E}} = \varepsilon_2 \dfrac{V_1 - V_2}{R_2 - R_1} \dfrac{R_1 R_2}{r^2} \vec{\mathbf{e}}_r & z < 0 \end{cases} \qquad [5.127]$$

Thus, we deduce that on the upper hemisphere $(r = R_1, z > 0)$ and the lower hemisphere $(r = R_1, z < 0)$, the surface charge densities of free charges are given, respectively, by:

$$\begin{aligned} \sigma_1 &= \vec{\mathbf{D}}\,(R_1, z > 0) \cdot \vec{\mathbf{e}}_r = \varepsilon_1 \frac{V_1 - V_2}{R_2 - R_1} \frac{R_2}{R_1} \\ \sigma_1' &= \vec{\mathbf{D}}\,(R_1, z < 0) \cdot \vec{\mathbf{e}}_r = \varepsilon_2 \frac{V_1 - V_2}{R_2 - R_1} \frac{R_2}{R_1} \end{aligned} \qquad [5.128]$$

The difference in permittivity results in a force acting between the two electrodes of the capacitor. In fact, the force experienced by the charges on a surface element dS of the central electrode is:

$$d\vec{\mathbf{F}}_1 = \begin{cases} \dfrac{\sigma_1^2}{2\varepsilon_1} dS \vec{\mathbf{e}}_r & z > 0 \\ \dfrac{\sigma_1'^2}{2\varepsilon_2} dS \vec{\mathbf{e}}_r & z < 0 \end{cases} \qquad [5.129]$$

We then obtain the total force exerted on the central electrode:

$$\begin{aligned} \vec{\mathbf{F}} &= \frac{R_1^2}{2} \left(\frac{\sigma_1^2}{\varepsilon_1} - \frac{\sigma_1'^2}{\varepsilon_2} \right) \int_0^{\pi/2} \sin\theta d\theta \int_0^{2\pi} d\varphi \vec{\mathbf{e}}_r \\ &= \pi R_1^2 \left(\frac{\sigma_1^2}{\varepsilon_1} - \frac{\sigma_1'^2}{\varepsilon_2} \right) \int_0^{\pi/2} \sin\theta \cos\theta d\theta \vec{\mathbf{k}} = \frac{Q^2}{8\pi R_1^2} \frac{\varepsilon_1 - \varepsilon_2}{(\varepsilon_1 + \varepsilon_2)^2} \vec{\mathbf{k}} \end{aligned} \qquad [5.130]$$

where $Q = 2\pi R_1^2 (\sigma_1 + \sigma_1')$ represents the charge of the capacitor.

EXAMPLE 5.10. (Semi-infinite dielectric slab in a uniform field).– An LIH dielectric material with permittivity ε and thickness h is assumed to occupy the entire region $0 \leq x \leq h$ and $-\infty < y, z < \infty$. The material is placed in an external uniform electrostatic field $\vec{\mathbf{E}}_0$.

1) $\vec{\mathbf{E}}_0$ *is perpendicular to the slab*: by symmetry, the electrostatic potential must not depend on variables y and z. The Laplace equation reduces to $\frac{d^2V}{dx^2} = 0$ in the three regions: $x < 0$, $0 \leq x \leq h$ and $h < x < \infty$. Its general solution is $V(x) = Ax + B$, where A and B are determined to ensure the continuity of the potential and the continuity of the normal component of $\vec{\mathbf{D}}$, since no free charge is present inside. For simplicity, we set the potential to zero at $x = 0$. It follows that:

$$V(x) = \begin{cases} A_1 x & x < 0 \\ A_2 x & 0 \leq x \leq h \\ A_3 x + B_3 & h < x < \infty \end{cases} \qquad [5.131]$$

The boundary conditions impose $A_2 h = A_3 h + B_3$, $\varepsilon_0 A_1 = \varepsilon A_2$ and $\varepsilon A_2 = \varepsilon_0 A_3$, which yield $A_1 = A_3 = A$, $A_2 = \frac{A}{\varepsilon_r}$ and $B_3 = Ah\left(\frac{1}{\varepsilon_r} - 1\right)$. To determine the constant A, we assume that far from the material ($|x| \gg h$), the field must coincide with the externally applied field, which leads to $A_3 = -E_0$. Thus, we obtain:

$$V(x) = \begin{cases} -E_0 x & x < 0 \\ -\dfrac{E_0}{\varepsilon_r} x & 0 \leq x \leq h \\ -E_0 x - E_0 h\left(\dfrac{1}{\varepsilon_r} - 1\right) & h < x < \infty \end{cases} \qquad [5.132]$$

and, consequently:

$$\vec{\mathbf{E}} = \begin{cases} \vec{\mathbf{E}}_0 & x < 0 \text{ and } x > h \\ \dfrac{\vec{\mathbf{E}}_0}{\varepsilon_r} x & 0 \leq x \leq h \end{cases} \qquad [5.133]$$

The polarization vector in the dielectric is therefore:

$$\vec{\mathbf{P}} = \varepsilon_0 (\varepsilon_r - 1) \vec{\mathbf{E}} = \varepsilon_0 \vec{\mathbf{E}}_0 - \varepsilon_0 \vec{\mathbf{E}} \qquad [5.134]$$

which leads to:

$$\vec{\mathbf{E}} = \vec{\mathbf{E}}_0 - \frac{\vec{\mathbf{P}}}{3} \qquad [5.135]$$

Since $\vec{\nabla} \cdot \vec{\mathbf{P}} = 0$, only surface polarization charges exist. The surface charge densities at the interfaces $x = h$ and $x = 0$ are, respectively, given by:

$$\begin{aligned} \sigma_{p,h} &= -\vec{\mathbf{P}} \cdot \vec{\mathbf{i}} = \varepsilon_0 \left(1 - \frac{1}{\varepsilon_r}\right) E_0 > 0 \\ \sigma_{p,0} &= -\vec{\mathbf{P}} \cdot \left(-\vec{\mathbf{i}}\right) = -\varepsilon_0 \left(1 - \frac{1}{\varepsilon_r}\right) E_0 < 0 \end{aligned} \qquad [5.136]$$

These surface charge distributions create a depolarizing field $-\frac{\vec{\mathbf{P}}}{3}$ within the dielectric, which opposes the inducing field $\vec{\mathbf{E}}_0$.

2) $\vec{\mathbf{E}}_0$ *is parallel to the slab*: let us take $\vec{\mathbf{E}}_0 = E_0 \vec{\mathbf{j}}$. The potential now depends only on the variable y in the three regions $x < 0$, $0 \leq x \leq h$ and $h < x < \infty$, that is, it is of the form $V(y) = Ay + B$. To ensure the continuity of the tangential component of the electric field at the lateral interfaces $x = 0$ and $x = h$, the constants A and B must be the same everywhere. Since $\vec{\mathbf{E}}$ must also be equal to $\vec{\mathbf{E}}_0$ far from the slab, that is, for $|x| \longrightarrow \infty$, we must set $A = -E_0$ and $B = 0$. We thus conclude that the external field completely penetrates the dielectric (no depolarizing field exists).

$$\vec{\mathbf{E}} = \vec{\mathbf{E}}_0 \qquad [5.137]$$

which leads to the polarization:

$$\vec{\mathbf{P}} = \varepsilon_0 (\varepsilon_r - 1) \vec{\mathbf{E}}_0 \qquad [5.138]$$

The absence of the depolarizing field can be explained as follows: the uniform polarization [5.138] generates surface charges located at the edges of the slab $|y| \longrightarrow \infty$, which creates a negligible field inside the dielectric (i.e. at points far from the edges). These results are consistent with Example 5.7.

EXAMPLE 5.11. (Sphere immersed in a uniform electric field).– Let us determine the potential in the vicinity of a sphere of radius R, made of a LIH dielectric material. The sphere is immersed in a uniform electric field E_0 directed along the Oz-axis. By solving the Laplace equation in spherical coordinates and noting that the system remains invariant under rotation around Oz (i.e. it is independent of the variable φ), we obtain:

$$V(r, \theta) = \begin{cases} \sum_{l=0}^{\infty} A_l r^l P_l(\cos\theta) & r < R \\ \sum_{l=0}^{\infty} \frac{B_l}{r^{l+1}} P_l(\cos\theta) - E_0 r\cos\theta & r > R \end{cases} \qquad [5.139]$$

The term $\frac{B_0}{r}$ induces an electric field $\frac{B_0 \vec{\mathbf{r}}}{r^3}$ and could be interpreted as a field resulting from a charge $Q = 4\pi\varepsilon_0 B_0$ distributed over the sphere. However, this term is zero because the sphere is uncharged. At the boundary of the sphere, we must have $D_\perp(R^-) = D_\perp(R^+)$ and $E_\parallel(R^-) = E_\parallel(R^+)$. In terms of potential V, these two equations can also be written as:

$$\begin{aligned} \varepsilon_1 \left.\frac{\partial V}{\partial r}\right|_{R^-} &= \varepsilon_0 \left.\frac{\partial V}{\partial r}\right|_{R^+} \\ \left.\frac{1}{R}\frac{\partial V}{\partial \theta}\right|_{R^-} &= \left.\frac{1}{R}\frac{\partial V}{\partial \theta}\right|_{R^+} \end{aligned} \qquad [5.140]$$

By substituting expressions [5.139] into [5.140], we obtain:

$$\varepsilon_1 A_1 \cos\theta + \varepsilon_1 \sum_{l=2}^{\infty} l A_l R^{l-1} P_l(\cos\theta) = -\varepsilon_0 \left(\frac{2B_1}{R^3} + E_0\right)\cos\theta - \varepsilon_0 \sum_{l=2}^{\infty} \frac{(l+1)B_l}{R^{l+2}} P_l(\cos\theta) \qquad [5.141]$$

$$A_1 - \sum_{l=2}^{\infty} A_l R^{l-1} P_l'(\cos\theta) = -\left(\frac{B_1}{R^3} - E_0\right) - \sum_{l=2}^{\infty} \frac{B_l}{R^{l+2}} P_l'(\cos\theta)$$

(we have simplified by $\sin\theta$ in the second equation), which leads to: $A_1 = -\frac{\varepsilon_0}{\varepsilon_1}\left(\frac{2B_1}{R^3} + E_0\right)$, $\varepsilon_1 l A_l R^{l-1} = -\frac{\varepsilon_0(l+1)B_l}{R^{l+2}}$ for all $l \geq 2$, $A_1 = \frac{B_1}{R^3} - E_0$, and $A_l R^{l-1} = \frac{B_l}{R^{l+2}}$ for all $l \geq 2$. Solving these constraints simultaneously gives: $B_1 = \left(\frac{\varepsilon_1 - \varepsilon_0}{\varepsilon_1 + 2\varepsilon_0}\right) R^3 E_0$, and $A_1 = -\left(\frac{3\varepsilon_0}{\varepsilon_1 + 2\varepsilon_0}\right) E_0$. All other parameters are zero, that is, $A_l = B_l = 0$, for all $l \geq 2$. Thus, the potential is given by:

$$V(r,\theta) = \begin{cases} -\left(\dfrac{3\varepsilon_0}{\varepsilon_1 + 2\varepsilon_0}\right) E_0 r \cos\theta & r < R \\ -E_0 r\cos\theta + \left(\dfrac{\varepsilon_1 - \varepsilon_0}{\varepsilon_1 + 2\varepsilon_0}\right) \dfrac{R^3}{r^3} E_0 r \cos\theta & r > R \end{cases} \qquad [5.142]$$

and, consequently:

$$\vec{\mathbf{E}} = \begin{cases} \left(\dfrac{3\varepsilon_0}{\varepsilon_1 + 2\varepsilon_0}\right) \vec{\mathbf{E}}_0 & r < R \\ \vec{\mathbf{E}}_0 + \left(\dfrac{\varepsilon_1 - \varepsilon_0}{\varepsilon_1 + 2\varepsilon_0}\right) R^3 E_0 \dfrac{(2\cos\theta\,\vec{\mathbf{e}}_r + \sin\theta\,\vec{\mathbf{e}}_\theta)}{r^3} & r > R \end{cases} \qquad [5.143]$$

Inside the sphere, the electric field appears as the superposition of the external electric field $\vec{\mathbf{E}}_0$ and a depolarizing electric field equal to $-\left(\frac{\varepsilon_1 - 2\varepsilon_0}{\varepsilon_1 + 2\varepsilon_0}\right)\vec{\mathbf{E}}_0$. On the other hand, outside the sphere, the electric field $\vec{\mathbf{E}}$ is the sum of the external field $\vec{\mathbf{E}}_0$ and the electric field created by a dipole moment $\vec{\mathbf{p}} = 4\pi\varepsilon_0 R^3 \left(\frac{\varepsilon_1 - \varepsilon_0}{\varepsilon_1 + 2\varepsilon_0}\right)\vec{\mathbf{E}}_0$ located in the center of the sphere.

EXAMPLE 5.12. (Point charge near an infinite parallelepiped dielectric medium located in the region $z \leq 0$).– Consider a point charge q placed at a distance d from an infinite dielectric medium with permittivity ε, occupying the region $z \leq 0$, as illustrated in Figure 5.17. The point charge q is located at $(0, 0, d)$ in the upper half-space. According to the image method, the potential at a point $M\left(r, z > 0\right)$ outside the dielectric (in cylindrical coordinates) is given by:

$$V\left(r, z > 0\right) = \frac{1}{4\pi\varepsilon_0}\left(\frac{q}{R_1} + \frac{q'}{R_2}\right) = \frac{1}{4\pi\varepsilon_0}\left(\frac{q}{\sqrt{r^2 + \left(d - z\right)^2}} + \frac{q'}{\sqrt{r^2 + \left(d + z\right)^2}}\right)$$
[5.144]

where q' is the image charge that mimics the surface charge induced at the dielectric interface due to the presence of q. Furthermore, the potential created by the charge q at a point $M\left(r, z < 0\right)$ inside the dielectric is given by:

$$V\left(r, z < 0\right) = \frac{1}{4\pi\varepsilon}\frac{q''}{\sqrt{r^2 + \left(d - z\right)^2}} \qquad [5.145]$$

where q'' is now the charge observed from inside the dielectric. In fact, the charge q is reduced within the dielectric by the dipoles of the medium (or, equivalently, by the surface charge induced at the interface). This phenomenon is known as the *screening effect*. To determine q' and q'', we must apply the boundary conditions at the interface: $E_r\left(r, z = 0^+\right) = E_r\left(r, z = 0^-\right)$ and $D_z\left(r, z = 0^+\right) = D_z\left(r, z = 0^-\right)$. Expressed in terms of the potential, these boundary conditions become:

$$\begin{aligned} \left.\frac{\partial V}{\partial r}\right|_{z=0+} &= \left.\frac{\partial V}{\partial r}\right|_{z=0-} \\ \varepsilon_0 \left.\frac{\partial V}{\partial z}\right|_{z=0+} &= \varepsilon \left.\frac{\partial V}{\partial z}\right|_{z=0-} \end{aligned} \qquad [5.146]$$

Substituting expressions [5.144] and [5.145] into [5.146], we obtain:

$$\begin{aligned} \frac{1}{4\pi\varepsilon_0}\frac{rq}{\left(r^2 + (d)^2\right)^{3/2}} + \frac{1}{4\pi\varepsilon_0}\frac{rq'}{\left(r^2 + (d)^2\right)^{3/2}} &= \frac{1}{4\pi\varepsilon}\frac{rq''}{\left(r^2 + (d)^2\right)^{3/2}} \\ \frac{\varepsilon_0}{4\pi\varepsilon_0}\left(\frac{dq}{\left(r^2 + (d)^2\right)^{3/2}} - \frac{dq'}{\left(r^2 + (d)^2\right)^{3/2}}\right) &= \frac{\varepsilon}{4\pi\varepsilon}\frac{dq''}{\left(r^2 + (d)^2\right)^{3/2}} \end{aligned} \qquad [5.147]$$

which implies that:

$$\begin{aligned} \frac{q + q'}{\varepsilon_0} &= \frac{q''}{\varepsilon} \\ q - q' &= q'' \end{aligned} \qquad [5.148]$$

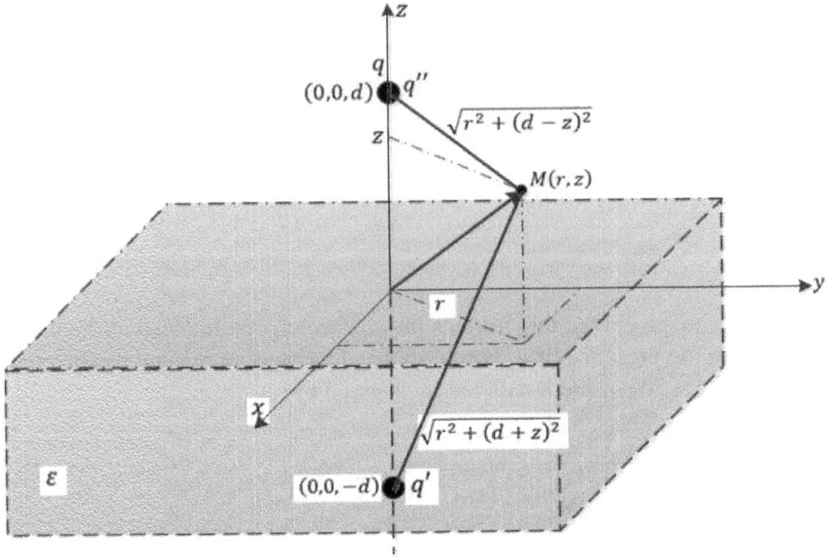

Figure 5.17. *Point charge near a dielectric occupying the lower half-space. For a color version of this figure, see www.iste.co.uk/abdesselam/electrodynamics2.zip*

The solution of the system of equations [5.148] is:

$$\begin{aligned} q' &= -\frac{\varepsilon-\varepsilon_0}{\varepsilon+\varepsilon_0}q \\ q'' &= \frac{2\varepsilon}{\varepsilon+\varepsilon_0}q \end{aligned} \qquad [5.149]$$

Thus, the potentials outside and inside the dielectric are, respectively, given by:

$$\begin{aligned} V_{\text{ext}}(\vec{\mathbf{r}}) &= \frac{q}{4\pi\varepsilon_0}\left[\frac{1}{\left|\vec{\mathbf{r}}-\vec{\mathbf{r}}'\right|}-\frac{(\varepsilon-\varepsilon_0)(\varepsilon+\varepsilon_0)}{\left|\vec{\mathbf{r}}-\left[\vec{\mathbf{r}}'-2\left(\vec{\mathbf{n}}\cdot\vec{\mathbf{r}}'\right)\vec{\mathbf{n}}\right]\right|}\right] \\ V_{\text{int}}(\vec{\mathbf{r}}) &= \frac{2}{4\pi(\varepsilon+\varepsilon_0)}\frac{q}{\left|\vec{\mathbf{r}}-\vec{\mathbf{r}}'\right|} \end{aligned} \qquad [5.150]$$

where $\vec{\mathbf{r}}'$ denotes the position of the charge q, and $\vec{\mathbf{n}}$ represents the normal to the dielectric interface. These expressions can be generalized for an arbitrary charge distribution as follows:

$$\begin{aligned} V_{\text{ext}}(\vec{\mathbf{r}}) &= \frac{1}{4\pi\varepsilon_0}\int_{\tau'}\rho(\vec{\mathbf{r}}')\left[\frac{1}{\left|\vec{\mathbf{r}}-\vec{\mathbf{r}}'\right|}-\frac{(\varepsilon-\varepsilon_0)/(\varepsilon+\varepsilon_0)}{\left|\vec{\mathbf{r}}-\left[\vec{\mathbf{r}}'-2\left(\vec{\mathbf{n}}\cdot\vec{\mathbf{r}}'\right)\vec{\mathbf{n}}\right]\right|}\right]d\tau' \\ V_{\text{int}}(\vec{\mathbf{r}}) &= \frac{2}{4\pi(\varepsilon+\varepsilon_0)}\int_{\tau'}\frac{\rho(\vec{\mathbf{r}}')}{\left|\vec{\mathbf{r}}-\vec{\mathbf{r}}'\right|}d\tau' \end{aligned} \qquad [5.151]$$

EXAMPLE 5.13. (Image of a line of charges near a cylindrical dielectric).– The same thing is also observed for a line charge λ placed near a cylindrical dielectric of radius R. The cylinder and charge line axes are parallel and separated by a distance h, as illustrated in Figure 5.18. The image line charge, which mimics the surface charge density distributed within the cylinder due to the presence of λ, is denoted by λ' and is located at a distance $b < R$. In addition, to keep the dielectric neutral, we add a line of $\hat{a}\lambda$ charges along its axis. According to the image method, the potential outside the cylinder is:

$$\begin{aligned} V\left(\vec{\mathbf{r}}\right) = &-\frac{1}{2\pi\varepsilon_0}\left(\lambda \ln\sqrt{r^2+h^2-2rh\cos\theta}\right. \\ &\left.+\lambda' \ln\sqrt{r^2+b^2-2rb\cos\theta} - \lambda' \ln r\right) \end{aligned} \quad [5.152]$$

whereas the potential inside the dielectric is given by:

$$V\left(\vec{\mathbf{r}}\right) = -\frac{1}{2\pi\varepsilon}\lambda'' \ln\sqrt{r^2+h^2-2rh\cos\theta} \quad [5.153]$$

Here, λ'' is the charge observed in place of λ from the inside (screening effect). To determine λ' and λ'', we must use the boundary conditions $\varepsilon_0 \left.\frac{\partial V}{\partial r}\right|_{R^+} = \varepsilon \left.\frac{\partial V}{\partial r}\right|_{R^-}$ and $\left.\frac{\partial V}{\partial \theta}\right|_{R^+} = \left.\frac{\partial V}{\partial \theta}\right|_{R^-}$, which yield:

$$\begin{aligned} -\frac{\varepsilon\lambda''}{2\pi\varepsilon}\frac{R-h\cos\theta}{R^2+h^2-2Rh\cos\theta} &= -\frac{\varepsilon_0}{2\pi\varepsilon_0}\left[\frac{\lambda\left(R-h\cos\theta\right)}{R^2+h^2-2Rh\cos\theta}+\right. \\ &\left.\frac{\lambda'\left(R-b\cos\theta\right)}{R^2+b^2-2Rb\cos\theta}-\frac{\lambda'}{R}\right] \\ -\frac{\lambda''}{2\pi\varepsilon}\frac{Rh\sin\theta}{R^2+h^2-2Rh\cos\theta} &= -\frac{R}{2\pi\varepsilon_0}\left[\frac{\lambda h\sin\theta}{R^2+h^2-2Rh\cos\theta}+\right. \\ &\left.\frac{\lambda' b\sin\theta}{R^2+b^2-2Rb\cos\theta}\right] \end{aligned} \quad [5.154]$$

By simplifying these expressions, we get:

$$\begin{aligned} &\lambda''-\lambda = \lambda'\frac{b}{R}\frac{\left(R\cos\theta-b\right)\left(R^2+h^2-2Rh\cos\theta\right)}{\left(R-h\cos\theta\right)\left(R^2+b^2-2Rb\cos\theta\right)} \\ &\left(\frac{\lambda''}{\varepsilon}-\frac{\lambda}{\varepsilon_0}\right)\frac{\varepsilon_0 h}{\lambda' b} = \frac{R^2+h^2-2Rh\cos\theta}{R^2+b^2-2Rb\cos\theta} \equiv K^2 \end{aligned} \quad [5.155]$$

These equations must be satisfied for all values of θ, which implies, from the second equation of system [5.155], that $2hR = K^2 2bR$ and $R^2+h^2 = K^2(R^2+b^2)$. We therefore conclude that $K^2 = \frac{h}{b}$ and, consequently, $R^2+h^2 = \frac{h}{b}(R^2+b^2)$. The solution to this last equation is:

$$h = \frac{a^2}{b} \quad [5.156]$$

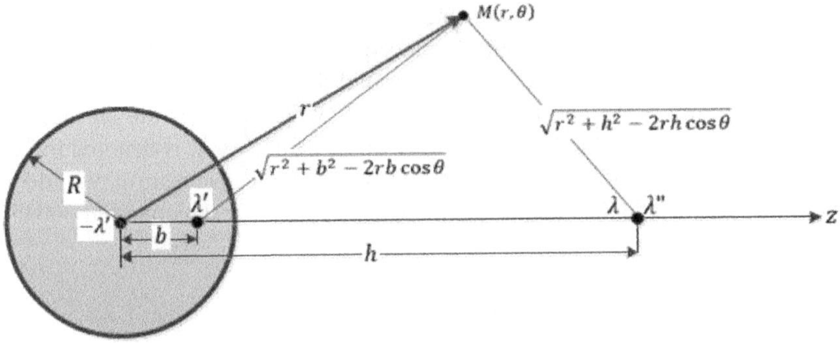

Figure 5.18. *Image method for a line charge λ near a cylindrical dielectric of permittivity ε. For a color version of this figure, see www.iste.co.uk/abdesselam/electrodynamics2.zip*

Based on these results, we obtain:

$$\begin{aligned} &\lambda'' - \lambda = -\lambda' \\ &\frac{\lambda''}{\varepsilon} - \frac{\lambda}{\varepsilon_0} = \frac{\lambda'}{\varepsilon_0} \end{aligned} \qquad [5.157]$$

The system [5.157] admits the following solutions:

$$\begin{aligned} \lambda' &= -\frac{\varepsilon - \varepsilon_0}{\varepsilon + \varepsilon_0}\lambda \\ \lambda'' &= \frac{2\varepsilon}{\varepsilon + \varepsilon_0}\lambda \end{aligned} \qquad [5.158]$$

Finally, we have the following:

$$V(r,\theta) = \begin{cases} -\dfrac{\lambda}{2\pi\varepsilon_0}\left(\ln\sqrt{r^2 + h^2 - 2rh\cos\theta} \right. \\ \left. -\dfrac{\varepsilon - \varepsilon_0}{\varepsilon + \varepsilon_0}\ln\dfrac{\sqrt{r^2 + b^2 - 2rb\cos\theta}}{r}\right) & r > R \\ -\dfrac{\lambda}{\pi(\varepsilon + \varepsilon_0)}\ln\sqrt{r^2 + h^2 - 2rh\cos\theta} & r < R \end{cases} \qquad [5.159]$$

It is worth pointing out that no image charge can be found for a point charge near a spherical dielectric. The image method does not work.

EXAMPLE 5.14. (Conducting sphere floating on a dielectric medium).– Consider a conducting sphere of radius R and mass m floating on the surface of a dielectric fluid with a relative permittivity ε_r. At zero potential (uncharged sphere), a quarter of its volume is immersed in equilibrium, as illustrated in Figure 5.19(a). When the sphere is at potential V_S, it is observed to be half-immersed in equilibrium (Figure 5.19(b)). Our goal is to determine V_S. We use the spherical coordinates attached to the center O of the sphere, with the Oz-axis perpendicular to the fluid–air interface. The potential $V_{\text{ext}}\,(r > R)$ induced outside the sphere satisfies the Laplace equation $\Delta V_{\text{ext}} = 0$, while the potential within the sphere is uniform and given by $V_{\text{int}}\,(r \leq R) = V_S$. For symmetry reasons, the potential V_{ext} depends only on r. The boundary conditions are as follows: (1) continuity of the potential on the surface of the sphere, $V_{\text{ext}}\,(R) = V_{\text{int}}\,(R) = V_S$; (2) at infinity, that is, as $r \longrightarrow \infty$, the potential V_{ext} tends to zero. Outside the sphere, the Laplace equation is written as $\Delta V_{\text{ext}} = \frac{1}{r^2}\frac{d}{dr}\left(r^2\frac{dV_{\text{ext}}(r)}{dr}\right) = 0$, which leads to the general solution $V_{\text{ext}}\,(r) = \frac{A}{r} + B$, where A and B are constants. Applying the boundary conditions, we obtain $V_{\text{ext}}\,(r \longrightarrow \infty) = B = 0$ and $V_{\text{ext}}\,(R) = V_S$, which also gives $A = V_S R$. It therefore follows that:

$$V_{\text{ext}}\,(r) = \frac{V_S R}{r} \qquad [5.160]$$

The electric field is obtained as follows:

$$\vec{\mathbf{E}}_{\text{ext}} = -\vec{\nabla} V_{\text{ext}} = -\frac{dV_{\text{ext}}}{dr}\vec{\mathbf{e}}_r = \frac{V_S R}{r^2}\vec{\mathbf{e}}_r \qquad [5.161]$$

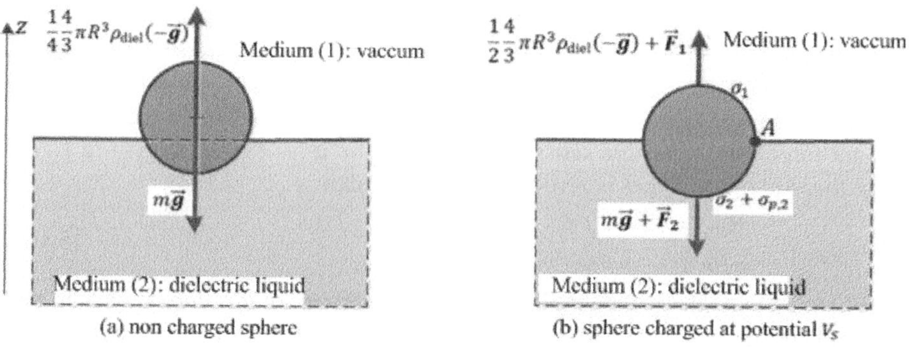

Figure 5.19. *Conducting sphere floating on the surface of a liquid. For a color version of this figure, see www.iste.co.uk/abdesselam/electrodynamics2.zip*

At the interface of the upper cap of the sphere, the perpendicular component $\left(\vec{\mathbf{D}}\right)_\perp$ (normal to the cap) is discontinuous and satisfies the relation $\left(\vec{\mathbf{D}}_{\text{ext},1}\,(R) - \vec{\mathbf{D}}_{\text{int}}\,(R)\right) \cdot \vec{\mathbf{e}}_r = \sigma_1$, where σ_1 is the surface charge density of the free charges on the upper cap. Since the electric field inside the conducting sphere is zero, it follows that:

$$\sigma_1 = \overrightarrow{\mathbf{D}}_{\text{ext},1}(R) \cdot \overrightarrow{\mathbf{e}}_r = \varepsilon_0 \overrightarrow{\mathbf{E}}_{\text{ext},1}(R) \cdot \overrightarrow{\mathbf{e}}_r = \frac{\varepsilon_0 V_S}{R} \qquad [5.162]$$

To determine the surface charge density of free charges on the lower cap of the sphere in contact with the dielectric medium, we apply the same relation:

$$\sigma_2 = \overrightarrow{\mathbf{D}}_{\text{ext},2}(R) \cdot \overrightarrow{\mathbf{e}}_r = \varepsilon_0 \varepsilon_r \overrightarrow{\mathbf{E}}_{\text{ext},2}(R) \cdot \overrightarrow{\mathbf{e}}_r = \frac{\varepsilon_0 \varepsilon_r V_S}{R} \qquad [5.163]$$

Thus, we get:

$$\sigma_2 = \varepsilon_r \varepsilon_0 \frac{V_S}{R} = \varepsilon_r \sigma_1 \qquad [5.164]$$

Note that we used the same expression for $\overrightarrow{\mathbf{E}}_{\text{ext}}$ on the lower cap. In fact, the continuity of the tangential component of the electric field at point A (the vacuum–dielectric–conductor interface) requires that: $\overrightarrow{\mathbf{E}}_{\text{ext},2}(R) \cdot \overrightarrow{\mathbf{e}}_r = \overrightarrow{\mathbf{E}}_{\text{ext},1}(R) \cdot \overrightarrow{\mathbf{e}}_r$, which implies that: $\overrightarrow{\mathbf{E}}_{\text{ext},2}(R) \cdot \overrightarrow{\mathbf{e}}_r = \frac{V_S}{R}$. The surface polarization charge densities can be derived from the polarization vector $\overrightarrow{\mathbf{P}}$. Since there is no polarization in medium (1), we have $\sigma_{p,1} = 0$. However, the surface polarization charge density in medium (2) is given by:

$$\sigma_{p,2} = -\overrightarrow{\mathbf{P}} \cdot \overrightarrow{\mathbf{e}}_r = -\varepsilon_0 \chi_e \frac{V_S}{R} = -\left(1 - \frac{1}{\varepsilon_r}\right) \sigma_2 = -\left(\varepsilon_r - 1\right) \sigma_1 \qquad [5.165]$$

In the absence of potential, that is, when $V_S = 0$, only a quarter of the sphere is submerged. Since the system is in equilibrium, the mass density of the fluid can be determined using Archimedes' principle. The forces acting (weight of the sphere and buoyant force) on the sphere satisfy the equation: $m\overrightarrow{\mathbf{g}} + \frac{1}{4}\frac{4}{3}\pi R^3 \rho_{\text{diel}} \left(-\overrightarrow{\mathbf{g}}\right) = 0$ (since the sum of the forces is zero), which implies that:

$$\rho_{\text{diel}} = \frac{3m}{\pi R^3} \qquad [5.166]$$

When the potential is increased to V_S, the immersed volume corresponds to half of the sphere's volume due to the electrostatic pressures now exerted on its surfaces:

$$m\overrightarrow{\mathbf{g}} + \frac{1}{2}\frac{4}{3}\pi R^3 \rho_{\text{diel}} \left(-\overrightarrow{\mathbf{g}}\right) + \overrightarrow{\mathbf{F}}_1 + \overrightarrow{\mathbf{F}}_2 = 0 \qquad [5.167]$$

where $\overrightarrow{\mathbf{F}}_1$ and $\overrightarrow{\mathbf{F}}_2$ correspond to the pressure forces exerted on each of the hemispheres:

$$\begin{aligned} \overrightarrow{\mathbf{F}}_1 &= \int_{\text{uppercap}} p_1 d\overrightarrow{\mathbf{S}} = \int_0^{2\pi} d\varphi \int_0^{\frac{\pi}{2}} p_1 \cos\theta R^2 \sin\theta d\theta \overrightarrow{\mathbf{k}} = \pi R^2 p_1 \overrightarrow{\mathbf{k}} \\ \overrightarrow{\mathbf{F}}_2 &= \int_{\text{lowercap}} p_2 d\overrightarrow{\mathbf{S}} = \int_0^{2\pi} d\varphi \int_{\frac{\pi}{2}}^{\pi} p_2 \cos\theta R^2 \sin\theta d\theta \overrightarrow{\mathbf{k}} = -\pi R^2 p_2 \overrightarrow{\mathbf{k}} \end{aligned} \qquad [5.168]$$

The electrostatic pressures p_1 and p_2 are given by:

$$\begin{aligned} p_1 &= \left|\frac{d\vec{\mathbf{F}}^1}{d\vec{\mathbf{S}}}\right| = \left|\frac{d\vec{\mathbf{E}}_1 dq}{d\vec{\mathbf{S}}}\right| = \frac{\sigma_1}{2\varepsilon_0}\frac{\sigma_1 dS}{dS} = \frac{\sigma_1^2}{2\varepsilon_0} \\ p_2 &= \left|\frac{d\vec{\mathbf{F}}^2}{d\vec{\mathbf{S}}}\right| = \left|\frac{d\vec{\mathbf{E}}_2 dq}{d\vec{\mathbf{S}}}\right| = \frac{\sigma_2 + \sigma_{p2}}{2\varepsilon_0}\frac{\sigma_2 dS}{dS} = \frac{\sigma_2^2}{2\varepsilon_r\varepsilon_0} = \frac{\varepsilon_r\sigma_1^2}{2\varepsilon_0} \end{aligned} \qquad [5.169]$$

By projecting [5.167] onto Oz, we finally obtain:

$$-mg + \frac{1}{2}\frac{4}{3}\pi R^3 \frac{3m}{\pi R^3} g + \frac{\sigma_1^2}{2\varepsilon_0}\pi R^2 - \frac{\varepsilon_r \sigma_1^2}{2\varepsilon_0}\pi R^2 = 0 \qquad [5.170]$$

and, consequently:

$$V_S = \sqrt{\frac{2mg}{\varepsilon_0 \pi (\varepsilon_r - 1)}} \qquad [5.171]$$

5.7. Polarization current

The presence of a local field $\vec{\mathbf{E}}_{\text{loc}}$ will cause a partial displacement of bound charges, and thus generate a current. In a dielectric medium, the polarization is:

$$\vec{\mathbf{P}} = -\sum_k n_k e \vec{\mathbf{r}}_k \qquad [5.172]$$

where $\vec{\mathbf{r}}_k$ and n_k denote the average position and number of electrons per unit volume of the k level, respectively. By definition, the current $\vec{\mathbf{J}}_p$, here called the *polarization current*, associated with the polarization charges, is given by:

$$\vec{\mathbf{J}}_p = -\sum_k n_k e \vec{\mathbf{v}}_k = -\sum_k n_k e \frac{\partial \vec{\mathbf{r}}_k}{\partial t} \qquad [5.173]$$

It therefore follows that:

$$\vec{\mathbf{J}}_p = \frac{\partial \vec{\mathbf{P}}}{\partial t} \qquad [5.174]$$

We can verify that this expression is consistent with the charge conservation equation. Indeed, since $\rho_p = -\vec{\nabla} \cdot \vec{\mathbf{P}}$, the divergence of [5.174] yields:

$$\vec{\nabla} \cdot \vec{\mathbf{J}}_p = \vec{\nabla} \cdot \left(\frac{\partial \vec{\mathbf{P}}}{\partial t}\right) = \frac{\partial}{\partial t}\left(\vec{\nabla} \cdot \vec{\mathbf{P}}\right) = -\frac{\partial \rho_p}{\partial t} \qquad [5.175]$$

This current density is on a macroscopic scale and represents an average of the microscopic currents associated with the slight displacements of bound charges. It produces the same magnetic effects as conventional currents. However, its distinctive feature is that it cannot leave the dielectric material (as it is associated with bound charges) and therefore cannot be used in an external circuit.

5.8. Kramers–Krönig relations

In the case of a time-dependent regime, the polarization at a given position $\vec{\mathbf{r}}$ and time t can only depend on the values of the field at neighboring positions $\vec{\mathbf{r}}'$ and at earlier times t': this is the principle of causality. In other words, the induction of polarization at $(\vec{\mathbf{r}},t)$ does not occur directly due to the electric field but rather indirectly. If we consider a linear dielectric, meaning that the polarization depends linearly on the electric field, the most general form of the induced polarization $\vec{\mathbf{P}}$ by an applied electric field $\vec{\mathbf{E}}$ is as follows:

$$P_i\left(\vec{\mathbf{r}},t\right) = \varepsilon_0 \int_{-\infty}^{t} dt' \int d\tau' \chi_{ij}\left(\vec{\mathbf{r}} - \vec{\mathbf{r}}',t-t'\right) E_j\left(\vec{\mathbf{r}}',t'\right) \qquad [5.176]$$

where $\chi_{ij}\left(\vec{\mathbf{r}} - \vec{\mathbf{r}}',t-t'\right)$ is the *electric response function* of the dielectric, also known as the *electric temporal susceptibility*. The term $\varepsilon_0\chi_{ij}\left(\vec{\mathbf{r}} - \vec{\mathbf{r}}',t-t'\right)$ represents the ith component of the polarization induced at $\vec{\mathbf{r}}$, at time t, by the jth component of a unit-strength electric field at $\vec{\mathbf{r}}'$, at time t'. In the most general case, it is a tensor. In an isotropic medium, this tensor simplifies to $\chi_{ij} = \delta_{ij}\chi$. Hence, we obtain:

$$\vec{\mathbf{P}}\left(\vec{\mathbf{r}},t\right) = \varepsilon_0 \int_{-\infty}^{t} dt' \int d\tau' \chi\left(\vec{\mathbf{r}} - \vec{\mathbf{r}}',t-t'\right) \vec{\mathbf{E}}\left(\vec{\mathbf{r}}',t'\right) \qquad [5.177]$$

The spatial dependence of dielectric susceptibility $\chi\left(\vec{\mathbf{r}},t\right)$ is generally negligible, since the function $\chi\left(\vec{\mathbf{r}},t\right)$ vanishes over a short distance. If $\vec{\mathbf{E}}\left(\vec{\mathbf{r}},t'\right)$ is the electric field applied at $\vec{\mathbf{r}}$ at an earlier time $t' < t$, then the polarization $\vec{\mathbf{P}}\left(\vec{\mathbf{r}},t\right)$, in accordance with the principle of causality and neglecting the spatial variation in [5.177], is given by:

$$\vec{\mathbf{P}}\left(\vec{\mathbf{r}},t\right) = \varepsilon_0 \int_{-\infty}^{t} dt' \chi\left(t-t'\right) \vec{\mathbf{E}}\left(\vec{\mathbf{r}},t'\right) \qquad [5.178]$$

where $\chi\left(t-t'\right)$ is the temporal susceptibility. (1) Equation [5.178] generalizes the expression valid in the static case, $\vec{\mathbf{P}}\left(\vec{\mathbf{r}}\right) = \varepsilon_0\chi\vec{\mathbf{E}}\left(\vec{\mathbf{r}}\right)$. (2) This is a local equation that relates the field in $\vec{\mathbf{r}}$ to polarization at the same point. (3) The integral is limited to $t' < t$ to ensure causality. (4) Equation [5.178] is time translation invariant, that is, whenever we impose a causal $\vec{\mathbf{E}}\left(\vec{\mathbf{r}},t'\right)$, the response $\vec{\mathbf{P}}\left(\vec{\mathbf{r}},t\right)$ depends only on the difference $t-t'$.

We agree in accordance with what has just been stated that $\chi(t) = 0$ for $t < 0$. Thus, we write:

$$\begin{aligned} &\vec{\mathbf{P}}\left(\vec{\mathbf{r}},t\right) = \varepsilon_0 \int_{-\infty}^{+\infty} dt' \chi\left(t-t'\right) \vec{\mathbf{E}}\left(\vec{\mathbf{r}},t'\right) \\ &\chi(t) = 0 \quad \text{fort} < 0 \end{aligned} \qquad [5.179]$$

In mathematical terms, the above formula represents a convolution. Polarization is therefore the convolution of the applied electric field $\overrightarrow{\mathbf{E}}\left(\overrightarrow{\mathbf{r}},t\right)$ and the susceptibility $\chi\left(t\right)$. The relation [5.179] becomes much simpler in the Fourier transform space:

$$\begin{aligned}\hat{\overrightarrow{\mathbf{P}}}\left(\overrightarrow{\mathbf{r}},\omega\right) &= \varepsilon_0 \int_{-\infty}^{+\infty}\left[\int_{-\infty}^{+\infty}\chi\left(t-t'\right)\overrightarrow{\mathbf{E}}\left(\overrightarrow{\mathbf{r}},t'\right)dt'\right]e^{i\omega t}dt \\ &= \varepsilon_0 \int_{-\infty}^{+\infty} e^{i\omega t'}\overrightarrow{\mathbf{E}}\left(\overrightarrow{\mathbf{r}},t'\right)dt'\left[\int_{-\infty}^{+\infty}\chi\left(t-t'\right)e^{i\omega\left(t-t'\right)}dt\right] \\ &= \varepsilon_0\left[\int_{-\infty}^{+\infty}\chi\left(t\right)e^{i\omega t}dt\right]\left[\int_{-\infty}^{+\infty}e^{i\omega t'}\overrightarrow{\mathbf{E}}\left(\overrightarrow{\mathbf{r}},t'\right)dt'\right] \\ &= \hat{\chi}\left(\omega\right)\hat{\overrightarrow{\mathbf{E}}}\left(\overrightarrow{\mathbf{r}},\omega\right)\end{aligned} \qquad [5.180]$$

where $\hat{\overrightarrow{\mathbf{E}}}\left(\overrightarrow{\mathbf{r}},\omega\right)$ and $\hat{\chi}\left(\omega\right)$ are the Fourier transforms of $\overrightarrow{\mathbf{E}}\left(\overrightarrow{\mathbf{r}},t\right)$ and $\chi\left(t\right)$, respectively. The linear relationship between polarization and the electric field is therefore straightforward in Fourier space. It follows that:

$$\hat{\varepsilon}\left(\omega\right) = \varepsilon_0\left(1+\hat{\chi}\left(\omega\right)\right) \qquad [5.181]$$

The dielectric constant $\hat{\varepsilon}(\omega)$ generally depends on the frequency. Since $\chi(t)=0$ for $t<0$, we have:

$$\hat{\chi}\left(\omega\right) = \int_{0}^{+\infty}\chi\left(t\right)e^{i\omega t}dt \qquad [5.182]$$

It is easy to verify that $\overline{\hat{\chi}\left(\omega\right)} = \hat{\chi}\left(-\omega\right)$. By decomposing $\hat{\chi}\left(\omega\right)$ into its real and imaginary parts, we obtain:

$$\begin{aligned}\hat{\chi}'\left(-\omega\right) &= \hat{\chi}'\left(\omega\right) \\ \hat{\chi}''\left(-\omega\right) &= -\hat{\chi}''\left(\omega\right)\end{aligned} \qquad [5.183]$$

Here are some properties of the functions $\chi\left(t\right)$ and $\hat{\chi}\left(\omega\right)$:

– If we assume that a constant field E has been applied since an infinitely long time in the past, it is reasonable to assume that equilibrium has been established, regardless of the time t at which $P\left(t\right)$ is measured, and therefore:

$$P = \varepsilon_0\int_{-\infty}^{0}\chi\left(-t'\right)Edt' = \varepsilon_0\chi E \qquad [5.184]$$

This remark therefore provides an integral relation for $\chi\left(t\right)$:

$$\int_0^{\infty}\chi\left(t\right)dt = \chi \qquad [5.185]$$

– After reaching equilibrium, the applied field E is switched off at $t = 0$. The system must then return to equilibrium, where $P = 0$:

$$P(t) = \varepsilon_0 \int_{-\infty}^{0} \chi(t - t')\, E dt' = \varepsilon_0 E \left(\chi - \int_0^t \chi(t')\, dt'\right) \qquad [5.186]$$

This expression shows that the susceptibility cannot have a mathematical singularity for $t > 0$ with a positive value, since, physically, P must be continuous.

– The function $\hat{\chi}(\omega)$ is always well-defined in the upper complex half-plane of ω (i.e. for $\Im(\omega) > 0$) because the integral over t necessarily converges, as it decreases exponentially (since the function $\chi(t)$ itself cannot exhibit singular behavior and must decay over time). Moreover, we observe that $\hat{\chi}(\omega)$ must tend to zero as $|\omega| \longrightarrow \infty$. Therefore, $\hat{\chi}(\omega)$ is analytic in the upper half-plane of the complex ω-plane.

Consider the contour of the complex ω'-plane illustrated in Figure 5.20. The radius of the outer semicircle C_1 must tend to infinity, while the radius of the inner semicircle C_2, centered at ω, must tend to zero. Next, let us examine the complex integral along the contour shown in Figure 5.20, $\oint_C \frac{\hat{\chi}(\omega')}{\omega' - \omega} d\omega'$, where ω is real. The function $\hat{\chi}(\omega')$ is analytic in the upper half-plane. By Cauchy's theorem, this integral must vanish because the contour does not enclose any poles of the integrand (the semicircle C_2 is used to circumvent the singularity at ω):

$$0 = \int_{C_1} \frac{\hat{\chi}(\omega')}{\omega' - \omega} d\omega' + \lim_{a \longrightarrow 0} \int_{-\infty}^{\omega - a} \frac{\hat{\chi}(\omega')}{\omega' - \omega} d\omega' + \lim_{a \longrightarrow 0} \int_{\omega + a}^{+\infty} \frac{\hat{\chi}(\omega')}{\omega' - \omega} d\omega' + \int_{C_2} \frac{\hat{\chi}(\omega')}{\omega' - \omega} d\omega' \qquad [5.187]$$

The contribution of the semicircle C_1 to the integral vanishes in the limit as its radius tends to infinity since the function $\hat{\chi}(\omega')$ approaches zero as $|\omega'| \longrightarrow \infty$. For the semicircle C_2, we obtain:

$$\begin{aligned} \int_{C_2} \frac{\hat{\chi}(\omega')}{\omega' - \omega} d\omega' &= \lim_{a \longrightarrow 0} \int_{\pi}^{0} \frac{\hat{\chi}\left(\omega + a e^{\mathrm{i}\theta}\right)}{a e^{\mathrm{i}\theta}} d\left(\omega + a e^{\mathrm{i}\theta}\right) \\ &= \lim_{a \longrightarrow 0} \mathrm{i} \int_{\pi}^{0} \hat{\chi}\left(\omega + a e^{\mathrm{i}\theta}\right) d\theta = -\mathrm{i}\pi \hat{\chi}(\omega) \end{aligned} \qquad [5.188]$$

Finally, the contribution of the two remaining integrals (along the real axis), in the limit $a \longrightarrow 0$, is, by definition, the principal value:

$$\mathrm{VP} \int_{-\infty}^{\infty} \frac{\hat{\chi}(\omega')}{\omega' - \omega} d\omega' = \lim_{a \longrightarrow 0} \left[\int_{-\infty}^{\omega - a} \frac{\hat{\chi}(\omega')}{\omega' - \omega} d\omega' + \int_{\omega + a}^{+\infty} \frac{\hat{\chi}(\omega')}{\omega' - \omega} d\omega'\right] \qquad [5.189]$$

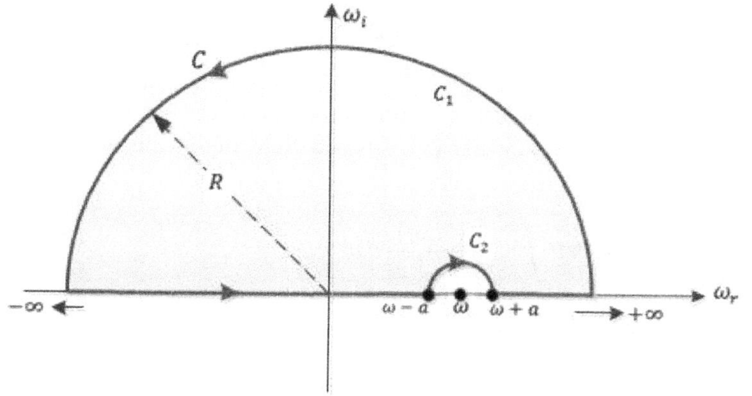

Figure 5.20. *Integration contour used to derive the Kramers–Krönig relations*

By substituting these results into [5.188], we obtain:

$$\mathrm{VP}\int_{-\infty}^{\infty}\frac{\hat{\chi}(\omega')}{\omega'-\omega}d\omega' = \mathrm{i}\pi\hat{\chi}(\omega) \quad [5.190]$$

By decomposing $\hat{\chi}(\omega')$ into its real and imaginary parts, it follows that:

$$\begin{aligned}\hat{\chi}''(\omega) &= -\frac{1}{\pi}\mathrm{VP}\int_{-\infty}^{\infty}\frac{\hat{\chi}'(\omega')}{\omega'-\omega}d\omega' \\ \hat{\chi}'(\omega) &= \frac{1}{\pi}\mathrm{VP}\int_{-\infty}^{\infty}\frac{\hat{\chi}''(\omega')}{\omega'-\omega}d\omega'\end{aligned} \quad [5.191]$$

These are the *Kramers–Krönig relations*. Finally, we can express these relations in terms of the dielectric constant $\hat{\varepsilon} = \varepsilon_0(1+\hat{\chi})$, by defining $\hat{\varepsilon} = \hat{\varepsilon}' + \mathrm{i}\hat{\varepsilon}''$, which gives $\hat{\varepsilon}' = \varepsilon_0(1+\hat{\chi}')$ and $\hat{\varepsilon}'' = \varepsilon_0\hat{\chi}''$. Given that $\mathrm{VP}\int_{-\infty}^{\infty}\frac{d\omega'}{\omega'-\omega} = 0$, we obtain the following relations:

$$\begin{aligned}\hat{\varepsilon}''(\omega) &= -\frac{1}{\pi}\mathrm{VP}\int_{-\infty}^{\infty}\frac{\hat{\varepsilon}'(\omega')}{\omega'-\omega}d\omega' \\ \hat{\varepsilon}'(\omega) &= \varepsilon_0 + \frac{1}{\pi}\mathrm{VP}\int_{-\infty}^{\infty}\frac{\hat{\varepsilon}''(\omega')}{\omega'-\omega}d\omega'\end{aligned} \quad [5.192]$$

The Kramers–Krönig relations [5.191] and [5.192] allow us to express the real part of $\hat{\varepsilon}$ (or equivalently $\hat{\chi}$) in terms of an integral of its imaginary part and vice versa. This implies that the electric susceptibility cannot always be purely real; its imaginary part must not be zero, at least within a certain frequency range. This also explains why, if the imaginary part exhibits Lorentzian behavior near a resonance, the real part takes the form of a dispersion curve. These relations have numerous implications in the theory of linear response.

Since the function $\chi(t)$ is real, it is straightforward to verify that $\overline{\hat{\chi}(\omega)} = \hat{\chi}(-\omega)$. In other words, $\hat{\chi}'(\omega)$ is even, while $\hat{\chi}'(\omega)$ is odd. Substituting this into the expression [5.191], we find:

$$\begin{aligned} \hat{\chi}''(\omega) &= -\frac{2\omega}{\pi}\text{VP}\int_{-\infty}^{\infty}\frac{\hat{\chi}'(\omega')}{\omega'^2-\omega^2}d\omega' \\ \hat{\chi}'(\omega) &= \frac{2}{\pi}\text{VP}\int_{-\infty}^{\infty}\frac{\omega'\hat{\chi}''(\omega')}{\omega'^2-\omega^2}d\omega' \end{aligned} \quad [5.193]$$

The same remark applies to the dielectric constant:

$$\begin{aligned} \hat{\varepsilon}''(\omega) &= -\frac{2}{\pi}\text{VP}\int_{-\infty}^{\infty}\frac{\hat{\varepsilon}'(\omega')}{\omega'^2-\omega^2}d\omega' \\ \hat{\varepsilon}'(\omega) &= \varepsilon_0+\frac{2}{\pi}\text{VP}\int_{-\infty}^{\infty}\frac{\hat{\varepsilon}''(\omega')}{\omega'^2-\omega^2}d\omega' \end{aligned} \quad [5.194]$$

EXAMPLE 5.15.– Let $\hat{\chi}''(\omega) = \frac{\omega}{1+\omega^2}$. We compute $\hat{\chi}'(\omega)$ and $\hat{\chi}(\omega)$ by integrating the complex function $f(\omega') = \frac{\hat{\chi}''(\omega)}{(\omega'-\omega)}$ on the contour illustrated in Figure 5.21. According to the residue theorem, this integral evaluates to:

$$\begin{aligned} &\oint_C \frac{\hat{\chi}''(\omega)}{(\omega'-\omega)}d\omega' = 2\pi\text{iRes}\left[\frac{\omega'}{(\omega'-\omega)(1+\omega'^2)},\text{i}\right] \\ &= 2\pi\text{i}\lim_{\omega'\longrightarrow\text{i}}\frac{\omega'(\omega'-\text{i})}{(\omega'-\omega)(1+\omega'^2)} = 2\pi\text{i}\lim_{\omega'\longrightarrow\text{i}}\frac{\omega'}{(\omega'-\omega)(\omega'+\text{i})} = \frac{\text{i}\pi}{(\text{i}-\omega)} \end{aligned} \quad [5.195]$$

Similarly, the contribution of the semicircle C_1 to the integral is zero since $f(\omega')$ vanishes at infinity. For the semicircle C_2, we get $-\text{i}\pi\hat{\chi}''(\omega) = -\frac{\text{i}\pi\omega}{1+\omega^2}$. Finally, the contribution along the real axis, in the limit $a \longrightarrow 0$, is simply VP $\int_{-\infty}^{\infty}\frac{\hat{\chi}''(\omega)}{(\omega'-\omega)}d\omega'$. Taking these results into account, we obtain:

$$\text{VP}\int_{-\infty}^{\infty}\frac{\hat{\chi}''(\omega)}{(\omega'-\omega)}d\omega' - \frac{\text{i}\pi\omega}{1+\omega^2} = \frac{\text{i}\pi}{(\text{i}-\omega)} \quad [5.196]$$

which gives:

$$\begin{aligned} \hat{\chi}'(\omega) &= \frac{1}{\pi}\text{VP}\int_{-\infty}^{\infty}\frac{\hat{\chi}''(\omega)}{(\omega'-\omega)}d\omega' = \frac{1}{1+\omega^2} \\ \hat{\chi}(\omega) &= \hat{\chi}'(\omega)+\text{i}\hat{\chi}''(\omega) = \frac{1}{1-\text{i}\omega} \end{aligned} \quad [5.197]$$

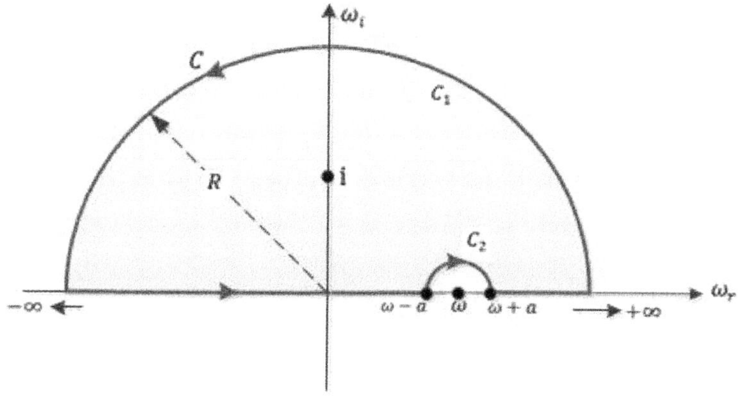

Figure 5.21. *Contour used for the calculation of the integral VP* $\int_{-\infty}^{\infty}\frac{\omega'}{\left(\left(1+\omega'^2\right)\omega'-\omega\right)}d\omega'$

EXAMPLE 5.16.– Let us also take $\hat{\chi}''(\omega)=\frac{\gamma\omega}{\left(\omega_0^2-\omega^2\right)^2+\gamma^2\omega^2}$. Let us compute $\hat{\chi}'(\omega)$ and $\hat{\chi}'(\omega)$. First, let us note that this function can be factored as:

$$\hat{\chi}''(\omega)=\frac{\gamma\omega}{(\omega-z_1)(\omega+z_1)(\omega-\overline{z_1})(\omega+\overline{z_1})} \quad [5.198]$$

where $z_1=\sqrt{\omega_0^2-\frac{\gamma^2}{4}}+\frac{i\gamma}{2}$. Applying the same procedure to the complex function $\frac{\hat{\chi}''(\omega)}{(\omega'-\omega)}$ along the contour shown in Figure 5.22, we immediately obtain:

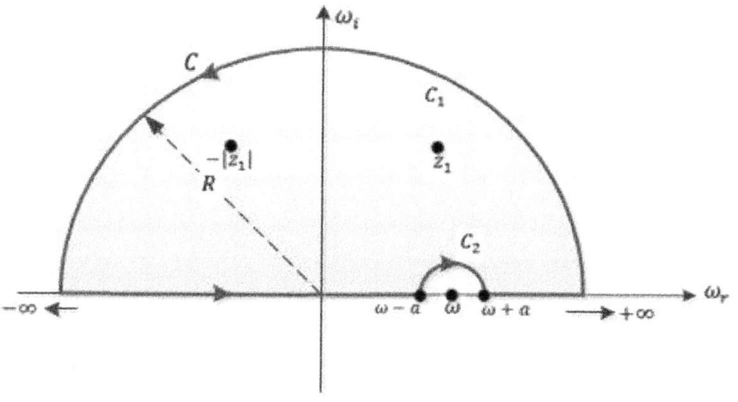

Figure 5.22. *Contour used for VP* $\int_{-\infty}^{\infty}\frac{\gamma\omega'}{\left[\left(\omega_0^2-\omega'^2\right)^2+\gamma^2\omega'^2\right](\omega'-\omega)}d\omega'$

$$\begin{aligned}&\frac{1}{\pi}\text{VP}\int_{-\infty}^{\infty}\frac{\hat{\chi}''(\omega')}{(\omega'-\omega)}d\omega' = \mathrm{i}\hat{\chi}''(\omega) + 2\mathrm{i}\text{Res}\left[\frac{\hat{\chi}''(z)}{(\omega'-\omega)}, z_1\right]\\ &+2\mathrm{i}\text{Res}\left[\frac{\hat{\chi}''(z)}{(\omega'-\omega)}, -\overline{z_1}\right] = \mathrm{i}\hat{\chi}''(\omega) + 2\mathrm{i}\lim_{\omega\longrightarrow z_1}(\omega'-z_1)\frac{\hat{\chi}''(\omega)}{\omega'-\omega}+\\ &2\mathrm{i}\lim_{\omega\longrightarrow -\overline{z_1}}(\omega'+\overline{z_1})\frac{\hat{\chi}''(\omega)}{\omega'-\omega} = \frac{z_1\overline{z_1}-\omega^2}{(\omega-z_1)(\omega+z_1)(\omega-\overline{z_1})(\omega+\overline{z_1})}\end{aligned} \quad [5.199]$$

which gives:

$$\begin{aligned}&\hat{\chi}'(\omega) = \frac{1}{\pi}\text{VP}\int_{-\infty}^{\infty}\frac{\hat{\chi}''(\omega)}{(\omega'-\omega)}d\omega' = \frac{\omega_0^2-\omega^2}{(\omega_0^2-\omega^2)^2+\gamma^2\omega^2}\\ &\hat{\chi}(\omega) = \hat{\chi}'(\omega) + \mathrm{i}\hat{\chi}''(\omega) = \frac{\omega_0^2-\omega^2+\mathrm{i}\gamma\omega}{(\omega^2-\omega_0^2)^2+\gamma^2\omega^2} = \frac{1}{\omega_0^2-\omega^2-\mathrm{i}\gamma\omega}\end{aligned} \quad [5.200]$$

The expressions [5.197] and [5.200] are consistent with the results previously obtained.

Chapter 6

Magnetization, Magnetized Media

6.1. Magnetic field generated by magnetized medium: macroscopic description

6.1.1. *Magnetization vector field*

In addition to electric dipoles, a material medium can also contain *magnetic (dipolar) moments* that are regularly distributed within its volume. These materials or *magnetized* media generate magnetic fields that are observed to be of the same nature as those produced by conventional currents. The origin of these magnetic fields is microscopic in nature. Using the *Bohr model*, it is easy to see that the atoms possess an intrinsic magnetic moment. The Bohr model for the hydrogen atom considers an electron with charge $-e$ moving in a uniform circular motion around a central nucleus (proton) with a period $T = \frac{2\pi}{\omega}$. This motion generates an electric current given by $I = \frac{e}{T} = \frac{e\omega}{2\pi}$. Thus, we have a circular current loop with a radius a_0, known as the *Bohr radius* (the average electron–proton distance). Consequently, the hydrogen atom acquires an intrinsic magnetic moment:

$$\overrightarrow{\mathrm{m}} = IS\overrightarrow{\mathrm{n}} = \frac{e\omega}{2} a_0^2 \overrightarrow{\mathrm{n}} = \frac{e}{2m_e}\left(m_e \omega a_0^2 \overrightarrow{\mathrm{n}}\right) = -\frac{e}{2m_e}\overrightarrow{\mathbf{L}}_e \qquad [6.1]$$

where m_e and $\overrightarrow{\mathbf{L}}_e$ are, respectively, the mass and the angular momentum of the electron. The minus sign comes from the fact that the current intensity I and the electron velocity $\overrightarrow{\mathbf{v}}$ are opposite. We will revisit this in more detail in the following sections. The factor $\gamma = \frac{e}{2m_e}$ is called the *gyromagnetic ratio*. This model can be generalized to complex atoms or molecules. In this case, we see that the resulting magnetic moment is proportional to the total angular momentum of the electrons inside the atom or molecule (it is also possible to imagine that the nuclei and electrons are spinning, that is, they possess spin). In this case, we refer to the spin magnetic moment.

Let us try to calculate the vector potential $\vec{\mathbf{A}}$ and the magnetic field $\vec{\mathbf{B}}$ generated by this microscopic magnetic moment. Since atomic dimensions are very small, we are only concerned with the magnetic field created far from the atom. Furthermore, the shape of the loop is not important; so, for simplicity, we will take a rectangular loop, as illustrated in Figure 6.1.

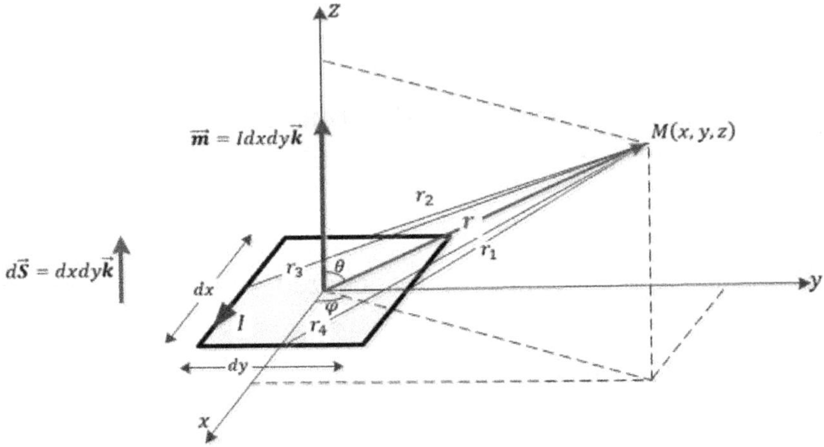

Figure 6.1. *Atomic magnetic dipole moment consisting of an infinitesimal current loop. For a color version of this figure, see www.iste.co.uk/abdesselam/electrodynamics2.zip*

The vector potential generated by the loop is then:

$$\vec{\mathbf{A}} = \frac{\mu_0 I}{4\pi}\left[dx\left(\frac{1}{r_3} - \frac{1}{r_1}\right)\vec{\mathbf{i}} + dy\left(\frac{1}{r_4} - \frac{1}{r_2}\right)\vec{\mathbf{j}}\right] \qquad [6.2]$$

where r_1, r_2, r_3 and r_4 are the distances measured from the point M to the centers of the four sides of the infinitesimal loop (the average distances). They are given by:

$$\begin{aligned}
r_1 &= r^2 + \left(\frac{dy}{2}\right)^2 - rdy\,\vec{\mathbf{e}}_r \cdot \vec{\mathbf{j}} = r^2 + \left(\frac{dy}{2}\right)^2 - rdy\sin\theta\sin\varphi \\
r_2 &= r^2 + \left(\frac{dx}{2}\right)^2 - rdx\,\vec{\mathbf{e}}_r \cdot \vec{\mathbf{i}} = r^2 + \left(\frac{dx}{2}\right)^2 - rdx\sin\theta\cos\varphi \\
r_3 &= r^2 + \left(\frac{dy}{2}\right)^2 + rdy\,\vec{\mathbf{e}}_r \cdot \vec{\mathbf{j}} = r^2 + \left(\frac{dy}{2}\right)^2 + rdy\sin\theta\sin\varphi \\
r_4 &= r^2 + \left(\frac{dx}{2}\right)^2 + rdx\,\vec{\mathbf{e}}_r \cdot \vec{\mathbf{i}} = r^2 + \left(\frac{dx}{2}\right)^2 + rdx\sin\theta\cos\varphi
\end{aligned} \qquad [6.3]$$

For distant points M, the expression [6.2] simplifies to:

$$\overrightarrow{\mathbf{A}} = \frac{\mu_0 I}{4\pi}\left[\frac{dx}{r}\left(\frac{1}{\sqrt{1+\frac{dy}{2r}\left(\frac{dy}{2r}+2\sin\theta\sin\varphi\right)}}-\right.\right.$$

$$\left.\frac{1}{\sqrt{1+\frac{dy}{2r}\left(\frac{dy}{2r}-2\sin\theta\sin\varphi\right)2}}\right)\overrightarrow{\mathbf{i}}+\frac{dy}{r}\left(\frac{1}{\sqrt{1+\frac{dx}{2r}\left(\frac{dx}{2r}-2\sin\theta\cos\varphi\right)}}\right. \quad [6.4]$$

$$\left.\left.-\frac{1}{\sqrt{1+\frac{dx}{2r}\left(\frac{dx}{2r}+2\sin\theta\cos\varphi\right)}}\right)\overrightarrow{\mathbf{j}}\right] \simeq -\frac{\mu_0 I}{4\pi r^2}dxdy\left[\sin\theta\sin\varphi\,\overrightarrow{\mathbf{i}}-\right.$$

$$\left.\sin\theta\cos\varphi\,\overrightarrow{\mathbf{j}}\right] = \frac{\mu_0 IdS}{4\pi r^2}\sin\theta\left[-\sin\varphi\,\overrightarrow{\mathbf{i}}+\cos\varphi\,\overrightarrow{\mathbf{j}}\right] = \frac{\mu_0 IdS}{4\pi r^2}\sin\theta\,\overrightarrow{\mathbf{e}}_\varphi$$

The *magnetic moment* $\overrightarrow{\mathbf{m}}$ is defined as the product of I and the vector $d\overrightarrow{\mathbf{S}}$:

$$\overrightarrow{\mathbf{m}} = Id\overrightarrow{\mathbf{S}} = IdS\,\overrightarrow{\mathbf{k}} \quad [6.5]$$

Thus, the vector potential is written (more generally) as:

$$\overrightarrow{\mathbf{A}} = \frac{\mu_0 m}{4\pi r^2}\sin\theta\,\overrightarrow{\mathbf{e}}_\varphi = \frac{\mu_0\left(\overrightarrow{\mathbf{m}}\times\overrightarrow{\mathbf{e}}_r\right)}{4\pi r^2} \quad [6.6]$$

with the associated magnetic field:

$$\begin{aligned}\overrightarrow{\mathbf{B}} = \overrightarrow{\nabla}\times\overrightarrow{\mathbf{A}} &= \frac{1}{r\sin\theta}\frac{\partial\left(A_\varphi\sin\theta\right)}{\partial\theta}\overrightarrow{\mathbf{e}}_r - \frac{1}{r}\frac{\partial\left(rA_\varphi\right)}{\partial r}\overrightarrow{\mathbf{e}}_\theta \\ &= \frac{\mu_0}{4\pi r^3}\left(2\cos\theta\,\overrightarrow{\mathbf{e}}_r+\sin\theta\,\overrightarrow{\mathbf{e}}_\theta\right)\end{aligned} \quad [6.7]$$

This field is identical to the electric dipole field (simply replace $\frac{p}{\varepsilon_0}$ with $\mu_0 m$).

According to the Ampère model, magnetic materials can be modeled as having a volume filled with infinitesimal current loops with a density n, where each loop is the result of the motion of electrons around the nuclei of the constituent atoms, as illustrated in Figure 6.2.

Thus, every volume element $d\tau$ surrounding a point $\overrightarrow{\mathbf{r}}'$ in a magnetized medium has a *macroscopic magnetic moment*:

$$d\overrightarrow{\mathbf{M}} = \overrightarrow{\mathbf{M}}\left(\overrightarrow{\mathbf{r}}'\right)d\tau \quad [6.8]$$

where $\overrightarrow{\mathbf{M}}$, referred to as the *magnetization vector* (or simply *magnetization*), is defined as the density of magnetic moments:

$$\overrightarrow{\mathbf{M}} = n\overrightarrow{\mathbf{m}} = nId\overrightarrow{\mathbf{S}} \quad [6.9]$$

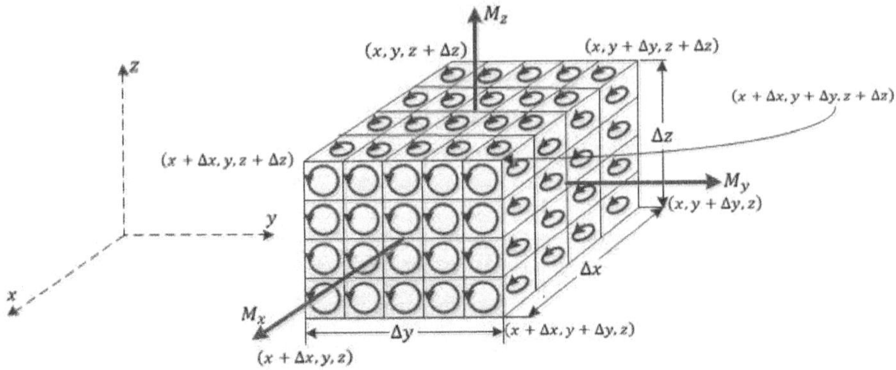

Figure 6.2. *The Ampère model for magnetized materials. For a color version of this figure, see www.iste.co.uk/abdesselam/electrodynamics2.zip*

The magnetization thus macroscopically describes the magnetic properties of magnetized materials.

6.1.2. *Magnetization current*

For the infinitesimal contour $ABCD$ located in the Oxy plane, as shown in Figure 6.3, only magnetic moments oriented along the Ox or Oy directions contribute to currents passing perpendicularly through the surface bounded by the contour. Moreover, magnetic moments entirely inside the contour do not contribute to a net current because the current crosses the surface twice: once in the positive direction of the Oz axis and once in the opposite direction. Only the magnetic moments located on the edges contribute to the net current passing through the surface (since the current crosses the surface once and then again outside of it). Since the sides Δx and Δy of the contour are very small, we assume that the magnetic moments neither change in number nor direction along each edge. The total net current enclosed by the contour near each side is equal to the product of the current I associated with each moment and the number of moments that pass only once through the contour (i.e. those cut by the edges of the contour). If $\overrightarrow{m}$ makes an angle θ with the Oy axis, the net current enclosed along the line AB is:

$$-IndS\Delta y\cos\theta|_x = -M_y(x)\,\Delta y \qquad [6.10]$$

The minus sign appears because the current inside the contour adjacent to the x coordinate flows in the negative z direction. Similarly, the net current enclosed perpendicularly by the contour near the $x + \Delta x$ coordinate is:

$$IndS\Delta y\cos\theta|_{x+\Delta x} = M_y(x+\Delta x)\,\Delta y \qquad [6.11]$$

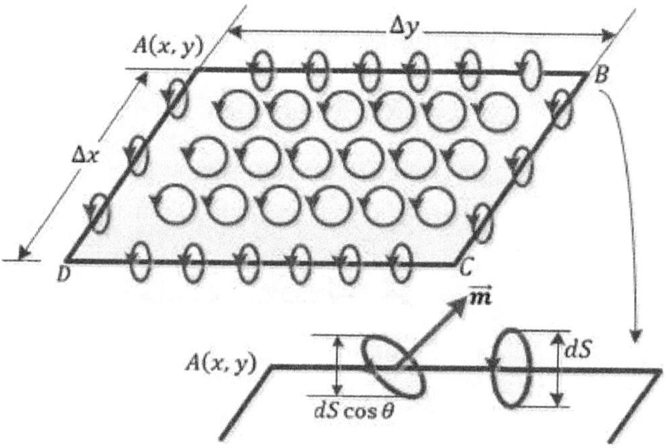

Figure 6.3. *Magnetization current enclosed by an arbitrary contour $ABCD$ within the magnetized medium. For a color version of this figure, see www.iste.co.uk/abdesselam/electrodynamics2.zip*

Along the edges located at y and $y + \Delta y$, the currents enclosed are given by:

$$\begin{aligned} & IndS\Delta x \cos\theta|_y = M_x(y)\,\Delta x \\ & -IndS\Delta x \cos\theta|_{y+\Delta y} = -M_x(y+\Delta y)\,\Delta x \end{aligned} \qquad [6.12]$$

The total current in the Oz direction enclosed by this contour is therefore the sum of contributions [6.10], [6.11] and [6.12]:

$$I_{\text{tot}} = \Delta x \Delta y \left(\frac{M_y(x+\Delta x) - M_y(x)}{\Delta x} - \frac{M_x(y+\Delta y) - M_x(y)}{\Delta y} \right) \qquad [6.13]$$

If the magnetization is uniform, the total net current is zero (the current flowing in one direction within the loop is canceled by the current flowing in the opposite direction on the other side). However, if the magnetization varies with position, there will be a net current leaving the surface of the loop. This is possible unless the current, surface, density or orientation angle of the magnetic moments depends on position. If Δx and Δy tend to zero, the terms on the right-hand side reduce to the partial derivatives $\frac{\partial M_y}{\partial x}$ and $\frac{\partial M_x}{\partial y}$. Thus, we obtain:

$$J_z = \lim_{\Delta x, \Delta y \longrightarrow 0} \frac{I_{\text{tot}}}{\Delta x \Delta y} = \left(\frac{\partial M_y}{\partial x} - \frac{\partial M_x}{\partial y} \right) = \left(\overrightarrow{\nabla} \times \overrightarrow{\mathbf{M}} \right)_z \qquad [6.14]$$

This is the volume current density measured in the Oz direction. If the loop is oriented in the Oxz or Oyz plane, we can similarly show that the volume currents in the Oy and Ox directions are, respectively:

$$J_y = \left(\frac{\partial M_x}{\partial z} - \frac{\partial M_z}{\partial x}\right) = \left(\vec{\nabla} \times \vec{\mathbf{M}}\right)_y$$
$$J_x = \left(\frac{\partial M_z}{\partial y} - \frac{\partial M_y}{\partial z}\right) = \left(\vec{\nabla} \times \vec{\mathbf{M}}\right)_x \quad [6.15]$$

In 3D, we therefore obtain:

$$\vec{\mathbf{J}}_m = \vec{\nabla} \times \vec{\mathbf{M}} \quad [6.16]$$

$\vec{\mathbf{J}}_m$ here denotes the *magnetization volume current*. This current also generates a magnetic field and can be included in the Maxwell–Ampère equation:

$$\vec{\nabla} \times \frac{\vec{\mathbf{B}}}{\mu_0} = \vec{\mathbf{J}} + \vec{\mathbf{J}}_m = \vec{\mathbf{J}} + \vec{\nabla} \times \vec{\mathbf{M}} \quad [6.17]$$

where $\vec{\mathbf{J}}$ is the conduction current present in the magnetized medium caused by the motion of free charges. The current $\vec{\mathbf{J}}_m$ results from the motion of bound charges within the material.

6.1.3. *Vector potential and macroscopic magnetic field*

The vector potential resulting from both the conduction current and the magnetic moments within the magnetized medium is:

$$\vec{\mathbf{A}}\left(\vec{\mathbf{r}}\right) = \frac{\mu_0}{4\pi}\int_{\tau'} \frac{\vec{\mathbf{J}}\left(\vec{\mathbf{r}}'\right) d\tau'}{\left|\vec{\mathbf{r}} - \vec{\mathbf{r}}'\right|} + \frac{\mu_0}{4\pi}\int_{\tau'} \frac{\vec{\mathbf{M}}\left(\vec{\mathbf{r}}'\right) \times \left(\vec{\mathbf{r}} - \vec{\mathbf{r}}'\right) d\tau'}{\left|\vec{\mathbf{r}} - \vec{\mathbf{r}}'\right|^3} \quad [6.18]$$

The second integral can also be expressed as:

$$\begin{aligned}\int_{\tau'} \frac{\vec{\mathbf{M}}\left(\vec{\mathbf{r}}'\right) \times \left(\vec{\mathbf{r}} - \vec{\mathbf{r}}'\right) d\tau'}{\left|\vec{\mathbf{r}} - \vec{\mathbf{r}}'\right|^3} &= \int_{\tau'} \vec{\mathbf{M}}\left(\vec{\mathbf{r}}'\right) \times \vec{\nabla}'\left(\frac{1}{\left|\vec{\mathbf{r}} - \vec{\mathbf{r}}'\right|}\right) d\tau' \\ &= \int_{\tau'} \frac{\vec{\nabla}' \times \vec{\mathbf{M}}\left(\vec{\mathbf{r}}'\right)}{\left|\vec{\mathbf{r}} - \vec{\mathbf{r}}'\right|} d\tau' - \int_{\tau'} \vec{\nabla}' \times \left(\frac{\vec{\mathbf{M}}\left(\vec{\mathbf{r}}'\right)}{\left|\vec{\mathbf{r}} - \vec{\mathbf{r}}'\right|}\right) d\tau' \\ &= \int_{\tau'} \frac{\vec{\nabla}' \times \vec{\mathbf{M}}\left(\vec{\mathbf{r}}'\right)}{\left|\vec{\mathbf{r}} - \vec{\mathbf{r}}'\right|} d\tau' - \oint_{S'} \frac{\vec{\mathbf{M}}\left(\vec{\mathbf{r}}'\right)}{\left|\vec{\mathbf{r}} - \vec{\mathbf{r}}'\right|} \times d\vec{\mathbf{S}}'\end{aligned} \quad [6.19]$$

where we have used the identity $\int_{\tau'} \vec{\nabla}' \times \vec{\mathbf{A}}' d\tau' = -\oint_{S'} \vec{\mathbf{A}}' \times d\vec{\mathbf{S}}'$. Here, S' represents the boundary of τ'. Thus, the vector potential in [6.18] takes the form:

$$\vec{\mathbf{A}}\left(\vec{\mathbf{r}}\right) = \frac{\mu_0}{4\pi}\int_{\tau'} \frac{\vec{\mathbf{J}}\left(\vec{\mathbf{r}}'\right) + \vec{\nabla}' \times \vec{\mathbf{M}}\left(\vec{\mathbf{r}}'\right)}{\left|\vec{\mathbf{r}} - \vec{\mathbf{r}}'\right|} d\tau' + \frac{\mu_0}{4\pi}\oint_{S'} \frac{\vec{\mathbf{M}}\left(\vec{\mathbf{r}}'\right) \times \vec{\mathbf{n}}'}{\left|\vec{\mathbf{r}} - \vec{\mathbf{r}}'\right|} dS' \quad [6.20]$$

This new expression can be immediately interpreted in terms of currents. The first term represents the vector potential generated by the superposition of the conduction current $\vec{\mathbf{J}}$ and the magnetization current $\vec{\mathbf{J}}_m = \vec{\nabla} \times \vec{\mathbf{M}}$ (due to magnetization) distributed throughout the entire volume τ', while the second term represents the vector potential generated by a surface current distribution spread over the surface S', called the *surface magnetization current*:

$$\vec{\mathbf{K}}_m = \vec{\mathbf{M}} \times \vec{\mathbf{n}} \qquad [6.21]$$

where $\vec{\mathbf{n}}$ is the normal to the surface. We thus conclude that, similarly to polarization, which contributes to the surface charge at the boundary of a dielectric, magnetization also acts as a surface current at the boundaries of a magnetized material. Describing magnetization using these equivalent currents is highly useful for calculating the magnetic field generated by such media. Taking the curl of [6.20], we obtain the expression for the magnetic field created by a magnetized material in terms of the equivalent currents:

$$\begin{aligned} \vec{\mathbf{B}}\left(\vec{\mathbf{r}}\right) &= \frac{\mu_0}{4\pi}\int_{\tau'} \frac{\vec{\mathbf{J}}\left(\vec{\mathbf{r}}'\right) \times \left(\vec{\mathbf{r}} - \vec{\mathbf{r}}'\right)}{\left|\vec{\mathbf{r}} - \vec{\mathbf{r}}'\right|^3} d\tau' \\ &+ \frac{\mu_0}{4\pi}\int_{\tau'} \frac{\vec{\mathbf{J}}_m\left(\vec{\mathbf{r}}'\right) \times \left(\vec{\mathbf{r}} - \vec{\mathbf{r}}'\right)}{\left|\vec{\mathbf{r}} - \vec{\mathbf{r}}'\right|^3} d\tau' \\ &+ \frac{\mu_0}{4\pi}\oint_{S'} \frac{\vec{\mathbf{K}}_m\left(\vec{\mathbf{r}}'\right) \times \left(\vec{\mathbf{r}} - \vec{\mathbf{r}}'\right)}{\left|\vec{\mathbf{r}} - \vec{\mathbf{r}}'\right|^3} dS' \end{aligned} \qquad [6.22]$$

The total magnetic field is therefore the vector sum of the magnetic field $\vec{\mathbf{B}}_c$ due to the conduction current (first term) and the magnetic field $\vec{\mathbf{B}}_m$ due to magnetization (second and third terms). The surface current $\vec{\mathbf{K}}_m$ is related to the discontinuity of the magnetization vector ($\vec{\mathbf{M}}$, which is non-zero inside the material and zero outside, undergoing an apparent discontinuity when crossing the material's surface). This discontinuity gives rise to a discontinuity in the field $\vec{\mathbf{B}}_m$ when crossing the material's surface given by:

$$\left(\vec{\mathbf{B}}_m^{\text{ext}} - \vec{\mathbf{B}}_m^{\text{int}}\right)_{\parallel} = -\mu_0 \vec{\mathbf{M}}_{\parallel} \qquad [6.23]$$

The magnetic field [6.22] obviously satisfies the first local equation of magnetostatics or the Maxwell–Thomson equation:

$$\vec{\nabla} \cdot \vec{\mathbf{B}} = 0 \qquad [6.24]$$

To obtain the modified Maxwell-Ampère equation governing $\vec{\nabla} \times \vec{\mathbf{B}}$, we use the identity $\vec{\nabla} \times \vec{\mathbf{B}} = \vec{\nabla} \times \left(\vec{\nabla} \times \vec{\mathbf{A}}\right) = \vec{\nabla}\left(\vec{\nabla} \cdot \vec{\mathbf{A}}\right) - \Delta\vec{\mathbf{A}}$. In the Coulomb gauge, we also impose the condition $\vec{\nabla} \cdot \vec{\mathbf{A}} = 0$, which simplifies the equation to:

$$\begin{aligned}\vec{\nabla} \times \vec{\mathbf{B}}\left(\vec{\mathbf{r}}\right) &= -\frac{\mu_0}{4\pi}\Delta \int_{\tau'} \frac{\vec{\mathbf{J}}\left(\vec{\mathbf{r}}'\right) + \vec{\nabla}' \times \vec{\mathbf{M}}\left(\vec{\mathbf{r}}'\right)}{\left|\vec{\mathbf{r}} - \vec{\mathbf{r}}'\right|} d\tau' \\ &\quad - \frac{\mu_0}{4\pi}\Delta \oint_{S'} \frac{\vec{\mathbf{M}}\left(\vec{\mathbf{r}}'\right) \times \vec{\mathbf{n}}'}{\left|\vec{\mathbf{r}} - \vec{\mathbf{r}}'\right|} dS' = \mu_0 \int_{\tau'} \left[\vec{\mathbf{J}}\left(\vec{\mathbf{r}}'\right)\right. \\ &\quad \left.+ \vec{\nabla}' \times \vec{\mathbf{M}}\left(\vec{\mathbf{r}}'\right)\right] \delta^3\left(\vec{\mathbf{r}} - \vec{\mathbf{r}}'\right) d\tau' \\ &\quad + \mu_0 \oint_{S'} \left[\vec{\mathbf{M}}\left(\vec{\mathbf{r}}'\right) \times \vec{\mathbf{n}}'\right] \delta^3\left(\vec{\mathbf{r}} - \vec{\mathbf{r}}'\right) dS' \\ &= \mu_0 \left[\vec{\mathbf{J}}\left(\vec{\mathbf{r}}\right) + \vec{\nabla} \times \vec{\mathbf{M}}\left(\vec{\mathbf{r}}\right)\right]\end{aligned} \quad [6.25]$$

for any point $\vec{\mathbf{r}}$ outside S'. We conclude that the field $\vec{\mathbf{B}}$ indeed satisfies the local equation of magnetostatics in magnetized media, $\vec{\nabla} \times \vec{\mathbf{B}} = \mu_0\left(\vec{\mathbf{J}} + \vec{\nabla} \times \vec{\mathbf{M}}\right)$ (already introduced in [6.17]). This equation remains valid even in the quasi-stationary regime, as long as propagation effects can be neglected.

6.1.4. *Magnetic excitation vector* $\vec{\mathbf{H}}$

By grouping the circulation terms on the left-hand side of equation [6.17], we obtain:

$$\vec{\nabla} \times \left(\frac{\vec{\mathbf{B}}}{\mu_0} - \vec{\mathbf{M}}\right) = \vec{\mathbf{J}} \quad [6.26]$$

The vector field

$$\vec{\mathbf{H}} = \frac{\vec{\mathbf{B}}}{\mu_0} - \vec{\mathbf{M}} \iff \vec{\mathbf{B}} = \mu_0\left(\vec{\mathbf{H}} + \vec{\mathbf{M}}\right) \quad [6.27]$$

is called the *magnetic excitation field* (sometimes referred to as the *magnetic intensity field*). $\vec{\mathbf{H}}$ satisfies the equation:

$$\vec{\nabla} \times \vec{\mathbf{H}} = \vec{\mathbf{J}} \quad [6.28]$$

We therefore conclude that (1) the conduction current $\vec{\mathbf{J}}$ is the source of the field $\vec{\mathbf{H}}$; (2) the magnetization current $\vec{\mathbf{J}}_m$ is the source of the field $\vec{\mathbf{M}}$; (3) the total current $\vec{\mathbf{J}} + \vec{\mathbf{J}}_m$ is the source of the field $\vec{\mathbf{B}}$. In a vacuum, equation [6.27] simplifies to $\vec{\mathbf{B}} = \mu_0\vec{\mathbf{H}}$ since $\vec{\mathbf{M}} = 0$. Note that in equation [6.27], the parameter μ_0 multiplies both $\vec{\mathbf{H}}$ and $\vec{\mathbf{M}}$, unlike the permittivity ε_0, which only multiplies $\vec{\mathbf{E}}$.

The circulation of the vector $\vec{\mathbf{H}}$ along a closed contour is therefore determined solely by the conduction current:

$$\int_S \left(\vec{\nabla} \times \vec{\mathbf{H}}\right) \cdot d\vec{\mathbf{S}} = \oint_\Gamma \vec{\mathbf{H}} \cdot d\vec{\mathbf{l}} = \int_S \vec{\mathbf{J}} \cdot d\vec{\mathbf{S}} \qquad [6.29]$$

This is the *Ampère modified theorem*. However, it would be incorrect to think that this vector depends only on this current, since its divergence indeed depends on the magnetization:

$$\vec{\nabla} \cdot \vec{\mathbf{H}} = -\vec{\nabla} \cdot \vec{\mathbf{M}} \qquad [6.30]$$

6.1.5. *Boundary conditions for the fields $\vec{\mathrm{B}}$ and $\vec{\mathrm{H}}$*

The boundary relations at the interface between two media (1) and (2) are derived from the local properties of the fields. These media can either be two magnetized materials or a magnetized material and a vacuum. Let $\vec{\mathrm{n}}_{12}$ denote the unit normal vector at the interface, oriented from medium (1) to medium (2). Since the magnetic field $\vec{\mathbf{B}}$ is solenoidal, its normal component remains continuous at every point on the surface. In fact, consider a cylinder of height Δz positioned at the interface between the two media, with its upper half in (2) and its lower half in (1). The cylinder axis is parallel to the normal vector $\vec{\mathrm{n}}_{12}$ and is denoted Oz, as shown in Figure 6.4. By integrating $\vec{\nabla} \cdot \vec{\mathbf{B}} = 0$ into the volume of the cylinder and letting $\Delta z \to 0$, we obtain:

$$\int_S B_{1,z} dS = \int_S B_{2,z} dS \qquad [6.31]$$

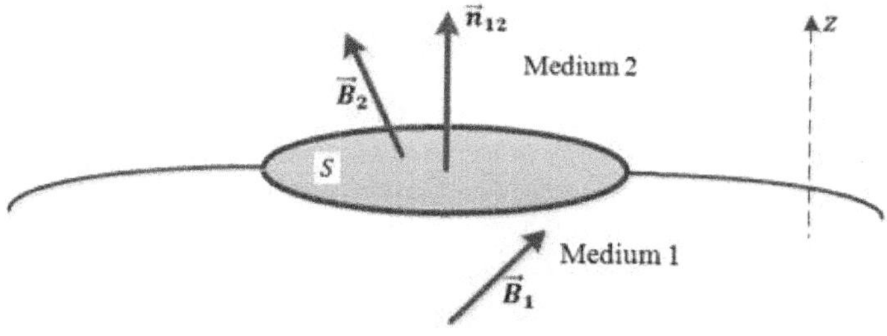

Figure 6.4. *Continuity of the normal component of the magnetic field $\vec{\mathrm{B}}$ across the interface between two media. For a color version of this figure, see www.iste.co.uk/abdesselam/electrodynamics2.zip*

Since S is arbitrary, it follows that $B_{1,z} = B_{2,z}$. This can also be written as $\vec{\mathrm{n}}_{12} \cdot \left(\vec{\mathbf{B}}_2 - \vec{\mathbf{B}}_1\right) = 0$ or, more simply:

$$\left(\vec{\mathbf{B}}_2\right)_\perp = \left(\vec{\mathbf{B}}_1\right)_\perp \qquad [6.32]$$

which means that the normal component of $\vec{\mathbf{B}}$ is continuous across the interface between the two media (1) and (2).

From $\vec{\nabla} \times \vec{\mathbf{H}} = \vec{\mathbf{J}}$, we deduce that if the interface carries a surface conduction current $\vec{\mathbf{K}}$, the tangential component of the vector $\vec{\mathbf{H}}$ undergoes a discontinuity:

$$\vec{\mathbf{n}}_{12} \times \left(\vec{\mathbf{H}}_2 - \vec{\mathbf{H}}_1\right) = \vec{\mathbf{K}} \qquad [6.33]$$

In fact, let us integrate the equation $\vec{\nabla} \times \vec{\mathbf{H}} = \vec{\mathbf{J}}$ along the contour $ABCDA$, as illustrated in Figure 6.5, and then take the limit $\Delta z \to 0$:

$$\begin{aligned} &\int_{y_0-\frac{L}{2}}^{y_0+\frac{L}{2}} H_{1,y}(x_0, y, z_0)\, dy - \int_{y_0-\frac{L}{2}}^{y_0+\frac{L}{2}} H_{2,y}(x_0, y, z_0)\, dy \\ &= \lim_{\Delta z \longrightarrow 0} \int_{y_0-\frac{L}{2}}^{y_0+\frac{L}{2}} \int_{z_0-\frac{\Delta z}{2}}^{z_0+\frac{\Delta z}{2}} J_x(x_0, y, z)\, dy dz \end{aligned} \qquad [6.34]$$

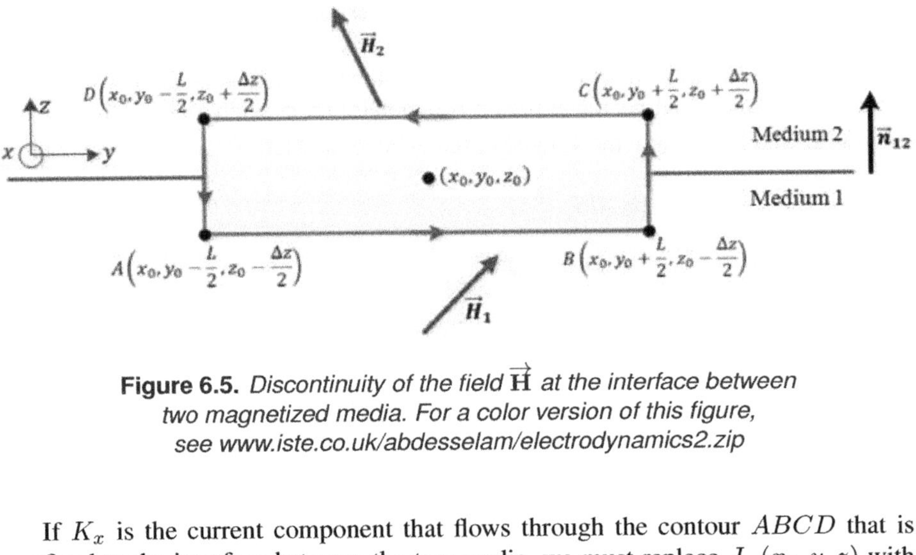

Figure 6.5. *Discontinuity of the field $\vec{\mathbf{H}}$ at the interface between two magnetized media. For a color version of this figure, see www.iste.co.uk/abdesselam/electrodynamics2.zip*

If K_x is the current component that flows through the contour $ABCD$ that is confined to the interface between the two media, we must replace $J_x(x_0, y, z)$ with $K_x(x_0, y)\,\delta(z)$. Thus, we write $\int_{y_0-\frac{L}{2}}^{y_0+\frac{L}{2}} [H_{1,y}(x_0, y, z_0) - H_{2,y}(x_0, y, z_0)]\, dy = \int_{y_0-\frac{L}{2}}^{y_0+\frac{L}{2}} K_x(x_0, y)\, dy$, which gives $H_{1,y} - H_{2,y} = K_x$, at every point on the interface. Similarly, by changing the orientation of the contour, we obtain $H_{1,x} - H_{2,x} = -K_y$. These two results can be combined into a single equation:

$$\left(\vec{\mathbf{H}}_2 - \vec{\mathbf{H}}_1\right)_{\parallel} = \vec{\mathbf{K}} \times \vec{\mathbf{n}}_{12} \qquad [6.35]$$

which means that the tangential component of $\vec{\mathbf{H}}$ is discontinuous by $\vec{\mathbf{K}}$ at the interface between the two media (1) and (2). We therefore conclude that the normal component $\left(\vec{\mathbf{B}}\right)_{\perp}$ is continuous, while the tangential component $\left(\vec{\mathbf{H}}\right)_{\parallel}$ is discontinuous by $\vec{\mathbf{K}}$. This last equation is equivalent to expression [6.33]. Since equation [6.16] has the same form as [6.28], we can therefore use [6.33] and [6.35] by replacing $\vec{\mathbf{H}}$ with $\vec{\mathbf{M}}$ and $\vec{\mathbf{J}}$ with $\vec{\mathbf{K}}_m$:

$$\vec{\mathbf{n}}_{12} \times \left(\vec{\mathbf{M}}_2 - \vec{\mathbf{M}}_1\right) = \vec{\mathbf{K}}_m \iff \left(\vec{\mathbf{M}}_2 - \vec{\mathbf{M}}_1\right)_{\parallel} = \vec{\mathbf{K}}_m \times \vec{\mathbf{n}}_{12} \qquad [6.36]$$

These results are consistent with expression [6.21]. Here are some examples of calculations when magnetization $\vec{\mathbf{M}}$ is known.

EXAMPLE 6.1. (Infinitely magnetized cylinder along its axis).– Consider a very long cylinder with its axis along Oz and a radius R (its length is assumed to be much greater than R so that the edge effects can be neglected). The magnetization of the cylinder is assumed to be uniform and directed along Oz, as illustrated in Figure 6.6. Since the magnetization $\vec{\mathbf{M}} = M\vec{\mathbf{k}}$ is uniform, the magnetic current $\vec{\mathbf{J}}_m = \vec{\nabla} \times \vec{\mathbf{M}}$ is zero. The cylinder is therefore equivalent to a sheet of azimuthal current distributed on its surface (this surface current is thus directed along $\vec{e}_\theta$):

$$\vec{\mathbf{K}}_m = \vec{\mathbf{M}} \times \vec{e}_r = M\vec{e}_\theta \qquad [6.37]$$

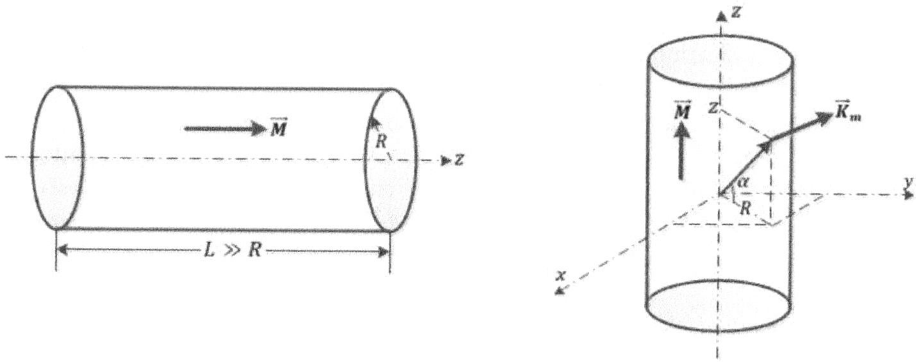

Figure 6.6. *Infinitely long cylinder of radius R magnetized along its axis. For a color version of this figure, see www.iste.co.uk/abdesselam/electrodynamics2.zip*

The calculation of the corresponding magnetic field is similar to that of a solenoid with the same geometry as the cylinder, and where $nI = M$. Far from the edges, we obtain:

$$\vec{\mathbf{B}}_m = \begin{cases} 0 & r > R \\ \mu_0 \vec{\mathbf{M}} & r < R \end{cases} \quad [6.38]$$

where r is the distance from the point being considered and the Oz axis. We therefore conclude that the magnetic field is uniform inside the cylinder and that the field $\vec{\mathbf{H}}$ is zero everywhere. These results can be confirmed through direct calculations. In fact, we know that (1) the magnetic current is zero inside the cylinder, that is, $\vec{\mathbf{J}}_m = 0$; and (2) the surface current $\vec{\mathbf{K}}_m$ is equal to $M\vec{\mathbf{e}}_\theta$ on the lateral surface and zero on the two bases. At the center of the cylinder, the magnetic field $\vec{\mathbf{B}}_m$ is equal to:

$$\begin{aligned} \vec{\mathbf{B}}_m &= \frac{\mu_0}{4\pi}\int_{\text{S.lat.}} \frac{\vec{\mathbf{M}}dS \times \left(-R\vec{\mathbf{e}}_r - z\vec{\mathbf{k}}\right)}{\left(R^2+z^2\right)^{3/2}} \\ &= \frac{\mu_0}{4\pi}\int_{\varphi=0}^{2\pi}\int_{z_1}^{z_2} \frac{M\vec{\mathbf{e}}_\theta R d\varphi dz \times \left(-R\vec{\mathbf{e}}_r - z\vec{\mathbf{k}}\right)}{\left(R^2+z^2\right)^{3/2}} \\ &= \frac{\mu_0 M R^2}{2}\vec{\mathbf{k}}\int_{z_1}^{z_2}\frac{dz}{\left(R^2+z^2\right)^{3/2}} = \frac{\mu_0 M}{2}\vec{\mathbf{k}}\int_{-\alpha_1}^{\alpha_2}\cos\alpha d\alpha \\ &= \frac{\mu_0 M}{2}\vec{\mathbf{k}}\left(\sin\alpha_2 + \sin\alpha_1\right) \end{aligned} \quad [6.39]$$

If the cylinder has the shape of a disk ($L \ll R$), that is, $\alpha_1 \approx \alpha_2 \approx 0$, then $\vec{\mathbf{B}}_m = 0$. If the cylinder is very long ($L \gg R$) (infinite), that is, $\alpha_1 \approx \alpha_2 \approx \frac{\pi}{2}$, we obtain $\vec{\mathbf{B}}_m = \mu_0 M\vec{\mathbf{k}}$.

EXAMPLE 6.2. (An infinitely long cylinder uniformly magnetized in a direction perpendicular to its axis).– To simplify, it is also assumed that magnetization is uniform and directed along the Ox axis, that is, $\vec{\mathbf{M}} = M\vec{\mathbf{i}}$, as shown in Figure 6.7. The magnetic current (inside) is given by $\vec{\mathbf{J}}_m = \vec{\nabla} \times \vec{\mathbf{M}} = 0$, since $\vec{\mathbf{M}}$ is uniform within the cylinder. The magnetic current on the surface is $\vec{\mathbf{K}}_m = \vec{\mathbf{M}} \times \vec{\mathbf{n}} = M\vec{\mathbf{i}} \times \vec{\mathbf{e}}_r = M\sin\varphi\vec{\mathbf{k}}$. The vector potential $\vec{\mathbf{A}}$ can be determined in several ways.

1) *Direct method*: the vector potential $\vec{\mathbf{A}}$ generated by the cylinder is given by:

$$\vec{\mathbf{A}} = \frac{\mu_0}{4\pi}\int_{\text{cylinder}} \frac{\vec{\mathbf{M}} \times (\vec{\mathbf{r}} - \vec{\mathbf{r}}')}{\left|\vec{\mathbf{r}} - \vec{\mathbf{r}}'\right|^3} d\tau = \frac{\mu_0}{4\pi}\vec{\mathbf{M}} \times \left[\int_{\text{cylinder}} \frac{\vec{\mathbf{r}} - \vec{\mathbf{r}}'}{\left|\vec{\mathbf{r}} - \vec{\mathbf{r}}'\right|^3} d\tau\right] \quad [6.40]$$

The term $\int_{\text{cylinder}} \frac{\vec{\mathbf{r}} - \vec{\mathbf{r}}'}{\left|\vec{\mathbf{r}} - \vec{\mathbf{r}}'\right|^3} d\tau$ is equivalent to the electrostatic field created by a volume charge distribution $\rho = 4\pi\varepsilon_0$ uniformly distributed within the cylinder (assumed here to be infinite to neglect edge effects). This field can also be calculated using Gauss's theorem, taking into account the symmetries of the system. It can be easily shown that:

$$\int_{\text{cylinder}} \frac{\overrightarrow{\mathbf{r}} - \overrightarrow{\mathbf{r}}'}{\left|\overrightarrow{\mathbf{r}} - \overrightarrow{\mathbf{r}}'\right|^3} d\tau = \begin{cases} 2\pi r\, \overrightarrow{\mathbf{e}}_r & r \leq R \\ 2\pi \dfrac{R^2}{r} \overrightarrow{\mathbf{e}}_r & r > R \end{cases} \qquad [6.41]$$

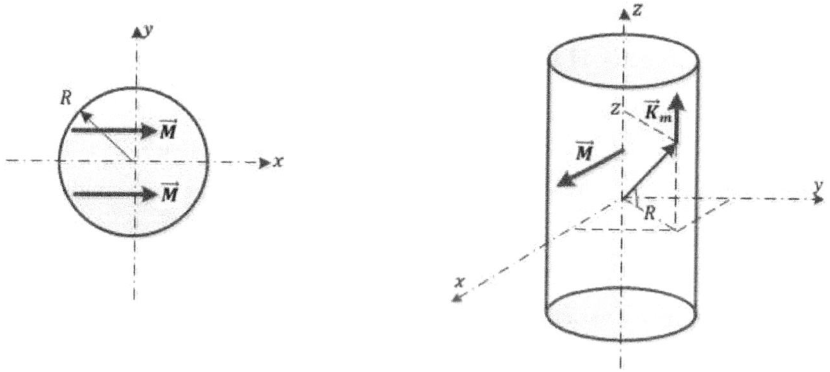

Figure 6.7. *Infinite cylinder of radius R magnetized parallel to its axis. For a color version of this figure, see www.iste.co.uk/abdesselam/electrodynamics2.zip*

Since $\overrightarrow{\mathbf{M}} = M\,\overrightarrow{\mathbf{i}}$, we can directly deduce the vector potential:

$$\overrightarrow{\mathbf{A}}_m\left(\overrightarrow{\mathbf{r}}\right) = \begin{cases} \dfrac{\mu_0 M y}{2} \overrightarrow{\mathbf{k}} & r \leq R \\ \dfrac{\mu_0 M y R^2}{2r^2} \overrightarrow{\mathbf{k}} & r > R \end{cases} \qquad [6.42]$$

and then the Cartesian components $B_{m,x} = \frac{\partial A_{m,z}}{\partial y}$ and $B_{m,y} = -\frac{\partial A_{m,z}}{\partial x}$ of the magnetic field:

$$B_{m,x} = \begin{cases} \dfrac{\mu_0 M}{2} & r \leq R \\ \dfrac{\mu_0 M R^2}{2r^4}\left(x^2 - y^2\right) & r > R \end{cases}$$
$$B_{m,y} = \begin{cases} 0 & r \leq R \\ \dfrac{\mu_0 M R^2}{r^4} xy & r > R \end{cases} \qquad [6.43]$$

In vector form, the magnetic field is written as:

$$\overrightarrow{\mathbf{B}}_m\left(\overrightarrow{\mathbf{r}}\right) = \begin{cases} \dfrac{\mu_0 \overrightarrow{\mathbf{M}}}{2} & r \leq R \\ \dfrac{\mu_0 M R^2}{2r^2}\left(\cos 2\varphi\, \overrightarrow{\mathbf{i}} + \sin 2\varphi\, \overrightarrow{\mathbf{j}}\right) & r > R \end{cases} \qquad [6.44]$$

It is clear that the magnetic field is uniform inside the cylinder. To determine $\vec{\mathbf{H}}$, it is sufficient to apply definition [6.27]:

$$\vec{\mathbf{H}}\left(\vec{\mathbf{r}}\right) = \begin{cases} -\dfrac{\vec{\mathbf{M}}}{2} & r \leq R \\ \dfrac{MR^2}{2r^2}\left(\cos 2\varphi\,\vec{\mathbf{i}} + \sin 2\varphi\,\vec{\mathbf{j}}\right) & r > R \end{cases} \qquad [6.45]$$

In this case, the vector $\vec{\mathbf{H}}$ is not zero inside the cylinder.

2) *Indirect method*: it is known that the vector potential can also be calculated using the equation $\vec{\mathbf{A}} = \frac{\mu_0}{4\pi}\int \frac{\vec{\mathbf{K}}_m dS}{r}$, which implies that $\vec{\mathbf{A}} = A_z\,\vec{\mathbf{k}}$, since the current $\vec{\mathbf{K}}_m$ is directed along the Oz axis. As $\vec{\mathbf{A}}$ satisfies the Coulomb gauge condition:

$$\vec{\nabla}\cdot\vec{\mathbf{A}} = \frac{1}{r}\frac{\partial\left(rA_r\right)}{\partial r} + \frac{1}{r}\frac{\partial A_\varphi}{\partial r} + \frac{\partial A_z}{\partial z} = \frac{\partial A_z}{\partial z} = 0 \qquad [6.46]$$

the variable z does not appear in the expression of A_z, that is, $A_z = A_z\left(r,\varphi\right)$. Moreover, the absence of a real current inside the cylinder allows us to write $\vec{\nabla}\times\vec{\mathbf{B}} = \mu_0\vec{\mathbf{J}} = 0$, which implies $\vec{\nabla}\times\left(\vec{\nabla}\times\vec{\mathbf{A}}\right) = 0$. Substituting $\vec{\nabla}\times\vec{\mathbf{A}} = \frac{1}{r}\frac{\partial A_z}{\partial\varphi}\vec{\mathbf{e}}_r - \frac{\partial A_z}{\partial r}\vec{\mathbf{e}}_\varphi$ into $\vec{\nabla}\times\left(\vec{\nabla}\times\vec{\mathbf{A}}\right) = 0$, we obtain:

$$\frac{1}{r}\frac{\partial}{\partial r}\left(r\frac{\partial A_z}{\partial r}\right) + \frac{1}{r^2}\frac{\partial^2 A_z}{\partial\varphi^2} = 0 \qquad [6.47]$$

To obtain a solution to equation [6.47], we proceed by separation of variables: $A_z\left(r,\varphi\right) = f\left(r\right)\Phi\left(\varphi\right)$, which gives $\frac{r}{f}\frac{d}{dr}\left(r\frac{df}{dr}\right) = -\frac{1}{\Phi}\frac{\partial^2\Phi}{\partial\varphi^2} \equiv \lambda^2$. This leads to the following two equations: $\frac{d}{dr}\left(r\frac{df}{dr}\right) = \lambda^2\frac{f}{r}$ and $\frac{\partial^2\Phi}{\partial\varphi^2} + \lambda^2\Phi = 0$. It then follows that:

$$\begin{aligned} f\left(r\right) &= C_1 r^{\lambda} + C_2 r^{-\lambda} \\ \Phi\left(\varphi\right) &= D_1\cos\lambda\varphi + D_2\sin\lambda\varphi \end{aligned} \qquad [6.48]$$

where λ must be an integer due to the periodicity condition $A_z(r,\varphi+2\pi) = A_z(r,\varphi)$. We conclude that: $A_z\left(r,\varphi\right) = \sum_{n=1}^{\infty}\left(C_{1,n}r^n + C_{2,n}r^{-n}\right)\left(D_{1,n}\cos n\varphi + D_{2,n}\sin n\varphi\right)$ is a solution of the partial differential equation [6.47]. Thus, for $r < R$, we have:

$$A_z^{\text{int}}\left(r,\varphi\right) = \sum_{n=1}^{\infty} r^n\left(D_{1,n}^{\text{int}}\cos n\varphi + D_{2,n}^{\text{int}}\sin n\varphi\right) \qquad [6.49]$$

($C_{2,n}$ are zero because the solution must not diverge as $r \to 0$) and for $r > R$:

$$A_z^{\text{ext}}\left(r,\varphi\right) = \sum_{n=1}^{\infty}\frac{1}{r^n}\left(D_{1,n}^{\text{ext}}\cos n\varphi + D_{2,n}^{\text{ext}}\sin n\varphi\right) \qquad [6.50]$$

($C_{1,n}$ are zero because the solution must remain finite as $r \to \infty$). The continuity of the vector potential at the boundary leads to the relations $R^n D_{1,n}^{\text{int}} = \frac{D_{1,n}^{\text{ext}}}{R^n}$ and $R^n D_{2,n}^{\text{int}} = \frac{D_{2,n}^{\text{ext}}}{R^n}$, $\forall n \geq 1$. The magnetic fields inside and outside of the cylinder are given by the equations:

$$
\begin{aligned}
&\overrightarrow{\mathbf{B}}_{\text{int}} = \frac{1}{r}\frac{\partial A_z^{\text{int}}}{\partial \varphi}\overrightarrow{\mathbf{e}}_r - \frac{\partial A_z^{\text{int}}}{\partial r}\overrightarrow{\mathbf{e}}_{\varphi} r = \sum_{n=1}^{\infty} n r^{n-1}\left[\left(-D_{1,n}^{\text{int}} \sin n\varphi + \right.\right.\\
&\left.D_{2,n}^{\text{int}} \cos n\varphi\right)\overrightarrow{\mathbf{e}}_r - \left.\left(D_{1,n}^{\text{int}} \cos n\varphi + D_{2,n}^{\text{int}} \sin n\varphi\right)\overrightarrow{\mathbf{e}}_{\varphi}\right]\\
&\overrightarrow{\mathbf{B}}_{\text{ext}} = \frac{1}{r}\frac{\partial A_z^{\text{ext}}}{\partial \varphi}\overrightarrow{\mathbf{e}}_r - \frac{\partial A_z^{\text{ext}}}{\partial r}\overrightarrow{\mathbf{e}}_{\varphi} = \sum_{n=1}^{\infty} \frac{n}{r^{n+1}}\left[\left(-D_{1,n}^{\text{ext}} \sin n\varphi + \right.\right.\\
&\left.D_{2,n}^{\text{ext}} \cos n\varphi\right)\overrightarrow{\mathbf{e}}_r + \left.\left(D_{1,n}^{\text{ext}} \cos n\varphi + D_{2,n}^{\text{ext}} \sin n\varphi\right)\overrightarrow{\mathbf{e}}_{\varphi}\right]
\end{aligned}
\qquad [6.51]
$$

Over the circumference of the cylinder, there is continuity of the normal component and discontinuity of the tangential component of the magnetic field $\overrightarrow{\mathbf{B}}$, that is, $B_{r,\text{ext}}(R) - B_{r,\text{int}}(R) = 0$ and $B_{\varphi,\text{ext}}(R) - B_{\varphi,\text{int}}(R) = \mu_0 K_m = \mu_0 M \sin\varphi$. Given the expressions [6.51], we deduce that $\sum_{n=1}^{\infty} n\left[\left(R^{n-1}D_{1,n}^{\text{int}} - \frac{D_{1,n}^{\text{ext}}}{R^{n+1}}\right)\sin n\varphi + \left(\frac{D_{2,n}^{\text{ext}}}{R^{n+1}} - R^{n-1}D_{2,n}^{\text{int}}\right)\cos n\varphi\right] = 0$ and $\sum_{n=1}^{\infty} n\left[\left(\frac{D_{1,n}^{\text{ext}}}{R^{n+1}} + R^{n-1}D_{1,n}^{\text{int}}\right)\cos n\varphi + \left(\frac{D_{2,n}^{\text{ext}}}{R^{n+1}} + R^{n-1}D_{2,n}^{\text{int}}\right)\sin n\varphi\right] = \mu_0 M \sin\varphi$, which implies that ($\forall n \geq 1$):

$$
\begin{aligned}
&R^{n-1}D_{1,n}^{\text{int}} - \frac{D_{1,n}^{\text{ext}}}{R^{n+1}} = \frac{D_{2,n}^{\text{ext}}}{R^{n+1}} - R^{n-1}D_{2,n}^{\text{int}} = 0\\
&n\left(\frac{D_{1,n}^{\text{ext}}}{R^{n+1}} + R^{n-1}D_{1,n}^{\text{int}}\right) = 0\\
&n\left(\frac{D_{2,n}^{\text{ext}}}{R^{n+1}} + R^{n-1}D_{2,n}^{\text{int}}\right) = \mu_0 M \delta_{n,1}
\end{aligned}
\qquad [6.52]
$$

The simultaneous solution of the constraints [6.52] gives $D_{1,n}^{\text{ext}} = D_{1,n}^{\text{int}} = 0$ $(n \geq 1)$, $D_{2,1}^{\text{ext}} = \frac{1}{2}\mu_0 M R^2$, $D_{2,1}^{\text{int}} = \frac{1}{2}\mu_0 M$ and $D_{2,n}^{\text{ext}} = D_{2,n}^{\text{int}} = 0$ $(n \geq 2)$. It follows that:

$$
\begin{aligned}
&\overrightarrow{\mathbf{A}}_{\text{int}} = \frac{1}{2}\mu_0 M r \sin\varphi\, \overrightarrow{\mathbf{k}}\\
&\overrightarrow{\mathbf{A}}_{\text{ext}} = \frac{1}{2}\mu_0 \frac{MR^2}{r}\sin\varphi\, \overrightarrow{\mathbf{k}}\\
&\overrightarrow{\mathbf{B}}_{\text{int}} = \frac{1}{2}\mu_0 M \cos\varphi\, \overrightarrow{\mathbf{e}}_r - \frac{1}{2}\mu_0 M \sin\varphi\, \overrightarrow{\mathbf{e}}_{\varphi}\\
&\overrightarrow{\mathbf{B}}_{\text{ext}} = \frac{\mu_0 MR^2}{2r^2}\cos\varphi\, \overrightarrow{\mathbf{e}}_r + \frac{\mu_0 MR^2}{2r^2}\sin\varphi\, \overrightarrow{\mathbf{e}}_{\varphi}
\end{aligned}
\qquad [6.53]
$$

These results are analogous to those obtained using the first method.

EXAMPLE 6.3. (Uniformly magnetized sphere along a diameter).– It is assumed that the magnetization of the sphere is uniform and directed along a diameter, as shown in Figure 6.8. We define the axis Oz so that $\overrightarrow{\mathbf{M}} = M\overrightarrow{\mathbf{k}}$, where M is a constant. The vector potential can be evaluated in two different ways:

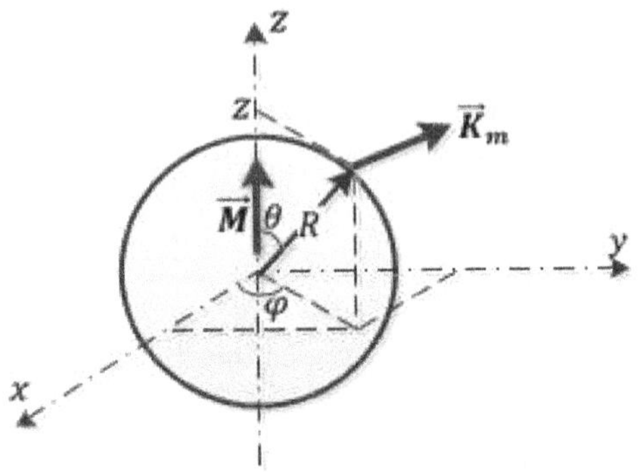

Figure 6.8. *Uniformly magnetized sphere along the Oz direction. For a color version of this figure, see www.iste.co.uk/abdesselam/electrodynamics2.zip*

1) *Direct method*: the vector potential is written in a form similar to [6.40], where this time, the integral $\int_{\text{sphere}} \frac{\vec{\mathbf{r}}-\vec{\mathbf{r}}'}{|\vec{\mathbf{r}}-\vec{\mathbf{r}}'|^3} d\tau'$ is extended to the sphere centered at O with radius R. This integral appears in the expression for the electrostatic field created by a distribution $\rho = 4\pi\varepsilon_0$ uniformly distributed inside a sphere of the same dimension. This field is calculated in a similar manner using Gauss's theorem:

$$\int_{\text{sphere}} \frac{\vec{\mathbf{r}}-\vec{\mathbf{r}}'}{\left|\vec{\mathbf{r}}-\vec{\mathbf{r}}'\right|^3} d\tau = \begin{cases} \frac{4\pi}{3} r \vec{\mathbf{e}}_r & r \le R \\ \frac{4\pi}{3}\frac{R^3}{r^2} \vec{\mathbf{e}}_r & r > R \end{cases} \quad [6.54]$$

and consequently:

$$\vec{\mathbf{A}}_m = \begin{cases} \frac{\mu_0 M}{3} r \sin\theta\, \vec{\mathbf{e}}_\varphi & r \le R \\ \frac{\mu_0 M}{3}\frac{R^3 \sin\theta}{r^2} \vec{\mathbf{e}}_\varphi & r > R \end{cases} \quad [6.55]$$

The spherical components $B_{m,r} = \frac{1}{r\sin\theta}\frac{\partial(\sin\theta A_{m,\varphi})}{\partial\theta}$, $B_{m,\theta} = -\frac{1}{r}\frac{\partial(rA_{m,\varphi})}{\partial r}$, and $B_{m,\varphi} = 0$ of the magnetic field are written as:

$$B_{m,r} = \begin{cases} \frac{2\mu_0 M}{3} \cos\theta & r \le R \\ \frac{2\mu_0 M R^3}{3r^3} \cos\theta & r > R \end{cases}$$

$$B_{m,\theta} = \begin{cases} -\dfrac{\mu_0 M}{3} \sin\theta & r \le R \\ \dfrac{\mu_0 M R^3}{3r^3} \sin\theta & r > R \end{cases} \qquad [6.56]$$

In vector form, we obtain:

$$\vec{\mathbf{B}}_m = \begin{cases} \dfrac{2\mu_0 M}{3} \left(\cos\theta \vec{\mathbf{e}}_r - \sin\theta \vec{\mathbf{e}}_\theta\right) \equiv \dfrac{2\mu_0 \vec{\mathbf{M}}}{3} & r \le R \\ \dfrac{\mu_0 M R^3}{3r^3} \left(2\cos\theta \vec{\mathbf{e}}_r + \sin\theta \vec{\mathbf{e}}_\theta\right) & r > R \end{cases} \qquad [6.57]$$

We thus conclude that inside the sphere, the magnetic field is uniform and equal to $\frac{2\mu_0 \vec{\mathbf{M}}}{3}$. Outside the sphere, it is equivalent to the magnetic field produced by a point magnetic dipole $\frac{4\pi R^3}{3} \vec{\mathbf{M}}$ placed at the center of the sphere.

2) *Indirect method*: since $\vec{\mathbf{M}} = M\vec{\mathbf{k}}$, the volume and surface magnetic currents are given, respectively, by $\vec{\mathbf{J}}_m = \vec{\nabla} \times \vec{\mathbf{M}} = 0$ and $\vec{\mathbf{K}}_m = \vec{\mathbf{M}} \times \vec{\mathbf{n}} = M\vec{\mathbf{k}} \times \vec{\mathbf{e}}_r = M\sin\theta \vec{\mathbf{e}}_\varphi$. From the expression $\vec{\mathbf{A}} = \frac{\mu_0}{4\pi} \int_{\text{sphere}} \frac{\vec{\mathbf{K}}_m dS}{r}$, we see that the vector potential has only one component along $\vec{\mathbf{e}}_\varphi$, that is, $\vec{\mathbf{A}} = A_\varphi(r,\theta,\varphi)\vec{\mathbf{e}}_\varphi$. Moreover, it satisfies the Coulomb gauge condition:

$$\vec{\nabla} \cdot \vec{\mathbf{A}} = \frac{1}{r\sin\theta} \frac{\partial A_\varphi}{\partial \varphi} = 0 \qquad [6.58]$$

The variable φ therefore does not appear in the expression for A_φ, that is, $\vec{\mathbf{A}} = A_\varphi(r,\theta)\vec{\mathbf{e}}_\varphi$. The absence of conduction current inside the sphere also allows us to write that $\vec{\nabla} \times \vec{\mathbf{B}} = 0$, which implies that $\vec{\nabla} \times \left(\vec{\nabla} \times \vec{\mathbf{A}}\right) = 0$. Taking into account that $\vec{\nabla} \times \vec{\mathbf{A}} = \frac{1}{r\sin\theta} \frac{\partial(\sin\theta A_\varphi)}{\partial\theta} \vec{\mathbf{e}}_r - \frac{1}{r} \frac{\partial(rA_\varphi)}{\partial r} \vec{\mathbf{e}}_\theta$, we then deduce the following partial differential equation:

$$\frac{1}{r} \frac{\partial^2 (rA_\varphi)}{\partial r^2} + \frac{1}{r} \frac{\partial}{\partial\theta} \left[\frac{1}{r\sin\theta} \frac{\partial(\sin\theta A_\varphi)}{\partial\theta}\right] = 0 \qquad [6.59]$$

which can be solved by separation of variables $A_\varphi(r,\theta) = f(r)\Theta(\theta)$. Substituting this expression into equation [6.59], we obtain $\frac{r}{f} \frac{d^2(rf)}{dr^2} = -\frac{1}{\Theta} \frac{d}{d\theta} \left[\frac{1}{\sin\theta} \frac{d(\sin\theta\Theta)}{d\theta}\right] \equiv \lambda(\lambda+1)$. This separates into the two differential equations $\frac{d^2(rf)}{dr^2} = \frac{\lambda(\lambda+1)f}{r}$ and $\frac{1}{\sin\theta} \frac{d}{d\theta} \left[\sin\theta \frac{d\Theta}{d\theta}\right] + \left[\lambda(\lambda+1) - \frac{1}{\sin^2\theta}\right]\Theta = 0$. These are well-known differential equations. To ensure finite solutions, we must choose λ to be an integer, that is, $\lambda = l$. In this case, we obtain $f(r) = Cr^l + \frac{D}{r+1}$ and $\Theta(\theta) = \frac{dP_l(\theta)}{d\theta}$, where P_l is the Legendre polynomial of degree l. It follows that the general solution to the partial differential equation [6.59] is: $A_\varphi(r,\theta)=\sum_{l=1}^{\infty}\left(C_l r^l+\frac{D_l}{r^{l+1}}\right)\frac{dP_l(\theta)}{d\theta}$. To ensure the solution remains finite near $r=0$ and as $r\longrightarrow\infty$, we take:

$$A_\varphi(r,\theta)=\begin{cases}\sum_{l=1}^{\infty}C_l^{\text{int}}r^l\dfrac{dP_l(\theta)}{d\theta} & r\leq R\\ \sum_{l=1}^{\infty}\dfrac{D_l^{\text{ext}}}{r^{l+1}}\dfrac{dP_l(\theta)}{d\theta} & r>R\end{cases} \qquad [6.60]$$

The continuity of the vector potential at the boundary of the sphere leads to:

$$C_l^{\text{int}}R^l=\frac{D_l^{\text{ext}}}{R^{l+1}},\ (\forall l\geq 1) \qquad [6.61]$$

Therefore, the magnetic field is:

$$\vec{\mathbf{B}}=\begin{cases}\sum_{l=1}^{\infty}\left[\dfrac{C_l^{\text{int}}r^{l-1}}{\sin\theta}\dfrac{d}{d\theta}\left(\sin\theta\dfrac{dP_l}{d\theta}\right)\vec{\mathbf{e}}_r-(l+1)\,C_l^{\text{int}}r^{l-1}\dfrac{dP_l}{d\theta}\vec{\mathbf{e}}_\theta\right] & r\leq R\\ \sum_{l=1}^{\infty}\left[\dfrac{D_l^{\text{ext}}}{r^{l+2}\sin\theta}\dfrac{d}{d\theta}\left(\sin\theta\dfrac{dP_l}{d\theta}\right)\vec{\mathbf{e}}_r+\dfrac{lD_l^{\text{ext}}}{r^{l+2}}\dfrac{dP_l}{d\theta}\vec{\mathbf{e}}_\theta\right] & r>R\end{cases}$$

[6.62]

We obviously verify that the normal component of the field $\vec{\mathbf{A}}$ is indeed continuous at the interface $r=R$. The discontinuity of the tangential component at the same interface, $B_\varphi^{\text{ext}}(R)-B_\varphi^{\text{int}}(R)=\mu_0 K_m=\mu_0 M\sin\theta$, requires that $\sum_{l=1}^{\infty}\left[\frac{lD_l^{\text{ext}}}{R^{l+2}}+(l+1)C_l^{\text{int}}R^{l-1}\right]\frac{dP_l}{d\theta}=\mu_0 M\sin\theta$, which implies that:

$$\begin{aligned}&\frac{lD_l^{\text{ext}}}{R^{l+2}}+(l+1)\,C_l^{\text{int}}R^{l-1}=0\ \ \forall l\geq 2\\ &-\frac{D_1^{\text{ext}}}{R^3}-2C_1^{\text{int}}=\mu_0 M\end{aligned} \qquad [6.63]$$

The simultaneous resolution of equations [6.61] and [6.63] yields $D_1^{\text{ext}}=-\frac{1}{3}\mu_0 MR^3$, $C_1^{\text{int}}=-\frac{1}{3}\mu_0 M$ and $C_l^{\text{int}}=D_l^{\text{ext}}=0$ $(l\geq 2)$. Therefore, we get:

$$\begin{aligned}\vec{\mathbf{A}}_{\text{int}}&=\frac{1}{3}\mu_0 Mr\sin\theta\,\vec{\mathbf{e}}_\varphi\\ \vec{\mathbf{A}}_{\text{ext}}&=\frac{1}{3}\mu_0\frac{MR^3}{r^2}\sin\theta\,\vec{\mathbf{e}}_\varphi\\ \vec{\mathbf{B}}_{\text{int}}&=\frac{2\mu_0 M}{3}\left(\cos\theta\,\vec{\mathbf{e}}_r-\sin\theta\,\vec{\mathbf{e}}_\theta\right)\\ \vec{\mathbf{B}}_{\text{ext}}&=\frac{\mu_0 MR^3}{3r^3}\left(2\cos\theta\,\vec{\mathbf{e}}_r+\sin\theta\,\vec{\mathbf{e}}_\theta\right)\end{aligned} \qquad [6.64]$$

6.2. Microscopic description of magnetization

6.2.1. *Problem statement*

There is a direct analogy between dielectric polarization and magnetization. Similarly, to establish the law linking magnetization $\vec{\mathbf{M}}$ to the magnetic field $\vec{\mathbf{B}}$, we must apply the Lorentz force to various atomic models.

Historically, Ampère suggested that the magnetic field produced by a material could originate from microscopic current loops associated with the structure of the medium. These current loops, considered as magnetic moments, help explain most of the macroscopic properties of magnetized materials. Recent developments in physics have largely confirmed Ampère's model. It has been established that the diamagnetism of materials arises from the motion of electrons within atoms. The existence of a link between angular momentum and magnetic moment has been demonstrated both experimentally and theoretically by equation [6.1]. We have already seen that theoretically a magnetic moment $\vec{\mathbf{m}}$ can be associated with any angular momentum $\vec{\mathbf{L}}$, to which it is proportional:

$$\vec{\mathbf{m}} = \gamma \vec{\mathbf{L}} \qquad [6.65]$$

where γ is the gyromagnetic ratio, generally given by $\gamma = \frac{q}{2m}$, and where q and m are the charge and mass of the moving object, respectively. The experimental observation of a negative sign for γ confirms the electronic origin of magnetism. However, it has been found that the measured value of $|\gamma|$ is twice the theoretical value $\gamma_e = -\frac{e}{2m_e}$ for the gyromagnetic ratio of the electron. This result cannot be explained by classical theory. Therefore, classical physics cannot accurately describe magnetism, and quantum mechanics must be used instead. According to quantum mechanics, the relationship between angular momentum and magnetic moment with the same gyromagnetic ratio as in classical theory holds only for *orbital magnetism*, which is associated with the motion of atomic electrons. However, quantum theory also predicts that elementary particles possess an intrinsic angular momentum, known as *spin*, which has no classical analogue. The spin angular momentum, denoted by $\vec{\mathbf{S}}$, gives the electron a magnetic moment $\vec{\mathbf{m}}$ equal to:

$$\vec{\mathbf{m}} = -\frac{e}{m_e}\vec{\mathbf{S}} \qquad [6.66]$$

We see that the gyromagnetic ratio is indeed twice that predicted by classical theory (which agrees with experimental results). This type of magnetism is known as *spin magnetism*. It often dominates over orbital magnetism. In quantum theory, the projection J_z of a quantum angular momentum $\vec{\mathbf{J}}$ (the quantum analogue of classical angular momentum) of a complex system (composed of the orbital angular momenta and spins of its constituents, that is, $\overrightarrow{\mathbf{J}} = \sum_i \overrightarrow{\mathbf{L}}_i + \sum_j \overrightarrow{\mathbf{S}}_j$), or of a single particle ($\overrightarrow{\mathbf{J}} = \overrightarrow{\mathbf{S}}$), onto an arbitrary axis Oz, is a quantized quantity:

$$J_z = \mathrm{m}\hbar \qquad [6.67]$$

where $\hbar$ is the *reduced Planck constant*, and m can take only one of the $2J+1$ possible values:

$$\mathrm{m} = -J,\ J+1,\ -J+2, \ldots,\ J-1,\ J \qquad [6.68]$$

J is a positive integer or half-integer. The numbers J and m are commonly referred to as the *intrinsic quantum number* and the *magnetic quantum number*, respectively. The magnitude of the angular momentum is given by:

$$\left|\overrightarrow{\mathbf{J}}\right| = \hbar\sqrt{J\,(J+1)} \qquad [6.69]$$

If J is associated with the intrinsic angular momentum of an elementary particle, it is called *spin* and can take only one value, which is characteristic of the given particle. Particles with an integer spin are called *bosons*, while those with a half-integer spin are known as *fermions*. Fermions obey the *Pauli exclusion principle*: two fermions cannot occupy the same quantum state simultaneously, unlike bosons. It follows that the magnetic moment of an atom or molecule originates primarily from its electrons. It is generally written as:

$$\overrightarrow{\mathrm{m}} = -g m_B \overrightarrow{\mathbf{J}} \qquad [6.70]$$

where $\overrightarrow{\mathbf{J}}$ is the total angular momentum of the atom, $m_B = \frac{e\hbar}{2m_e}$ is the *Bohr magneton*, and g is the *Landé g-factor*. The total angular momentum $\overrightarrow{\mathbf{J}}$ is composed of the orbital angular momentum $\overrightarrow{\mathbf{L}}$ and the spin angular momentum $\overrightarrow{\mathbf{S}}$, both arising from the electrons. The factor g lies between 1 and 2, depending on which of these two contributions is larger.

The majority of materials only manifest magnetic properties when they are placed in an external magnetic field. In this case, they are said to acquire *induced magnetization*, which will then be the source of a new magnetic field that will modify the applied external field. Materials can be classified into two categories. In the case of one category, comprising *diamagnetic* and *paramagnetic*, materials or media, magnetic effects are small and the applied field experiences negligible perturbation. The other category, ferromagnetic and ferrimagnetic, can modify the external field to a considerable extent.

6.2.2. *Diamagnetism*

Electrons orbiting the atomic nucleus can be seen as current loops (with the current flowing in the direction opposite to the electron's velocity). If an electron rotates with an angular velocity ω along an orbit of radius R, as illustrated in Figure 6.9, the current and the magnetic moment are given, respectively, by:

$$I = \frac{e\omega}{2\pi}$$
$$m = I\pi R^2 = \frac{e\omega}{2}R^2 \qquad [6.71]$$

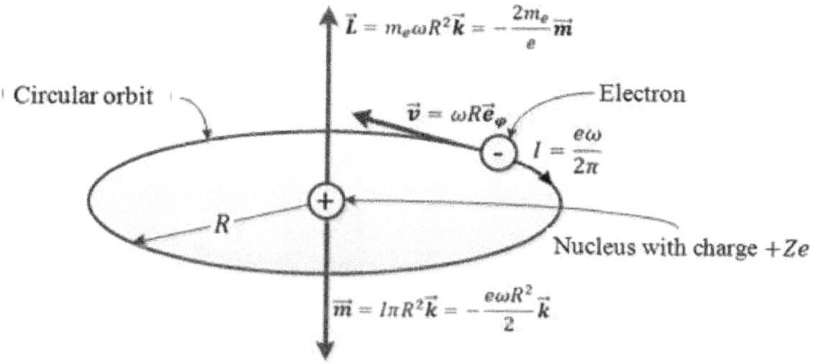

Figure 6.9. *Bohr atomic model. For a color version of this figure, see www.iste.co.uk/abdesselam/electrodynamics2.zip*

It should be noted that the angular momentum $\vec{\mathbf{L}}$ and the magnetic moment $\vec{\mathrm{m}}$ are directed in opposite directions. They are related by the following equation:

$$\vec{\mathbf{L}} = m_e R\vec{\mathbf{e}}_r \times \vec{\mathbf{v}} = m_e\omega R^2\vec{\mathbf{k}} = -\frac{2m_e}{e}\vec{\mathrm{m}} \qquad [6.72]$$

where m_e is the mass of the electron. Since quantum mechanics requires that angular momentum is quantized (the smallest value of L is $\hbar$), we see that the smallest value of the magnetic moment, known as the *Bohr magneton*, is equal to:

$$m_B = \frac{e\hbar}{2m_e} \qquad [6.73]$$

Within a homogeneous material, these moments are randomly distributed, so that each electron orbiting in a given direction has another nearby electron orbiting in the opposite direction. It can thus be concluded that in the absence of an external magnetic field, there is no net magnetization.

To determine R and ω, it should be noted that the attractive force between the orbiting electron and the atomic nucleus with a charge of $+Ze$ is balanced by the centrifugal force, that is:

$$m_e\omega^2 R = \frac{Ze^2}{4\pi\varepsilon_0 R^2} \qquad [6.74]$$

Since the left-hand side is proportional to the square of the quantized angular momentum, it follows that the radius R is also quantized. The smallest value of R is determined by writing $m_e\left(\frac{\hbar}{m_e R^2}\right)^2 R = \frac{Ze^2}{4\pi\varepsilon_0 R^2}$, which leads to the following:

$$R = \frac{4\pi\varepsilon_0\hbar^2}{m_e Ze^2} \qquad [6.75]$$

and consequently:

$$\omega_0 = \frac{\hbar}{m_e R^2} = \frac{Z^2 e^4 m_e}{(4\pi\varepsilon_0)^2\,\hbar^3} \qquad [6.76]$$

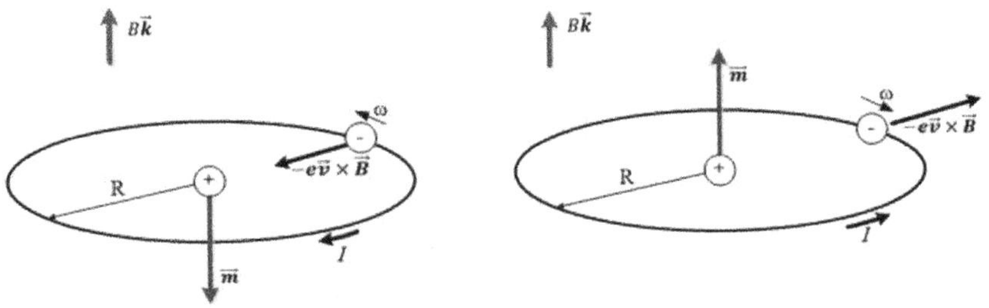

Figure 6.10. *Diamagnetic effects in materials. For a color version of this figure, see www.iste.co.uk/abdesselam/electrodynamics2.zip*

When a magnetic field $B\overrightarrow{\mathbf{k}}$ is applied, as shown in Figure 6.10, the electron loop with a magnetic moment opposite to the field experiences an additional radial force directed inward, while the loop with a magnetic moment collinear to the field experiences a radial force directed outward. Since the orbital radius R cannot change because it is quantized, this magnetic force causes a change in the angular velocity of $\Delta\omega$:

$$\begin{aligned} m_e\left(\omega_0+\Delta\omega_1\right)^2 R &= e\left(\frac{Ze}{4\pi\varepsilon_0 R^2}+(\omega_0+\Delta\omega_1)RB\right)\\ m_e\left(\omega_0+\Delta\omega_2\right)^2 R &= e\left(\frac{Ze}{4\pi\varepsilon_0 R^2}-(\omega_0+\Delta\omega_2)RB\right)\end{aligned} \qquad [6.77]$$

where the first electron speeds up, while the second slows down. Since the change in angular velocity $\Delta\omega$ is much smaller than the initial velocity ω_0, we have, approximately:

$$\begin{aligned}\Delta\omega_1 &= \frac{e\omega_0 B}{2m_e\omega_0 - eB}\\ \Delta\omega_2 &= \frac{-e\omega_0 B}{2m_e\omega_0 + eB}\end{aligned} \qquad [6.78]$$

where the terms of order $(\Delta\omega)^2$ have been neglected. However, even under a strong magnetic field, it is generally observed that $eB \ll 2m_e\omega_0$, and consequently, the previous equations reduce to:

$$\Delta\omega_1 \simeq -\Delta\omega_2 \simeq \frac{eB}{2m_e} \qquad [6.79]$$

The net magnetic moment for this pair of current loops is:

$$m = \frac{eR^2}{2}(\omega_2 - \omega_1) = -eR^2\Delta\omega_1 = -\frac{e^2R^2}{2m_e}B \qquad [6.80]$$

which is oriented opposite to the applied magnetic field. If there are n such pairs of loops per unit volume, the magnetization vector is:

$$\vec{\mathbf{M}} = n\vec{\mathbf{m}} = -\frac{ne^2R^2}{2m_e}B\vec{\mathbf{k}} \qquad [6.81]$$

which also points in the direction opposite to the applied field. Since magnetization is linearly related to the magnetic field, we define the *diamagnetic susceptibility* $(\chi_m)_{\text{diam}}$ as:

$$\vec{\mathbf{M}}_{\text{diam}} = \frac{(\chi_m)_{\text{diam}}}{\mu_0}\vec{\mathbf{B}} \qquad [6.82]$$

which immediately gives:

$$(\chi_m)_{\text{diam}} = -\frac{ne^2\mu_0R^2}{2m_e} \qquad [6.83]$$

This shows that $(\chi_m)_{\text{diam}}$ is negative. By substituting equation [6.75] into [6.83], we obtain:

$$(\chi_m)_{\text{diam}} = -\frac{8\pi^2\varepsilon_0^2n\mu_0\hbar^4}{m_e^3Z^2e^2} \qquad [6.84]$$

The magnetic excitation field is then given by:

$$\vec{\mathbf{H}} = \frac{\vec{\mathbf{B}}}{\mu_0} - \vec{\mathbf{M}} = \frac{1-\chi_m}{\mu_0}\vec{\mathbf{B}} = \frac{1}{\mu_0\mu_r}\vec{\mathbf{B}} = \frac{1}{\mu}\vec{\mathbf{B}} \qquad [6.85]$$

where $\mu_r = \frac{1}{1-\chi_m}$ is called the *relative permeability*, and μ is the *magnetic permeability* of the medium. In vacuum: $\chi_m = 0$, hence, $\mu_r = 1$ and $\mu = \mu_0$. In diamagnetic media (with negative magnetic susceptibility), we have $\mu_r < 1$ and $\mu < \mu_0$. Furthermore, since $|\chi_m| \ll 1$, magnetic effects in diamagnetic materials are very weak. In this case, we can write with good approximation:

$$\mu_r \simeq 1 + \chi_m \qquad [6.86]$$

There is another way to derive these results, often found in the literature. Let us consider the Bohr model of the atom. In the absence of a magnetic field, the electron describes a circular orbit with angular velocity ω_0. The force $\overrightarrow{\mathbf{F}}$ acting on the electron is simply the Coulomb electrostatic force, which is proportional to the electric field $\overrightarrow{\mathbf{E}}$ and is given by:

$$\overrightarrow{\mathbf{F}} = -e\overrightarrow{\mathbf{E}} = -\frac{Ze^2}{4\pi\varepsilon_0 r^2}\overrightarrow{\mathbf{e}}_r = e\overrightarrow{\nabla}V \qquad [6.87]$$

where Z is the number of protons in the nucleus, $-e$ is the charge of the electron, r is the distance to the nucleus and V is the potential created by the nucleus. Electrons orbiting the nucleus induce magnetic moments, but since these are randomly oriented in space, the net average magnetization is zero. When a magnetic field $\overrightarrow{\mathbf{B}}$ is gradually applied (for example, along the Oz axis) between times $t = 0$ and t_1 (with $\overrightarrow{\mathbf{B}}$ remaining constant after t_1), the electron orbit is modified: its plane becomes perpendicular to the magnetic field, as shown in Figure 6.11. A new force $\overrightarrow{\mathbf{F}}'$ appears and acts on the electron with velocity $\overrightarrow{\mathbf{v}}$:

$$\overrightarrow{\mathbf{F}}' = -e\left(\overrightarrow{\mathbf{E}}' + \overrightarrow{\mathbf{v}} \times \overrightarrow{\mathbf{B}}\right) = -e\left(-\overrightarrow{\nabla}V - \frac{\partial\overrightarrow{\mathbf{A}}}{\partial t} + \overrightarrow{\mathbf{v}} \times \overrightarrow{\mathbf{B}}\right) \qquad [6.88]$$

where $\overrightarrow{\mathbf{A}}$ is the vector potential from which $\overrightarrow{\mathbf{B}}$ is derived. Since $\overrightarrow{\mathbf{B}}$ is constant, we choose $\overrightarrow{\mathbf{A}} = \frac{1}{2}\overrightarrow{\mathbf{B}} \times \overrightarrow{\mathbf{r}} = \frac{1}{2}Br\overrightarrow{\mathbf{e}}_\varphi$, as $\overrightarrow{\mathbf{B}}$ points in the $\overrightarrow{\mathbf{k}}$ direction. As $\overrightarrow{\mathbf{v}}$ remains aligned with $\overrightarrow{\mathbf{e}}_\varphi$ as long as the perturbation is small, we write:

$$\overrightarrow{\mathbf{F}}' = -e\left[\frac{Ze}{4\pi\varepsilon_0 r^2}\overrightarrow{\mathbf{e}}_r - \frac{\partial\left(\frac{1}{2}Br\overrightarrow{\mathbf{e}}_\varphi\right)}{\partial t} + r\frac{d\varphi}{dt}B\overrightarrow{\mathbf{e}}_r\right] \qquad [6.89]$$

Using Newton's second law and assuming that the acceleration remains confined to the orbit plan ($\frac{d^2z}{dt^2} = 0$), we also have $\overrightarrow{\mathbf{F}}' = m_e\overrightarrow{\gamma} = m_e\left[\left(\frac{d^2r}{dt^2} - r\left(\frac{d\varphi}{dt}\right)^2\right)\overrightarrow{\mathbf{e}}_r + \left(2\frac{dr}{dt}\frac{d\varphi}{dt} + r\frac{d^2\varphi}{dt^2}\right)\overrightarrow{\mathbf{e}}_\varphi\right]$. If we assume that the magnetic field perturbs only the angular velocity of the electron and not its distance from the proton, then both $\frac{dr}{dt}$ and $\frac{d^2r}{dt^2}$ must be zero. Comparing the components along $\overrightarrow{\mathbf{e}}_\varphi$, we get $m_e r\frac{d^2\varphi}{dt^2} = e\frac{\partial\left(\frac{1}{2}Br\right)}{\partial t}$, which leads to:

$$m_e\frac{d\omega}{dt} = e\frac{\partial\left(\frac{1}{2}B\right)}{\partial t} \qquad [6.90]$$

Integrating with respect to time gives:

$$\omega - \omega_0 = \frac{eB}{2m_e} \qquad [6.91]$$

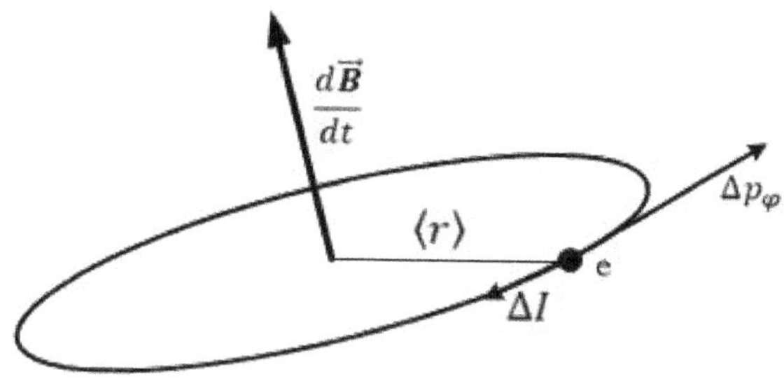

Figure 6.11. *Magnetic moment of an orbiting electron induced by the variation of an external magnetic field*

The angular velocity therefore varies by an amount $\omega_L = \frac{eB}{2m_e}$, which is called the *Larmor frequency*. This additional angular velocity corresponds to an additional electric current:

$$\Delta I = -\frac{e}{T} = \frac{e}{2\pi}\omega_L = -\frac{e^2 B}{4\pi m_e} \qquad [6.92]$$

and the magnetic moment of the electron will also increase by an additional quantity $\Delta\vec{\mathbf{m}} = \Delta I \vec{\mathbf{S}}_\perp = \frac{-e^2 B(x^2+y^2)}{4m_e}\vec{\mathbf{k}}$, where $\vec{\mathbf{S}}_\perp = \pi\left(x^2 + y^2\right)\vec{\mathbf{k}}$ is the area enclosed by the projection of the Bohr orbit onto the Oxy plane, perpendicular to the direction of the magnetic field $\vec{\mathbf{B}}$. On average, this becomes:

$$\langle \Delta\vec{\mathbf{m}} \rangle = \frac{-e^2 B\left(\langle x^2 \rangle + \langle y^2 \rangle\right)}{4m_e}\vec{\mathbf{k}} = \frac{-e^2 B\langle r^2 \rangle}{6m_e}\vec{\mathbf{k}} \qquad [6.93]$$

If the atom has Z independent electrons, the total contribution is:

$$\langle \Delta\vec{\mathbf{m}} \rangle_T = \sum_{i=1}^{Z} \frac{-e^2 B\langle r_i^2 \rangle}{6m_e}\vec{\mathbf{k}} \qquad [6.94]$$

Assuming there are n atoms per unit volume, the resulting magnetization is given by:

$$\vec{\mathbf{M}}_{\text{diam}} = \sum_{i=1}^{Z} \frac{-ne^2 B\langle r_i^2 \rangle}{6m_e}\vec{\mathbf{k}} \qquad [6.95]$$

By definition, $\vec{\mathbf{M}}_{\text{diam}} = \frac{(\chi_m)_{\text{diam}}}{\mu_0}\vec{\mathbf{B}}$, which immediately gives:

$$\chi_m = -\sum_{i=1}^{Z} \frac{\mu_0 n e^2}{6m_e}\langle r_i^2 \rangle \qquad [6.96]$$

It is also possible to use the Lenz law to derive a formula similar to [6.93] (up to a constant factor). According to this law, the electromotive force induced around the electron orbit is $\mathcal{E} = \frac{1}{e}\oint \vec{\mathbf{F}} \cdot d\vec{\mathbf{l}} = -\int \frac{d\vec{\mathbf{B}}}{dt} \cdot d\vec{\mathbf{S}}$, which implies $\mathcal{E} = \frac{1}{e}\oint \frac{d\vec{\mathbf{p}}}{dt} \cdot d\vec{\mathbf{l}} = -\int \frac{d\vec{\mathbf{B}}}{dt} \cdot d\vec{\mathbf{S}}$. Since the orbits are circular, it follows that $\frac{dp_\varphi}{dt} 2\pi \langle r \rangle_\perp = e\pi \langle r^2 \rangle_\perp \frac{dB_z}{dt}$, and therefore $\Delta p_\varphi = \frac{e\langle r^2\rangle_\perp \Delta B_z}{2\langle r \rangle_\perp}$. On the other hand, we know that $\vec{\mathbf{m}} = -\frac{e}{2m_e}\vec{\mathbf{L}}_e = -\frac{e}{2m_e}\vec{\mathbf{r}} \times \vec{\mathbf{p}}$, which gives $\Delta m_z = -\frac{e\langle r \rangle_\perp}{2m_e}\Delta p_\varphi = -\frac{e^2\langle r^2\rangle_\perp}{4m_e}\Delta B_z$. By generalizing these results, we write: $\Delta\vec{\mathbf{m}} = -\frac{e^2\langle r^2\rangle_\perp}{4m_e}\Delta\vec{\mathbf{B}}$. We finally obtain $\chi_m = -\frac{ne^2\langle r^2\mu_0\rangle_\perp}{12m_e}$. Here, $\langle r \rangle_\perp$ and $\langle r^2 \rangle_\perp$ represent the average radius and the average area of the orbit projected onto the plane perpendicular to $\vec{\mathbf{B}}$.

6.2.3. *Paramagnetism*

6.2.3.1. *Langevin model*

As with orientation polarization, an external magnetic field applied to a magnetized medium (referred to here as *paramagnetic*) exerts a torque on each magnetic moment, tending to align it with the field. For simplicity, we assume that the current loop associated with the magnetic moment is rectangular in shape, as illustrated in Figure 6.12. The forces acting on the different sides of the loop are:

$$\begin{aligned}
\vec{\mathbf{F}}_1 &= I\Delta x\,\vec{\mathbf{i}} \times \vec{\mathbf{B}} = I\Delta x\left[B_y\vec{\mathbf{k}} - B_z\vec{\mathbf{j}}\right] \\
\vec{\mathbf{F}}_2 &= I\Delta x\left(-\vec{\mathbf{i}}\right) \times \vec{\mathbf{B}} = -I\Delta x\left[B_y\vec{\mathbf{k}} - B_z\vec{\mathbf{j}}\right] \\
\vec{\mathbf{F}}_3 &= I\Delta y\,\vec{\mathbf{j}} \times \vec{\mathbf{B}} = I\Delta y\left[-B_x\vec{\mathbf{k}} + B_z\vec{\mathbf{i}}\right] \\
\vec{\mathbf{F}}_4 &= I\Delta y\left(-\vec{\mathbf{j}}\right) \times \vec{\mathbf{B}} = -I\Delta y\left[-B_x\vec{\mathbf{k}} + B_z\vec{\mathbf{i}}\right]
\end{aligned} \qquad [6.97]$$

When $\vec{\mathbf{B}}$ is uniform, the forces acting on opposite sides of the loop are equal in magnitude but opposite in direction. Therefore, the net force on the loop is zero. However, the loop experiences a torque given by:

$$\begin{aligned}
\vec{\mathbf{N}} &= \sum_{k=1}^{4} \vec{\mathbf{r}} \times d\vec{\mathbf{F}}_k = \frac{\Delta y}{2}\left(-\vec{\mathbf{j}} \times d\vec{\mathbf{F}}_1 + \vec{\mathbf{j}} \times d\vec{\mathbf{F}}_2\right) \\
&+ \frac{\Delta x}{2}\left(\vec{\mathbf{i}} \times d\vec{\mathbf{F}}_3 - \vec{\mathbf{i}} \times d\vec{\mathbf{F}}_4\right) \\
&= I\Delta x\Delta y\left(B_x\vec{\mathbf{j}} - B_y\vec{\mathbf{i}}\right) = \vec{\mathbf{m}} \times \vec{\mathbf{B}}
\end{aligned} \qquad [6.98]$$

Let $\vec{\mathbf{L}}$ be the internal angular momentum of the particle producing the current loop. In the absence of any other external torques, we write:

$$\frac{d\vec{\mathbf{L}}}{dt} = \vec{\mathbf{m}} \times \vec{\mathbf{B}} \qquad [6.99]$$

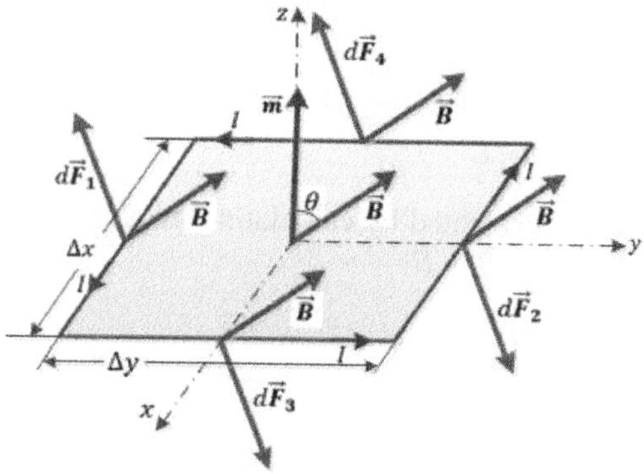

Figure 6.12. *Forces and torque acting on an infinitesimal loop. For a color version of this figure, see www.iste.co.uk/abdesselam/electrodynamics2.zip*

In addition, we know that $\vec{\mathbf{m}} = \gamma \vec{\mathbf{L}}$, where γ is the gyromagnetic ratio of the particle. It follows that:

$$\frac{d\vec{\mathbf{m}}}{dt} = -\gamma \vec{\mathbf{B}} \times \vec{\mathbf{m}} \qquad [6.100]$$

Since $\vec{\mathbf{B}} \times \vec{\mathbf{m}}$ is perpendicular to both $\vec{\mathbf{m}}$ and $\vec{\mathbf{B}}$, the time derivative of $\vec{\mathbf{m}}$ is also perpendicular to $\vec{\mathbf{m}}$. Therefore, the magnitude $\left|\vec{\mathbf{m}}\right|$ and the scalar product $\vec{\mathbf{m}} \cdot \vec{\mathbf{B}}$ also remain constant. As a result, the magnetic moment of the particle undergoes a rotational motion (precession) around the direction of the magnetic field, with angular velocity $\vec{\boldsymbol{\Omega}}_L = -\gamma \vec{\mathbf{B}}$. If the magnetic moment originates from an electronic orbital motion, that is, $\gamma = -\frac{e}{2m_e}$, then we recover the Larmor frequency $\omega_L = \frac{eB}{2m_e}$.

The work required to rotate a magnetic moment by a small angle $d\theta$ is:

$$dW = Nd\theta = mB \sin\theta d\theta \qquad [6.101]$$

and therefore, the total work needed to rotate it from $\theta = 0$ to an angle θ is given by the integral:

$$W = \int_{\theta=0}^{\theta} Nd\theta = -mB \cos\theta|_0^\theta = mB\,(1 - \cos\theta) \qquad [6.102]$$

The magnetic field $\vec{\mathbf{B}}$ is assumed to be directed along the z-axis. This work is stored as potential energy, since if the magnetic moment is released, it will align parallel to the field. Thermal agitation counteracts this alignment. According to the Boltzmann distribution, the number of magnetic moments per unit volume with energy W is:

$$
\begin{aligned}
n(\theta) &= n(0)\exp\left(-\frac{W}{k_BT}\right) = n(0)\exp\left(-\frac{mB(1-\cos\theta)}{k_BT}\right) \\
&\equiv n_0 \exp\left(\frac{mB\cos\theta}{k_BT}\right)
\end{aligned}
\qquad [6.103]
$$

The quantity n_0 can be evaluated by calculating the average number of magnetic moments within a sphere of radius R:

$$
\begin{aligned}
N &= \int_{r=0}^{R}\int_{\theta=0}^{\pi}\int_{\varphi=0}^{2\pi} n_0 \exp\left(\frac{mB\cos\theta}{k_BT}\right) r^2 \sin\theta dr d\theta d\varphi \\
&= \frac{2\pi R^3 n_0}{3}\int_0^{\pi} e^{a\cos\theta}\sin\theta d\theta
\end{aligned}
\qquad [6.104]
$$

where $a = \frac{mB}{k_BT}$. This integral reduces, using the change of variable $u = a\cos\theta$, to $N = \frac{-2\pi R^3 n_0}{3a}\int_a^{-a} e^u du = \frac{4\pi R^3 n_0}{3a}\sinh a$. Thus, the average number density is $n_{\text{avg}} = \frac{N}{\frac{4}{3}\pi R^3} = \frac{n_0}{a}\sinh a$, which leads to:

$$
n_0 = \frac{n_{\text{avg}} a}{\sinh a} \qquad [6.105]
$$

Clearly, n_{avg} is independent of R; it represents the average number of magnetic moments per unit volume, regardless of their orientation. Substituting equation [6.105] into [6.103], we find:

$$
n(\theta) = \frac{n_{\text{avg}} a}{\sinh a} e^{a\cos\theta} \qquad [6.106]
$$

Magnetic moments located between angles θ and $\theta + d\theta$ contribute to a net magnetization in the direction of the applied magnetic field by an amount $dM(\theta) = \frac{1}{\frac{4}{3}\pi R^3} \times m \times n(\theta) \times \cos\theta \times r^2 \sin\theta dr d\theta d\varphi$. Therefore, the total magnetization due to all of the moments inside the sphere is:

$$
\begin{aligned}
M &= \frac{m n_{\text{avg}} a}{2\sinh a}\int_0^{\pi}\sin\theta\cos\theta e^{a\cos\theta} d\theta = \frac{-m n_{\text{avg}}}{2a\sinh a}\int_a^{-a} u e^u du \\
&= \frac{-m n_{\text{avg}}}{2a\sinh a}\left[(-a-1)e^{-a} - (a-1)e^{a}\right] = m n_{\text{avg}}\left[\coth a - \frac{1}{a}\right]
\end{aligned}
\qquad [6.107]
$$

This formula, called the *Langevin equation*, is represented in Figure 6.13. At low temperatures and/or for intense fields (high a), it is observed that the magnetization tends to $M = n_{\text{avg}} m$, because all of the moments are aligned with the field. This is the saturation phenomenon or effect. At high temperatures ($a \ll 1$), magnetization increases linearly with the field. At ambient temperatures, the parameter a is also generally low. The following approximation can then be made:

$$M \simeq mn_{\text{avg}}\left[\frac{1+\frac{a^2}{2}}{a+\frac{a^3}{6}}-\frac{1}{a}\right] \simeq mn_{\text{avg}}\left[\frac{\left(1+\frac{a^2}{2}\right)\left(1-\frac{a^2}{6}\right)}{a}-\frac{1}{a}\right]$$
$$\simeq \frac{mn_{\text{avg}}a}{3} = \frac{m^2 n_{\text{avg}}}{3k_B T}B \quad [6.108]$$

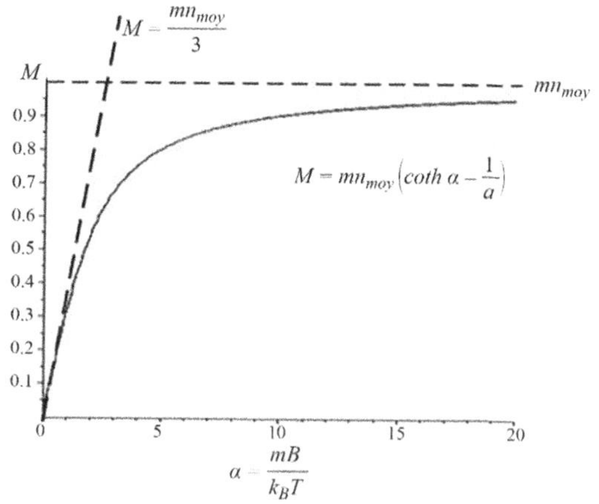

Figure 6.13. *Langevin equation for paramagnetic materials*

This shows that *paramagnetic susceptibility* is a positive quantity, inversely proportional to temperature T. In fact, according to the relation $\vec{\mathbf{M}}_{\text{para}} = \frac{(\chi_m)_{\text{para}}}{\mu_0}\vec{\mathbf{B}}$, we deduce:

$$(\chi_m)_{\text{para}} = \frac{\mu_0 m^2 n_{\text{avg}}}{3k_B T} \quad [6.109]$$

This is the *Curie law*, and the constant $\frac{\mu_0 m^2 n_{\text{avg}}}{3k_B}$ is known as the *Curie constant.*

6.2.3.2. *Contribution from quantum mechanics*

Quantum mechanics corrects the Langevin model due to the quantization of the angular momentum. For an atom with total angular momentum $L = \hbar\sqrt{J(J+1)}$, this means that the angle θ is no longer a continuous variable but instead takes on discrete values given by:

$$\cos\theta_M = \frac{\mathrm{m}}{\sqrt{J(J+1)}} \quad [6.110]$$

where $-J \leq \mathrm{m} \leq J$. The expression [6.102] for the potential energy ($W \sim -mB\cos\theta$) must therefore be replaced by:

$$W_{\mathrm{m}} \sim -\mathrm{m}gm_B B \quad [6.111]$$

It follows that the probability of finding a magnetic moment with this energy, according to the Boltzmann distribution, is given by $\frac{e^{ma}}{\mathcal{Z}}$, where $a = \frac{gm_B B}{k_B T}$. The quantity $\mathcal{Z}$, called the *partition function*, ensures the normalization of the probabilities and is given by:

$$\mathcal{Z} = \sum_{m=-J}^{J} e^{ma} = \frac{\sinh\left(\frac{a}{2}(2J+1)\right)}{\sinh\left(\frac{a}{2}\right)} \quad [6.112]$$

The average magnetization along the field is obtained from the relation:

$$\begin{aligned} M &= n_{avg}\frac{\sum_{m=-J}^{J} gm_B m e^{ma}}{\mathcal{Z}} = n_{avg} gm_B \frac{\partial \ln \mathcal{Z}}{\partial a} \\ &= n_{avg} gm_B J\left[\frac{2J+1}{2J}\coth\left(\frac{a(2J+1)}{2}\right) - \frac{1}{2J}\coth\left(\frac{a}{2}\right)\right] \end{aligned} \quad [6.113]$$

where n_{avg} is the average number of atoms per unit volume. This is the *quantum Langevin equation* of order J, also known as the *Brillouin function*. The function [6.113] is represented in Figure 6.14.

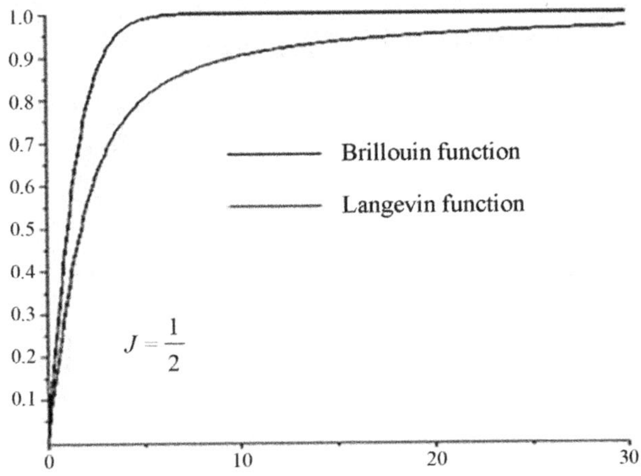

Figure 6.14. *Comparison of the Brillouin and Langevin models in the case $J = \frac{1}{2}$. For a color version of this figure, see www.iste.co.uk/abdesselam/electrodynamics2.zip*

For $a \ll 1$, equation [6.113] leads to:

$$M \simeq \frac{n_{avg} m_{eff}^2 B}{3k_B T} \quad [6.114]$$

where:

$$m_{\text{eff}} = gm_B\sqrt{J\left(J+1\right)} \qquad [6.115]$$

If we identify m_{eff} with m, both the classical and quantum theories yield the same result. At low temperatures $(a \gg 1)$, the magnetization reaches saturation:

$$M = n_{\text{avg}}gm_BJ \qquad [6.116]$$

Equation [6.113] is in better agreement with the experimental data. We can see that when $J \longrightarrow \infty$, the Langevin and Brillouin models converge.

6.2.3.3. *Paramagnetism in a molecular field*

Up to now, we have considered that the magnetic field acting on the constituents of matter was limited to the external field $\overrightarrow{\mathbf{B}}$. However, similar to the case of dielectrics, atoms generate a molecular field $\overrightarrow{\mathbf{B}}_{\text{mol}}$, which adds to the external field. This molecular field $\overrightarrow{\mathbf{B}}_{\text{mol}}$ is especially significant in very dense media. As a result, the local magnetic field within the material is given by:

$$\overrightarrow{\mathbf{B}}_{\text{loc}} = \overrightarrow{\mathbf{B}} + \overrightarrow{\mathbf{B}}_{\text{mol}} \qquad [6.117]$$

On the one hand, the molecular field $\overrightarrow{\mathbf{B}}_{\text{mol}}$ can be approximated as proportional to its source, that is, the local magnetization:

$$\overrightarrow{\mathbf{B}}_{\text{mol}} = \lambda\overrightarrow{\mathbf{M}} \qquad [6.118]$$

On the other hand, we have previously shown that at ambient temperature, the magnetization is given by:

$$\overrightarrow{\mathbf{M}} = \frac{m^2 n_{\text{avg}}}{3k_BT}\overrightarrow{\mathbf{B}}_{\text{loc}} \qquad [6.119]$$

this time using the local field. Substituting, we get $\overrightarrow{\mathbf{M}} = \frac{m^2 n_{\text{avg}}}{3k_BT}\left(\overrightarrow{\mathbf{B}} + \lambda\overrightarrow{\mathbf{M}}\right)$, which leads to:

$$\overrightarrow{\mathbf{M}} = \frac{m^2 n_{\text{avg}}}{3k_B\left(T - \frac{\lambda m^2 n_{\text{avg}}}{3k_B}\right)}\overrightarrow{\mathbf{B}} \qquad [6.120]$$

This gives a very simple form for the paramagnetic susceptibility:

$$(\chi_m)_{\text{para}} = \frac{m^2\mu_0 n_{\text{avg}}}{3k_B\left(T - \frac{\lambda m^2 n_{\text{avg}}}{3k_B}\right)} \equiv \frac{C}{T - T_c} \qquad [6.121]$$

where $C = \frac{m^2 \mu_0 n_{\text{avg}}}{3k_B}$ is the *Curie constant*, and $T_c = \frac{\lambda m^2 n_{\text{avg}}}{3k_B}$ is the *critical temperature (Curie)* for paramagnetism. This is the *Curie–Weiss law*. In practice, this equation is found to be valid only for temperatures above the critical temperature T_c, where the susceptibility becomes infinite. Below T_c, the material enters a different ordered magnetic phase known as the *ferromagnetic phase*, which will be discussed in the next section.

6.2.4. *Ferromagnetism*

Ferromagnetic materials or media have the property of acquiring a strong magnetization under the influence of an external magnetic field and retaining this magnetization even after the field is removed. In this sense, they become permanent *magnets*. Their magnetization is much greater than that of paramagnetic materials, and as a result, they can significantly modify the external magnetic field.

Ferromagnetism is temperature dependent. It is observed that a ferromagnetic material loses its ferromagnetic properties when its temperature T exceeds a certain threshold T_c', known as the *Curie temperature of the ferromagnetic or the critical temperature of the ferromagnetic*. In other words, a phase transition occurs from an ordered ferromagnetic state to a disordered state, in which the material loses its spontaneous magnetization and behaves like a paramagnetic material. This transition is reversible. At temperatures well above T_c', the paramagnetic susceptibility of the material follows the Curie–Weiss equation:

$$(\chi_m)_{\text{para}} = \frac{C}{T - T_c} \qquad [6.122]$$

where T_c is the *Curie temperature for paramagnetism*, which is slightly higher than T_c'.

The interpretation of ferromagnetism is described by the Weiss model. According to this model, a ferromagnetic material is divided into regions called *Weiss domains*, each of which possesses its own magnetization, called *spontaneous magnetization*. This magnetization results from the collective alignment of all of the elementary magnetic moments within the domain, along a preferred direction known as the *easy axis of magnetization*. In the absence of an external field, the magnetizations of the various domains are randomly oriented, leading to zero net macroscopic magnetization. The boundaries between domains, called *Bloch walls*, are transition zones, where the elementary magnetic moments gradually rotate from the orientation of one domain to that of a neighboring one, thus changing the direction of magnetization smoothly across the interface. The alignment of elementary moments in the same direction ($\uparrow\uparrow$) within a domain cannot be explained by classical physics. In fact, classical magnetic dipole interactions would instead favor antiparallel alignment (↑↓). It was Heisenberg, within the framework of quantum mechanics, who proposed an explanation for this phenomenon by introducing a coupling between neighboring spins through atomic wave functions – an interaction known as the exchange interaction. This interaction is described by a coupling energy between two neighboring spins (↑↓) of the form:

$$W_e = -2J_e \vec{\mathbf{S}}_i \cdot \vec{\mathbf{S}}_j \qquad [6.123]$$

where J_e is a quantity called the *exchange integral*, which depends not only on the shape of the atomic wavefunctions but, more importantly, on their overlap. Moreover, J_e also depends on the interatomic distance. In particular, this distance must be sufficiently small for J_e to be positive, resulting in a coupling that favors spin alignment. It should also be noted that the formation of multiple domains with different magnetization directions helps to reduce the magnetic field generated by the magnetization itself and, consequently, the associated positive magnetostatic energy until a stable equilibrium is reached. The smaller the domains, the more stable the equilibrium tends to be.

When an external magnetic field is applied to a ferromagnetic material, the domains whose magnetization is oriented close to the direction of the field are energetically favored and tend to grow at the expense of the others, whose volume decreases. In other words, this leads to a displacement of the domain walls that separate adjacent regions. As long as the applied field remains weak, the movement of domain walls is reversible, provided it is not too extensive and does not encounter structural obstacles within the material. In such a case, when the field is removed, the domain walls return to their original positions. However, if the magnetic field is sufficiently strong, the *favorably oriented domains* expand significantly, resulting in a non-zero net magnetization. In this regime, the motion of the domain wall becomes irreversible, generally due to imperfections or internal stresses in the material structure. Under intense magnetic fields, the number of domains decreases and the remaining ones become larger. Consequently, the magnetizations of the various domains tend to align with the direction of the applied field. In a permanent magnet, the domain walls are pinned, preventing the sample from returning to a demagnetized state. This domain wall pinning is the physical origin of magnetic *hysteresis*.

To illustrate the previous discussion, let us apply a magnetic field to a material that initially has no magnetization. We obtain a curve $M = M(H)$, known as the *initial magnetization curve*, whose general shape is shown in Figure 6.15(a), where the successive phases of the magnetization process are indicated. Phase 1 corresponds to reversible displacements of the walls of the Weiss domain that occur when the applied magnetic field H is still relatively weak. Note that the slope at the origin of the curve is not zero. In phase 2, magnetization increases rapidly with H, reflecting irreversible displacements of the domain walls due to structural pinning or other imperfections in the material. In phase 3, the domain walls disappear and the individual magnetic moments reorient themselves in the direction of the applied field. For large values of H, the magnetization curve approaches a horizontal asymptote – the magnetization tends toward a limiting value M_{sat}, known as the *saturation magnetization*. This saturation magnetization depends on temperature: it decreases with increasing material temperature and disappears at the ferromagnetic Curie temperature T_c'. Above this temperature, the material ceases to be ferromagnetic and becomes paramagnetic. The curve representing the variation of the magnetic field B as a function of H, that is, $B(H)$, is obtained from the curve $\mu_0 M(H)$ by adding the linear contribution $\mu_0 H$, as illustrated in Figure 6.15(b). For large H, this curve tends toward the asymptotic line described by equation $B = \mu_0 H + \mu_0 M_{\text{sat}}$.

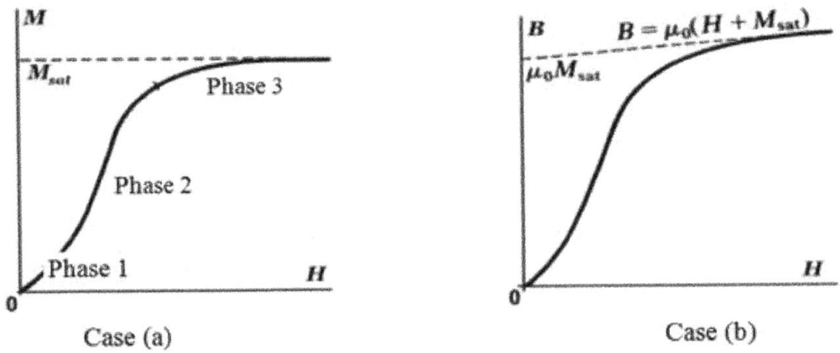

Figure 6.15. *Initial magnetization curves of a ferromagnetic material*

Let us suppose that an initially unmagnetized material is made to describe its first magnetization curve, up to a value H_M of the field H for which the magnetization is close to saturation magnetization. If we gradually decrease H, we see that the magnetization decreases but takes on higher values (for the same value of H) than it did on the first magnetization curve. In other words, the demagnetization curve is above the first magnetization curve, as shown in Figure 6.16(a). This behavior is referred to as *hysteresis*. This phenomenon results from the irreversibility of wall displacements between Weiss domains.

When H cancels out, the magnetization does not: it retains a value M_r, called *remanent magnetization*. This property of a retaining magnetization in the absence of an exciting field characterizes ferromagnetic materials. If we continue toward negative values of H, the magnetization cancels out for the value $H = -H_c$, known as the *coercive excitation of the material*. By further decreasing the values of H, M takes negative values. If H is varied in alternative ways between the two opposite values $+H_M$ and $-H_M$, the characteristic point describes indefinitely the two symmetrical curves 2 and 3 of Figure 6.16(b), forming a cycle called the material *hysteresis cycle*.

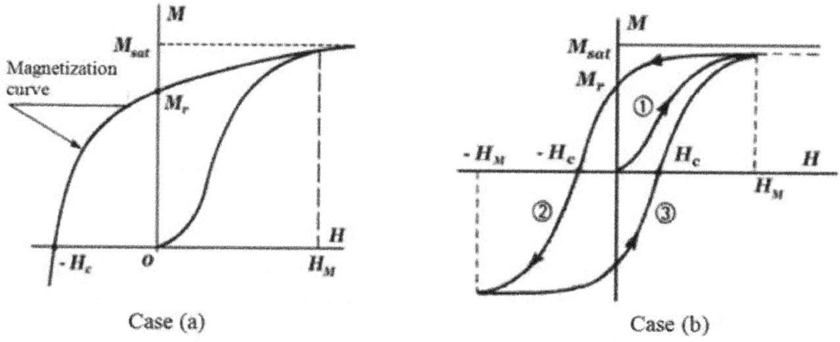

Figure 6.16. *Ferromagnetic materials exhibit hysteresis, where the magnetization saturates under high magnetic fields and retains a non-zero remanent magnetization M_r even when the applied field H is reduced to zero. A coercive field $-H_c$ is required to bring the magnetization back to zero*

6.3. Constitutive relations

6.3.1. *Linear magnetic medium*

We have already seen that for diamagnetic materials, the relationship between magnetization $\overrightarrow{\mathbf{M}}$ and the magnetic field $\overrightarrow{\mathbf{B}}$ is linear. Excluding situations involving intense magnetic fields and/or very low temperatures, this relation also remains approximately linear for paramagnetic materials. Similarly to dielectrics, we can then write, for diamagnetic and paramagnetic materials:

$$\overrightarrow{\mathbf{M}} = \frac{\chi_m}{\mu_0}\overrightarrow{\mathbf{B}} \qquad [6.124]$$

For anisotropic materials, the previous relation must be replaced by $M_i = \frac{(\chi_m)_{ij}}{\mu_0}B_j$, where $(\chi_m)_{ij}$ is a matrix 3×3. The quantity χ_m is a dimensionless parameter called the *magnetic susceptibility*. For both diamagnetic and paramagnetic materials, we observe that $|\chi_m| \ll 1$. We can therefore use the approximation $\overrightarrow{\mathbf{H}} = \frac{1}{\mu_0}\overrightarrow{\mathbf{B}} - \overrightarrow{\mathbf{M}} \approx \frac{1}{\mu_0}\overrightarrow{\mathbf{B}}$, which leads to:

$$\overrightarrow{\mathbf{M}} = \chi_m \overrightarrow{\mathbf{H}} \qquad [6.125]$$

Historically, magnetic susceptibility was introduced through relation [6.125] and not relation [6.124]. With this definition, we now have:

$$\overrightarrow{\mathbf{B}} = \mu_0\left(\overrightarrow{\mathbf{M}} + \overrightarrow{\mathbf{H}}\right) = \mu_0\left(1 + \chi_m\right)\overrightarrow{\mathbf{H}} = \mu\overrightarrow{\mathbf{H}} \qquad [6.126]$$

The quantity $\mu = \mu_0(1 + \chi_m)$ is the *magnetic permeability* of the material and the dimensionless ratio $\mu_r = \frac{\mu}{\mu_0} = 1 + \chi_m$ is its *relative permeability*. The relation between magnetization and magnetic field can then be expressed as:

$$\vec{\mathbf{M}} = \left(\frac{1}{\mu_0} - \frac{1}{\mu}\right)\vec{\mathbf{B}} \qquad [6.127]$$

For diamagnetic and paramagnetic materials, we have already seen that the susceptibility χ_m does not depend on the magnetic field $\vec{\mathbf{H}}$. For ferromagnetic materials, we use the same definitions for quantities χ_m, μ and μ_r. However, the above relation cannot be considered linear, since χ_m depends on H.

6.3.2. *Calculation of induced magnetization*

Determining the magnetization acquired by a material subjected to an external magnetic field is essentially a feedback problem. In fact, if the distribution of magnetization within the sample is known, the induced magnetic field $\vec{\mathbf{B}}\mathrm{m}$ can be calculated using the procedure already described in section 6.1. This induced field then adds to the external magnetic field $\vec{\mathbf{B}}\mathrm{ext}$ that initially causes magnetization. As a result, the total magnetic field is given by:

$$\vec{\mathbf{B}} = \vec{\mathbf{B}}_{\mathrm{ext}} + \vec{\mathbf{B}}_{\mathrm{m}} \qquad [6.128]$$

However, magnetization $\vec{\mathbf{M}}$ itself depends on the total field through the constitutive relation $\vec{\mathbf{M}} = \vec{\mathbf{M}}(H)$ of the material. Combining this relation with $\vec{\mathbf{B}} = \mu_0\left(\vec{\mathbf{M}} + \vec{\mathbf{H}}\right)$ should, in principle, allow us to successively determine $\vec{\mathbf{H}}$, then $\vec{\mathbf{M}}$, and finally $\vec{\mathbf{B}}$. For diamagnetic and paramagnetic media, the problem is simpler. As the magnetic susceptibility of these materials is very low, their induced magnetic field can be neglected compared to the external field. We then have $\vec{\mathbf{B}} = \mu_0\left(1 + \chi_m\right)\vec{\mathbf{H}} \approx \vec{\mathbf{B}}_{\mathrm{ext}} \approx \mu_0\vec{\mathbf{H}}$, and consequently, $\vec{\mathbf{M}} \approx \frac{\chi_m}{\mu_0}\vec{\mathbf{B}}_{\mathrm{ext}}$. In contrast, for ferromagnetic materials, solving the coupled equations can be more complicated. To address this, we may begin by making a plausible assumption regarding the orientation of the magnetization and then verify its consistency with the set of coupled field equations. Below are a few examples (each time, we will, as far as possible, introduce a different calculation method – these methods are equivalent; it is just the technique that changes):

EXAMPLE 6.4. (Ferromagnetic sphere in a uniform magnetic field (I)).– Let us consider a homogeneous and isotropic ferromagnetic sphere placed in a uniform external magnetic field $\vec{\mathbf{B}}_0 = B_0\vec{\mathbf{k}}$. We assume that the magnetization acquired by the sphere is uniform and aligned with the external field: $\vec{\mathbf{M}} = M\vec{\mathbf{k}}$, where M is a constant. As shown in example 6.3, this magnetization produces an induced magnetic field within the sphere equal to $\frac{2}{3}\mu_0\overrightarrow{\mathbf{M}}$. It follows that the total magnetic field inside the sphere is also uniform $\overrightarrow{\mathbf{B}} = B\overrightarrow{\mathbf{k}}$, with $B = B_0 + \frac{2}{3}\mu_0 M$, which gives the magnetic field strength $\overrightarrow{\mathbf{H}} = H\overrightarrow{\mathbf{k}}$ with $H = \frac{B_0}{\mu_0} - \frac{M}{3}$. To express M as a function of B_0, we solve the equation

$$M = M(H) = 3\left(\frac{B_0}{\mu_0} - H\right) \quad [6.129]$$

One of two cases may arise:

1) If the constitutive relation $M(H)$ is known, the solution can, for example, be obtained graphically. The assumption of a uniform magnetization is therefore compatible with the coupled equations, which can then be satisfied at every point within the sphere.

2) If the constitutive relation is linear (i.e. with a susceptibility χ_m independent of H), that is, $M = \chi_m H$, the solution is given by:

$$H = \frac{3B_0}{\mu_0(3+\chi_m)} \quad [6.130]$$

and consequently:

$$\left\{\begin{array}{l} M = \dfrac{3\chi_m B_0}{\mu_0(3+\chi_m)} = \dfrac{3(\mu_r-1)}{\mu_r+2}\dfrac{B_0}{\mu_0} \\ B_{\text{int}} = B_0 + \dfrac{2}{3}\mu_0 M = \dfrac{3\mu_r}{\mu_r+2}B_0 \\ H_{\text{int}} = B_0 + \dfrac{2}{3}\mu_0 M = \dfrac{3}{\mu_r+2}\dfrac{B_0}{\mu_0} \end{array}\right. \Longrightarrow \left\{\begin{array}{l} \overrightarrow{\mathbf{M}} = \dfrac{3\chi_m \overrightarrow{\mathbf{B}}_0}{\mu_0(3+\chi_m)} = \dfrac{3(\mu_r-1)}{\mu_r+2}\dfrac{\overrightarrow{\mathbf{B}}_0}{\mu_0} \\ \overrightarrow{\mathbf{B}}_{\text{int}} = \overrightarrow{\mathbf{B}}_0 + \dfrac{2}{3}\mu_0\overrightarrow{\mathbf{M}} = \dfrac{3\mu_r}{\mu_r+2}\overrightarrow{\mathbf{B}}_0 \\ \overrightarrow{\mathbf{H}}_{\text{int}} = \dfrac{3}{\mu_r+2}\dfrac{\overrightarrow{\mathbf{B}}_0}{\mu_0} \end{array}\right.$$

[6.131]

We can now also calculate the magnetic field induced outside the sphere. As previously discussed, this field, which adds to the external field $B_0\overrightarrow{\mathbf{k}}$, is equivalent to the field produced by a point magnetic dipole located at the center of the sphere:

$$\begin{aligned} \overrightarrow{\mathbf{B}}_{\text{ext}} &= B_0\overrightarrow{\mathbf{k}} + \frac{\mu_r-1}{\mu_r+2}\frac{R^3}{r^3}B_0\left(2\cos\theta\,\overrightarrow{\mathbf{e}}_r + \sin\theta\,\overrightarrow{\mathbf{e}}_\theta\right) \\ &= \left[1 + 2\frac{\mu_r-1}{\mu_r+2}\frac{R^3}{r^3}\right]B_0\cos\theta\,\overrightarrow{\mathbf{e}}_r - \left[1 - \frac{\mu_r-1}{\mu_r+2}\frac{R^3}{r^3}\right]B_0\sin\theta\,\overrightarrow{\mathbf{e}}_\theta \\ \overrightarrow{\mathbf{H}}_{\text{ext}} &= \left[1 + 2\frac{\mu_r-1}{\mu_r+2}\frac{R^3}{r^3}\right]\frac{B_0}{\mu_0}\cos\theta\,\overrightarrow{\mathbf{e}}_r - \left[1 - \frac{\mu_r-1}{\mu_r+2}\frac{R^3}{r^3}\right]\frac{B_0}{\mu_0}\sin\theta\,\overrightarrow{\mathbf{e}}_\theta \end{aligned} \quad [6.132]$$

We observe that $H_{\theta,\text{ext}} - H_{\theta,\text{int}} = 0$, which indicates that no current is observed at the surface of the sphere.

EXAMPLE 6.5. (Ferromagnetic sphere in a uniform magnetic field (II)).– There exists an alternative, simpler method to solve this problem, based on theoretical calculations without making any initial assumptions. Since there is no free current, we have, on the one hand, $\vec{\nabla} \times \vec{\mathbf{H}} = 0$, inside and outside the sphere, which implies that $\vec{\mathbf{H}} = -\vec{\nabla} V_m$, where V_m is a scalar magnetic potential to be determined. On the other hand, equation $\vec{\nabla} \cdot \vec{\mathbf{B}} = 0$ leads to $0 = \vec{\nabla} \cdot \left(\mu \vec{\mathbf{H}}\right) = \vec{\nabla} \cdot \left(-\mu \vec{\nabla} V_m\right) = -\mu \Delta V_m$, since μ is constant both inside and outside the sphere. It follows that the magnetic potential satisfies the Laplace equation in each region: $\Delta V_m = 0$. We can therefore use the same techniques developed for the electric field by trying a scalar potential of the form:

$$V_m(r,\theta) = \begin{cases} Ar\cos\theta & r < R \\ \left(Dr + \dfrac{C}{r^2}\right)\cos\theta & r > R \end{cases} \quad [6.133]$$

Quite simply speaking, we should use the entire series $\sum_{l=0}^{\infty} A_l r^l P_l(\cos\theta)$ for $r < R$ and $\sum_{l=0}^{\infty} \left(C_l r^l + \frac{D_l}{r^{l+1}}\right) P_l(\cos\theta)$ for $r > R$, but this leads to the same results. The fields $\vec{\mathbf{H}}$ and $\vec{\mathbf{B}}$ associated with V_m are:

$$\vec{\mathbf{H}} = -\vec{\nabla} V_m = \begin{cases} -A\left(\cos\theta\vec{\mathbf{e}}_r - \sin\theta\vec{\mathbf{e}}_\theta\right) = -A\vec{\mathbf{k}} & r < R \\ -\left(D - \dfrac{2C}{r^3}\right)\cos\theta\vec{\mathbf{e}}_r + \left(D + \dfrac{C}{r^3}\right)\sin\theta\vec{\mathbf{e}}_\theta & r > R \end{cases}$$

$$\vec{\mathbf{B}} = \begin{cases} -A\mu\left(\cos\theta\vec{\mathbf{e}}_r - \sin\theta\vec{\mathbf{e}}_\theta\right) = -A\mu\vec{\mathbf{k}} & r < R \\ -\mu_0\left(D - \dfrac{2C}{r^3}\right)\cos\theta\vec{\mathbf{e}}_r + \mu_0\left(D + \dfrac{C}{r^3}\right)\sin\theta\vec{\mathbf{e}}_\theta & r > R \end{cases} \quad [6.134]$$

Far from the sphere (as $r \longrightarrow \infty$), the field $\vec{\mathbf{B}}$ must approach the uniform applied field, that is, $\vec{\mathbf{B}}(r \longrightarrow \infty) = B_0\vec{\mathbf{k}} = -\mu_0 D\left(\cos\theta\vec{\mathbf{e}}_r - \sin\theta\vec{\mathbf{e}}_\theta\right)$, which leads to:

$$D = -\frac{B_0}{\mu_0} \quad [6.135]$$

The magnetic field inside the sphere is uniform and aligned with the direction of the applied field. The external field is equivalent to the one created by a magnetic dipole with moment $m = 4\pi C$, placed at the center of the sphere. The other parameters, A and C, are determined using the boundary conditions on the surface of the sphere, that is, at $r = R$. In fact, the tangential components of $\vec{\mathbf{H}}$ and the normal components of $\vec{\mathbf{B}}$ are continuous throughout the surface of the sphere. They therefore obey the conditions $H_\theta(R^+) = H_\theta(R^-)$ and $B_r(R^+) = B_r(R^-)$. This leads to the following system of equations $A = D + \frac{C}{R^3}$ and $A\mu_0 = \mu\left(D - \frac{2C}{R^3}\right)$. Solving this system gives:

$$C = \frac{\mu_0 - \mu}{\mu_0 + 2\mu} R^3 \frac{B_0}{\mu_0}$$
$$A = -\frac{3\mu}{\mu_0 + 2\mu}\frac{B_0}{\mu_0} \quad [6.136]$$

By substituting equations [6.135] and [6.136] into [6.134], we recover the same results as previously obtained, namely, [6.131] and [6.132].

EXAMPLE 6.6. (Perfectly conducting sphere in a uniform magnetic field $B_0 \overrightarrow{\mathbf{k}}$).– If the sphere with magnetic permeability μ is replaced by a perfectly conducting sphere of the same radius R, the same reasoning still applies, except that this time, the magnetic field $\overrightarrow{\mathbf{H}}$ is zero inside the sphere, that is, $A = 0$. The normal component of the magnetic field $\overrightarrow{\mathbf{B}}$ is then also zero just outside the sphere, that is, $B_r(R^+) = 0$, which leads to:

$$C = \frac{B_0 R^3}{\mu_0} \quad [6.137]$$

As a result, we obtain:

$$\begin{aligned} \overrightarrow{\mathbf{H}}_{\text{ext}} &= \frac{B_0}{\mu_0}\left[\left(1 - \frac{R^3}{r^3}\right)\cos\theta\,\overrightarrow{\mathbf{e}}_r - \left(1 + \frac{2R^3}{r^3}\right)\sin\theta\,\overrightarrow{\mathbf{e}}_\theta\right] \\ \overrightarrow{\mathbf{B}}_{\text{ext}} &= B_0\left[\left(1 - \frac{R^3}{r^3}\right)\cos\theta\,\overrightarrow{\mathbf{e}}_r - \left(1 + \frac{2R^3}{r^3}\right)\sin\theta\,\overrightarrow{\mathbf{e}}_\theta\right] \end{aligned} \quad [6.138]$$

The surface current at the interface $r = R$ is obtained from the discontinuity of the tangential component of the field $\overrightarrow{\mathbf{H}}$:

$$K = H_\theta(R+0) = -\frac{3}{2}\frac{B_0}{\mu_0}\sin\theta \quad [6.139]$$

EXAMPLE 6.7. (Sphere with magnetization $\overrightarrow{\mathbf{M}}_2 = M_2 \overrightarrow{\mathbf{k}}$ surrounded by a medium with magnetization $\overrightarrow{\mathbf{M}}_1 = M_1 \overrightarrow{\mathbf{k}}$).– Let us consider a sphere with a magnetization $\overrightarrow{\mathbf{M}}_2 = M_2 \overrightarrow{\mathbf{k}}$, immersed in a surrounding medium that also has a uniform magnetization $\overrightarrow{\mathbf{M}}_1 = M_1 \overrightarrow{\mathbf{k}}$. If a uniform magnetic field $B_0 \overrightarrow{\mathbf{k}}$ is applied, the same calculations remain valid, except that the following boundary conditions must be satisfied: (1) $H_\theta(R^+) = H_\theta(R^-) \Longrightarrow A = D + \frac{C}{R^3}$; (2) $B_r(R^+) = B_r(R^-) \Longrightarrow H_r(R^+) + M_1\cos\theta = H_r(R^-) + M_2\cos\theta \Longrightarrow -\left(D - \frac{2C}{R^3}\right) + M_1 = -A + M_2$. From these, we obtain the solutions:

$$\begin{aligned} A &= -\frac{B_0}{\mu_0} + \frac{1}{3}(M_2 - M_1) \\ C &= \frac{R^3}{3}(M_2 - M_1) \end{aligned} \quad [6.140]$$

and consequently:

$$\vec{\mathbf{H}} = \begin{cases} \left[\frac{B_0}{\mu_0} + \frac{1}{3}(M_1 - M_2)\right]\left(\cos\theta\,\vec{\mathbf{e}}_r - \sin\theta\,\vec{\mathbf{e}}_\theta\right) & r < R \\ \left[\frac{B_0}{\mu_0} + \frac{2R^3}{3r^3}(M_2 - M_1)\right]\cos\theta\,\vec{\mathbf{e}}_r & \\ -\left[\frac{B_0}{\mu_0} + \frac{R^3}{3r^3}(M_1 - M_2)\right]\sin\theta\,\vec{\mathbf{e}}_\theta & r > R \end{cases} \quad [6.141]$$

EXAMPLE 6.8. (Infinitely long magnetically permeable cylinder in a uniform magnetic field).– Consider an infinite, homogeneous, isotropic ferromagnetic cylinder of radius R placed in an external magnetic field $\vec{\mathbf{B}}_0$.

1) *Magnetic field* $\vec{\mathbf{B}}_0$ *along the axis of the cylinder*: suppose that the external magnetic field is along the axis of the cylinder, that is, $\vec{\mathbf{B}}_0 = B_0\,\vec{\mathbf{k}}$. The induced magnetization within the cylinder is also uniform and in the same direction, that is, $\vec{\mathbf{M}} = M\,\vec{\mathbf{k}}$. According to example 6.1, this magnetization produces a uniform magnetic field inside the cylinder, $\mu_0 M\,\vec{\mathbf{k}}$. Therefore, the total magnetic field within the cylinder is also uniform and is equal to $\vec{\mathbf{B}} = (B_0 + \mu_0 M)\,\vec{\mathbf{k}}$, which leads to $\vec{\mathbf{H}} = \frac{\vec{\mathbf{B}}}{\mu_0} - \vec{\mathbf{M}} = \frac{\vec{\mathbf{B}}_0}{\mu_0}$. If the constitutive relation is linear, we conclude that $\vec{\mathbf{M}} = \chi_m\vec{\mathbf{H}} = \frac{\chi_m}{\mu_0}\vec{\mathbf{B}}_0 = \frac{\mu_r - 1}{\mu_0}\vec{\mathbf{B}}_0$ and $\vec{\mathbf{B}} = \mu_r\vec{\mathbf{B}}_0$.

2) *Magnetic field* $\vec{\mathbf{B}}_0$ *perpendicular to the axis of the cylinder*: if the magnetic field $\vec{\mathbf{B}}_0$ is perpendicular to the axis of the cylinder, that is, $\vec{\mathbf{B}}_0 = B_0\,\vec{\mathbf{i}}$, then $\vec{\mathbf{M}} = M\,\vec{\mathbf{i}}$. According to example 6.2, a magnetic field $\frac{\mu_0 M}{2}\vec{\mathbf{i}}$ will superpose to the field $\vec{\mathbf{B}}_0$, so $\vec{\mathbf{B}} = \left(B_0 + \frac{\mu_0 M}{2}\right)\vec{\mathbf{i}}$, which gives $\vec{\mathbf{H}} = \frac{\vec{\mathbf{B}}}{\mu_0} - \vec{\mathbf{M}} = \left(\frac{B_0}{\mu_0} - \frac{M}{2}\right)\vec{\mathbf{i}}$. Therefore, $\vec{\mathbf{M}} = 2\left(\frac{\vec{\mathbf{B}}_0}{\mu_0} - \vec{\mathbf{H}}\right)$. If $\vec{\mathbf{M}} = \chi_m\vec{\mathbf{H}}$, then $\vec{\mathbf{H}} = \frac{2}{\chi_m + 2}\frac{\vec{\mathbf{B}}_0}{\mu_0} = \frac{2}{\mu_r + 1}\frac{\vec{\mathbf{B}}_0}{\mu_0}$, which gives $\vec{\mathbf{M}} = \frac{2\chi_m}{\chi_m + 2}\frac{\vec{\mathbf{B}}_0}{\mu_0} = \frac{2(\mu_r - 1)}{\mu_r + 1}\frac{\vec{\mathbf{B}}_0}{\mu_0}$. Finally, $\vec{\mathbf{B}} = \frac{2(\chi_m + 1)}{\chi_m + 2}\vec{\mathbf{B}}_0 = \frac{2\mu_r}{\mu_r + 1}\vec{\mathbf{B}}_0$.

EXAMPLE 6.9. (Magnetically permeable spherical shell in a uniform field).– A spherical shell, with an inner radius R_1, an outer radius R_2 and a permeability μ_1, is placed in an initially uniform magnetic field $\vec{\mathbf{B}}_0$, oriented in the direction of the Oz axis that passes through the center O of the shell, as shown in Figure 6.17. Let us calculate the magnetization $\vec{\mathbf{M}}$ inside the shell. In the absence of free current, we have $\vec{\nabla} \times \vec{\mathbf{H}} = 0 \Longrightarrow \vec{\mathbf{H}}$ is the gradient of a scalar potential V_m, called the *magnetic scalar potential*, so $\vec{\mathbf{H}} = -\vec{\nabla}V_m$. In addition, we have $\vec{\nabla}\cdot\vec{\mathbf{B}} = \vec{\nabla}\cdot\left(\mu\vec{\mathbf{H}}\right) = \vec{\nabla}\cdot\left(-\mu\vec{\nabla}V_m\right) = 0$. Since ν is constant in regions I, II and III, we conclude that $\Delta V_m = 0$ in each region. In terms of V_m, the magnetic field is expressed as $\overrightarrow{\mathbf{B}} = -\mu\vec{\nabla}V_m$. At a sufficiently large distance from the shell, the magnetic field $\overrightarrow{\mathbf{B}}$ must be equal to $B_0\overrightarrow{\mathbf{k}} \Longrightarrow V_m(r \longrightarrow \infty) = -\frac{B_0}{\mu_0}z = -\frac{B_0}{\mu_0}r\cos\theta$. In spherical coordinates, the potential V_m is given by (in the three regions):

$$V_m(r) = \begin{cases} -\dfrac{B_0}{\mu_0}r\cos\theta + \displaystyle\sum_{l=0}^{\infty}\dfrac{A_l}{r^{l+1}}P_l(\cos\theta) & R_2 < r < \infty \\ \displaystyle\sum_{l=0}^{\infty}\left(C_l r^l + \dfrac{D_l}{r^{l+1}}\right)P_l(\cos\theta) & R_1 < r < R_2 \\ \displaystyle\sum_{l=0}^{\infty}F_l r^l P_l(\cos\theta) & 0 < r < R_1 \end{cases} \quad [6.142]$$

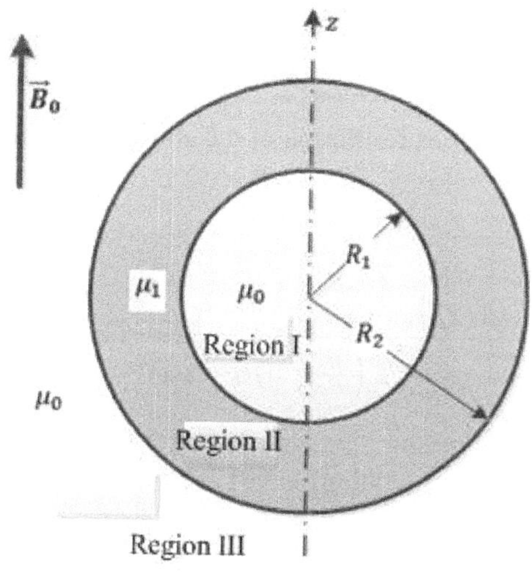

Figure 6.17. *Spherical shell with permeability* μ_1

At the spherical boundaries of radii R_1 and R_2, the following conditions must hold:

$$\begin{cases} B_r(R_1^+) = B_r(R_1^-) \\ B_r(R_2^+) = B_r(R_2^-) \\ H_\theta(R_1^+) = H_\theta(R_1^-) \\ H_\theta(R_2^+) = H_\theta(R_2^-) \end{cases} \Longrightarrow \begin{cases} \mu_1 \left.\dfrac{\partial V_m}{\partial r}\right|_{R_1^+} = \mu_0 \left.\dfrac{\partial V_m}{\partial r}\right|_{R_1^-} \\ \mu_0 \left.\dfrac{\partial V_m}{\partial r}\right|_{R_2^+} = \mu_1 \left.\dfrac{\partial V_m}{\partial r}\right|_{R_2^-} \\ \left.\dfrac{\partial V_m}{\partial \theta}\right|_{R_1^+} = \left.\dfrac{\partial V_m}{\partial \theta}\right|_{R_1^-} \\ \left.\dfrac{\partial V_m}{\partial \theta}\right|_{R_2^+} = \left.\dfrac{\partial V_m}{\partial \theta}\right|_{R_2^-} \end{cases} \quad [6.143]$$

Substituting [6.142] into [6.143], we get:

$$\begin{aligned}
&B_0\cos\theta + \sum_{l=0}^{\infty}\frac{A_l\mu_0\,(l+1)}{R_2^{l+2}}P_l(\cos\theta) = \mu_1\sum_{l=0}^{\infty}\left(\frac{D_l\,(l+1)}{R_2^{l+2}} - lC_l b^{l-1}\right)\\
&\qquad\times P_l(\cos\theta)\\
&\mu_1\sum_{l=0}^{\infty}\left(lC_lR_1^{l-1} - \frac{D_l\,(l+1)}{R_1^{l+2}}\right)P_l(\cos\theta) = \mu_0\sum_{l=0}^{\infty} lF_lR_1^{l-1}P_l(\cos\theta)\\
&\frac{B_0}{\mu_0}R_2\sin\theta + \sum_{l=0}^{\infty}\frac{A_l}{R_2^{l+1}}\frac{dP_l(\cos\theta)}{d\theta} = \sum_{l=0}^{\infty}\left(C_lR_2^l + \frac{D_l}{R_2^{l+1}}\right)\frac{dP_l(\cos\theta)}{d\theta}\\
&\sum_{l=0}^{\infty}\left(C_lR_1^l + \frac{D_l}{R_1^{l+1}}\right)\frac{dP_l(\cos\theta)}{d\theta} = \sum_{l=0}^{\infty}F_lR_1^l\frac{dP_l(\cos\theta)}{d\theta}
\end{aligned} \qquad [6.144]$$

It is clear that $A_0 = C_0 = D_0 = F_0 = 0$. For $l = 1$, we find $-B_0 - \frac{2A_1\mu_0}{R_2^3} = \mu_1\left(C_1 - \frac{2D_1}{R_2^3}\right)$, $\mu_1\left(C_1 - \frac{2D_1}{R_1^3}\right) = \mu_0F_1$, $\frac{B_0R_2}{\mu_0} - \frac{A_1}{R_2^2} = -C_1R_2 - \frac{D_1}{R_2^2}$ and $C_1R_1 + \frac{D_1}{R_1^2} = F_1R_1$. The simultaneous resolution of the above equations yields:

$$\begin{aligned}
A_1 &= \frac{(2\mu_1+\mu_0)(\mu_1-\mu_0)}{(2\mu_1+\mu_0)(2\mu_0+\mu_1) - 2\left(\frac{R_1}{R_2}\right)^3(\mu_1-\mu_0)^2}\left(R_2^3 - R_1^3\right)\frac{B_0}{\mu_0}\\
C_1 &= \frac{-3(2\mu_1+\mu_0)}{(2\mu_1+\mu_0)(2\mu_0+\mu_1) - 2\left(\frac{R_1}{R_2}\right)^3(\mu_1-\mu_0)^2}B_0\\
D_1 &= \frac{-3(\mu_1-\mu_0)R_1^3}{(2\mu_1+\mu_0)(2\mu_0+\mu_1) - 2\left(\frac{R_1}{R_2}\right)^3(\mu_1-\mu_0)^2}B_0\\
F_1 &= \frac{-9\mu_1}{(2\mu_1+\mu_0)(2\mu_0+\mu_1) - 2\left(\frac{R_1}{R_2}\right)^3(\mu_1-\mu_0)^2}B_0
\end{aligned} \qquad [6.145]$$

All other parameters are zero, that is, $A_l = C_l = D_l = F_l = 0, \forall l \geq 2$. Thus, we obtain:

$$V_m = \begin{cases} -\frac{B_0}{\mu_0}r\cos\theta + \\ \frac{(2\mu_1+\mu_0)(\mu_1-\mu_0)\left(R_2^3-R_1^3\right)}{\Xi}\frac{B_0}{\mu_0}\frac{\cos\theta}{r^2} & r > R_2\\ -3\left[(2\mu_1+\mu_0)r + \frac{(\mu_1-\mu_0)R_1^3}{r^2}\right]\frac{B_0\cos\theta}{\Xi} & R_1 < r < R_2\\ -\frac{9\mu_1}{\Xi}B_0r\cos\theta & r < R_1 \end{cases} \qquad [6.146]$$

where $\Xi = (2\mu_1+\mu_0)(2\mu_0+\mu_1) - 2\left(\frac{R_1}{R_2}\right)^3(\mu_1-\mu_0)^2$.

The induced fields inside the shell are:

$$\overrightarrow{\mathbf{H}}_{II} = 3\frac{(2\mu_1 + \mu_0)\overrightarrow{\mathbf{B}}_0 - (\mu_1 - \mu_0)\left(\frac{R_1}{r}\right)^3 B_0\left(2\cos\theta\overrightarrow{\mathbf{e}}_r + \sin\theta\overrightarrow{\mathbf{e}}_\theta\right)}{(2\mu_1 + \mu_0)(2\mu_0 + \mu_1) - 2\left(\frac{R_1}{R_2}\right)^3(\mu_1 - \mu_0)^2}$$
$$\overrightarrow{\mathbf{B}}_{II} = 3\mu_1\frac{(2\mu_1 + \mu_0)\overrightarrow{\mathbf{B}}_0 - (\mu_1 - \mu_0)\left(\frac{R_1}{r}\right)^3 B_0\left(2\cos\theta\overrightarrow{\mathbf{e}}_r + \sin\theta\overrightarrow{\mathbf{e}}_\theta\right)}{(2\mu_1 + \mu_0)(2\mu_0 + \mu_1) - 2\left(\frac{R_1}{R_2}\right)^3(\mu_1 - \mu_0)^2} \quad [6.147]$$

It follows that the magnetization within the shell is equal to:

$$\overrightarrow{\mathbf{M}}_{II} = 3\left(\frac{\mu_1}{\mu_0} - 1\right)\frac{(2\mu_1 + \mu_0)\overrightarrow{\mathbf{B}}_0 - (\mu_1 - \mu_0)\left(\frac{R_1}{r}\right)^3 B_0\left(2\cos\theta\overrightarrow{\mathbf{e}}_r + \sin\theta\overrightarrow{\mathbf{e}}_\theta\right)}{(2\mu_1 + \mu_0)(2\mu_0 + \mu_1) - 2\left(\frac{R_1}{R_2}\right)^3(\mu_1 - \mu_0)^2} \quad [6.148]$$

It is clear that $\overrightarrow{\mathbf{M}}$ is not collinear with $\overrightarrow{\mathbf{B}}_0$. In regions I and III, we obtain:

$$\overrightarrow{\mathbf{H}}_I = \frac{9\mu_1}{(2\mu_1 + \mu_0)(2\mu_0 + \mu_1) - 2\left(\frac{R_1}{R_2}\right)^3(\mu_1 - \mu_0)^2}\overrightarrow{\mathbf{B}}_0 \quad [6.149]$$

$$\overrightarrow{\mathbf{B}}_I = \frac{9\mu_0\mu_1}{(2\mu_1 + \mu_0)(2\mu_0 + \mu_1) - 2\left(\frac{R_1}{R_2}\right)^3(\mu_1 - \mu_0)^2}\overrightarrow{\mathbf{B}}_0$$

$$\overrightarrow{\mathbf{H}}_{III} = \frac{\overrightarrow{\mathbf{B}}_0}{\mu_0} + \frac{(2\mu_1 + \mu_0)(\mu_1 - \mu_0)\frac{\left(R_2^3 - R_1^3\right)}{r^3}\frac{B_0}{\mu_0}\left(2\cos\theta\overrightarrow{\mathbf{e}}_r + \sin\theta\overrightarrow{\mathbf{e}}_\theta\right)}{(2\mu_1 + \mu_0)(2\mu_0 + \mu_1) - 2\left(\frac{R_1}{R_2}\right)^3(\mu_1 - \mu_0)^2}$$

$$\overrightarrow{\mathbf{B}}_{III} = \overrightarrow{\mathbf{B}}_0 + \frac{(2\mu_1 + \mu_0)(\mu_1 - \mu_0)\frac{\left(R_2^3 - R_1^3\right)}{r^3}B_0\left(2\cos\theta\overrightarrow{\mathbf{e}}_r + \sin\theta\overrightarrow{\mathbf{e}}_\theta\right)}{(2\mu_1 + \mu_0)(2\mu_0 + \mu_1) - 2\left(\frac{R_1}{R_2}\right)^3(\mu_1 - \mu_0)^2}$$

In particular, if $R_1 = 0$, we recover the results from [6.131]. If $\mu_1 \gg \mu_0$, we deduce that:

$$\overrightarrow{\mathbf{H}}_I = \frac{9}{2\mu_1}\frac{1}{1 - \left(\frac{R_1}{R_2}\right)^3}\overrightarrow{\mathbf{B}}_0$$

$$\overrightarrow{\mathbf{B}}_I = \frac{9\mu_0}{2\mu_1}\frac{1}{1 - \left(\frac{R_1}{R_2}\right)^3}\overrightarrow{\mathbf{B}}_0 \ll \overrightarrow{\mathbf{B}}_0$$

$$\vec{\mathbf{H}}_{II} = \frac{3}{\mu_1}\frac{\vec{\mathbf{B}}_0 - \frac{1}{2}\left(\frac{R_1}{r}\right)^3 B_0\left(2\cos\theta\,\vec{\mathbf{e}}_r + \sin\theta\,\vec{\mathbf{e}}_\theta\right)}{1-\left(\frac{R_1}{R_2}\right)^3}$$

$$\vec{\mathbf{B}}_{II} = 3\frac{\vec{\mathbf{B}}_0 - \frac{1}{2}\left(\frac{R_1}{r}\right)^3 B_0\left(2\cos\theta\,\vec{\mathbf{e}}_r + \sin\theta\,\vec{\mathbf{e}}_\theta\right)}{1-\left(\frac{R_1}{R_2}\right)^3}$$

$$\vec{\mathbf{H}}_{III} = \frac{\vec{\mathbf{B}}_0}{\mu_0} + \frac{R_2^3}{r^3}\frac{B_0}{\mu_0}\left(2\cos\theta\,\vec{\mathbf{e}}_r + \sin\theta\,\vec{\mathbf{e}}_\theta\right)$$

$$\vec{\mathbf{B}}_{III} = \vec{\mathbf{B}}_0 + \frac{R_2^3}{r^3}B_0\left(2\cos\theta\,\vec{\mathbf{e}}_r + \sin\theta\,\vec{\mathbf{e}}_\theta\right)$$

$$\vec{\mathbf{M}}_{II} = \frac{3}{2\mu_0}\frac{2\vec{\mathbf{B}}_0 - \left(\frac{R_1}{r}\right)^3 B_0\left(2\cos\theta\,\vec{\mathbf{e}}_r + \sin\theta\,\vec{\mathbf{e}}_\theta\right)}{1-\left(\frac{R_1}{R_2}\right)^3} \qquad [6.150]$$

EXAMPLE 6.10. (Infinitely long magnetically permeable cylinder carrying a linear electric current).– Let us consider an infinitely long, homogeneous, isotropic cylinder of permeability μ and radius R, carrying a linear current I along its axis, as shown in Figure 6.18. We aim to compute the magnetic field $\vec{\mathbf{B}}$, the magnetic field intensity $\vec{\mathbf{H}}$, the magnetization $\vec{\mathbf{M}}$ and the magnetization current throughout the cylinder. We choose a circular path of radius r around the current. Using the integral form of the modified Ampère law, we get:

$$\oint \vec{\mathbf{H}}\cdot d\vec{\mathbf{l}} = H_\varphi 2\pi r = I \implies H_\varphi = \frac{I}{2\pi r} \qquad [6.151]$$

and therefore:

$$B_\varphi = \begin{cases} \mu H_\varphi = \dfrac{\mu I}{2\pi r} & 0<r<R \\ \mu_0 H_\varphi = \dfrac{\mu_0 I}{2\pi r} & r>R \end{cases} \qquad [6.152]$$

Magnetization is calculated using the relation:

$$\vec{\mathbf{M}} = \frac{\vec{\mathbf{B}}}{\mu_0} - \vec{\mathbf{H}} \implies M_\varphi = \begin{cases} \left(\dfrac{\mu}{\mu_0}-1\right)H_\varphi = \dfrac{\mu-\mu_0}{\mu_0}\dfrac{I}{2\pi r} & 0<r<R \\ 0 & r>R \end{cases} \qquad [6.153]$$

This magnetization gives rise to a volume magnetization current inside the cylinder (i.e. for $0<r<R$), given by:

$$\vec{\mathbf{J}}_m = \vec{\nabla}\times\vec{\mathbf{M}} = -\frac{\partial M_\varphi}{\partial z}\vec{\mathbf{e}}_r + \frac{1}{r}\frac{\partial(rM_\varphi)}{\partial r}\vec{\mathbf{k}} = 0 \qquad [6.154]$$

At the surface of the cylinder, the surface magnetization current density is:

$$\vec{\mathbf{K}}_m = \vec{\mathbf{M}} \times \vec{\mathbf{n}} = \vec{\mathbf{M}} \times \vec{\mathbf{e}}_r = -\frac{\mu - \mu_0}{\mu_0}\frac{I}{2\pi R}\vec{\mathbf{k}} \quad [6.155]$$

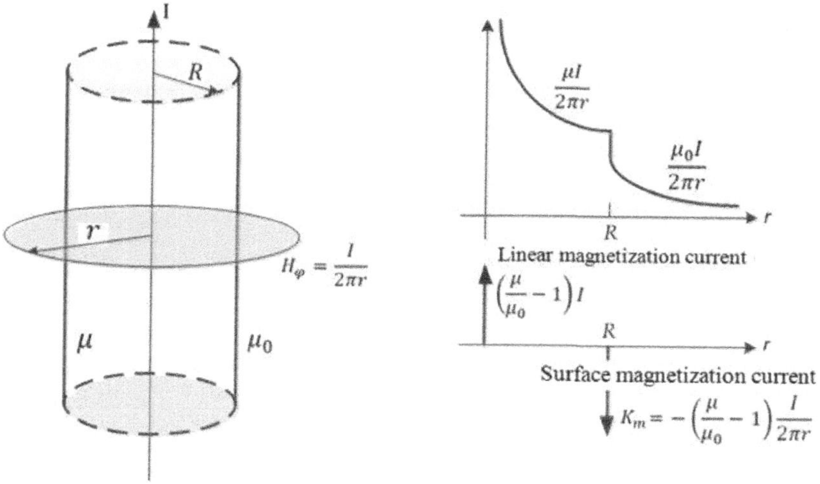

Figure 6.18. *Magnetically permeable cylinder carrying a conduction current I*

Clearly, the volume magnetization current density is zero. However, there exists a line magnetization current at $r = 0$, which can be found using the expression $\int_S \left(\vec{\nabla} \times \vec{\mathbf{M}}\right) \cdot d\vec{\mathbf{S}} = \oint_L \vec{\mathbf{M}} \cdot d\vec{\mathbf{l}} = \int_S \vec{\mathbf{J}}_m \cdot d\vec{\mathbf{S}}$. Choosing a path around the cylinder axis with $r < R$, we obtain:

$$M_\varphi 2\pi r = \frac{\mu - \mu_0}{\mu_0} I = I_m \quad [6.156]$$

where I_m is the equivalent magnetization line current.

6.3.3. *Calculation of $\vec{\mathrm{H}}$ and $\vec{\mathrm{B}}$ in a magnetized medium (given $\vec{\mathrm{M}}$)*

Equation $\vec{\nabla} \cdot \vec{\mathbf{B}} = 0$ implies that, in principle, magnetic charges or monopoles do not exist in nature. Nevertheless, the formulation involving magnetic monopoles is of great theoretical importance. From the definition of the magnetic field $\vec{\mathbf{H}}$, namely, $\vec{\mathbf{B}} = \mu_0 \left(\vec{\mathbf{H}} + \vec{\mathbf{M}}\right)$ and condition $\vec{\nabla} \cdot \vec{\mathbf{B}} = 0$, we can write:

$$\vec{\nabla} \cdot \vec{\mathbf{H}} = -\vec{\nabla} \cdot \vec{\mathbf{M}} \quad [6.157]$$

Note that in a region of space where the current density $\overrightarrow{\mathbf{J}}$ is zero ($\overrightarrow{\mathbf{J}} = 0$), we also have $\overrightarrow{\nabla} \times \overrightarrow{\mathbf{H}} = 0$, which implies that the magnetic field $\overrightarrow{\mathbf{H}}$ can be expressed as the gradient of a scalar magnetic potential V_m, that is, $\overrightarrow{\mathbf{H}} = -\overrightarrow{\nabla} V_m$. We therefore conclude that, in the absence of free current density, the scalar potential satisfies the equation:

$$\Delta V_m = -\overrightarrow{\nabla} \cdot \overrightarrow{\mathbf{M}} \qquad [6.158]$$

It is clear that $\overrightarrow{\nabla} \cdot \overrightarrow{\mathbf{M}}$ acts as a source for the scalar magnetic potential V_m, in a way analogous to how $\overrightarrow{\nabla} \cdot \overrightarrow{\mathbf{P}}$ acts as a source for the electric potential. According to the well-known results for the Poisson equation, the solution for V_m is given by:

$$V_m\left(\overrightarrow{\mathbf{r}}\right) = -\frac{1}{4\pi} \int_{\tau'} \frac{\overrightarrow{\nabla}' \cdot \overrightarrow{\mathbf{M}}\left(\overrightarrow{\mathbf{r}}'\right)}{\left|\overrightarrow{\mathbf{r}} - \overrightarrow{\mathbf{r}}'\right|} d\tau' \qquad [6.159]$$

where τ' is the volume containing all of the magnetization $\overrightarrow{\mathbf{M}}$. Using the identity $\overrightarrow{\nabla}' \cdot \left(\frac{\overrightarrow{\mathbf{M}}(\overrightarrow{\mathbf{r}}')}{|\overrightarrow{\mathbf{r}}-\overrightarrow{\mathbf{r}}'|}\right) = \frac{\overrightarrow{\nabla}' \cdot \overrightarrow{\mathbf{M}}(\overrightarrow{\mathbf{r}}')}{|\overrightarrow{\mathbf{r}}-\overrightarrow{\mathbf{r}}'|} + \overrightarrow{\mathbf{M}}\left(\overrightarrow{\mathbf{r}}'\right) \cdot \overrightarrow{\nabla}' \left(\frac{1}{|\overrightarrow{\mathbf{r}}-\overrightarrow{\mathbf{r}}'|}\right) = \frac{\overrightarrow{\nabla}' \cdot \overrightarrow{\mathbf{M}}(\overrightarrow{\mathbf{r}}')}{|\overrightarrow{\mathbf{r}}-\overrightarrow{\mathbf{r}}'|} + \frac{\overrightarrow{\mathbf{M}}(\overrightarrow{\mathbf{r}}') \cdot (\overrightarrow{\mathbf{r}}-\overrightarrow{\mathbf{r}}')}{|\overrightarrow{\mathbf{r}}-\overrightarrow{\mathbf{r}}'|^3}$ and the divergence theorem, formula [6.159] reduces to:

$$\begin{aligned} V_m &= -\frac{1}{4\pi} \int_{\tau'} \overrightarrow{\nabla}' \cdot \frac{\overrightarrow{\mathbf{M}}\left(\overrightarrow{\mathbf{r}}'\right)}{\left|\overrightarrow{\mathbf{r}} - \overrightarrow{\mathbf{r}}'\right|} d\tau' + \frac{1}{4\pi} \int_{\tau'} \frac{\overrightarrow{\mathbf{M}}\left(\overrightarrow{\mathbf{r}}'\right) \cdot \left(\overrightarrow{\mathbf{r}} - \overrightarrow{\mathbf{r}}'\right)}{\left|\overrightarrow{\mathbf{r}} - \overrightarrow{\mathbf{r}}'\right|} d\tau' \\ &= -\frac{1}{4\pi} \oint_{S'} \frac{\overrightarrow{\mathbf{M}}\left(\overrightarrow{\mathbf{r}}'\right) \cdot d\overrightarrow{\mathbf{S}}'}{\left|\overrightarrow{\mathbf{r}} - \overrightarrow{\mathbf{r}}'\right|} + \frac{1}{4\pi} \int_{\tau'} \frac{\overrightarrow{\mathbf{M}}\left(\overrightarrow{\mathbf{r}}'\right) \cdot \left(\overrightarrow{\mathbf{r}} - \overrightarrow{\mathbf{r}}'\right)}{\left|\overrightarrow{\mathbf{r}} - \overrightarrow{\mathbf{r}}'\right|} d\tau' \end{aligned} \qquad [6.160]$$

where S' is a closed surface that contains all magnetization. The integral $\oint_{S'} \frac{\overrightarrow{\mathbf{M}}(\overrightarrow{\mathbf{r}}') \cdot d\overrightarrow{\mathbf{S}}'}{|\overrightarrow{\mathbf{r}}-\overrightarrow{\mathbf{r}}'|}$ is obviously zero. Consequently, the previous expression simplifies to:

$$V_m\left(\overrightarrow{\mathbf{r}}\right) = \frac{1}{4\pi} \int_{\tau'} \frac{\overrightarrow{\mathbf{M}}\left(\overrightarrow{\mathbf{r}}'\right) \cdot \left(\overrightarrow{\mathbf{r}} - \overrightarrow{\mathbf{r}}'\right)}{\left|\overrightarrow{\mathbf{r}} - \overrightarrow{\mathbf{r}}'\right|} d\tau' \qquad [6.161]$$

and can also be written as:

$$\begin{aligned} V_m\left(\overrightarrow{\mathbf{r}}\right) &= -\frac{1}{4\pi} \int_{\tau'} \overrightarrow{\mathbf{M}}\left(\overrightarrow{\mathbf{r}}'\right) \cdot \overrightarrow{\nabla} \left(\frac{1}{\left|\overrightarrow{\mathbf{r}} - \overrightarrow{\mathbf{r}}'\right|}\right) d\tau' \\ &= -\frac{1}{4\pi} \overrightarrow{\nabla} \cdot \int_{\tau'} \frac{\overrightarrow{\mathbf{M}}\left(\overrightarrow{\mathbf{r}}'\right)}{\left|\overrightarrow{\mathbf{r}} - \overrightarrow{\mathbf{r}}'\right|} d\tau' \end{aligned} \qquad [6.162]$$

EXAMPLE 6.11.– Let us compute the magnetic scalar potential of a sphere of radius R with a magnetization of the form $\overrightarrow{\mathbf{M}}\left(\overrightarrow{\mathbf{r}}\right) = M_0 z \overrightarrow{\mathbf{i}}$. According to [6.162], we obtain:

$$V_m\left(\vec{\mathbf{r}}\right) = -\frac{1}{4\pi}\vec{\nabla}\cdot\int_{\tau'}\frac{M_0 z'\vec{\mathbf{i}}}{\left|\vec{\mathbf{r}}-\vec{\mathbf{r}}'\right|}d\tau' = -\frac{M_0}{4\pi}\frac{\partial}{\partial x}\int_{\tau'}\frac{z'}{\left|\vec{\mathbf{r}}-\vec{\mathbf{r}}'\right|}d\tau' \qquad [6.163]$$

This integral is evaluated using spherical harmonics through the identity $\frac{1}{\left|\vec{\mathbf{r}}-\vec{\mathbf{r}}'\right|} = \sum_{l=0}^{\infty}\sum_{m=-l}^{l}\frac{4\pi}{2l+1}Y_l^m(\theta,\varphi)\overline{Y_l^m(\theta',\varphi')}\frac{r_<^l}{r_>^{l+1}}$ and noting that $z' = \sqrt{\frac{4\pi}{3}}r'Y_1^0(\theta',\varphi')$. Substituting these expressions into [6.163] and using the orthogonality relations of the spherical harmonics, we obtain:

$$\begin{aligned}V_m\left(\vec{\mathbf{r}}\right) &= -\frac{M_0}{4\pi}\frac{\partial}{\partial x}\left[\sum_{l=0}^{\infty}\sum_{m=-l}^{l}\frac{4\pi}{2l+1}Y_l^m(\theta,\varphi)\int_0^{\pi}\int_0^{2\pi}\sqrt{\frac{4\pi}{3}}Y_1^0(\theta',\varphi')\right.\\ &\left.\times\overline{Y_l^m(\theta',\varphi')}d\Omega'\int_0^R\frac{r_<^l}{r_>^{l+1}}r'^3dr'\right] = -\frac{M_0}{4\pi}\frac{\partial}{\partial x}\left[\sum_{l=0}^{\infty}\sum_{m=-l}^{l}\frac{4\pi}{2l+1}Y_l^m(\theta,\varphi)\right.\\ &\left.\times\sqrt{\frac{4\pi}{3}}\delta_{l1}\delta_{m0}\int_0^R\frac{r_<^l}{r_>^{l+1}}r'^3dr'\right] = -\frac{M_0}{3}\frac{\partial}{\partial x}\left[\sqrt{\frac{4\pi}{3}}Y_1^0(\theta,\varphi)\int_0^R\frac{r_<}{r_>^2}r'^3dr'\right]\\ &= -\frac{M_0}{3}\frac{\partial}{\partial x}\left[\frac{z}{r}\int_0^R\frac{r_<}{r_>^2}r'^3dr'\right]\end{aligned} \qquad [6.164]$$

There are therefore two possible cases:

– for $r > R \Longrightarrow r_< = r'$ and $r_> = r$:

$$V_m(r>R) = -\frac{M_0 z}{3}\frac{\partial}{\partial x}\int_0^R\frac{r'^4}{r^3}dr' = -\frac{M_0 z}{3}\frac{\partial}{\partial x}\left(\frac{R^5}{5r^3}\right) = \frac{M_0 zxR^5}{5r^5} \qquad [6.165]$$

$$\begin{cases}\vec{\mathbf{H}}(r>R) = M_0R^5\left[\frac{\left(x^2-\frac{1}{5}r^2\right)z}{r^7}\vec{\mathbf{i}} + \frac{xyz}{r^7}\vec{\mathbf{j}} + \frac{\left(z^2-\frac{1}{5}r^2\right)x}{r^7}\vec{\mathbf{k}}\right]\\ \vec{\mathbf{B}}(r>R) = \mu_0M_0R^5\left[\frac{\left(x^2-\frac{1}{5}r^2\right)z}{r^7}\vec{\mathbf{i}} + \frac{xyz}{r^7}\vec{\mathbf{j}} + \frac{\left(z^2-\frac{1}{5}r^2\right)x}{r^7}\vec{\mathbf{k}}\right]\end{cases} \qquad [6.166]$$

– for $r < R$:

$$\begin{aligned}V_m(r<R) &= -\frac{M_0 z}{3}\frac{\partial}{\partial x}\left[\int_0^r\frac{r'^4}{r^3}dr' + \int_r^R r'dr'\right]\\ &= -\frac{M_0}{3}\frac{\partial}{\partial x}\left[-\frac{3r^2}{10}+\frac{R^2}{2}\right] = \frac{M_0 zx}{5}\end{aligned} \qquad [6.167]$$

$$\begin{cases}\vec{\mathbf{H}}(r<R) = -\frac{M_0}{5}\left(z\vec{\mathbf{i}} + x\vec{\mathbf{k}}\right)\\ \vec{\mathbf{B}}(r<R) = \frac{\mu_0M_0}{5}\left(4z\vec{\mathbf{i}} - x\vec{\mathbf{k}}\right)\end{cases} \qquad [6.168]$$

■

If $\overrightarrow{\mathbf{M}}$ is uniform except for discontinuities at the material boundaries, a slightly different approach is used to calculate V_m. By analogy with dielectrics, it is expected that the ends of magnetic moments located at the boundaries behave like surface magnetic charges (monopoles). At the boundaries, where M is discontinuous, the divergence becomes singular and formula [6.159] cannot be integrated in this form. To handle the discontinuity of $\overrightarrow{\mathbf{M}}$ at the material boundaries, the integration volume τ must be divided into two regions, as illustrated in Figure 6.19: (1) τ_1' ends at a small infinitesimal distance ϵ within the material; (2) τ_2' starts within the material and extends beyond the outer boundary of the original volume τ'. In this case, we write:

$$V_m\left(\overrightarrow{\mathbf{r}}\right) = \int_{\tau_1'} \frac{-\overrightarrow{\nabla}'\cdot\overrightarrow{\mathbf{M}}\left(\overrightarrow{\mathbf{r}}'\right)}{4\pi\left|\overrightarrow{\mathbf{r}}-\overrightarrow{\mathbf{r}}'\right|}d\tau_1' + \int_{\tau_2'} \frac{-\overrightarrow{\nabla}'\cdot\overrightarrow{\mathbf{M}}\left(\overrightarrow{\mathbf{r}}'\right)}{4\pi\left|\overrightarrow{\mathbf{r}}-\overrightarrow{\mathbf{r}}'\right|}d\tau_2' \qquad [6.169]$$

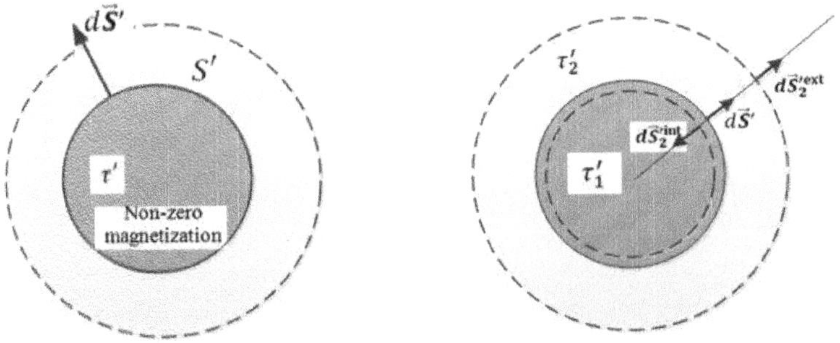

Figure 6.19. *The integration volume τ' is divided into a region τ_1' with uniform magnetization, and an outer region τ_2' containing only the boundaries of the magnetized material*

Using the same reasoning to establish formula [6.160], we obtain:

$$\int_{\tau_2'} \frac{\overrightarrow{\nabla}'\cdot\overrightarrow{\mathbf{M}}\left(\overrightarrow{\mathbf{r}}'\right)}{4\pi\left|\overrightarrow{\mathbf{r}}-\overrightarrow{\mathbf{r}}'\right|}d\tau_2' = \oint_{S_2^{\text{ext}}} \frac{\overrightarrow{\mathbf{M}}\cdot d\overrightarrow{\mathbf{S}}'^{\text{ext}}_2}{4\pi\left|\overrightarrow{\mathbf{r}}-\overrightarrow{\mathbf{r}}'\right|} + \oint_{S_2^{\text{int}}} \frac{\overrightarrow{\mathbf{M}}\cdot d\overrightarrow{\mathbf{S}}'^{\text{ext}}_2}{4\pi\left|\overrightarrow{\mathbf{r}}-\overrightarrow{\mathbf{r}}'\right|} - \int_{\tau_2'} \frac{\overrightarrow{\mathbf{M}}\cdot\left(\overrightarrow{\mathbf{r}}-\overrightarrow{\mathbf{r}}'\right)}{4\pi\left|\overrightarrow{\mathbf{r}}-\overrightarrow{\mathbf{r}}'\right|^3}d\tau_2' \qquad [6.170]$$

The integral over S_2^{ext} disappears because M is zero on this surface. As ϵ tends to zero, the volume integral over τ_2 also vanishes, and the surface S_2^{int} tends to the actual boundary S' of the magnetized material. It is important to note that the normal to the surface S_2^{int} and the normal to S' are in opposite directions, that is, $d\overrightarrow{\mathbf{S}}'^{\text{int}}_2 \longrightarrow d\overrightarrow{\mathbf{S}}'$. We then obtain:

$$V_m\left(\overrightarrow{\mathbf{r}}\right) = \int_{\tau_1'} \frac{-\overrightarrow{\nabla}'\cdot\overrightarrow{\mathbf{M}}\left(\overrightarrow{\mathbf{r}}'\right)}{4\pi\left|\overrightarrow{\mathbf{r}}-\overrightarrow{\mathbf{r}}'\right|}d\tau_1' + \oint_{S'} \frac{\overrightarrow{\mathbf{M}}\cdot d\overrightarrow{\mathbf{S}}'}{4\pi\left|\overrightarrow{\mathbf{r}}-\overrightarrow{\mathbf{r}}'\right|} \qquad [6.171]$$

where the surface integral extends over all boundaries and the volume integral excludes the boundaries. The surface integral exactly represents the contribution to the scalar potential of a magnetic charge $\overrightarrow{\mathbf{M}} \cdot \overrightarrow{n}$ on the surface of the material. To illustrate these results, let us give a few examples.

EXAMPLE 6.12. (A uniformly magnetized sphere of radius R, with magnetization $\overrightarrow{\mathbf{M}} = M_0 z \overrightarrow{\mathbf{i}}$).– Using [6.171] to calculate the scalar potential, we observe that the divergence of the magnetization within the sphere vanishes, and the term $\overrightarrow{\mathbf{i}} \cdot d\overrightarrow{\mathbf{S}}$ in the second integral reduces to $\sin\theta' \sin\varphi' R^2 d\Omega'$. It then follows that:

$$
\begin{aligned}
V_m\left(\overrightarrow{\mathbf{r}}\right) &= \frac{M_0}{4\pi} \oint \frac{z' \sin\theta' \sin\varphi' R^2 d\Omega'}{\left|\overrightarrow{\mathbf{r}} - \overrightarrow{\mathbf{r}}'\right|} \qquad [6.172] \\
&= M_0 R^3 \sum_{l=0}^{\infty} \sum_{m=-l}^{l} \frac{Y_l^m(\theta,\varphi)}{2l+1} \oint \overline{Y_l^m(\theta',\varphi')} \cos\theta' \sin\theta' \sin\varphi' R^2 \frac{r_<^l}{r_>^{l+1}} d\Omega'
\end{aligned}
$$

Since $\cos\theta' \sin\theta' \sin\varphi' = \sqrt{\frac{2\pi}{15}} \left(Y_2^{-1}(\theta,\varphi) - Y_2^1(\theta,\varphi)\right)$, we obtain:

$$
\begin{aligned}
V_m\left(\overrightarrow{\mathbf{r}}\right) &= \frac{M_0 R^3 r_<^2 \sqrt{\frac{2\pi}{15}} \left[Y_2^{-1}(\theta,\varphi) - Y_2^1(\theta,\varphi)\right]}{5 r_>^3} \\
&= \frac{M_0 R^3 \cos\theta \sin\theta \sin\varphi r_<^2}{5 r_>^3} = \frac{M_0 R^3 z x r_<^2}{5 r^2 r_>^3}
\end{aligned} \qquad [6.173]
$$

Two cases can be considered: (1) $r < R \Longrightarrow r_> = R \Longrightarrow V_m(r < R) = \frac{M_0 z x}{5}$; (2) $r > R \Longrightarrow r_< = R \Longrightarrow V_m(r > R) = \frac{M_0 z x R^5}{5 r^5}$. These results are similar to those previously obtained.

EXAMPLE 6.13. (A uniformly magnetized cylinder of length L and radius R, with magnetization aligned along its axis).– The uniform magnetization inside the cylinder has zero divergence and therefore does not contribute to the scalar magnetic potential. The only contribution comes from the surface integrals at the ends of the cylinder. For simplicity, we fix the origin at the center of the cylinder, with the z-axis aligned along its axis, as shown in Figure 6.20. Using equation [6.171], we obtain:

$$
\begin{aligned}
V_m(0,0,z) &= \frac{1}{4\pi} \int_{\text{rightf.}} \frac{M dS}{\sqrt{\left(z - \frac{L}{2}\right)^2 + r^2}} + \frac{1}{4\pi} \int_{\text{leftf.}} \frac{-M dS}{\sqrt{\left(z + \frac{L}{2}\right)^2 + r^2}}, \\
&= \frac{M}{2} \int_0^R \frac{r dr}{\sqrt{\left(z - \frac{L}{2}\right)^2 + r^2}} - \frac{M}{2} \int_0^R \frac{r dr}{\sqrt{\left(z + \frac{L}{2}\right)^2 + r^2}}, \\
&= -\frac{M}{2} \left[\sqrt{\left(z + \frac{L}{2}\right)^2 + r^2} - \sqrt{\left(z - \frac{L}{2}\right)^2 + r^2} - 2z\right] \qquad [6.174]
\end{aligned}
$$

and, therefore:

$$\begin{aligned}\vec{\mathbf{H}} &= -\vec{\nabla} V_m = -\frac{\vec{\mathbf{M}}}{2}\left[\frac{z+\frac{L}{2}}{\sqrt{\left(z+\frac{L}{2}\right)^2+r^2}}+\frac{\frac{L}{2}-z}{\sqrt{\left(z-\frac{L}{2}\right)^2+r^2}}-2\right] \\ \vec{\mathbf{B}} &= \mu_0\left(\vec{\mathbf{H}}+\vec{\mathbf{M}}\right) = \frac{\mu_0\vec{\mathbf{M}}}{2}\left[\frac{z+\frac{L}{2}}{\sqrt{\left(z+\frac{L}{2}\right)^2+r^2}}+\frac{\frac{L}{2}-z}{\sqrt{\left(z-\frac{L}{2}\right)^2+r^2}}\right]\end{aligned} \quad [6.175]$$

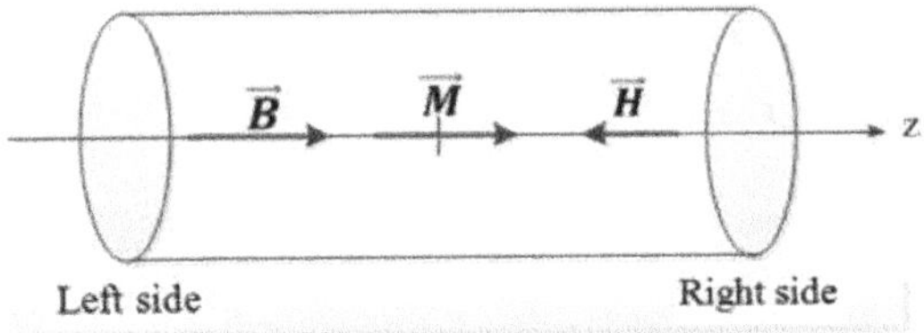

Figure 6.20. *A cylinder uniformly magnetized along its axis*

The field $\vec{\mathbf{H}}$ opposes magnetization, whereas the magnetic field $\vec{\mathbf{B}}$ is in the same direction as $\vec{\mathbf{M}}$.

6.4. Method of images

6.4.1. *Image of a magnetic moment in an infinitely permeable medium*

Although magnetic charges or monopoles do not exist in nature, they provide a powerful tool for dealing with problems involving induced magnetization. Suppose that magnetic charges q_m do exist and generate a scalar magnetic potential given by:

$$V_m\left(\vec{\mathbf{r}}\right) = \frac{q_m}{4\pi\left|\vec{\mathbf{r}}-\vec{\mathbf{r}}'\right|} \quad [6.176]$$

As with the method of images used for electric charges, we assume that the magnetic charge induced at the interface of a magnetically permeable material can be modeled by an image charge q'_m located inside the material, as seen from the outside. When observing the charge q_m from within the material, a reduced charge q''_m is perceived due to the screening effect of the medium. This situation closely resembles the phenomenon observed for electric charges. For simplicity, we assume that the plane Oxy lies at the interface of the material, with the charge q_m located a distance dd from the origin along the Oz axis, as shown in Figure 6.21. Inside the material, the potential is given by:

$$V_m(z<0) = \frac{q''_m}{4\pi R_1} = \frac{q''_m}{4\pi\sqrt{(d-z)^2+r^2}} \qquad [6.177]$$

Outside the material, it takes the form:

$$V_m(z>0) = \frac{q_m}{4\pi R_1} + \frac{q'_m}{4\pi R_2} = \frac{q_m}{4\pi\sqrt{(d-z)^2+r^2}} + \frac{q'_m}{4\pi\sqrt{(a+z)^2+r^2}} \qquad [6.178]$$

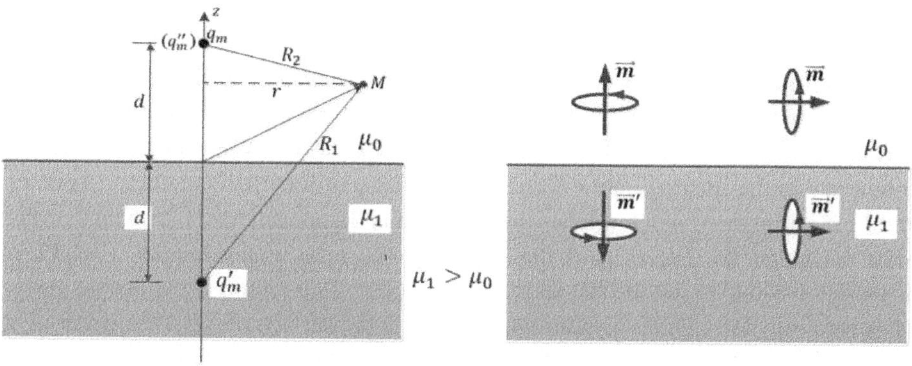

Figure 6.21. *Image of a magnetic charge q_m near a permeable medium with permeability μ_1*

At the interface $z=0$, the tangential component of the magnetic field strength $H_{\|} = -\frac{\partial V_m}{\partial r}$ and the normal component of the magnetic field $B_{\perp} = -\mu\frac{\partial V_m}{\partial z}$ must be continuous. This leads to the conditions $\frac{rq_m}{(d^2+r^2)^{3/2}} + \frac{rq'_m}{(d^2+r^2)^{3/2}} = \frac{rq''_m}{(d^2+r^2)^{3/2}} \Longrightarrow q_m + q'_m = q''_m$ and $-\mu_0\frac{dq_m}{(d^2+r^2)^{3/2}} + \mu_0\frac{dq'_m}{(d^2+r^2)^{3/2}} = -\mu_1\frac{dq''_m}{(d^2+r^2)^{3/2}} \Longrightarrow -\mu_0 q_m + \mu_0 q'_m = -\mu_1 q''_m$. Solving this system yields:

$$\begin{aligned} q'_m &= \frac{\mu_0-\mu_1}{\mu_0+\mu_1}q_m \\ q''_m &= \frac{2\mu_0}{\mu_0+\mu_1}q_m \end{aligned} \qquad [6.179]$$

Since magnetic monopoles do not exist, magnetic dipoles are modeled as pairs of opposite magnetic charges. Let q_m and $-q_m$ be located at positions $\vec{\mathbf{r}}_+$ and $\vec{\mathbf{r}}_-$, respectively, so that $q_m\left(\vec{\mathbf{r}}_+ - \vec{\mathbf{r}}_-\right) = \vec{\mathbf{m}}$. As with electric charges, q_m has an image q'_m located at $\vec{\mathbf{r}}_+ - 2\left(\vec{\mathbf{r}}_+\cdot\vec{\mathbf{n}}\right)\vec{\mathbf{n}}$, while $-q_m$ has an image $-q'_m$ located at $\vec{\mathbf{r}}_- - 2\left(\vec{\mathbf{r}}_-\cdot\vec{\mathbf{n}}\right)\vec{\mathbf{n}}$, where $\vec{\mathbf{n}}$ is the unit normal to the interface. As a result, the image of the magnetic moment is:

$$\begin{aligned}\vec{\mathrm{m}}' &= q'_m\left(\vec{\mathrm{r}}'_+ - \vec{\mathrm{r}}'_-\right) = q_m\left[\vec{\mathrm{r}}_+ - \vec{\mathrm{r}}_- - 2\vec{\mathrm{n}}\left(\vec{\mathrm{r}}_+ - \vec{\mathrm{r}}_-\right)\cdot\vec{\mathrm{n}}\right] \\ &= \frac{\mu_0 - \mu_1}{\mu_0 + \mu_1}\left[\vec{\mathrm{m}} - 2\vec{\mathrm{n}}\left(\vec{\mathrm{m}}\cdot\vec{\mathrm{n}}\right)\right]\end{aligned} \qquad [6.180]$$

Similarly, the magnetic moment as seen from inside the material is:

$$\vec{\mathrm{m}}'' = \frac{2\mu_0 - \mu_1}{\mu_0 + \mu_1}\vec{\mathrm{m}} \qquad [6.181]$$

6.4.2. *Image of a current line*

An infinite current line in the Oz direction is located at a distance d above a perfectly conducting or infinitely permeable plane, as shown in Figure 6.22. In both cases, the magnetic field $\vec{\mathbf{H}}$ inside the material must be zero, but the boundary conditions at the interface are different. In the case of a perfect conductor, both $\vec{\mathbf{B}}$ and $\vec{\mathbf{H}}$ must be zero, which implies that the normal component of $\vec{\mathbf{B}}$ (and thus of $\vec{\mathbf{H}}$) must vanish at the interface of the material. The tangential component of $\vec{\mathbf{H}}$ is discontinuous due to the surface current. In the case of an infinitely permeable plane, $\vec{\mathbf{H}}$ is zero, but $\vec{\mathbf{B}}$ is finite. No current can flow at the surface because the material is not conductive, which indicates that the tangential component of $\vec{\mathbf{H}}$ is continuous and therefore zero. As a result, the field $\vec{\mathbf{B}}$ is normal to the interface.

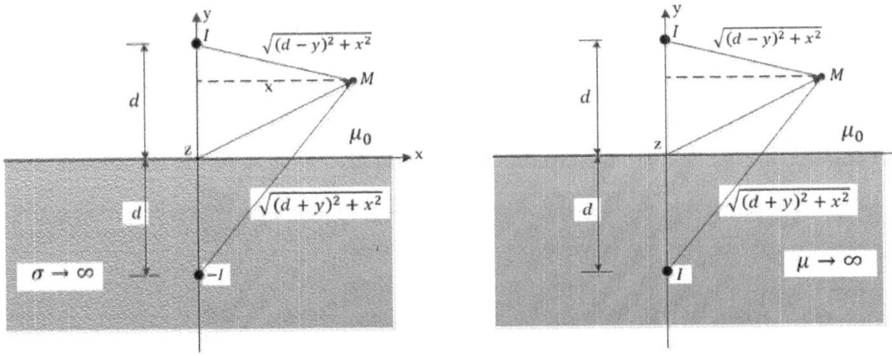

Figure 6.22. *Image of a current line I near (1) a perfect conductor and (2) a medium with infinite permeability*

Both sets of boundary conditions can be satisfied by placing an image current line I at $y = d$, flowing in the opposite direction for the conducting plane and in the same direction for the infinitely permeable plane. In both cases, the vector potential due to the two currents is found by superimposing the vector potentials of each line separately:

$$A_z = -\frac{\mu_0 I}{2\pi}\left[\ln\left[x^2+(y-d)^2\right]^{1/2} \mp \ln\left[x^2+(y+d)^2\right]^{1/2}\right]$$
$$= -\frac{\mu_0 I}{4\pi}\left[\ln\left[x^2+(y-d)^2\right] \mp \ln\left[x^2+(y+d)^2\right]\right] \qquad [6.182]$$

where the minus sign is used for the conducting plane and the plus sign for the infinitely permeable plane. The components A_x and A_y are zero. The resulting magnetic field $\overrightarrow{\mathbf{H}}$ is:

$$\overrightarrow{\mathbf{H}} = \frac{1}{\mu_0}\overrightarrow{\nabla}\times\overrightarrow{\mathbf{H}} = \frac{1}{\mu_0}\left(\frac{\partial A_z}{\partial y}\overrightarrow{\mathbf{i}} - \frac{\partial A_z}{\partial x}\overrightarrow{\mathbf{j}}\right)$$
$$= -\frac{I}{2\pi}\left[\frac{(y-d)\,\overrightarrow{\mathbf{i}} - x\,\overrightarrow{\mathbf{j}}}{x^2+(y-d)^2} \mp \frac{(y+d)\,\overrightarrow{\mathbf{i}} - x\,\overrightarrow{\mathbf{j}}}{x^2+(y+d)^2}\right] \qquad [6.183]$$

For the conducting plane, the surface current distribution is given by the discontinuity in the tangential component of $\overrightarrow{\mathbf{H}}$, $K_z = -H_x\,(y=0) = -\frac{Id}{\pi(d^2+x^2)}$. This current density leads to a total current:

$$I_T = \int_{-\infty}^{\infty} K_z dx = -\frac{Id}{\pi}\int_{-\infty}^{\infty}\frac{dx}{x^2+d^2} = -I \qquad [6.184]$$

which is exactly equal to the image current.

The force per unit length experienced by the current in each case is just due to the magnetic field from its image:

$$\overrightarrow{\mathbf{F}} = \pm\frac{\mu_0 I^2}{4\pi d}\overrightarrow{\mathbf{j}} \qquad [6.185]$$

This force is repulsive for the conducting plane and attractive for the magnetically permeable plane.

In particular, suppose that the half-space $y < 0$ is now occupied by a linear, homogeneous and isotropic magnetic medium with permeability μ. The half-space $y > 0$ remains empty, as shown in Figure 6.23. We aim to calculate the magnetic field at every point in space. The magnetically permeable half-space is mimicked by a current I' passing through the point $(0, -d, 0)$ and flowing in the same direction as I. From within the half-plane $y < 0$, we observe a current I'' instead of I, due to the screening effect. The total vector potential is given by:

$$A_z\,(y>0) = -\frac{\mu_0 I}{2\pi}\ln\left[x^2+(y-d)^2\right]^{1/2} - \frac{\mu_0 I'}{2\pi}\ln\left[x^2+(y+d)^2\right]^{1/2}$$
$$A_z\,(y<0) = -\frac{\mu_0 I''}{2\pi}\ln\left[x^2+(y-d)^2\right]^{1/2} \qquad [6.186]$$

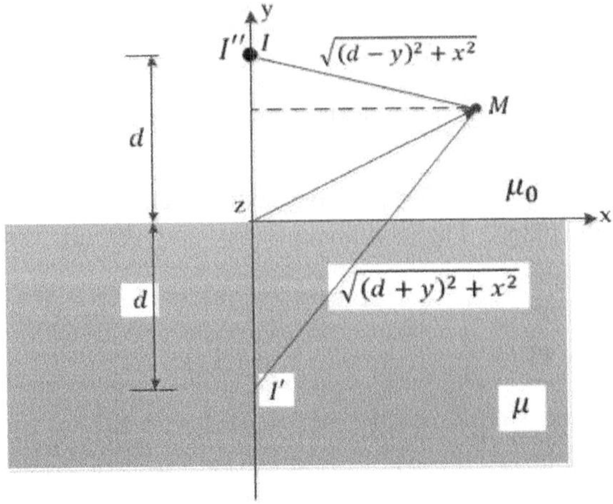

Figure 6.23. *Image of a current line located near a magnetically permeable plane with permeability* μ

The fields $\vec{\mathbf{B}}$ $\left(\vec{\mathbf{B}} = \vec{\nabla} \times \vec{\mathbf{A}}\right)$ and $\vec{\mathbf{H}}$, induced by the potential $\vec{\mathbf{A}}$, are:

$$\begin{aligned}
\vec{\mathbf{B}}\,(y>0) &= -\frac{\mu_0}{2\pi}\left[\frac{y-d}{x^2+(y-d)^2}I + \frac{y+d}{x^2+(y+d)^2}I'\right]\vec{\mathbf{i}} + \\
&\quad \frac{\mu_0}{2\pi}\left[\frac{x}{x^2+(y-d)^2}I + \frac{x}{x^2+(y+d)^2}I'\right]\vec{\mathbf{j}} \\
\vec{\mathbf{B}}\,(y<0) &= -\frac{\mu_0 I''}{2\pi}\left[\frac{y-d}{x^2+(y-d)^2}\right]\vec{\mathbf{i}} + \frac{\mu_0 I''}{2\pi}\left[\frac{x}{x^2+(y-d)^2}\right]\vec{\mathbf{j}} \\
\vec{\mathbf{H}}\,(y>0) &= -\frac{1}{2\pi}\left[\frac{y-d}{x^2+(y-d)^2}I + \frac{y+d}{x^2+(y+d)^2}I'\right]\vec{\mathbf{i}} + \\
&\quad \frac{1}{2\pi}\left[\frac{x}{x^2+(y-d)^2}I + \frac{x}{x^2+(y+d)^2}I'\right]\vec{\mathbf{j}} \\
\vec{\mathbf{B}}\,(y<0) &= -\frac{\mu_0 I''}{2\pi\mu}\left[\frac{y-d}{x^2+(y-d)^2}\right]\vec{\mathbf{i}} + \frac{\mu_0 I''}{2\pi\mu}\left[\frac{x}{x^2+(y-d)^2}\right]\vec{\mathbf{j}}
\end{aligned} \qquad [6.187]$$

At the interface $y = 0$, the components $\vec{\mathbf{B}}_{\perp}$ and $\vec{\mathbf{H}}_{\parallel}$ must be continuous, that is, we must have $\frac{\mu_0}{2\pi}\frac{x(I+I')}{x^2+d^2}I = \frac{\mu_0}{2\pi}\frac{xI''}{x^2+d^2}$ and $-\frac{1}{2\pi}\frac{d\,(I'-I)}{x^2+d^2} = \frac{\mu_0}{2\pi\mu}\frac{dI''}{x^2+d^2}$, which implies that $I + I' = I''$ and $-I + I' = -\frac{\mu_0}{\mu}I''$. The system of equations obtained allows us to calculate I and I'. Thus, we find:

$$\begin{aligned} I' &= \frac{\mu - \mu_0}{\mu + \mu_0} I \\ I'' &= \frac{2\mu}{\mu + \mu_0} I \end{aligned} \qquad [6.188]$$

It follows that:

$$\begin{aligned}
\vec{\mathbf{B}}\,(y>0) &= -\frac{\mu_0 I}{2\pi}\left[\frac{y-d}{x^2+(y-d)^2}+\frac{\mu-\mu_0}{\mu+\mu_0}\frac{y+d}{x^2+(y+d)^2}\right]\vec{\mathbf{i}}+ \\
&\quad \frac{\mu_0 I}{2\pi}\left[\frac{x}{x^2+(y-d)^2}+\frac{\mu-\mu_0}{\mu+\mu_0}\frac{x}{x^2+(y+d)^2}\right]\vec{\mathbf{j}} \\
\vec{\mathbf{B}}\,(y<0) &= -\frac{I}{\pi}\frac{\mu_0\mu}{\mu+\mu_0}\left[\frac{y-d}{x^2+(y-d)^2}\right]\vec{\mathbf{i}}+\frac{I}{\pi}\frac{\mu_0\mu}{\mu+\mu_0}\left[\frac{x}{x^2+(y-d)^2}\right]\vec{\mathbf{j}} \\
\vec{\mathbf{H}}\,(y>0) &= -\frac{I}{2\pi}\left[\frac{y-d}{x^2+(y-d)^2}+\frac{\mu-\mu_0}{\mu+\mu_0}\frac{y+d}{x^2+(y+d)^2}\right]\vec{\mathbf{i}}+ \\
&\quad \frac{I}{2\pi}\left[\frac{x}{x^2+(y-d)^2}+\frac{\mu-\mu_0}{\mu+\mu_0}\frac{x}{x^2+(y+d)^2}\right]\vec{\mathbf{j}} \\
\vec{\mathbf{H}}\,(y<0) &= -\frac{I}{\pi}\frac{\mu_0}{\mu+\mu_0}\left[\frac{y-d}{x^2+(y-d)^2}\right]\vec{\mathbf{i}}+\frac{I}{\pi}\frac{\mu_0}{\mu+\mu_0}\left[\frac{x}{x^2+(y-d)^2}\right]\vec{\mathbf{j}}
\end{aligned} \qquad [6.189]$$

If $\mu \longrightarrow \infty$, the previous results are recovered. The force per unit length experienced by the current line I is:

$$\vec{\mathbf{F}} = -\frac{\mu_0 I^2}{4\pi d}\frac{\mu-\mu_0}{\mu+\mu_0}\vec{\mathbf{j}} \qquad [6.190]$$

6.5. Magnetic field and force

A magnetizable medium in which a free current $\vec{\mathbf{J}}$ flows is placed in a magnetic field $\vec{\mathbf{B}}$ that depends on the position. In addition to the Lorentz force, the medium experiences forces due to the interactions of the field $\vec{\mathbf{B}}$ with its magnetic moments. For simplicity, imagine a rectangular magnetic moment in the plane Oxy, as shown in Figure 6.24.

On the sides AB, BC, CD and DA, the following forces are observed:

$$\begin{aligned}
\vec{\mathbf{f}}_{AB} &= I\Delta x\,\vec{\mathbf{i}} \times \left[B_x(x,y,z)\,\vec{\mathbf{i}} + B_y(x,y,z)\,\vec{\mathbf{j}} + B_z(x,y,z)\,\vec{\mathbf{k}}\right] \\
&= I\Delta x\left[-B_z(x,y,z)\,\vec{\mathbf{j}} + B_y(x,y,z)\,\vec{\mathbf{k}}\right] \\
\vec{\mathbf{f}}_{BC} &= I\Delta y\,\vec{\mathbf{j}} \times \left[B_x(x+\Delta x,y,z)\,\vec{\mathbf{i}} + B_y(x+\Delta x,y,z)\,\vec{\mathbf{j}}\right. \\
&\quad \left. + B_z(x+\Delta x,y,z)\,\vec{\mathbf{k}}\right] \\
&= I\Delta y\left[B_z(x+\Delta x,y,z)\,\vec{\mathbf{i}} - B_x(x+\Delta x,y,z)\,\vec{\mathbf{k}}\right] \\
\vec{\mathbf{f}}_{CD} &= -I\Delta x\,\vec{\mathbf{i}} \times \left[B_x(x+\Delta x,y+\Delta y,z)\,\vec{\mathbf{i}} + \right. \\
&\quad \left. + B_y(x+\Delta x,y+\Delta y,z)\,\vec{\mathbf{j}} + B_z(x+\Delta x,y+\Delta y,z)\,\vec{\mathbf{k}}\right] \\
&\simeq -I\Delta x\left[-B_z(x,y+\Delta y,z)\,\vec{\mathbf{j}} + B_y(x,y+\Delta y,z)\,\vec{\mathbf{k}}\right] \\
\vec{\mathbf{f}}_{DA} &= -I\Delta y\,\vec{\mathbf{j}} \times \left[B_x(x,y+\Delta y,z)\,\vec{\mathbf{i}} + B_y(x,y+\Delta y,z)\,\vec{\mathbf{j}}\right. \\
&\quad \left. + B_z(x,y+\Delta y,z)\,\vec{\mathbf{k}}\right] \\
&\simeq -I\Delta y\left[B_z(x,y,z)\,\vec{\mathbf{j}} - B_x(x,y,z)\,\vec{\mathbf{k}}\right]
\end{aligned} \qquad [6.191]$$

where terms of order $(\Delta x)^2$ and $(\Delta y)^2$ have been neglected. Therefore, the total force acting on the dipole is:

$$\begin{aligned}
\vec{\mathbf{f}} &= \vec{\mathbf{f}}_{AB} + \vec{\mathbf{f}}_{BC} + \vec{\mathbf{f}}_{CD} + \vec{\mathbf{f}}_{DA} \simeq I\left[\frac{B_z(x+\Delta x,y,z) - B_z(x,y,z)}{\Delta x}\vec{\mathbf{i}}\right. \\
&\quad + \frac{B_z(x,y+\Delta y,z) - B_z(x,y,z)}{\Delta y}\vec{\mathbf{j}} - \frac{B_x(x+\Delta x,y,z) - B_x(x,y,z)}{\Delta x}\vec{\mathbf{k}} \\
&\quad \left. - \frac{B_y(x,y+\Delta y,z) - B_y(x,y,z)}{\Delta y}\vec{\mathbf{k}}\right]\Delta x\Delta y
\end{aligned} \qquad [6.192]$$

When Δx and Δy tend to zero, the terms in brackets reduce to partial derivatives, while the term $I\Delta x\Delta y$ simply becomes the magnetic moment $\vec{\mathbf{m}} = I\Delta x\Delta y\,\vec{\mathbf{k}}$:

$$\vec{\mathbf{f}} = m_z\left[\frac{\partial B_z}{\partial x}\vec{\mathbf{i}} - \left(\frac{\partial B_x}{\partial x} + \frac{\partial B_y}{\partial y}\right)\vec{\mathbf{k}} + \frac{\partial B_z}{\partial y}\vec{\mathbf{j}}\right] \qquad [6.193]$$

Now, since $\vec{\nabla}\cdot\vec{\mathbf{B}} = 0$, it follows that $\frac{\partial B_z}{\partial z} = -\left(\frac{\partial B_x}{\partial x} + \frac{\partial B_y}{\partial y}\right)$, and using $\vec{\nabla}\times\vec{\mathbf{B}} = \mu_0\left(\vec{\mathbf{J}} + \vec{\nabla}\times\vec{\mathbf{M}}\right) = \mu_0\vec{\mathbf{J}}_T$, we obtain: (1) $\frac{\partial B_z}{\partial y} = \frac{\partial B_y}{\partial z} + \mu_0 J_{T,x}$; (2) $\frac{\partial B_z}{\partial x} = \frac{\partial B_x}{\partial z}\mu_0 J_{T,y}$. It follows that:

$$\begin{aligned}
\vec{\mathbf{f}} &= m_z\left[\frac{\partial B_x}{\partial z}\vec{\mathbf{i}} + \frac{\partial B_y}{\partial z}\vec{\mathbf{j}} + \frac{\partial B_z}{\partial z}\vec{\mathbf{k}} - \mu_0\left(J_{T,y}\vec{\mathbf{i}} - J_{T,x}\vec{\mathbf{j}}\right)\right] \\
&= \left(\vec{\mathbf{m}}\cdot\vec{\nabla}\right)\vec{\mathbf{B}} + \mu_0\vec{\mathbf{m}}\times\vec{\mathbf{J}}_T
\end{aligned} \qquad [6.194]$$

where $\vec{\mathbf{J}}_T$ represents the sum of the free current density and the magnetization current. If the medium contains n dipoles per unit volume, the force density acting on both the dipoles and the free current is:

$$\begin{aligned}\vec{\mathbf{F}} &= n\vec{\mathbf{f}} = \left(\vec{\mathbf{M}}\cdot\vec{\nabla}\right)\vec{\mathbf{B}} + \mu_0\vec{\mathbf{M}}\times\vec{\mathbf{J}}_T + \vec{\mathbf{J}}\times\vec{\mathbf{B}} \\ &= \mu_0\left(\vec{\mathbf{M}}\cdot\vec{\nabla}\right)\left(\vec{\mathbf{H}}+\vec{\mathbf{M}}\right) + \mu_0\vec{\mathbf{M}}\times\left(\vec{\mathbf{J}}+\vec{\nabla}\times\vec{\mathbf{M}}\right) \\ &\quad + \mu_0\vec{\mathbf{J}}\times\left(\vec{\mathbf{H}}+\vec{\mathbf{M}}\right)\end{aligned} \qquad [6.195]$$

Using the vector identity $\vec{\mathbf{M}}\times\left(\vec{\nabla}\times\vec{\mathbf{M}}\right) = -\left(\vec{\mathbf{M}}\cdot\vec{\nabla}\right)\vec{\mathbf{M}} + \frac{1}{2}\vec{\nabla}\left(\vec{\mathbf{M}}\cdot\vec{\mathbf{M}}\right)$, equation [6.195] simplifies to:

$$\vec{\mathbf{F}} = \mu_0\left(\vec{\mathbf{M}}\cdot\vec{\nabla}\right)\vec{\mathbf{H}} + \mu_0\vec{\mathbf{J}}\times\vec{\mathbf{H}} + \vec{\nabla}\left(\frac{\mu_0}{2}\vec{\mathbf{M}}\cdot\vec{\mathbf{M}}\right) \qquad [6.196]$$

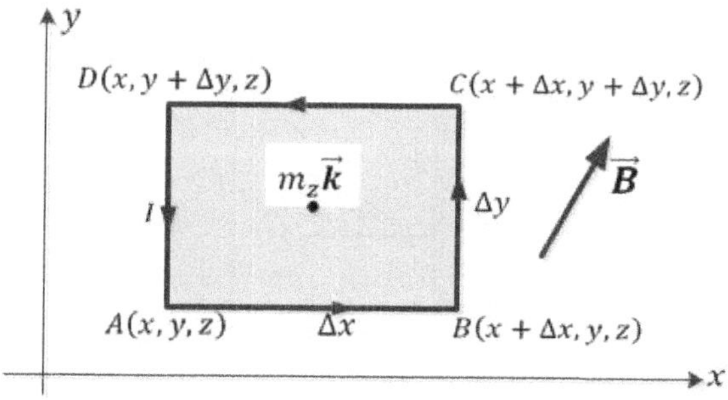

Figure 6.24. *Interaction of an infinitesimal rectangular magnetic dipole $ABCD$ with a position-dependent external magnetic field*

The total force experienced by the entire material is simply the integral of equation [6.196] over its volume:

$$\vec{\mathbf{F}}_T = \int_\tau \mu_0\left(\vec{\mathbf{M}}\cdot\vec{\nabla}\right)\vec{\mathbf{H}} + \mu_0\vec{\mathbf{J}}\times\vec{\mathbf{H}} + \vec{\nabla}\left(\frac{\mu_0}{2}\vec{\mathbf{M}}\cdot\vec{\mathbf{M}}\right)d\tau \qquad [6.197]$$

Applying the divergence theorem, the last term can be transformed into a surface integral:

$$\int_\tau \vec{\nabla}\left(\frac{\mu_0}{2}\vec{\mathbf{M}}\cdot\vec{\mathbf{M}}\right)d\tau = \oint_S \frac{\mu_0}{2}\vec{\mathbf{M}}\cdot\vec{\mathbf{M}}d\vec{\mathbf{S}} \qquad [6.198]$$

where S is a closed surface enclosing all of the magnetization. This integral vanishes, as $\overrightarrow{\mathbf{M}}$ is zero on S. For this reason, we retain only the first two terms in expression [6.197]:

$$\overrightarrow{\mathbf{F}} = \mu_0 \left(\overrightarrow{\mathbf{M}} \cdot \overrightarrow{\nabla}\right) \overrightarrow{\mathbf{H}} + \mu_0 \overrightarrow{\mathbf{J}} \times \overrightarrow{\mathbf{H}} \qquad [6.199]$$

It is the term $\mu_0 \left(\overrightarrow{\mathbf{M}} \cdot \overrightarrow{\nabla}\right) \overrightarrow{\mathbf{H}}$ that explains the attraction of a magnetizable object by a magnet. Magnetizable materials are drawn toward regions where $\overrightarrow{\mathbf{H}}$ is high.

EXAMPLE 6.14. (Motion on two parallel conducting rails (I)).– Consider two parallel wires connected by a third wire of length l, which is free to move, as shown in Figure 6.25. The two fixed wires act as rails for the mobile wire. A current I flows through the entire circuit, which is placed in a uniform external magnetic field $B_0 \overrightarrow{\mathbf{i}}$. As a result, the mobile wire moves along the rails. Indeed, the Lorentz force acting on the mobile wire is:

$$\overrightarrow{\mathbf{F}} = Il \overrightarrow{\mathbf{k}} \times B_0 \overrightarrow{\mathbf{i}} = IlB_0 \overrightarrow{\mathbf{j}} \implies F_y = IB_0 l \qquad [6.200]$$

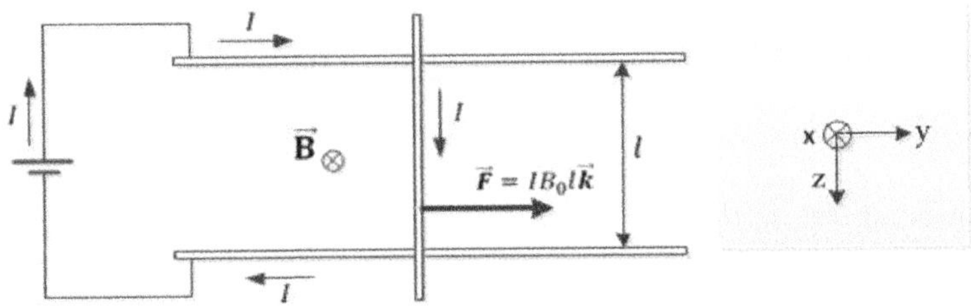

Figure 6.25. *Current-carrying wire sliding on two conductor rails under the action of an external magnetic field*

The magnetic field generated by the current is neglected, assuming it is much smaller than the imposed field B_0.

EXAMPLE 6.15. (Motion on two parallel conducting rails (II)).– The movable wire is now surrounded by a hollow, infinitely permeable cylinder with inner radius R_1 and outer radius R_2, both much smaller than the length l of the wire, as shown in Figure 6.26.

The magnetic field appears as the superposition of the field created by the current-carrying wire (which is approximately the same as that of an infinitely long wire) and the field generated by the cylinder. Since there is no current in the three regions, $0 < r < R_1$, $R_1 < r < R_2$ and $R_2 < r < \infty$, both the curl and the divergence of the magnetic field are zero in each medium. Consequently, it is possible to express $\vec{\mathbf{B}}$ in terms of a scalar magnetic potential V_m that satisfies Laplace's equation $\vec{\mathbf{B}} = -\vec{\nabla} V_m$. Note that we have used $\vec{\mathbf{B}} = -\vec{\nabla} V_m$ instead of $\vec{\mathbf{H}} = -\vec{\nabla} V_m$, since the media are linear.

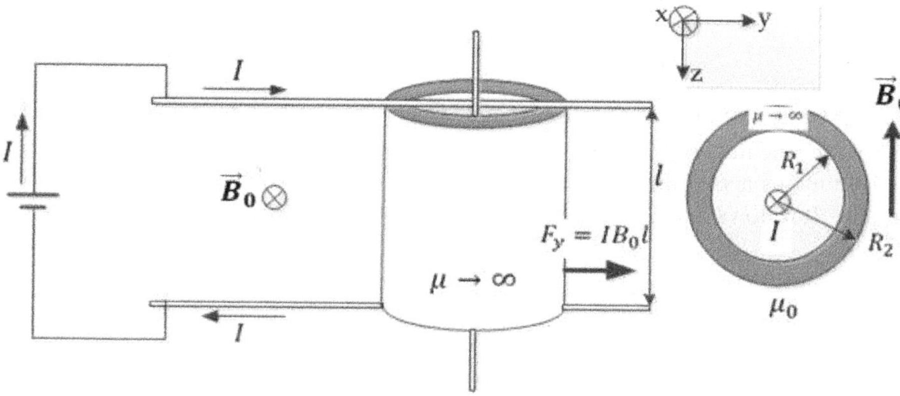

Figure 6.26. *The current-carrying wire is surrounded by a hollow cylinder with permeability $\mu \longrightarrow \infty$*

Since we are using cylindrical coordinates here, we will try a potential of the form:

$$V_m = \begin{cases} Ar\cos\varphi & 0 < r < R_1 \\ \left(Cr + \dfrac{D}{r}\right)\cos\varphi & R_1 < r < R_2 \\ \left(Er + \dfrac{F}{r}\right)\cos\varphi & R_2 < r < \infty \end{cases} \quad [6.201]$$

Thus, we obtain:

$$\begin{aligned} \vec{\mathbf{B}}\,(0 < r < R_1) &= -A\cos\varphi\,\vec{\mathbf{e}}_r + A\sin\varphi\,\vec{\mathbf{e}}_\varphi + \frac{\mu_0 I}{2\pi r}\vec{\mathbf{e}}_\varphi \\ \vec{\mathbf{B}}\,(R_1 < r < R_2) &= -\left(C - \frac{D}{r^2}\right)\cos\varphi\,\vec{\mathbf{e}}_r + \left(C + \frac{D}{r^2}\right)\sin\varphi\,\vec{\mathbf{e}}_\varphi \\ &\quad + \frac{\mu I}{2\pi r}\vec{\mathbf{e}}_\varphi \\ \vec{\mathbf{B}}\,(R_2 < r < \infty) &= -\left(E - \frac{F}{r^2}\right)\cos\varphi\,\vec{\mathbf{e}}_r + \left(E + \frac{F}{r^2}\right)\sin\varphi\,\vec{\mathbf{e}}_\varphi \\ &\quad + \frac{\mu_0 I}{2\pi r}\vec{\mathbf{e}}_\varphi \end{aligned} \quad [6.202]$$

$$\begin{aligned}
\vec{\mathbf{H}}\,(0<r<R_1) &= -\frac{A}{\mu_0}\cos\varphi\,\vec{\mathbf{e}}_r + \frac{A}{\mu_0}\sin\varphi\,\vec{\mathbf{e}}_\varphi + \frac{I}{2\pi r}\vec{\mathbf{e}}_\varphi \\
\vec{\mathbf{H}}\,(R_1<r<R_2) &= \frac{I}{2\pi r}\vec{\mathbf{e}}_\varphi \\
\vec{\mathbf{H}}\,(R_2<r<\infty) &= -\frac{1}{\mu_0}\left(E-\frac{F}{r^2}\right)\cos\varphi\,\vec{\mathbf{e}}_r + \\
&\quad \frac{1}{\mu_0}\left(E+\frac{F}{r^2}\right)\sin\varphi\,\vec{\mathbf{e}}_\varphi + \frac{I}{2\pi r}\vec{\mathbf{e}}_\varphi
\end{aligned} \qquad [6.203]$$

The constants A, B, C, D, E and F are determined by requiring that $\vec{\mathbf{B}}$ approaches the imposed field $B_0\,\vec{\mathbf{i}}$ as $r \longrightarrow \infty$, and that the components $B_\perp$ and $H_\parallel$ are continuous across the interfaces $r = R_1$ and $r = R_2$ of the infinitely permeable cylinder. It follows that $E = -B_0$, and:

$$\begin{aligned}
&B_r\,(R_1+0) = B_r\,(R_1-0) \Longrightarrow C - \frac{D}{R_1^2} = A \\
&B_r\,(R_2+0) = B_r\,(R_2-0) \Longrightarrow E - \frac{F}{R_2^2} = C - \frac{D}{R_2^2} \\
&H_\varphi\,(R_1+0) = H_\varphi\,(R_1-0) \Longrightarrow \frac{I}{2\pi R_1} = \frac{I}{2\pi R_1} + A\sin\varphi \Longrightarrow A = 0 \\
&H_\varphi\,(R_2+0) = H_\varphi\,(R_2-0) \Longrightarrow -\frac{1}{\mu_0}\left(E+\frac{F}{R_2^2}\right) + \frac{I}{2\pi R_2} = \frac{I}{2\pi R_2} \\
&\qquad\qquad \Longrightarrow E + \frac{F}{R_2^2} = 0
\end{aligned} \qquad [6.204]$$

Therefore, it follows that $C = -\frac{2B_0R_2^2}{R_2^2-R_1^2}$, $D = -\frac{2B_0R_1^2R_2^2}{R_2^2-R_1^2}$ and $F = B_0R_2^2$. Substituting these solutions into [6.202], we obtain:

$$\begin{aligned}
\vec{\mathbf{B}}\,(r<R_1) &= \frac{\mu_0 I}{2\pi r}\vec{\mathbf{e}}_\varphi \\
\vec{\mathbf{B}}\,(R_1<r<R_2) &= \frac{2B_0R_2^2}{R_2^2-R_1^2}\left[\left(1-\frac{R_1^2}{r^2}\right)\cos\varphi\,\vec{\mathbf{e}}_r \right.\\
&\quad \left. -\left(1+\frac{R_1^2}{r^2}\right)\sin\varphi\,\vec{\mathbf{e}}_\varphi\right] + \frac{\mu I}{2\pi r}\vec{\mathbf{e}}_\varphi \\
\vec{\mathbf{B}}\,(r>R_2) &= B_0\left[\left(1+\frac{R_2^2}{r^2}\right)\cos\varphi\,\vec{\mathbf{e}}_r - \left(1-\frac{R_2^2}{r^2}\right)\sin\varphi\,\vec{\mathbf{e}}_\varphi\right] \\
&\quad + \frac{\mu_0 I}{2\pi r}\vec{\mathbf{e}}_\varphi
\end{aligned}$$

$$\vec{\mathbf{H}}(r<R_1)=\frac{I}{2\pi r}\vec{\mathbf{e}}_\varphi$$

$$\vec{\mathbf{H}}(R_1<r<R_2)=\frac{I}{2\pi r}\vec{\mathbf{e}}_\varphi$$

$$\vec{\mathbf{H}}(r>R_2)=\frac{B_0}{\mu_0}\left[\left(1+\frac{R_2^2}{r^2}\right)\cos\varphi\,\vec{\mathbf{e}}_r-\left(1-\frac{R_2^2}{r^2}\right)\sin\varphi\,\vec{\mathbf{e}}_\varphi\right]+\frac{I}{2\pi r}\vec{\mathbf{e}}_\varphi \qquad [6.205]$$

It is clear that the imposed magnetic field B_0 does not affect the regions $0<r<R_1$ and $R_1<r<R_2$, as they are shielded by the infinitely permeable cylindrical shell. We therefore conclude that the action of the Lorentz force on the current-carrying wire is zero because of the screening effect of the cylinder. However, there is a magnetization force acting on the cylindrical shell, where the internal magnetic field $\vec{\mathbf{H}}$ is entirely due to the line current, $\vec{\mathbf{H}}=\frac{I}{2\pi r}\vec{\mathbf{e}}_\varphi$:

$$\begin{cases} H_r=0 \\ H_\varphi=\dfrac{I}{2\pi r}\end{cases}\Longrightarrow\begin{cases}\mu_0M_r=B_r-\mu_0H_r=\dfrac{2B_0R_2^2}{R_2^2-R_1^2}\left(1-\dfrac{R_1^2}{r^2}\right)\cos\varphi \\ \mu_0M_\varphi=B_\varphi-\mu_0H_\varphi=-\dfrac{2B_0R_2^2}{R_2^2-R_1^2}\left(1+\dfrac{R_1^2}{r^2}\right) \\ \qquad\times\sin\varphi+\dfrac{\mu_0}{2\pi r}I\end{cases} \qquad [6.206]$$

Since H_r depends only on r, and $\vec{\mathbf{e}}_\varphi=-\sin\varphi\,\vec{\mathbf{i}}+\cos\varphi\,\vec{\mathbf{j}}$, it follows that:

$$\begin{aligned}\vec{\mathbf{F}}&=-\frac{B_rI}{2\pi r^2}\vec{\mathbf{e}}_\varphi+\frac{(B_\varphi-\mu_0H_\varphi)I}{2\pi r^2}\frac{d\vec{\mathbf{e}}_\varphi}{d\varphi}\\&=\frac{I}{2\pi r^2}\left[-\frac{2B_0R_2^2}{R_2^2-R_1^2}\left(-\sin 2\varphi\,\vec{\mathbf{i}}+\cos 2\varphi\,\vec{\mathbf{j}}-\frac{R_1^2}{r^2}\vec{\mathbf{j}}\right)+\right.\\&\left.\frac{(\mu_0-\mu_0)I}{2\pi r}\left(\cos\varphi\,\vec{\mathbf{i}}+\sin\varphi\,\vec{\mathbf{j}}\right)\right]\end{aligned} \qquad [6.207]$$

The total force exerted on the cylinder is obtained by integrating [6.206] over r and φ:

$$\begin{aligned}\vec{\mathbf{F}}_T&=\int_{R_1}^{R_2}\int_0^{2\pi}\vec{\mathbf{F}}\,lrdrd\varphi=\int_{R_1}^{R_2}\int_0^{2\pi}\frac{1}{2\pi}\frac{2IlB_0R_1^2R_2^2}{R_2^2-R_1^2}\frac{rdrd\varphi}{r^4}\vec{\mathbf{j}}\\&=\frac{2IlB_0R_1^2R_2^2}{R_2^2-R_1^2}\int_{r=R_1}^{R_2}\frac{dr}{r^3}\vec{\mathbf{j}}=IlB_0\,\vec{\mathbf{j}}\end{aligned} \qquad [6.208]$$

The other trigonometric integrals are all zero. The force exerted on the cylinder is the same as in [6.200].

EXAMPLE 6.16. (Motion on two parallel conducting rails (III)).– If the wire is replaced by a solid cylinder of radius R and finite permeability, with a uniformly distributed current, as shown in Figure 6.27, the fields $\vec{\mathbf{B}}$ and $\vec{\mathbf{H}}$ are given by:

$$\begin{aligned}
\vec{\mathbf{B}}\,(r<R) &= -A\cos\varphi\,\vec{\mathbf{e}}_r + A\sin\varphi\,\vec{\mathbf{e}}_\varphi + \frac{\mu I r}{2\pi R^2}\vec{\mathbf{e}}_\varphi \\
\vec{\mathbf{B}}\,(r>R) &= \left(B_0 + \frac{F}{r^2}\right)\cos\varphi\,\vec{\mathbf{e}}_r - \left(B_0 - \frac{F}{r^2}\right)\sin\varphi\,\vec{\mathbf{e}}_\varphi + \frac{\mu_0 I}{2\pi r}\vec{\mathbf{e}}_\varphi \\
\vec{\mathbf{H}}\,(r<R) &= -\frac{A}{\mu}\cos\varphi\,\vec{\mathbf{e}}_r + \frac{A}{\mu}\sin\varphi\,\vec{\mathbf{e}}_\varphi + \frac{I r}{2\pi R^2}\vec{\mathbf{e}}_\varphi \\
\vec{\mathbf{H}}\,(r>R) &= \frac{1}{\mu_0}\left(B_0 + \frac{F}{r^2}\right)\cos\varphi\,\vec{\mathbf{e}}_r - \frac{1}{\mu_0}\left(B_0 - \frac{F}{r^2}\right)\sin\varphi\,\vec{\mathbf{e}}_\varphi + \frac{I}{2\pi r}\vec{\mathbf{e}}_\varphi
\end{aligned}$$

[6.209]

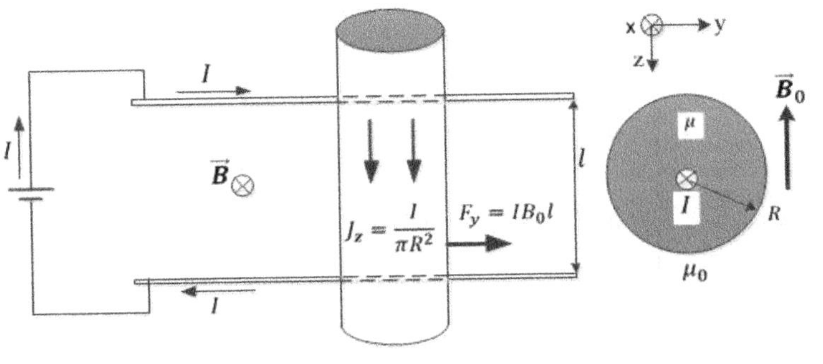

Figure 6.27. *If the current-carrying wire is replaced by a cylinder of radius R and finite permeability μ, the total force (Lorentz force + magnetization force) experienced remains unchanged*

The following boundary conditions must be imposed: $-A = B_0 + \frac{F}{R^2}$ and $\frac{A}{\mu} = -\frac{1}{\mu_0}\left(B_0 - \frac{F}{R^2}\right)$, which leads to $A = -\frac{2\mu}{\mu+\mu_0}B_0$ and $F = -\frac{\mu-\mu_0}{\mu+\mu_0}B_0R^2$. Thus, we find:

$$\begin{aligned}
\vec{\mathbf{B}}\,(r<R) &= \frac{2\mu B_0}{\mu+\mu_0}\left(\cos\varphi\,\vec{\mathbf{e}}_r - \sin\varphi\,\vec{\mathbf{e}}_\varphi\right) + \frac{\mu I r}{2\pi R^2}\vec{\mathbf{e}}_\varphi \\
\vec{\mathbf{B}}\,(r>R) &= B_0\left[\left(1 + \frac{\mu-\mu_0}{\mu+\mu_0}\frac{R^2}{r^2}\right)\cos\varphi\,\vec{\mathbf{e}}_r - \right. \\
&\quad \left.\left(1 - \frac{\mu-\mu_0}{\mu+\mu_0}\frac{R^2}{r^2}\right)\sin\varphi\,\vec{\mathbf{e}}_\varphi\right] + \frac{\mu_0 I}{2\pi r}\vec{\mathbf{e}}_\varphi
\end{aligned}$$

$$\begin{aligned}\vec{\mathbf{H}}(r<R) &= \frac{2B_0}{\mu+\mu_0}\left(\cos\varphi\,\vec{\mathbf{e}}_r-\sin\varphi\,\vec{\mathbf{e}}_\varphi\right)+\frac{Ir}{2\pi R^2}\vec{\mathbf{e}}_\varphi\\ \vec{\mathbf{H}}(r>R) &= \frac{B_0}{\mu_0}\left[\left(1+\frac{\mu-\mu_0}{\mu+\mu_0}\frac{R^2}{r^2}\right)\cos\varphi\,\vec{\mathbf{e}}_r-\right.\\ &\left.\left(1-\frac{\mu-\mu_0}{\mu+\mu_0}\frac{R^2}{r^2}\right)\sin\varphi\,\vec{\mathbf{e}}_\varphi\right]+\frac{I}{2\pi r}\vec{\mathbf{e}}_\varphi\end{aligned} \quad [6.210]$$

or equivalently:

$$\begin{aligned}\vec{\mathbf{B}}(0<r<R) &= \frac{2\mu B_0}{\mu+\mu_0}\vec{\mathbf{i}}+\frac{\mu I}{2\pi R^2}\left(-y\,\vec{\mathbf{i}}+x\,\vec{\mathbf{j}}\right)\\ \vec{\mathbf{H}}(0<r<R) &= \frac{2B_0}{\mu+\mu_0}\vec{\mathbf{i}}+\frac{I}{2\pi R^2}\left(-y\,\vec{\mathbf{i}}+x\,\vec{\mathbf{j}}\right)\end{aligned} \quad [6.211]$$

To calculate the force density experienced, it is simpler to express the field $\vec{\mathbf{H}}$ inside the cylinder in terms of Cartesian coordinates $\vec{\mathbf{H}} = H_x\vec{\mathbf{i}} + H_y\vec{\mathbf{j}}$, where $H_x = \frac{2B_0}{\mu+\mu_0} - \frac{Iy}{2\pi R^2}$ and $H_y = \frac{Ix}{2\pi R^2}$. The force density exerted on the cylinder is written as:

$$\begin{aligned}\vec{\mathbf{F}} &= \mu_0\left(\vec{\mathbf{M}}\cdot\vec{\nabla}\right)\vec{\mathbf{H}}+\mu_0\vec{\mathbf{J}}\times\vec{\mathbf{H}}\\ &= (\mu-\mu_0)\left(\vec{\mathbf{H}}\cdot\vec{\nabla}\right)\vec{\mathbf{H}}+\frac{\mu_0 I}{\pi R^2}\vec{\mathbf{k}}\times\vec{\mathbf{H}}\\ &= (\mu-\mu_0)\left(H_x\frac{\partial}{\partial x}+H_y\frac{\partial}{\partial y}\right)\left(H_x\vec{\mathbf{i}}+H_y\vec{\mathbf{j}}\right)\\ &+\frac{\mu_0 I}{\pi R^2}\left(H_x\vec{\mathbf{j}}-H_y\vec{\mathbf{i}}\right)\end{aligned} \quad [6.212]$$

Since $\frac{\partial H_x}{\partial x} = \frac{\partial H_y}{\partial y} = 0$ and $\frac{\partial H_x}{\partial y} = -\frac{\partial H_y}{\partial x} = \frac{I}{2\pi R^2}$, equation [6.211] simplifies to:

$$\begin{aligned}\vec{\mathbf{F}} &= (\mu-\mu_0)\left(H_x\frac{\partial H_y}{\partial x}\vec{\mathbf{j}}+H_y\frac{\partial H_x}{\partial y}\vec{\mathbf{i}}\right)+\frac{\mu_0 I}{\pi R^2}\left(H_x\vec{\mathbf{j}}-H_y\vec{\mathbf{i}}\right)\\ &= \frac{I(\mu+\mu_0)}{2\pi R^2}\left(H_x\vec{\mathbf{j}}-H_y\vec{\mathbf{i}}\right)\\ &= \frac{I(\mu+\mu_0)}{2\pi R^2}\left[\left(\frac{2B_0}{\mu+\mu_0}-\frac{Iy}{2\pi R^2}\right)\vec{\mathbf{j}}-\frac{Ix}{2\pi R^2}\vec{\mathbf{i}}\right]\\ &= \frac{IB_0}{\pi R^2}\vec{\mathbf{j}}-\frac{I^2(\mu+\mu_0)}{(2\pi R^2)^2}r\left(\cos\varphi\,\vec{\mathbf{i}}+\sin\varphi\,\vec{\mathbf{j}}\right)\end{aligned}$$

[6.213]

The total force experienced by the cylinder is:

$$\vec{\mathbf{F}}_T = \int_0^{2\pi}\int_0^R \vec{\mathbf{F}}\,lrdrd\varphi = \frac{IB_0 l}{\pi R^2}\vec{\mathbf{j}}\int_0^{2\pi}\int_0^R rdrd\varphi = IB_0 l\,\vec{\mathbf{j}} \quad [6.214]$$

The other trigonometric integrals are zero. We see that $\vec{\mathbf{F}}_T$ is independent of the permeability of the cylinder.

Chapter 7

Electromagnetic Fields in Matter

7.1. Maxwell equations

Historically, charge and its electric field were studied separately from current and its magnetic field in the early developments of electromagnetic theory. Until Faraday showed that a time-varying magnetic field generates an electric field, it was believed that the electric field and the magnetic field were distinct and decoupled. Faraday also believed that a time-varying electric field should generate a magnetic field, but he was unable to prove this hypothesis. This is the property of *duality*. It was Maxwell who later demonstrated that Faraday's hypothesis was correct and that, without this correction, the Ampère equation would be incompatible with the charge conservation equation:

$$\begin{aligned}&\textit{Ampère equation:} \quad \vec{\nabla} \times \vec{\mathbf{H}} = \vec{\mathbf{J}} \Longrightarrow \vec{\nabla} \cdot \vec{\mathbf{J}} = 0 \\ &\textit{Charge conservation equation:} \quad \vec{\nabla} \cdot \vec{\mathbf{J}} + \frac{\partial \rho}{\partial t} = 0\end{aligned} \qquad [7.1]$$

where $\vec{\mathbf{J}}$ and ρ are the conduction current density and the free volume charge density, respectively. It is evident that taking the divergence of Ampère's law implies that the current density must have zero divergence, because the divergence of the curl of any vector field is identically zero. This result contradicts the charge conservation equation when the charge density varies with time. Maxwell realized that adding the displacement current term to the right-hand side of the Ampère law would ensure compatibility with the charge conservation equation because of the Maxwell–Gauss equation, which relates $\vec{\mathbf{D}}$ to ρ, $\vec{\mathbf{D}} = \rho$. This seemingly simple correction had profound consequences. In particular, it enabled the prediction of electromagnetic waves propagating at the speed of light. As a result, it was established that light is, in fact, an electromagnetic wave. Due to the significance of this correction, the set of equations that govern electromagnetic fields became known as *Maxwell equations*.

1) *Maxwell–Faraday equation*:

$$\vec{\nabla} \times \vec{\mathbf{E}} = -\frac{\partial \vec{\mathbf{B}}}{\partial t} \Longrightarrow \oint_{\Gamma} \vec{\mathbf{E}} \cdot d\vec{\mathbf{l}} = -\frac{d}{dt}\int_{S} \vec{\mathbf{B}} \cdot d\vec{\mathbf{S}} \qquad [7.2]$$

2) *Ampère's law* with Maxwell's correction (displacement current term) or *Maxwell–Ampère equation*:

$$\vec{\nabla} \times \vec{\mathbf{H}} = \vec{\mathbf{J}} + \frac{\partial \vec{\mathbf{D}}}{\partial t} \Longrightarrow \oint_{\Gamma} \vec{\mathbf{H}} \cdot d\vec{\mathbf{l}} = \int_{S} \vec{\mathbf{J}} \cdot d\vec{\mathbf{S}} + \frac{d}{dt}\int_{S} \vec{\mathbf{D}} \cdot d\vec{\mathbf{S}} \qquad [7.3]$$

3) *Maxwell–Gauss equation*:

$$\vec{\nabla} \cdot \vec{\mathbf{D}} = \rho \Longrightarrow \oint_{S} \vec{\mathbf{D}} \cdot d\vec{\mathbf{S}} = \int_{\tau} \rho d\tau \qquad [7.4]$$

4) *Magnetic flux conservation equation or Maxwell–Thomson equation (for* $\vec{\mathbf{B}}$*)*:

$$\vec{\nabla} \cdot \vec{\mathbf{B}} = 0 \Longrightarrow \oint_{S} \vec{\mathbf{B}} \cdot d\vec{\mathbf{S}} = 0 \qquad [7.5]$$

5) *Charge conservation equation*:

$$\vec{\nabla} \cdot \vec{\mathbf{J}} + \frac{\partial \rho}{\partial t} = 0 \Longrightarrow \oint_{S} \vec{\mathbf{J}} \cdot d\vec{\mathbf{S}} + \frac{d}{dt}\int_{\tau} \rho d\tau = 0 \qquad [7.6]$$

It is important to note the following:

– Equation [7.6] can also be derived by computing the divergence of [7.3] and using [7.4].

– Equation [7.5] is not independent of [7.2]. In fact, taking the divergence of the Maxwell–Faraday equation yields $\frac{\partial}{\partial t}\left(\vec{\nabla} \cdot \vec{\mathbf{B}}\right) = 0$, which implies that $\vec{\nabla} \cdot \vec{\mathbf{B}}$ is a function independent of time. Since we can assume that at a certain moment, $\vec{\mathbf{B}} = 0$ (e.g. for $t < 0$, $\vec{\mathbf{B}}$ is considered switched off), this function must be identically null.

– The Maxwell equations become fully symmetric when a magnetic charge (magnetic monopole density) is added to the right-hand side of equation [7.5], along with an associated magnetic current due to the motion of magnetic charges on the right-hand side of equation [7.2]. So far, the existence of magnetic monopoles (and thus of magnetic current) has not been experimentally confirmed, despite major advances in physics. If such charges do exist, equations [7.2] and [7.5] must be modified accordingly, and an additional equation, similar to [7.6], must be introduced to represent the conservation of magnetic charge.

– The fields $\overrightarrow{\mathbf{D}}$ and $\overrightarrow{\mathbf{H}}$ are related to the fields $\overrightarrow{\mathbf{E}}$ and $\overrightarrow{\mathbf{B}}$ through the following relations $\overrightarrow{\mathbf{D}} = \varepsilon_0\overrightarrow{\mathbf{E}} + \overrightarrow{\mathbf{P}}$ and $\overrightarrow{\mathbf{B}} = \mu_0\left(\overrightarrow{\mathbf{H}} + \overrightarrow{\mathbf{M}}\right)$, where $\overrightarrow{\mathbf{P}}$ and $\overrightarrow{\mathbf{M}}$ are the *polarization* and *magnetization* of the medium, respectively. In terms of fields $\overrightarrow{\mathbf{E}}$, $\overrightarrow{\mathbf{B}}$, $\overrightarrow{\mathbf{P}}$ and $\overrightarrow{\mathbf{M}}$, equations [7.3] and [7.4] become:

$$\begin{aligned}&\overrightarrow{\nabla} \times \overrightarrow{\mathbf{B}} = \mu_0\left(\overrightarrow{\mathbf{J}} + \overrightarrow{\nabla} \times \overrightarrow{\mathbf{M}} + \varepsilon_0\frac{\partial\overrightarrow{\mathbf{E}}}{\partial t} + \frac{\partial\overrightarrow{\mathbf{P}}}{\partial t}\right) \Longrightarrow \oint_\Gamma \overrightarrow{\mathbf{B}} \cdot d\overrightarrow{\mathbf{l}} \\ &= \mu_0\left(\int_S \overrightarrow{\mathbf{J}} \cdot d\overrightarrow{\mathbf{S}} + \oint_\Gamma \overrightarrow{\mathbf{M}} \cdot d\overrightarrow{\mathbf{l}} + \varepsilon_0\frac{d}{dt}\int_S \overrightarrow{\mathbf{E}} \cdot d\overrightarrow{\mathbf{S}} + \frac{d}{dt}\int_S \overrightarrow{\mathbf{P}} \cdot d\overrightarrow{\mathbf{S}}\right) \\ &\overrightarrow{\nabla} \cdot \overrightarrow{\mathbf{E}} = \frac{\rho}{\varepsilon_0} - \overrightarrow{\nabla} \cdot \overrightarrow{\mathbf{P}} \Longrightarrow \oint_S \overrightarrow{\mathbf{E}} \cdot d\overrightarrow{\mathbf{S}} = \frac{1}{\varepsilon_0}\int_\tau \rho d\tau - \oint_S \overrightarrow{\mathbf{P}} \cdot d\overrightarrow{\mathbf{S}}\end{aligned} \qquad [7.7]$$

In the static case, magnetization acts as an additional source of magnetic field taken into account in the Maxwell–Ampère equation by adding the term $\overrightarrow{\nabla} \times \overrightarrow{\mathbf{M}}$ to the current density $\overrightarrow{\mathbf{J}}$. This magnetization current is not associated with any magnetic charge, since $\oint_S \left(\overrightarrow{\nabla} \times \overrightarrow{\mathbf{M}}\right) \cdot d\overrightarrow{\mathbf{S}} = 0$ (monopoles do not exist!). In the non-static case, an additional term appears. When a molecular dipole changes orientation or magnitude, a microscopic current is generated. If the polarization is uniform, the resulting transient current corresponds to the transport of bound surface charge from one side of the medium to the other. However, if the polarization is not uniform, charge concentrations appear in the medium in the form of a bound volume charge of polarization. This new bound charge, represented by the term $-\overrightarrow{\nabla} \cdot \overrightarrow{\mathbf{P}}$, leads to an additional electric field in the Maxwell–Gauss equation; and, on the other hand, to a polarization current $\frac{\partial\overrightarrow{\mathbf{P}}}{\partial t}$, which also acts as a source of magnetic field in the Maxwell–Ampère equation. The bound charge $\rho_p = -\overrightarrow{\nabla} \cdot \overrightarrow{\mathbf{P}}$ and the polarization current $\overrightarrow{\mathbf{J}}_p = \frac{\partial\overrightarrow{\mathbf{P}}}{\partial t}$ are related by a conservation law analogous to [7.6]:

$$\overrightarrow{\nabla} \cdot \overrightarrow{\mathbf{J}}_p + \frac{\partial\rho_p}{\partial t} = 0 \Longrightarrow \oint_S \overrightarrow{\mathbf{J}}_p \cdot d\overrightarrow{\mathbf{S}} + \frac{d}{dt}\int_\tau \rho_p d\tau = 0 \qquad [7.8]$$

– For linear, isotropic and homogeneous (LIH) media, we also have the following constitutive relations $\overrightarrow{\mathbf{P}} = \chi_e\varepsilon_0\overrightarrow{\mathbf{E}}$, $\overrightarrow{\mathbf{M}} = \chi_m\overrightarrow{\mathbf{H}}$ and $\overrightarrow{\mathbf{J}} = \gamma\overrightarrow{\mathbf{E}}$ (if the medium is ohmic, i.e. with losses).

– The interaction of matter with the electromagnetic field occurs entirely through the *Lorentz force* ($\overrightarrow{\mathbf{f}}$ is the force per unit volume):

$$\overrightarrow{\mathbf{f}} = \rho\overrightarrow{\mathbf{E}} + \overrightarrow{\mathbf{J}} \times \overrightarrow{\mathbf{B}} \qquad [7.9]$$

– The circuit theory assumes that the electric and magnetic fields are highly localized within the circuit elements. When the frequency of the fields is sufficiently low, we can still neglect the displacement current $\frac{\partial\overrightarrow{\mathbf{D}}}{\partial t}$ compared to the term $\mu_0\overrightarrow{\mathbf{J}}$

(except in the case of the capacitor, where $\overrightarrow{\mathbf{J}}$ is zero). The Ampère equation $\overrightarrow{\nabla} \times \overrightarrow{\mathbf{B}} = \mu_0 \overrightarrow{\mathbf{J}}$ allows us to deduce that the divergence of the current $\overrightarrow{\mathbf{J}}$ is therefore zero. Thus, we obtain *Kirchhoff's first law*, which states that the algebraic sum of all currents entering (or leaving) a node is zero:

$$\overrightarrow{\nabla} \cdot \overrightarrow{\mathbf{J}} = 0 \Longrightarrow \oint_S \overrightarrow{\mathbf{J}} \cdot d\overrightarrow{\mathbf{S}} = 0 \Longrightarrow \sum_k i_k = 0 \quad [7.10]$$

Similarly, we can neglect the term $\frac{\partial \overrightarrow{\mathbf{B}}}{\partial t}$ in the Maxwell–Faraday equation, which gives $\overrightarrow{\nabla} \times \overrightarrow{\mathbf{E}} = 0$. Thus, we obtain Kirchhoff's voltage law, or the *second Kirchhoff law*, which states that the algebraic sum of the voltage drops (or rises) around any closed loop in a circuit is zero:

$$\overrightarrow{\nabla} \cdot \overrightarrow{\mathbf{E}} = 0 \Longrightarrow \overrightarrow{\mathbf{E}} = -\overrightarrow{\nabla} V \Longrightarrow \oint_L \overrightarrow{\mathbf{E}} \cdot d\overrightarrow{\mathbf{l}} = 0 \Longrightarrow \sum_k V_k = 0 \quad [7.11]$$

This is the approximation of the *quasistatic or quasistationary regime.*

– Equations of continuity: let us summarize the different results. In most electromagnetism problems, we are faced with several media. It is essential to understand how fields behave at the interface of these media: some components are continuous, meaning their values are identical on both sides of the interface, whereas other components are discontinuous. These continuity equations come into play, for example, in problems of reflection or transmission of electromagnetic waves between media, etc. Figures 7.1(a) and (b) show the geometries considered for the study of continuity equations. The separating surface between the two media is assumed to be locally flat. Half of the contour (or the cylinder) lies in medium 1, and the other half in medium 2, as indicated in Figure 7.1(a) (or 7.1(b)). On the segments (l_1 and l_2) of the loop and on the bases (S_1 and S_2) of the cylinder parallel to the interface, uniform fields are assumed. This assumption becomes increasingly accurate as l_1, l_2, S_1 and S_2 are chosen to be very small. We then take the limit $\epsilon \longrightarrow 0$ to make (1) the surface enclosed by the contour vanish and (2) the volume enclosed by the cylinder vanish. The fields involved in the continuity equations are measured on both sides of the interface.

a) *Tangential component of* $\overrightarrow{\mathbf{E}}$: to derive the continuity equation for the tangential component of $\overrightarrow{\mathbf{E}}$, we apply the Maxwell–Faraday equation along the contour Γ:

$$\lim_{\epsilon \longrightarrow 0} \oint_\Gamma \overrightarrow{\mathbf{E}} \cdot d\overrightarrow{\mathbf{l}} = -\lim_{\epsilon \longrightarrow 0} \frac{d}{dt} \int_{S_\Gamma} \overrightarrow{\mathbf{B}} \cdot d\overrightarrow{\mathbf{S}} \quad [7.12]$$

Clearly, since the surface S_Γ tends to zero, the right-hand side vanishes. On the left-hand side, the line integral is reduced to the segments l_1 and l_2, where the fields are assumed to be uniform. We then obtain:

$$0 = \int_{l_2} \vec{\mathbf{E}}_2 \cdot \left(\vec{\mathbf{t}} \times \vec{\mathbf{n}}_{12}\right) dl + \int_{l_1} \vec{\mathbf{E}}_1 \cdot \left(-\vec{\mathbf{t}} \times \vec{\mathbf{n}}_{12}\right) dl = E_{2\|} l - E_{1\|} l \quad [7.13]$$

where $\vec{\mathbf{t}} \times \vec{\mathbf{n}}_{12}$ is a unit vector in the direction of segment l_2, that is, parallel to the interface. This implies that $E_{1\|}$ and $E_{2\|}$ must have the same value on both sides of the interface:

$$E_{2\|} = E_{1\|} \implies E_{2\|} - E_{1\|} = 0 \quad [7.14]$$

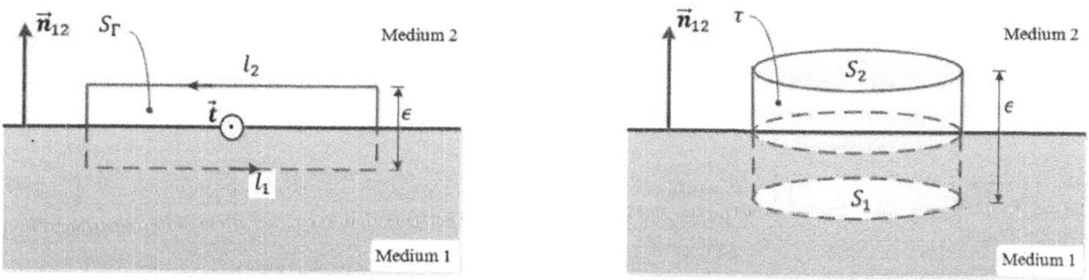

(a) Contour Γ at the interface between two media bounding a surface S_Γ

(b) Cylinder at the interface between tow media bounding a volume τ

Figure 7.1. *Different geometries used to establish the continuity equations for the fields and currents involved in the Maxwell equations. For a color version of this figure, see www.iste.co.uk/abdesselam/electrodynamics2.zip*

In other words, the tangential components of the electric field at the interface (denoted $E_{1\|}$ in medium 1 and $E_{2\|}$ in medium 2) are continuous. In vector form, the continuity equation [7.14] is written as:

$$\vec{\mathbf{n}}_{12} \times \left(\vec{\mathbf{E}}_2 - \vec{\mathbf{E}}_1\right) = 0 \quad [7.15]$$

Indeed, we have: $0 = \left(\vec{\mathbf{E}}_2 - \vec{\mathbf{E}}_1\right) \cdot \left(\vec{\mathbf{t}} \times \vec{\mathbf{n}}_{12}\right) = \vec{\mathbf{t}} \cdot \left[\vec{\mathbf{n}}_{12} \times \left(\vec{\mathbf{E}}_2 - \vec{\mathbf{E}}_1\right)\right]$. Since $\vec{\mathbf{t}}$ is arbitrary (as the loop can be oriented in any direction), it follows that: $\vec{\mathbf{n}}_{12} \times \left(\vec{\mathbf{E}}_2 - \vec{\mathbf{E}}_1\right) = 0$.

b) *Tangential component of* $\vec{\mathbf{H}}$: we proceed in the same way, but starting from the Maxwell–Ampère equation:

$$\lim_{\epsilon \longrightarrow 0} \oint_\Gamma \vec{\mathbf{H}} \cdot d\vec{\mathbf{l}} = \lim_{\epsilon \longrightarrow 0} \int_{S_\Gamma} \vec{\mathbf{J}} \cdot d\vec{\mathbf{S}} + \frac{d}{dt} \int_{S_\Gamma} \vec{\mathbf{D}} \cdot d\vec{\mathbf{S}} \quad [7.16]$$

This time, when the surface S_Γ tends to zero, the right-hand side is not necessarily zero, since a surface current density $\vec{\mathbf{K}}$ may exist at the interface, crossing the contour.

Reducing the surface of the loop to zero eliminates all contributions except that of the surface current. Thus, the right-hand side becomes:

$$\lim_{\epsilon \longrightarrow 0} \int_{S_\Gamma} \vec{\mathbf{J}} \cdot d\vec{\mathbf{S}} + \frac{d}{dt} \int_{S_\Gamma} \vec{\mathbf{D}} \cdot d\vec{\mathbf{S}} = \vec{\mathbf{K}} l \cdot \vec{\mathbf{t}} \qquad [7.17]$$

As for the left-hand side, it is again limited to the integration over segments l_1 and l_2, on which the fields are uniform. We therefore obtain:

$$\int_{l_2} \vec{\mathbf{H}}_2 \cdot \left(\vec{\mathbf{t}} \times \vec{\mathbf{n}}_{12}\right) dl + \int_{l_1} \vec{\mathbf{H}}_1 \cdot \left(-\vec{\mathbf{t}} \times \vec{\mathbf{n}}_{12}\right) dl = \vec{\mathbf{K}} l \cdot \vec{\mathbf{t}} \Longrightarrow \left(\vec{\mathbf{H}}_2 - \vec{\mathbf{H}}_1\right) \cdot \left(\vec{\mathbf{t}} \times \vec{\mathbf{n}}_{12}\right) = \vec{\mathbf{K}} \cdot \vec{\mathbf{t}} \Longrightarrow \vec{\mathbf{t}} \cdot \left[\vec{\mathbf{n}}_{12} \times \left(\vec{\mathbf{H}}_2 - \vec{\mathbf{H}}_1\right)\right] = \vec{\mathbf{K}} \cdot \vec{\mathbf{t}} \qquad [7.18]$$

Since $\vec{\mathbf{t}}$ is arbitrary, we deduce:

$$\vec{\mathbf{n}}_{12} \times \left(\vec{\mathbf{H}}_2 - \vec{\mathbf{H}}_1\right) = \vec{\mathbf{K}} \qquad [7.19]$$

Therefore, the tangential components of the magnetic field $\vec{\mathbf{H}}$ across the interface are discontinuous. This discontinuity depends on the surface current. The continuity equation [7.19] can also be written as:

$$H_{2\parallel} - H_{1\parallel} = K \qquad \Longleftrightarrow \qquad \left(\vec{\mathbf{H}}_2 - \vec{\mathbf{H}}_1\right)_\parallel = \vec{\mathbf{K}} \times \vec{\mathbf{n}}_{12} \qquad [7.20]$$

c) *Normal component of* $\vec{\mathbf{B}}$: the continuity equation for the normal component of the magnetic field $\vec{\mathbf{B}}$ is derived from equation [7.5], where S is the surface of the cylinder shown in Figure 7.1(b):

$$\lim_{\epsilon \longrightarrow 0} \oint_S \vec{\mathbf{B}} \cdot d\vec{\mathbf{S}} = 0 \qquad [7.21]$$

It is evident that this surface integral is limited to the two bases S_1 and S_2 (which have equal area). Since the fields are assumed to be uniform over these surfaces, the left-hand side becomes:

$$0 = \int_{S_2} \vec{\mathbf{B}}_2 \cdot \vec{\mathbf{n}}_{12} dS + \int_{S_1} \vec{\mathbf{B}}_1 \cdot \left(-\vec{\mathbf{n}}_{12}\right) dS = \left(B_{2\perp} - B_{1\perp}\right) S \qquad [7.22]$$

We thus conclude that the normal component of $\vec{\mathbf{B}}$ is continuous across the interface, that is:

$$B_{2\perp} - B_{1\perp} = 0 \Longleftrightarrow \vec{\mathbf{n}}_{12} \cdot \left(\vec{\mathbf{B}}_2 - \vec{\mathbf{B}}_1\right) = 0 \qquad [7.23]$$

d) *Normal component of* $\vec{\mathbf{D}}$: the continuity equation for the normal component $\left(\vec{\mathbf{D}}\right)_\perp$ is obtained from the Maxwell–Gauss equation by integrating over the volume of the cylinder and then taking the limit $\epsilon \longrightarrow 0$:

$$\lim_{\epsilon \longrightarrow 0} \oint_S \vec{\mathbf{D}} \cdot d\vec{\mathbf{S}} = \lim_{\epsilon \longrightarrow 0} \int_\tau \rho d\tau \qquad [7.24]$$

As the thickness ϵ of the cylinder tends to zero, the charge inside the infinitesimal volume may still be non-zero if there is a surface charge present on the interface. Thus, we write:

$$\lim_{\epsilon \longrightarrow 0} \oint_S \vec{\mathbf{D}} \cdot d\vec{\mathbf{S}} = \sigma S \qquad [7.25]$$

Integrating the left-hand side over the bases S_1 and S_2 (which are parallel to the interface and equal in area, $S_1 = S_2$), we obtain $(D_{2\perp} - D_{1\perp})\, S = \sigma S$, which leads to:

$$D_{2\perp} - D_{1\perp} = \sigma \implies \vec{\mathbf{n}}_{12} \cdot \left(\vec{\mathbf{D}}_2 - \vec{\mathbf{D}}_1\right) = \sigma \qquad [7.26]$$

The normal component of $\vec{\mathbf{D}}$ is thus discontinuous at the interface. It remains continuous only if there is no surface charge on the interface.

e) *Normal component of the free current density* $\vec{\mathbf{J}}$: using the same procedure, we obtain a continuity equation for the normal component of $\vec{\mathbf{J}}$ by integrating the charge conservation equation over the volume of the cylinder and letting $\epsilon \to 0$: $\lim_{\epsilon \longrightarrow 0} \oint_S \vec{\mathbf{J}} \cdot d\vec{\mathbf{S}} = -\lim_{\epsilon \longrightarrow 0} \int_\tau \frac{\partial \rho}{\partial t} d\tau = -\frac{\partial \sigma}{\partial t} S$. This leads to $(J_{2\perp} - J_{1\perp})\, S = -\frac{\partial \sigma}{\partial t} S$. By simplifying both sides by S, we get:

$$J_{2\perp} - J_{1\perp} = -\frac{\partial \sigma}{\partial t} \iff \vec{\mathbf{n}}_{12} \cdot \left(\vec{\mathbf{J}}_2 - \vec{\mathbf{J}}_1\right) = -\frac{\partial \sigma}{\partial t} \qquad [7.27]$$

where σ is the surface charge density at the interface. Assuming $\sigma = \sigma_0 e^{-\mathrm{i}\omega t}$ (oscillating field in complex notation) and $\vec{\mathbf{J}} = \gamma \vec{\mathbf{E}}$ (ohmic medium), equation [7.27] becomes $\gamma_2 E_{2\perp} - \gamma_1 E_{1\perp} = \mathrm{i}\omega\sigma$. In the same way, equation [7.26] gives $D_{2\perp} - D_{1\perp} = \sigma$, which results in $\varepsilon_2 E_{2\perp} - \varepsilon_1 E_{1\perp} = \sigma$. These two equations are compatible when $\sigma = 0$, unless $\frac{\gamma_1}{\varepsilon_1} = \frac{\gamma_2}{\varepsilon_2}$. Eliminating σ between these two equations, we get:

$$\left(\varepsilon_2 + \mathrm{i}\frac{\gamma_2}{\omega}\right) E_{2\perp} = \left(\varepsilon_1 + \mathrm{i}\frac{\gamma_1}{\omega}\right) E_{1\perp} \qquad [7.28]$$

In summary, the five continuity equations are expressed as follows:

$$\begin{aligned}
&\vec{\mathbf{n}}_{12} \times \left(\vec{\mathbf{E}}_2 - \vec{\mathbf{E}}_1\right) = 0 \\
&\vec{\mathbf{n}}_{12} \times \left(\vec{\mathbf{H}}_2 - \vec{\mathbf{H}}_1\right) = \vec{\mathbf{K}} \\
&\vec{\mathbf{n}}_{12} \cdot \left(\vec{\mathbf{B}}_2 - \vec{\mathbf{B}}_1\right) = 0 \\
&\vec{\mathbf{n}}_{12} \cdot \left(\vec{\mathbf{D}}_2 - \vec{\mathbf{D}}_1\right) = \sigma \\
&\vec{\mathbf{n}}_{12} \cdot \left(\vec{\mathbf{J}}_2 - \vec{\mathbf{J}}_1\right) = -\frac{\partial \sigma}{\partial t}
\end{aligned} \qquad [7.29]$$

where $\vec{\mathbf{n}}_{12}$ is a unit vector normal to the interface surface, directed from medium 1 toward medium 2. It is important to reiterate that *in reality a surface current exists only on a perfect conductor*. The same observation holds for the *surface charges*. In fact, to accumulate current or charge on a surface, the charges must be free to migrate there. This requires infinite conductivity to allow for such mobility. Another important property of perfect conductors is the absence of an electric field inside. The reason is as follows: with infinite conductivity ($\gamma \to \infty$), the current density (and hence the current) becomes infinite, which is physically impossible. It follows that the *electric field* $\vec{\mathbf{E}}$ *must be zero inside a perfect conductor*. Similarly, the magnetic field $\vec{\mathbf{B}}$ cannot be infinite in a magnetically perfect medium ($\mu \to \infty$). Therefore, the *field* $\vec{\mathbf{H}}$ *must vanish within such materials*. As an illustration, we can distinguish the following particular cases (assuming in all cases that $\vec{\mathbf{J}} = \gamma\vec{\mathbf{E}}$, $\vec{\mathbf{D}} = \varepsilon\vec{\mathbf{E}}$, $\vec{\mathbf{B}} = \mu\vec{\mathbf{H}}$, $\vec{\mathbf{H}} = \vec{\mathbf{H}}_0 e^{-i\omega t}$ and $\vec{\mathbf{E}} = \vec{\mathbf{E}}_0 e^{-i\omega t}$):

i) Both media are perfect dielectrics ($\gamma_1 = \gamma_2 = 0$). Neither volume nor surface currents can flow. Furthermore, no free charges are present. At the interface, the continuity equations are applied with $\vec{\mathbf{K}} = 0$ and $\sigma = 0$:

$$\begin{aligned}
&E_{1\parallel} = E_{2\parallel} \implies \vec{\mathbf{n}} \times \vec{\mathbf{E}}_1 = \vec{\mathbf{n}} \times \vec{\mathbf{E}}_2 \\
&H_{1\parallel} = H_{2\parallel} \implies \vec{\mathbf{n}} \times \vec{\mathbf{H}}_1 = \vec{\mathbf{n}} \times \vec{\mathbf{H}}_2 \\
&B_{1\perp} = B_{2\perp} \implies \vec{\mathbf{n}} \cdot \vec{\mathbf{B}}_1 = \vec{\mathbf{n}} \cdot \vec{\mathbf{B}}_2 \\
&D_{1\perp} = D_{2\perp} \implies \varepsilon_1 E_{1\perp} = \varepsilon_2 E_{2\perp} \implies \varepsilon_1 \vec{\mathbf{n}} \cdot \vec{\mathbf{E}}_1 = \varepsilon_2 \vec{\mathbf{n}} \cdot \vec{\mathbf{E}}_2
\end{aligned} \qquad [7.30]$$

It follows that $E_\parallel$, $H_\parallel$, $B_\perp$ and $D_\perp$ are all continuous across the interface. If both media have finite but non-zero conductivities (ohmic media), no surface current can occur in such materials. Consequently, $E_\parallel$, $\left(\varepsilon + i\frac{\gamma}{\omega}\right) E_\perp$, $H_\parallel$, and $B_\perp$ are all continuous.

ii) Medium 1 is a perfect conductor ($\gamma_1 \to \infty$). In this case, we know that $\vec{\mathbf{E}}_1 = 0$. Then, using equation $\vec{\nabla} \times \vec{\mathbf{E}}_1 = -\frac{\partial \vec{\mathbf{B}}_1}{\partial t} = i\mu\omega\vec{\mathbf{H}}_1 = 0$, we deduce that

$\vec{\mathbf{H}}_1 = 0$ (and therefore $\vec{\mathbf{B}}_1 = 0$). The boundary conditions are:

$$\begin{aligned} &E_{2\parallel} = 0 \Longrightarrow \vec{\mathbf{n}}_{12} \times \vec{\mathbf{E}}_2 = 0 \\ &\varepsilon_2 E_{2\perp} = \sigma \Longrightarrow \varepsilon_1 \vec{\mathbf{n}}_{12} \cdot \vec{\mathbf{E}}_2 = \sigma \\ &\vec{\mathbf{n}}_{12} \times \vec{\mathbf{H}}_2 = \vec{\mathbf{K}} \\ &B_{2\perp} = 0 \Longrightarrow \vec{\mathbf{n}}_{12} \cdot \vec{\mathbf{B}}_2 = 0 \end{aligned} \qquad [7.31]$$

iii) Medium 1 is perfectly magnetic ($\mu_1 \to \infty$). We then have $\vec{\mathbf{H}}_1 = 0$. From equation $\vec{\nabla} \times \vec{\mathbf{H}}_1 = (\gamma_1 - \mathrm{i}\varepsilon_1\omega)\,\vec{\mathbf{E}}_1 = 0$, it follows that $\vec{\mathbf{E}}_1 = 0$. The boundary conditions are:

$$\begin{aligned} &E_{2\parallel} = 0 \Longrightarrow \vec{\mathbf{n}}_{12} \times \vec{\mathbf{E}}_2 = 0 \\ &\varepsilon_2 E_{2\perp} = \sigma \Longrightarrow \varepsilon_2 \vec{\mathbf{n}}_{12} \cdot \vec{\mathbf{E}}_2 = \sigma \\ &H_{2\parallel} = 0 \Longrightarrow \vec{\mathbf{n}}_{12} \cdot \vec{\mathbf{H}}_2 = 0 \\ &B_{2\perp} = B_{1\perp} \Longrightarrow \vec{\mathbf{n}}_{12} \cdot \vec{\mathbf{B}}_2 = \vec{\mathbf{n}}_{12} \cdot \vec{\mathbf{B}}_1 \end{aligned} \qquad [7.32]$$

7.2. Energy conservation

7.2.1. *Electrostatic energy stored in a charge distribution*

We have already seen that the potential energy of a charged particle is stored in the fields generated by the particle itself. In what follows, we will see that this energy is also stored in the polarization and magnetization when dealing with a material medium. To demonstrate this, let us calculate the work δW required to add a small amount of charge δq to a charge distribution localized within a finite volume τ, without disturbing its potential V when it is added. By writing $\delta q = \int_\tau \delta\rho d\tau$, the work performed during the addition is:

$$\delta W = V\delta q = \int_\tau V\,(\delta\rho)\,d\tau = \int_\tau V\delta\left(\vec{\nabla}\cdot\vec{\mathbf{D}}\right)d\tau = \int_\tau V\vec{\nabla}\cdot\left(\delta\vec{\mathbf{D}}\right)d\tau \qquad [7.33]$$

Using the identity $\vec{\nabla}\cdot\left(V\delta\vec{\mathbf{D}}\right) = \left(\vec{\nabla}V\right)\cdot\delta\vec{\mathbf{D}} + V\vec{\nabla}\cdot\left(\delta\vec{\mathbf{D}}\right)$, the previous equation reduces to:

$$\delta W = \int_\tau \vec{\nabla}\cdot\left(V\delta\vec{\mathbf{D}}\right)d\tau - \int_\tau \vec{\nabla}V\cdot\delta\vec{\mathbf{D}}d\tau = \oint_S V\delta\vec{\mathbf{D}}\cdot d\vec{\mathbf{S}} + \int_\tau \vec{\mathbf{E}}\cdot\delta\vec{\mathbf{D}}d\tau \qquad [7.34]$$

These integrals can be extended to the entire space, that is, beyond the volume τ:

$$\delta W = \int_{\text{Space}} V\,(\delta\rho)\,d\tau = \lim_{R\longrightarrow\infty}\oint_S V\delta\vec{\mathbf{D}}\cdot d\vec{\mathbf{S}} + \int_{\text{Space}} \vec{\mathbf{E}}\cdot\delta\vec{\mathbf{D}}d\tau \qquad [7.35]$$

where R is the radius of a spherical surface S centered on the volume τ and located at infinity. On this imaginary sphere, the potential V and the field $\vec{\mathbf{D}}$ vary as $\frac{1}{R}$ and $\frac{1}{R^2}$, respectively. Since the surface area $S \sim R^2$, it follows that $\oint_S V\delta\vec{\mathbf{D}} \cdot d\vec{\mathbf{S}} \sim \frac{1}{R} \to 0$ as $R \to 0$. We thus conclude that the first integral vanishes at very large distances, and therefore:

$$\delta W = \int_{\text{Space}} \vec{\mathbf{E}} \cdot \delta\vec{\mathbf{D}} d\tau \qquad [7.36]$$

For an LIH medium, where $\vec{\mathbf{D}} = \varepsilon\vec{\mathbf{E}}$, the total work required to assemble the full charge distribution is:

$$\begin{aligned} W &= \int_0^D \delta W = \int_0^D \int_{\text{Space}} \vec{\mathbf{E}} \cdot \delta\vec{\mathbf{D}} d\tau = \varepsilon \int_{\text{Space}} \int_0^D \vec{\mathbf{E}} \cdot \delta\vec{\mathbf{E}} d\tau \\ &= \frac{\varepsilon}{2} \int_{\text{Space}} \vec{\mathbf{E}}^2 d\tau = \frac{1}{2} \int_{\text{Space}} \vec{\mathbf{E}} \cdot \vec{\mathbf{D}} d\tau \end{aligned} \qquad [7.37]$$

In the case of a linear anisotropic material, we write $D^i = \varepsilon^i_j E^j$, which leads to:

$$\begin{aligned} W &= \int_{\text{Space}} \int_0^D E_i \varepsilon^i_j \delta E^j d\tau = \int_{\text{Space}} \frac{\varepsilon^i_j E_i E^j}{2} d\tau \\ &= \frac{1}{2} \int_{\text{Space}} D_i E^i d\tau = \frac{1}{2} \int_{\text{Space}} \vec{\mathbf{E}} \cdot \vec{\mathbf{D}} d\tau \end{aligned} \qquad [7.38]$$

Note that the expression for the energy density in a linear anisotropic material, $U = \frac{1}{2}\varepsilon^i_j E_i E^j = \frac{1}{2}\varepsilon^j_i E_j E^i = \frac{1}{2}\varepsilon_{ij} E^i E^j$, shows that the permittivity tensor must be symmetric, that is, $\varepsilon_{ij} = \varepsilon_{ji}$.

7.2.2. *Poynting vector*

When both fields $\vec{\mathbf{E}}$ and $\vec{\mathbf{H}}$ are present, we proceed as follows: we expand the vector quantity:

$$\begin{aligned} \vec{\nabla} \cdot \left(\vec{\mathbf{E}} \times \vec{\mathbf{H}}\right) &= \vec{\mathbf{H}} \cdot \left(\vec{\nabla} \times \vec{\mathbf{E}}\right) - \vec{\mathbf{E}} \cdot \left(\vec{\nabla} \times \vec{\mathbf{H}}\right) \\ &= -\vec{\mathbf{H}} \cdot \frac{\partial \vec{\mathbf{B}}}{\partial t} - \vec{\mathbf{E}} \cdot \frac{\partial \vec{\mathbf{D}}}{\partial t} - \vec{\mathbf{E}} \cdot \vec{\mathbf{J}} \end{aligned} \qquad [7.39]$$

where the terms $\vec{\nabla} \times \vec{\mathbf{E}}$ and $\vec{\nabla} \times \vec{\mathbf{H}}$ were replaced using the Maxwell–Faraday and Maxwell–Ampère equations. For an LIH medium, we also have $\vec{\mathbf{D}} = \varepsilon\vec{\mathbf{E}}$ and $\vec{\mathbf{B}} = \mu\vec{\mathbf{H}}$. It follows that equation [7.39] can be rewritten as:

$$\vec{\nabla} \cdot \left(\vec{\mathbf{E}} \times \vec{\mathbf{H}}\right) + \frac{\partial}{\partial t}\left(\frac{1}{2}\varepsilon E^2 + \frac{1}{2}\mu H^2\right) = -\vec{\mathbf{E}} \cdot \vec{\mathbf{J}} \qquad [7.40]$$

This is the differential form of the *Poynting theorem*. Upon integrating [7.40] over a closed volume and then using the divergence theorem to convert the first term into a surface integral:

$$\oint_S \left(\vec{\mathbf{E}} \times \vec{\mathbf{H}}\right) \cdot d\vec{\mathbf{S}} + \frac{d}{dt}\int_\tau \left(\frac{1}{2}\varepsilon E^2 + \frac{1}{2}\mu H^2\right) d\tau = -\int_\tau \vec{\mathbf{E}} \cdot \vec{\mathbf{J}}\, d\tau \quad [7.41]$$

Here, $\frac{1}{2}\varepsilon E^2 + \frac{1}{2}\mu H^2$ is recognized as the electromagnetic energy density, suggesting the following interpretation of [7.41]:

$$P_{\text{out}} + \frac{dW_{\text{em}}}{dt} = -P_d \quad [7.42]$$

where P_{out} is the *total electromagnetic power* of density:

$$\vec{\mathbf{\Pi}} = \vec{\mathbf{E}} \times \vec{\mathbf{H}} \quad [7.43]$$

flowing out of the volume. In other words, P_{out} represents the outward flux of the vector $\vec{\mathbf{\Pi}}$ across the surface S surrounding the volume τ. $\vec{\mathbf{\Pi}}$ is called the *Poynting vector* (also called the *power density vector*). The quantities W_{em} and P_d are, respectively, the *stored electromagnetic energy* and the *dissipated power* in the volume:

$$\begin{aligned} P_{\text{out}} &= \oint_S \left(\vec{\mathbf{E}} \times \vec{\mathbf{H}}\right) \cdot d\vec{\mathbf{S}} = \oint_S \vec{\mathbf{\Pi}} \cdot d\vec{\mathbf{S}} \\ W_{\text{em}} &= \int_\tau \left(\frac{1}{2}\varepsilon E^2 + \frac{1}{2}\mu H^2\right) d\tau \\ P_d &= \int_\tau \vec{\mathbf{E}} \cdot \vec{\mathbf{J}}\, d\tau = \int_\tau \rho \vec{\mathbf{E}} \cdot \vec{\mathbf{v}}\, d\tau = \int_\tau \vec{\mathbf{f}} \cdot \vec{\mathbf{v}}\, d\tau \end{aligned} \quad [7.44]$$

If $\vec{\mathbf{E}}$ and $\vec{\mathbf{J}}$ are in the same direction as in an ohmic conductor (i.e. $\vec{\mathbf{E}} \cdot \vec{\mathbf{J}} = \gamma E^2$), then P_d is positive, representing power dissipation, as the right-hand side of [7.42] is negative. In the case of a source that supplies energy to the volume, $\vec{\mathbf{E}}$ and $\vec{\mathbf{J}}$ point in opposite directions, which implies that P_d is negative.

EXAMPLE 7.1. (Capacitor with losses).– Consider a cylindrical capacitor (with radius a and length l) with losses excited by a voltage source that varies over time, as illustrated in Figure 7.2. The electric field inside the cylinder is given by $E_z = \frac{V}{l}$ (or, in vector form, $\vec{\mathbf{E}} = \frac{V}{l}\vec{\mathbf{k}}$), while the field $\vec{\mathbf{H}}$ on the outer surface of the cylinder (generated by both the conduction current and the displacement current) is determined from $\oint_L \vec{\mathbf{H}} \cdot d\vec{\mathbf{l}} = \int_S \left(J_z + \varepsilon\frac{\partial E_z}{\partial t}\right) dS_z$, which gives $H_\varphi 2\pi a = \frac{\gamma S V}{l} + \frac{\varepsilon S}{l}\frac{dV}{dt} \equiv I$. Thus, we obtain:

$$H_\varphi = \frac{I}{2\pi a} \quad [7.45]$$

where:

$$I = \frac{V}{R} + C\frac{dV}{dt} \quad [7.46]$$

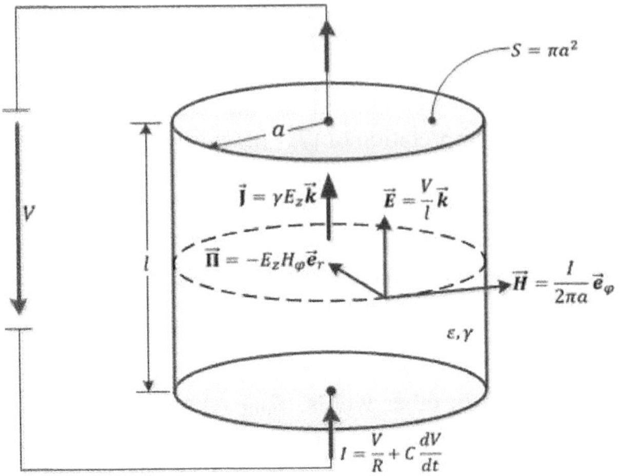

Figure 7.2. *Cylindrical capacitor with losses excited by a time-varying source. For a color version of this figure, see www.iste.co.uk/abdesselam/electrodynamics2.zip*

Here, I is the total current flowing through the circuit (ohmic current + displacement current). The quantities $R = \frac{l}{\gamma S}$ and $C = \frac{\varepsilon S}{l}$ represent the resistance and capacitance of the cylinder, respectively. Multiplying [7.46] by V, we get:

$$VI = \frac{V^2}{R} + \frac{d}{dt}\left(\frac{1}{2}CV^2\right) \quad [7.47]$$

The power delivered to the cylinder through the circuit is dissipated by the resistance and stored in the capacitor as electrical energy. The same result can be obtained using the Poynting theorem. The outward flux of the Poynting vector through the cylinder is:

$$\begin{aligned} \oint_S \left(\vec{\mathbf{E}} \times \vec{\mathbf{H}}\right) \cdot d\vec{\mathbf{S}} &= -\int_0^{2\pi}\int_0^l E_z H_\varphi a d\varphi dz \\ &= -\int_0^{2\pi} d\varphi \int_0^l dz \frac{VI}{2\pi l} = -VI \end{aligned} \quad [7.48]$$

where the negative sign arises from the fact that the Poynting vector $\vec{\mathbf{\Pi}}$ is directed inward. The power dissipated by the cylinder is given by:

$$P_d = \int_{\text{cylinder}} \vec{\mathbf{E}} \cdot \vec{\mathbf{J}}\, d\tau = \int_{\text{cylinder}} \gamma E_z^2 d\tau = \frac{\gamma S}{l} V^2 = \frac{V^2}{R} \quad [7.49]$$

Similarly, the energy stored inside the cylinder is:

$$W = \int_{\text{cylinder}} \frac{1}{2}\varepsilon E_z^2 d\tau = \frac{\varepsilon S}{2l} V^2 = \frac{1}{2}\mathbf{C}V^2 \qquad [7.50]$$

By substituting these results into the Poynting theorem [7.42], we once again arrive at equation [7.47].

Let us generalize equation [7.48]. This equation shows that the flux of $\vec{\mathbf{\Pi}}$ entering a circuit element is exactly equal to VI. In general, this result can be extended to a circuit within an N-branch network using the quasi-static approximation, which is valid outside of the circuit elements: (1) $\vec{\nabla} \times \vec{\mathbf{E}} = 0 \Longrightarrow \vec{\mathbf{E}} = -\vec{\nabla} V$; (2) $\vec{\nabla} \times \vec{\mathbf{H}} = \vec{\mathbf{J}} \Longrightarrow \vec{\nabla} \cdot \vec{\mathbf{J}} = 0$. Consider a sphere enclosing the circuit, as shown in Figure 7.3.

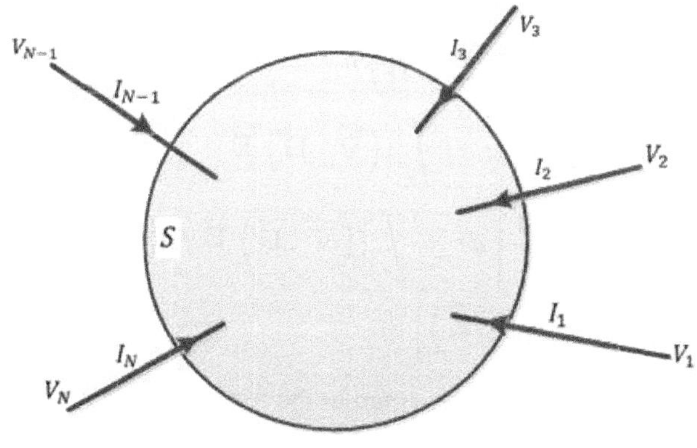

Figure 7.3. *The power $\sum_{k=1}^{N} V_k I_k$ received by a circuit in an N-branch network is equal to the inward flux of the Poynting vector through an imaginary surface enclosing the circuit, $-\oint_S \left(\vec{\mathbf{E}} \times \vec{\mathbf{H}}\right) \cdot d\vec{\mathbf{S}}$. For a color version of this figure, see www.iste.co.uk/abdesselam/electrodynamics2.zip*

The electromagnetic power received through the surface S, through the Poynting vector, is given by:

$$\begin{aligned} P_{\text{inward}} &= -\oint_S \left(\vec{\mathbf{E}} \times \vec{\mathbf{H}}\right) \cdot d\vec{\mathbf{S}} = -\int_\tau \vec{\nabla} \cdot \left(\vec{\mathbf{E}} \times \vec{\mathbf{H}}\right) d\tau \\ &= \int_\tau \vec{\nabla} \cdot \left(\vec{\nabla} V \times \vec{\mathbf{H}}\right) d\tau \end{aligned} \qquad [7.51]$$

(The negative sign appears because this is the power received by the sphere). Using the identity $\vec{\nabla} \cdot \left(\vec{\nabla} V \times \vec{\mathbf{H}}\right) = \vec{\mathbf{H}} \cdot \left(\vec{\nabla} \times \vec{\nabla} V\right) - \vec{\nabla} V \cdot \left(\vec{\nabla} \times \vec{\mathbf{H}}\right) = -\vec{\mathbf{J}} \cdot \vec{\nabla} V = -\vec{\nabla} \cdot \left(V \vec{\mathbf{J}}\right)$, it can be derived that:

$$P_{\text{inward}} = -\int_{\tau} \vec{\nabla} \cdot \left(V\vec{\mathbf{J}}\right) d\tau = -\oint_{S} V\vec{\mathbf{J}} \cdot d\vec{\mathbf{S}} \qquad [7.52]$$

On the surface S, the potential V is simply equal to the voltages across each branch. Since $d\vec{\mathbf{S}}$ is directed outward, it follows that:

$$P_{\text{inward}} = -\sum_{k=1}^{N} V_k \oint_{S} \vec{\mathbf{J}}_k \cdot d\vec{\mathbf{S}} = \sum_{k=1}^{N} V_k I_k \qquad [7.53]$$

7.2.3. *Electromagnetic momentum and Maxwell stress tensor*

A (charged) distribution spread over a volume τ is placed in an electromagnetic field. The total force experienced by the distribution is given by (where $\vec{\mathcal{P}}$ is the *mechanical momentum density*):

$$\begin{aligned}\vec{\mathbf{F}} &= \frac{d}{dt}\int_{\tau} \vec{\mathcal{P}} d\tau = \int_{\tau} \left(\rho\vec{\mathbf{E}} + \vec{\mathbf{J}} \times \vec{\mathbf{B}}\right) d\tau = \int_{\tau} \Bigg[\left(\vec{\nabla} \cdot \vec{\mathbf{D}}\right) \vec{\mathbf{E}} \\ &+ \left(\vec{\nabla} \times \vec{\mathbf{H}} - \frac{\partial \vec{\mathbf{D}}}{\partial t}\right) \times \vec{\mathbf{B}}\Bigg] d\tau = \int_{\tau} \Bigg[\left(\vec{\nabla} \cdot \vec{\mathbf{D}}\right) \vec{\mathbf{E}} + \left(\vec{\nabla} \times \vec{\mathbf{H}}\right) \times \vec{\mathbf{B}} \\ &- \frac{\partial}{\partial t}\left(\vec{\mathbf{D}} \times \vec{\mathbf{B}}\right) + \vec{\mathbf{D}} \times \frac{\partial \vec{\mathbf{B}}}{\partial t}\Bigg] d\tau = \int_{\tau} \Bigg[\left(\vec{\nabla} \cdot \vec{\mathbf{D}}\right) \vec{\mathbf{E}} + \left(\vec{\nabla} \times \vec{\mathbf{H}}\right) \times \vec{\mathbf{B}} \\ &- \frac{\partial}{\partial t}\left(\vec{\mathbf{D}} \times \vec{\mathbf{B}}\right) - \vec{\mathbf{D}} \times \left(\vec{\nabla} \times \vec{\mathbf{E}}\right)\Bigg] d\tau\end{aligned} \qquad [7.54]$$

This last equation can also be rewritten in the form:

$$\begin{aligned}\frac{d}{dt}\int_{\tau} \left(\vec{\mathcal{P}} + \vec{\mathbf{D}} \times \vec{\mathbf{B}}\right) d\tau = \int_{\tau} \Big[&\left(\vec{\nabla} \cdot \vec{\mathbf{D}}\right) \vec{\mathbf{E}} - \vec{\mathbf{B}} \times \left(\vec{\nabla} \times \vec{\mathbf{H}}\right) \\ &- \vec{\mathbf{D}} \times \left(\vec{\nabla} \times \vec{\mathbf{E}}\right)\Big] d\tau\end{aligned} \qquad [7.55]$$

If the fields are uniform, all of the spatial derivatives on the right-hand side vanish. The conservation of momentum suggests that $\vec{\mathcal{P}}_{\text{em}} = \vec{\mathbf{D}} \times \vec{\mathbf{B}}$ should be interpreted as the *momentum associated with the fields*, or the *electromagnetic momentum density*.

Let us express the right-hand side of [7.55] as the divergence of a second-order tensor, by setting:

$$\begin{aligned}f^j &= \left(\partial_k D^k\right) E^j - \epsilon^{jkl} B_k \epsilon_{lmn} \partial^m H^n - \epsilon^{jkl} D_k \epsilon_{lmn} \partial^m E^n \\ &= \left(\partial_k D^k\right) E^j - \epsilon^{ljk} \epsilon_{lmn} B_k \partial^m H^n - \epsilon^{ljk} \epsilon_{lmn} D_k \partial^m E^n \\ &= \left(\partial_k D^k\right) E^j - \left(\delta^j_m \delta^k_n - \delta^k_m \delta^j_n\right) B_k \partial^m H^n - \left(\delta^j_m \delta^k_n - \delta^k_m \delta^j_n\right) D_k \partial^m E^n \\ &= \left(\partial_k D^k\right) E^j - B_k \partial^j H^k + B_k \partial^k H^j - D_k \partial^j E^k + D_k \partial^k E^j\end{aligned} \qquad [7.56]$$

To give the previous equation a symmetric form, we add the term $0 = \left(\overrightarrow{\nabla} \cdot \overrightarrow{\mathbf{B}}\right) H^j = \left(\partial_k B^k\right) H^j$ to the right-hand side:

$$f^j = \partial_k D^k E^j + D_k \partial^k E^j + B_k \partial^k H^j + \partial^k B_k H^j - \left(B_k \partial^j H^k + D_k \partial^j E^k\right) \quad [7.57]$$

For linear media (not necessarily isotropic), the last two terms yield $B_k \partial^j H^k + D_k \partial^j E^k = \frac{1}{2} \partial^j (\overrightarrow{\mathbf{B}} \cdot \overrightarrow{\mathbf{H}} + \overrightarrow{\mathbf{D}} \cdot \overrightarrow{\mathbf{E}}) = \frac{1}{2} \delta^{jk} \partial_k (\overrightarrow{\mathbf{B}} \cdot \overrightarrow{\mathbf{H}} + \overrightarrow{\mathbf{D}} \cdot \overrightarrow{\mathbf{E}})$, and consequently:

$$f^j = \partial_k \left[D^k E^j + B^k H^j - \frac{1}{2} \delta^{jk} \left(\overrightarrow{\mathbf{B}} \cdot \overrightarrow{\mathbf{H}} + \overrightarrow{\mathbf{D}} \cdot \overrightarrow{\mathbf{E}}\right)\right] = -\partial_k T^{kj} \quad [7.58]$$

where:

$$T^{ij} = -D^i E^j - B^i H^j + \frac{1}{2} \delta^{ij} \left(\overrightarrow{\mathbf{B}} \cdot \overrightarrow{\mathbf{H}} + \overrightarrow{\mathbf{D}} \cdot \overrightarrow{\mathbf{E}}\right) \quad [7.59]$$

This is the *electromagnetic Maxwell stress tensor*.

EXAMPLE 7.2. (Stefan's equation for blackbodies).– Let us use the stress tensor to compute the pressure exerted by isotropic radiation on the walls of a cavity, assumed to be cubic and perfectly absorbent (a blackbody). We calculate the force $d\overrightarrow{\mathbf{F}}$ acting on a surface element $d\overrightarrow{\mathbf{S}}$ of the wall. According to the divergence theorem, we have $d\overrightarrow{\mathbf{F}} = \overleftrightarrow{\mathbf{T}} \otimes d\overrightarrow{\mathbf{S}}$. To simplify the calculation, we take $d\overrightarrow{\mathbf{S}}$ as an element of the right-hand wall, with a normal vector $-\overrightarrow{\mathbf{i}}$. Consequently, the only non-zero component of the force is:

$$\begin{aligned} dF^x &= -T^{xj} dS_j = T^{xx} dS_x \\ &= -\left[D^x E^x + B^x H^x - \frac{1}{2} \left(\overrightarrow{\mathbf{B}} \cdot \overrightarrow{\mathbf{H}} + \overrightarrow{\mathbf{D}} \cdot \overrightarrow{\mathbf{E}}\right)\right] dS_x \end{aligned} \quad [7.60]$$

For isotropic radiation, we have $E_x D_x = E_y D_y = E_z D_z = \frac{1}{3} \left(\overrightarrow{\mathbf{D}} \cdot \overrightarrow{\mathbf{E}}\right)$ and $B_x H_x = \frac{1}{3} \left(\overrightarrow{\mathbf{H}} \cdot \overrightarrow{\mathbf{B}}\right)$. Substituting these expressions into [7.60], we deduce that:

$$dF^x = \frac{1}{3} \left(\frac{\overrightarrow{\mathbf{D}} \cdot \overrightarrow{\mathbf{E}} + \overrightarrow{\mathbf{H}} \cdot \overrightarrow{\mathbf{B}}}{2}\right) dS_x = \frac{1}{3} U dS_x \quad [7.61]$$

We conclude that the radiation pressure $\mathsf{P} = \frac{dF^x}{dS_x}$ exerted on the wall is equal to $\frac{1}{3}$ of the energy density. This result forms the basis of the *Stefan equation* derived for blackbodies. Let us try to derive this equation. According to the first law of thermodynamics, the change in internal energy u during an infinitesimal transformation of a closed system satisfies:

$$du = \delta W + \delta Q \quad [7.62]$$

where δW is the work done by the pressure forces (the work induced by pressure forces corresponds to the most common form of work encountered in thermodynamics; it is expressed in the following differential form: $\delta W = -\mathsf{P}d\tau$) and δQ is the amount of heat exchanged with the external environment. For a blackbody, no heat is exchanged with the outside, which means $0 = \delta Q = du + \mathsf{P}d\tau$. Writing $u = U\tau = 3\mathsf{P}\tau$, we get $du = 3\mathsf{P}d\tau + 3\tau d\mathsf{P}$ and therefore $0 = 4\mathsf{P}d\tau + 3\tau d\mathsf{P}$, which leads to:

$$\mathsf{P}\tau^{\frac{4}{3}} = A \qquad [7.63]$$

where A is a constant of integration. The body therefore behaves like a perfect gas with $\gamma = \frac{4}{3}$. To establish the relationship between internal energy u and absolute temperature T, we just need to combine the two thermodynamic relations (here, F is the enthalpy of the system) $\mathsf{P} = -\frac{\partial \mathsf{F}}{\partial \tau}$ and $u = \mathsf{F} - T\frac{\partial \mathsf{F}}{\partial T}$, which gives $\frac{\partial u}{\partial \tau} = T\frac{\partial \mathsf{P}}{\partial T} - \mathsf{P}$. Thus, we obtain $U = \frac{T}{3}\frac{U}{\partial T} - \frac{U}{3}$, which leads to $4U = T\frac{\partial U}{\partial T}$. Integrating this last equation, we find the Stefan equation:

$$U = aT^4 \qquad [7.64]$$

Thus, the energy density of the radiation is proportional to T^4. It follows that the radiation pressure is given by:

$$\mathsf{P} = \frac{1}{3}aT^4 \qquad [7.65]$$

7.2.4. *Complex Poynting theorem*

If the fields $\vec{\mathbf{E}}$ and $\vec{\mathbf{H}}$ vary sinusoidally with time:

$$\begin{aligned} \vec{\mathbf{E}}\left(\vec{\mathbf{r}},t\right) &= \Re\left[\vec{\hat{\mathbf{E}}}\left(\vec{\mathbf{r}}\right)e^{-i\omega t}\right] = \frac{1}{2}\left[\vec{\hat{\mathbf{E}}}\left(\vec{\mathbf{r}}\right)e^{-i\omega t} + \vec{\hat{\mathbf{E}}}^*\left(\vec{\mathbf{r}}\right)e^{i\omega t}\right] \\ \vec{\mathbf{H}}\left(\vec{\mathbf{r}},t\right) &= \Re\left[\vec{\hat{\mathbf{H}}}\left(\vec{\mathbf{r}}\right)e^{-i\omega t}\right] = \frac{1}{2}\left[\vec{\hat{\mathbf{H}}}\left(\vec{\mathbf{r}}\right)e^{-i\omega t} + \vec{\hat{\mathbf{H}}}^*\left(\vec{\mathbf{r}}\right)e^{i\omega t}\right] \end{aligned} \qquad [7.66]$$

where $\vec{\hat{\mathbf{E}}}^*\left(\vec{\mathbf{r}}\right)$ and $\vec{\hat{\mathbf{H}}}^*\left(\vec{\mathbf{r}}\right)$ are the complex conjugates of $\vec{\hat{\mathbf{E}}}\left(\vec{\mathbf{r}}\right)$ and $\vec{\hat{\mathbf{H}}}\left(\vec{\mathbf{r}}\right)$, respectively, the Poynting vector (power density vector) is written as:

$$\begin{aligned} &\vec{\mathbf{\Pi}}\left(\vec{\mathbf{r}},t\right) = \vec{\mathbf{E}}\left(\vec{\mathbf{r}},t\right) \times \vec{\mathbf{H}}\left(\vec{\mathbf{r}},t\right) = \frac{1}{4}\left[\vec{\hat{\mathbf{E}}}\left(\vec{\mathbf{r}}\right)e^{-i\omega t} + \vec{\hat{\mathbf{E}}}^*\left(\vec{\mathbf{r}}\right)e^{i\omega t}\right] \\ &\times \left[\vec{\hat{\mathbf{H}}}\left(\vec{\mathbf{r}}\right)e^{-i\omega t} + \vec{\hat{\mathbf{H}}}^*\left(\vec{\mathbf{r}}\right)e^{i\omega t}\right] = \frac{1}{4}\left[\vec{\hat{\mathbf{E}}}\left(\vec{\mathbf{r}}\right) \times \vec{\hat{\mathbf{H}}}\left(\vec{\mathbf{r}}\right)e^{-2i\omega t} + \right. \\ &\left. \vec{\hat{\mathbf{E}}}^*\left(\vec{\mathbf{r}}\right) \times \vec{\hat{\mathbf{H}}}^*\left(\vec{\mathbf{r}}\right)e^{2i\omega t} + \vec{\hat{\mathbf{E}}}\left(\vec{\mathbf{r}}\right) \times \vec{\hat{\mathbf{H}}}^*\left(\vec{\mathbf{r}}\right) \vec{\hat{\mathbf{E}}}^*\left(\vec{\mathbf{r}}\right) \times \vec{\hat{\mathbf{H}}}\left(\vec{\mathbf{r}}\right)\right] \end{aligned} \qquad [7.67]$$

In practice, it is useful to compute the average of the Poynting vector over a period $T = \frac{2\pi}{\omega}$:

$$\langle\vec{\Pi}\rangle = \frac{1}{T}\int_0^T \vec{\Pi}\left(\vec{r}, t\right) dt = \frac{1}{4}\left[\vec{\hat{E}}\left(\vec{r}\right) \times \vec{\hat{H}}^*\left(\vec{r}\right) + \vec{\hat{E}}^*\left(\vec{r}\right) \times \vec{\hat{H}}\left(\vec{r}\right)\right] = \frac{1}{2}\Re\left[\vec{\hat{E}}\left(\vec{r}\right) \times \vec{\hat{H}}^*\left(\vec{r}\right)\right] \quad [7.68]$$

The integrals of the first two terms are zero. Based on equation [7.68], let us define the *complex Poynting vector*:

$$\vec{\hat{\Pi}}\left(\vec{r}\right) = \frac{1}{2}\left[\vec{\hat{E}}\left(\vec{r}\right) \times \vec{\hat{H}}^*\left(\vec{r}\right)\right] \quad [7.69]$$

Thus, the real part of the complex Poynting vector gives the time average of the Poynting vector (i.e. the time average of the power density vector):

$$\langle\vec{\Pi}\rangle = \Re\left[\vec{\hat{\Pi}}\right] \quad [7.70]$$

To derive a complex form of the Poynting theorem, let us rewrite the Maxwell equations (valid for sinusoidal fields in an LIH medium):

$$\begin{cases} \vec{\nabla} \times \vec{\hat{E}}\left(\vec{r}\right) = \mathrm{i}\omega\mu\vec{\hat{H}}\left(\vec{r}\right) \\ \vec{\nabla} \times \vec{\hat{H}}\left(\vec{r}\right) = \vec{\hat{J}}\left(\vec{r}\right) - \mathrm{i}\omega\varepsilon\vec{\hat{E}}\left(\vec{r}\right) \\ \vec{\nabla} \cdot \vec{\hat{E}}\left(\vec{r}\right) = \dfrac{\hat{\rho}\left(\vec{r}\right)}{\varepsilon} \\ \vec{\nabla} \cdot \vec{\hat{B}}\left(\vec{r}\right) = 0 \end{cases} \Longleftrightarrow \begin{cases} \vec{\nabla} \times \vec{\hat{E}}^*\left(\vec{r}\right) = -\mathrm{i}\omega\mu\vec{\hat{H}}^*\left(\vec{r}\right) \\ \vec{\nabla} \times \vec{\hat{H}}^*\left(\vec{r}\right) = \vec{\hat{J}}^*\left(\vec{r}\right) \\ \qquad + \mathrm{i}\omega\varepsilon\vec{\hat{E}}^*\left(\vec{r}\right) \\ \vec{\nabla} \cdot \vec{\hat{E}}^*\left(\vec{r}\right) = \dfrac{\hat{\rho}^*\left(\vec{r}\right)}{\varepsilon} \\ \vec{\nabla} \cdot \vec{\hat{B}}^*\left(\vec{r}\right) = 0 \end{cases} \quad [7.71]$$

and then expand the product:

$$\vec{\nabla} \cdot \vec{\hat{\Pi}}\left(\vec{r}\right) = \vec{\nabla} \cdot \left[\frac{1}{2}\vec{\hat{E}}\left(\vec{r}\right) \times \vec{\hat{H}}^*\left(\vec{r}\right)\right] = \frac{1}{2}\left[\vec{\hat{H}}^*\left(\vec{r}\right) \cdot \left(\vec{\nabla} \times \vec{\hat{E}}\left(\vec{r}\right)\right) - \vec{\hat{E}}\left(\vec{r}\right) \cdot \left(\vec{\nabla} \times \vec{\hat{H}}^*\left(\vec{r}\right)\right)\right] = \frac{\mathrm{i}\omega}{2}\left[\mu\vec{\hat{H}}\left(\vec{r}\right) \cdot \vec{\hat{H}}^*\left(\vec{r}\right) - \varepsilon\vec{\hat{E}}\left(\vec{r}\right) \cdot \vec{\hat{E}}^*\left(\vec{r}\right)\right] - \frac{1}{2}\vec{\hat{E}}\left(\vec{r}\right) \cdot \vec{\hat{J}}^*\left(\vec{r}\right) \quad [7.72]$$

Thus, we obtain:

$$\vec{\nabla} \cdot \vec{\hat{\Pi}} + 2\mathrm{i}\omega\left[\langle w_e\rangle - \langle w_m\rangle\right] = -\hat{p}_d \quad [7.73]$$

where:

$$\begin{aligned}
\langle w_e\rangle &= \frac{\varepsilon}{4}\vec{\hat{\mathbf{E}}}\left(\vec{\mathbf{r}}\right)\cdot\vec{\hat{\mathbf{E}}}^*\left(\vec{\mathbf{r}}\right) = \frac{\varepsilon}{4}\left|\vec{\hat{\mathbf{E}}}\left(\vec{\mathbf{r}}\right)\right|^2 \\
\langle w_m\rangle &= \frac{\mu}{4}\vec{\hat{\mathbf{H}}}\left(\vec{\mathbf{r}}\right)\cdot\vec{\hat{\mathbf{H}}}^*\left(\vec{\mathbf{r}}\right) = \frac{\mu}{4}\left|\vec{\hat{\mathbf{H}}}\left(\vec{\mathbf{r}}\right)\right|^2 \\
\hat{p}_d &= \frac{1}{2}\vec{\hat{\mathbf{E}}}\left(\vec{\mathbf{r}}\right)\cdot\vec{\hat{\mathbf{J}}}^*\left(\vec{\mathbf{r}}\right)
\end{aligned} \qquad [7.74]$$

This is *the complex Poynting theorem.* The quantities $\langle w_e\rangle$ and $\langle w_m\rangle$ represent the time-averaged energy densities of the electric and magnetic fields, respectively. In fact, we have: $\langle w_e\rangle = \frac{1}{T}\frac{\varepsilon}{2}\int_0^T\left|\vec{\mathbf{E}}\left(\vec{\mathbf{r}},t\right)\right|^2 dt = \frac{\varepsilon}{4}\vec{\hat{\mathbf{E}}}\left(\vec{\mathbf{r}}\right)\cdot\vec{\hat{\mathbf{E}}}^*\left(\vec{\mathbf{r}}\right)$. The complex Poynting theorem depends on the difference between the quantities $\langle w_e\rangle$ and $\langle w_m\rangle$, rather than their sum. The real part of $\hat{p}_d$ represents the average power density dissipated within the ohmic medium over a period. By integrating on a volume τ, we obtain the integral form of the complex Poynting theorem:

$$\oint_S \vec{\hat{\mathbf{\Pi}}}\cdot d\vec{\mathbf{S}} + 2\mathrm{i}\omega\left[\langle W_e\rangle - \langle W_m\rangle\right] = -\hat{P}_d \qquad [7.75]$$

where:

$$\begin{aligned}
\langle W_e\rangle &= \frac{\varepsilon}{4}\int_\tau\left|\vec{\hat{\mathbf{E}}}\left(\vec{\mathbf{r}}\right)\right|^2 d\tau \\
\langle W_m\rangle &= \frac{\mu}{4}\int_\tau\left|\vec{\hat{\mathbf{H}}}\left(\vec{\mathbf{r}}\right)\right|^2 d\tau \\
\hat{P}_d &= \frac{1}{2}\int_\tau \vec{\hat{\mathbf{E}}}\left(\vec{\mathbf{r}}\right)\cdot\vec{\hat{\mathbf{J}}}^*\left(\vec{\mathbf{r}}\right) d\tau
\end{aligned} \qquad [7.76]$$

7.3. Electromagnetic potentials

The introduction of matter has no effect on equations $\vec{\nabla}\cdot\vec{\mathbf{B}} = 0$ and $\vec{\nabla}\times\vec{\mathbf{E}} = -\frac{\partial\vec{\mathbf{B}}}{\partial t}$, nor on their consequences. In fact, equation $\vec{\nabla}\cdot\vec{\mathbf{B}} = 0$ implies that $\vec{\mathbf{B}} = \vec{\nabla}\times\vec{\mathbf{A}}$. Combining this with the Maxwell–Faraday equation, $\vec{\nabla}\times\vec{\mathbf{E}} = -\frac{\partial\vec{\mathbf{B}}}{\partial t}$, we deduce $\vec{\nabla}\times\left(\vec{\mathbf{E}} + \frac{\partial\vec{\mathbf{A}}}{\partial t}\right) = 0$, which leads to $\vec{\mathbf{E}} + \frac{\partial\vec{\mathbf{A}}}{\partial t} = -\vec{\nabla}V$ and consequently to $\vec{\mathbf{E}} = -\vec{\nabla}V - \frac{\partial\vec{\mathbf{A}}}{\partial t}$. As before, the vector potential $\vec{\mathbf{A}}$ is not unique: the potentials $\vec{\mathbf{A}}$ and $\vec{\mathbf{A}} + \vec{\nabla}f$ (with f being any scalar function) produce the same $\vec{\mathbf{B}}$. This is known as gauge freedom. This freedom in the choice of $\vec{\mathbf{A}}$ allows us to choose $\vec{\nabla}\cdot\vec{\mathbf{A}}$ at our convenience. By substituting $\vec{\mathbf{B}}$ with $\vec{\nabla}\times\vec{\mathbf{A}}$ and $\vec{\mathbf{E}}$ with $-\vec{\nabla}V - \frac{\partial\vec{\mathbf{A}}}{\partial t}$ into the other two Maxwell equations, we obtain:

1) Equation $\vec{\nabla} \times \vec{\mathbf{H}} = \vec{\mathbf{J}} + \varepsilon\frac{\partial \vec{\mathbf{E}}}{\partial t}$ becomes $\frac{1}{\mu}\vec{\nabla} \times \left[\vec{\nabla} \times \vec{\mathbf{A}}\right] = \vec{\mathbf{J}} + \varepsilon\frac{\partial}{\partial t}\left(-\vec{\nabla}V - \frac{\partial \vec{\mathbf{A}}}{\partial t}\right) = \frac{1}{\mu}\left[\vec{\nabla}\left(\vec{\nabla} \cdot \vec{\mathbf{A}}\right) - \Delta\vec{\mathbf{A}}\right]$. Thus, it follows that:

$$\Delta\vec{\mathbf{A}} - \mu\varepsilon\frac{\partial^2 \vec{\mathbf{A}}}{\partial t^2} = -\mu\vec{\mathbf{J}} + \vec{\nabla}\left[\vec{\nabla} \cdot \vec{\mathbf{A}} + \mu\varepsilon\frac{\partial V}{\partial t}\right] \quad [7.77]$$

2) The Maxwell–Gauss equation $\vec{\nabla} \cdot \vec{\mathbf{D}} = \rho$ can also be written as $\vec{\nabla} \cdot \vec{\mathbf{D}} = \vec{\nabla} \cdot \left(\varepsilon\vec{\mathbf{E}}\right) = \varepsilon\vec{\nabla} \cdot \left(-\vec{\nabla}V - \frac{\partial \vec{\mathbf{A}}}{\partial t}\right) = -\varepsilon\Delta V - \varepsilon\frac{\partial}{\partial t}\left(\vec{\nabla} \cdot \vec{\mathbf{A}}\right) = \rho$, which implies that:

$$\Delta V - \mu\varepsilon\frac{\partial^2 V}{\partial t^2} = -\frac{\rho}{\varepsilon} - \frac{\partial}{\partial t}\left[\vec{\nabla} \cdot \vec{\mathbf{A}} + \mu\varepsilon\frac{\partial V}{\partial t}\right] \quad [7.78]$$

By fixing $\vec{\nabla} \cdot \vec{\mathbf{A}} = -\mu\varepsilon\frac{\partial V}{\partial t}$ (Lorentz gauge condition), equations [7.77] and [7.78] reduce to:

$$\begin{aligned} \Delta\vec{\mathbf{A}} - \mu\varepsilon\frac{\partial^2 \vec{\mathbf{A}}}{\partial t^2} &= -\mu\vec{\mathbf{J}} \\ \Delta V - \mu\varepsilon\frac{\partial^2 V}{\partial t^2} &= -\frac{\rho}{\varepsilon} \end{aligned} \quad [7.79]$$

In other words, the vector potential $\vec{\mathbf{A}}$ and the scalar electric potential V satisfy inhomogeneous wave equations, with source terms $-\mu\vec{\mathbf{J}}$ and $-\frac{\rho}{\varepsilon}$, respectively. For time-harmonic (sinusoidal) fields, the above equations in complex notation become:

$$\begin{aligned} \Delta\vec{\hat{\mathbf{A}}} + \mu\varepsilon\omega^2\vec{\hat{\mathbf{A}}} &= -\mu\vec{\hat{\mathbf{J}}} \\ \Delta\hat{V} + \mu\varepsilon\omega^2\hat{V} &= -\frac{\hat{\rho}}{\varepsilon} \\ \vec{\nabla} \cdot \vec{\hat{\mathbf{A}}} - \mathrm{i}\mu\varepsilon\omega\hat{V} &= 0 \end{aligned} \quad [7.80]$$

7.4. Plane wave in material medium

Let us try to derive the wave equations for the fields $\vec{\mathbf{E}}$ and $\vec{\mathbf{B}}$ in an LIH medium with permittivity ε and permeability μ. To this end, we decouple the Maxwell equations by applying the operator $\vec{\nabla}$ to the Maxwell–Faraday equation:

$$\vec{\nabla} \times \left(\vec{\nabla} \times \vec{\mathbf{E}}\right) = -\frac{\partial}{\partial t}\left(\vec{\nabla} \times \vec{\mathbf{B}}\right) = -\mu\frac{\partial}{\partial t}\left(\vec{\mathbf{J}} + \frac{\partial \vec{\mathbf{D}}}{\partial t}\right) \quad [7.81]$$

We substitute $\vec{\mathbf{J}} = \gamma\vec{\mathbf{E}}$ (Ohmic medium) and $\vec{\mathbf{D}} = \varepsilon\vec{\mathbf{E}}$ (LIH medium):

$$\vec{\nabla}\left(\vec{\nabla} \cdot \vec{\mathbf{E}}\right) - \Delta\vec{\mathbf{E}} = -\mu\frac{\partial}{\partial t}\left(\gamma\vec{\mathbf{E}} + \varepsilon\frac{\partial \vec{\mathbf{E}}}{\partial t}\right) \quad [7.82]$$

In vacuum (or in a region without free charges), where $\vec{\nabla} \cdot \vec{\mathbf{D}} = 0$, and hence $\vec{\nabla} \cdot \vec{\mathbf{E}} = 0$, the preceding equation reduces to:

$$\Delta \vec{\mathbf{E}} - \mu\varepsilon \frac{\partial^2 \vec{\mathbf{E}}}{\partial t^2} = \gamma\mu \frac{\partial \vec{\mathbf{E}}}{\partial t} \quad [7.83]$$

In particular, for a time-harmonic (sinusoidal) field, the equation becomes:

$$\Delta \vec{\mathbf{E}} + \mu\varepsilon\omega^2 \left(1 + \mathrm{i}\frac{\gamma}{\omega\varepsilon}\right) \vec{\mathbf{E}} = 0 \quad [7.84]$$

In a similar fashion, we apply the same operator to the magnetic excitation field, $\vec{\nabla} \times \left(\vec{\nabla} \times \vec{\mathbf{H}}\right) = \vec{\nabla} \times \left(\vec{\mathbf{J}} + \frac{\partial \vec{\mathbf{D}}}{\partial t}\right) = \vec{\nabla} \times \left(\gamma \vec{\mathbf{E}} + \varepsilon \frac{\partial \vec{\mathbf{E}}}{\partial t}\right) = -\gamma \frac{\partial \vec{\mathbf{B}}}{\partial t} - \varepsilon \frac{\partial^2 \vec{\mathbf{B}}}{\partial t^2} \equiv -\frac{1}{\mu} \Delta \vec{\mathbf{B}}$, which gives the wave equation:

$$\Delta \vec{\mathbf{B}} - \mu\varepsilon \frac{\partial^2 \vec{\mathbf{B}}}{\partial t^2} = \gamma\mu \frac{\partial \vec{\mathbf{B}}}{\partial t} \quad [7.85]$$

Also, for a time-harmonic magnetic field $\vec{\mathbf{B}}$, we obtain:

$$\Delta \vec{\mathbf{B}} + \mu\varepsilon\omega^2 \left(1 + \mathrm{i}\frac{\gamma}{\omega\varepsilon}\right) \vec{\mathbf{B}} = 0 \quad [7.86]$$

For a nonlinear medium, the propagation equations for the fields $\vec{\mathbf{E}}$ and $\vec{\mathbf{B}}$ include the polarization vector $\vec{\mathbf{P}}$ and the magnetization vector $\vec{\mathbf{M}}$. In fact, by applying the curl operator to the Maxwell–Faraday equation, $\vec{\nabla} \times \left(\vec{\nabla} \times \vec{\mathbf{E}}\right) = -\mu_0 \frac{\partial}{\partial t} \left(\vec{\nabla} \times \vec{\mathbf{H}}\right) - \mu_0 \frac{\partial}{\partial t} \left(\vec{\nabla} \times \vec{\mathbf{M}}\right) = -\varepsilon_0\mu_0 \frac{\partial^2 \vec{\mathbf{E}}}{\partial t^2} - \mu_0 \frac{\partial^2 \vec{\mathbf{P}}}{\partial t^2} - \mu_0 \frac{\partial}{\partial t} \left(\vec{\mathbf{J}} + \vec{\nabla} \times \vec{\mathbf{M}}\right)$. This yields the wave equation:

$$\vec{\nabla} \times \left(\vec{\nabla} \times \vec{\mathbf{E}}\right) + \varepsilon_0\mu_0 \frac{\partial^2 \vec{\mathbf{E}}}{\partial t^2} = -\mu_0 \frac{\partial^2 \vec{\mathbf{P}}}{\partial t^2} - \mu_0 \frac{\partial}{\partial t} \left(\vec{\mathbf{J}} + \vec{\nabla} \times \vec{\mathbf{M}}\right) \quad [7.87]$$

The terms on the right-hand side of equation [7.87] act as sources of electromagnetic waves. A similar procedure for the magnetic field $\vec{H}$ gives $\vec{\nabla} \times \left(\vec{\nabla} \times \vec{\mathbf{H}}\right) = \vec{\nabla} \times \vec{\mathbf{J}} + \frac{\partial}{\partial t} \left(\vec{\nabla} \times \vec{\mathbf{D}}\right) = \vec{\nabla} \times \vec{\mathbf{J}} + \varepsilon_0 \frac{\partial}{\partial t} \left(\vec{\nabla} \times \vec{\mathbf{E}}\right) + \frac{\partial}{\partial t} \left(\vec{\nabla} \times \vec{\mathbf{P}}\right) = -\varepsilon_0\mu_0 \frac{\partial^2 \vec{\mathbf{H}}}{\partial t^2} - \varepsilon_0\mu_0 \frac{\partial^2 \vec{\mathbf{M}}}{\partial t^2} + \vec{\nabla} \times \left(\vec{\mathbf{J}} + \frac{\partial \vec{\mathbf{P}}}{\partial t}\right)$. It leads to:

$$\vec{\nabla} \times \left(\vec{\nabla} \times \vec{\mathbf{H}}\right) + \varepsilon_0\mu_0 \frac{\partial^2 \vec{\mathbf{H}}}{\partial t^2} = -\varepsilon_0\mu_0 \frac{\partial^2 \vec{\mathbf{M}}}{\partial t^2} + \vec{\nabla} \times \left(\vec{\mathbf{J}} + \frac{\partial \vec{\mathbf{P}}}{\partial t}\right) \quad [7.88]$$

In most practical situations, magnetization terms can be neglected. For sinusoidal (time-harmonic) variations, equations [7.87] and [7.88] take the form:

$$\begin{aligned} &\vec{\nabla} \times \left(\vec{\nabla} \times \vec{\hat{\mathbf{E}}}\right) - \varepsilon_0\mu_0\omega^2 \vec{\hat{\mathbf{E}}} = \mu_0\omega^2 \vec{\hat{\mathbf{P}}} + \mathrm{i}\mu_0\omega \left(\vec{\hat{\mathbf{J}}} + \vec{\nabla} \times \vec{\hat{\mathbf{M}}}\right) \\ &\vec{\nabla} \times \left(\vec{\nabla} \times \vec{\hat{\mathbf{H}}}\right) - \varepsilon_0\mu_0\omega^2 \vec{\hat{\mathbf{H}}} = \varepsilon_0\mu_0\omega^2 \vec{\hat{\mathbf{M}}} + \vec{\nabla} \times \left(\vec{\hat{\mathbf{J}}} - \mathrm{i}\omega \vec{\hat{\mathbf{P}}}\right) \end{aligned} \quad [7.89]$$

7.4.1. *Plane wave in a linear and isotropic dielectric*

Let us find the solutions to the Maxwell equations that depend only on the coordinate z and the time t in an LIH dielectric medium with permittivity ε and permeability μ. In this source-free medium, that is, with $\overrightarrow{\mathbf{J}} = 0$ and $\rho = 0$, the Maxwell equations reduce to:

$$\begin{aligned} &-\frac{\partial E_y}{\partial z}\overrightarrow{\mathbf{i}} + \frac{\partial E_x}{\partial z}\overrightarrow{\mathbf{j}} = -\mu\frac{\partial \overrightarrow{\mathbf{H}}}{\partial t} \\ &-\frac{\partial H_y}{\partial z}\overrightarrow{\mathbf{i}} + \frac{\partial H_x}{\partial z}\overrightarrow{\mathbf{j}} = \varepsilon\frac{\partial \overrightarrow{\mathbf{E}}}{\partial t} \\ &\varepsilon\frac{\partial E_z}{\partial z} = \mu\frac{\partial H_z}{\partial z} = 0 \end{aligned} \qquad [7.90]$$

These relations indicate that the components E_z and H_z are constant in space and time. Since constant fields do not correspond to propagating waves, we set $E_z = H_z = 0$. For the remaining components, we obtain:

$$\begin{aligned} &\frac{\partial E_x}{\partial z} = -\mu\frac{\partial H_y}{\partial t} \\ &\frac{\partial H_x}{\partial z} = \varepsilon\frac{\partial E_y}{\partial t} \\ &\frac{\partial E_y}{\partial z} = \mu\frac{\partial H_x}{\partial t} \\ &-\frac{\partial H_y}{\partial z} = \varepsilon\frac{\partial E_x}{\partial t} \end{aligned} \qquad [7.91]$$

These equations form two independent sets of equations. Each solution is characterized by electric and magnetic excitation fields $\overrightarrow{\mathbf{E}}$ and $\overrightarrow{\mathbf{H}}$ that are mutually perpendicular. The Poynting vector $\overrightarrow{\mathbf{\Pi}} = \overrightarrow{\mathbf{E}} \times \overrightarrow{\mathbf{H}}$ points in the Oz direction for each solution. Since the fields and the Poynting vector are mutually orthogonal, these are known as *transverse electromagnetic waves (TEM waves)*. By decoupling the equations above, we obtain one-dimensional wave equations:

$$\frac{\partial^2 E_x}{\partial z^2} - \frac{1}{v_\phi^2}\frac{\partial^2 E_x}{\partial t^2} = \frac{\partial^2 E_y}{\partial z^2} - \frac{1}{v_\phi^2}\frac{\partial^2 E_y}{\partial t^2} = 0 \qquad [7.92]$$

where v_ϕ is the *phase velocity* of the wave, given by:

$$v_\phi = \frac{1}{\sqrt{\varepsilon\mu}} = \frac{1}{\sqrt{\varepsilon_0\mu_0}\sqrt{\varepsilon_r\mu_r}} = \frac{c}{\sqrt{\varepsilon_r\mu_r}} \qquad [7.93]$$

In vacuum, we have $v_\phi = c$ (since $\varepsilon_r = \mu_r = 1$), which explains why light is a transverse electromagnetic wave. Similarly, for the remaining components of $\vec{\mathbf{H}}$, we obtain the following one-dimensional wave equations:

$$\frac{\partial^2 H_x}{\partial z^2} - \frac{1}{v_\phi^2}\frac{\partial^2 H_x}{\partial t^2} = \frac{\partial^2 H_y}{\partial z^2} - \frac{1}{v_\phi^2}\frac{\partial^2 H_y}{\partial t^2} = 0 \qquad [7.94]$$

Equations [7.92] and [7.94] are well known in physics. They admit the following general solutions:

$$\begin{aligned} E_x(z,t) &= E_{x,+}\left(t - \frac{z}{v_\phi}\right) + E_{x,-}\left(t + \frac{z}{v_\phi}\right) \\ H_y(z,t) &= H_{y,+}\left(t - \frac{z}{v_\phi}\right) + H_{y,-}\left(t + \frac{z}{v_\phi}\right) \\ E_y(z,t) &= E_{y,+}\left(t - \frac{z}{v_\phi}\right) + E_{y,-}\left(t + \frac{z}{v_\phi}\right) \\ H_x(z,t) &= H_{x,+}\left(t - \frac{z}{v_\phi}\right) + H_{x,-}\left(t + \frac{z}{v_\phi}\right) \end{aligned} \qquad [7.95]$$

where $E_{x,+}$, $E_{x,-}$, $H_{y,+}$, $H_{y,-}$, $E_{y,+}$, $E_{y,-}$, $H_{x,+}$ and $H_{x,-}$ depend on the initial and boundary conditions in space. These solutions can be easily found by introducing the new variables $\alpha = t - \frac{z}{v_\phi}$ and $\beta = t + \frac{z}{v_\phi}$, which gives $\frac{\partial \alpha}{\partial t} = 1$, $\frac{\partial \alpha}{\partial z} = -\frac{1}{v_\phi}$, $\frac{\partial \beta}{\partial t} = 1$ and $\frac{\partial \beta}{\partial z} = \frac{1}{v_\phi}$. It is easy to verify that:

$$\begin{cases} \dfrac{\partial^2 E_x}{\partial t^2} - v_\phi^2\dfrac{\partial^2 E_x}{\partial z^2} = \dfrac{\partial^2 E_x}{\partial\alpha\partial\beta} = 0 \\ \dfrac{\partial^2 H_y}{\partial t^2} - v_\phi^2\dfrac{\partial^2 H_y}{\partial z^2} = \dfrac{\partial^2 H_y}{\partial\alpha\partial\beta} = 0 \\ \dfrac{\partial^2 E_y}{\partial t^2} - v_\phi^2\dfrac{\partial^2 E_y}{\partial z^2} = \dfrac{\partial^2 E_y}{\partial\alpha\partial\beta} = 0 \\ \dfrac{\partial^2 H_x}{\partial t^2} - v_\phi^2\dfrac{\partial^2 H_x}{\partial z^2} = \dfrac{\partial^2 H_x}{\partial\alpha\partial\beta} = 0 \end{cases} \Longrightarrow \begin{cases} E_x = E_{x,+}(\alpha) + E_{x,-}(\beta) \\ H_y = H_{y,+}(\alpha) + H_{y,-}(\beta) \\ E_y = E_{y,+}(\alpha) + E_{y,-}(\beta) \\ H_x = H_{x,+}(\alpha) + H_{x,-}(\beta) \end{cases} \qquad [7.96]$$

The pairs $(H_{x,+}, E_{y,+})$, $(H_{x,-}, E_{y,-})$, $(H_{y,+}, E_{x,+})$ and $(H_{y,-}, E_{x,-})$ are not independent, as seen by substituting the solutions [7.95] into [7.91]:

$$\begin{aligned} \frac{\partial E_x}{\partial z} = -\mu\frac{\partial H_y}{\partial t} &\Longrightarrow \left(-\frac{dE_{x,+}}{d\alpha} + \frac{dE_{x,-}}{d\beta}\right) = -\mu v_\phi\left(\frac{dH_{y,+}}{d\alpha} + \frac{dH_{y,-}}{d\beta}\right) \\ \frac{\partial E_y}{\partial z} = \mu\frac{\partial H_x}{\partial t} &\Longrightarrow \left(-\frac{dE_{y,+}}{d\alpha} + \frac{dE_{y,-}}{d\beta}\right) = \mu v_\phi\left(\frac{dH_{x,+}}{d\alpha} + \frac{dH_{x,-}}{d\beta}\right) \\ \frac{\partial H_x}{\partial z} = \varepsilon\frac{\partial E_y}{\partial t} &\Longrightarrow \left(-\frac{dH_{x,+}}{d\alpha} + \frac{dH_{x,-}}{d\beta}\right) = \varepsilon v_\phi\left(\frac{dE_{y,+}}{d\alpha} + \frac{dE_{y,-}}{d\beta}\right) \\ \frac{\partial H_y}{\partial z} = -\varepsilon\frac{\partial E_x}{\partial t} &\Longrightarrow \left(-\frac{dH_{y,+}}{d\alpha} + \frac{dH_{y,-}}{d\beta}\right) = -\varepsilon v_\phi\left(\frac{dE_{x,+}}{d\alpha} + \frac{dE_{x,-}}{d\beta}\right) \end{aligned} \qquad [7.97]$$

This leads to the following relations:

$$\begin{aligned}\frac{d}{d\alpha}(E_{x,+} - \mu v_\phi H_{y,+}) &= \frac{d}{d\beta}(E_{x,-} + \mu v_\phi H_{y,-}) = 0 \\ \frac{d}{d\alpha}(E_{y,+} + \mu v_\phi H_{x,+}) &= \frac{d}{d\beta}(E_{y,-} - \mu v_\phi H_{x,-}) = 0\end{aligned} \qquad [7.98]$$

These equations immediately yield:

$$\begin{aligned}E_{x,+} &= ZH_{y,+} \qquad & E_{x,-} &= -ZH_{y,-} \\ E_{y,+} &= -ZH_{x,+} \qquad & E_{y,-} &= ZH_{x,-}\end{aligned} \qquad [7.99]$$

where $Z = \sqrt{\frac{\mu}{\varepsilon}}$ is called the *impedance of the medium*. Note that the solutions [7.95] represent waves propagating at phase velocity v_ϕ. As long as $t - \frac{z}{v_\phi}$ = const, we see that $E_{x,+}\left(t - \frac{z}{v_\phi}\right)$ remains unchanged. As time increases, z must vary so that $dt - \frac{dz}{v_\phi} = 0$, which means that $\frac{dz}{dt} = v_\phi$ for $E_{x,+}$ to stay constant. We conclude that $E_{x,+}\left(t - \frac{z}{v_\phi}\right)$ represents a wave that propagates at the speed v_ϕ in the positive direction of Oz. For similar reasons, $E_{x,-}\left(t + \frac{z}{v_\phi}\right)$ represents a wave propagating at the speed v_ϕ in the negative direction of the Oz axis, because an observer must move in that direction to keep the argument $t + \frac{z}{v_\phi}$ constant (i.e. $dt + \frac{dz}{v_\phi} = 0 \Longrightarrow \frac{dz}{dt} = -v_\phi$). These solutions are called *plane waves*, because at any constant z-plane, the fields are constant and do not vary with the x- and y-coordinates.

For a TEM wave, the Poynting vector is given by:

$$\begin{aligned}\overrightarrow{\Pi} = \overrightarrow{\mathbf{E}}(z,t) \times \overrightarrow{\mathbf{H}}(z,t) &= [(E_{x,+} + E_{x,-})(H_{y,+} + H_{y,-}) \\ &\quad - (E_{y,+} + E_{y,-})(H_{x,+} + H_{x,-})]\,\overrightarrow{\mathbf{k}} \\ &= Z^{-1}\left[E_{x,+}^2 - E_{x,-}^2 + E_{y,+}^2 - E_{y,-}^2\right]\overrightarrow{\mathbf{k}}\end{aligned} \qquad [7.100]$$

The total power flux is therefore equal to the difference between the power fluxes of the ascending and descending waves along Oz.

EXAMPLE 7.3. (TEM wave generated by a current sheet).– The simplest source of a plane wave is an infinite and uniform time-varying current sheet, which is assumed to be directed in the positive direction of the Ox axis. From the boundary conditions on $\overrightarrow{\mathbf{H}}$ (discontinuity of the tangential component), we observe that a current sheet directed along Ox gives rise to a field $\overrightarrow{\mathbf{H}}$ directed along Oy:

$$H_y(0^+) - H_y(0^-) = -K_x(t) \qquad [7.101]$$

In general, a uniform current sheet gives rise to a field $\overrightarrow{\mathbf{H}}$ perpendicular to the direction of current flow but in the plane of the sheet. Thus, to generate a $\overrightarrow{\mathbf{H}}$–field in the

Oy (respectively, Ox) direction, a current surface pointing toward Ox (respectively, Oy) is required. If there are no other sources, the waves will propagate away from the sheet in both directions to infinity, as shown in Figure 7.4.

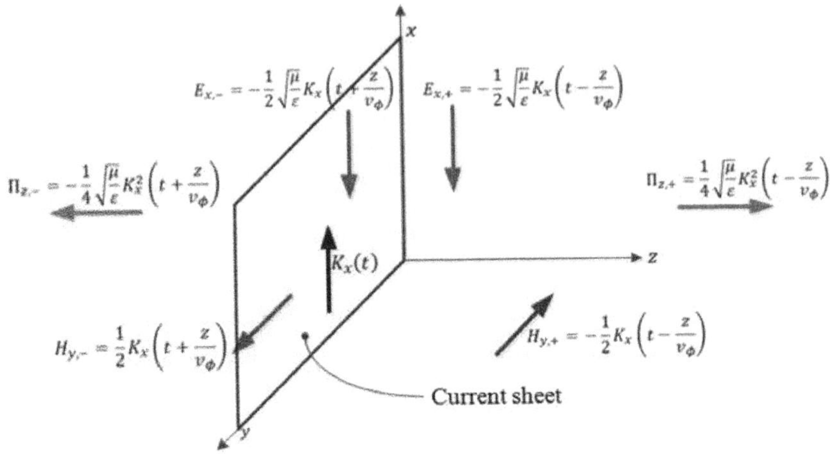

Figure 7.4. *Wave generated by a current sheet oriented along the Ox axis. For a color version of this figure, see www.iste.co.uk/abdesselam/electrodynamics2.zip*

The solutions on each side of the sheet are written as:

$$H_y(z,t) = \begin{cases} H_{y,+}\left(t - \dfrac{z}{v_\phi}\right) & z > 0 \\ H_{y,-}\left(t + \dfrac{z}{v_\phi}\right) & z < 0 \end{cases}$$
$$E_x(z,t) = \begin{cases} ZH_{y,+}\left(t - \dfrac{z}{v_\phi}\right) & z > 0 \\ -ZH_{y,-}\left(t + \dfrac{z}{v_\phi}\right) & z < 0 \end{cases} \quad [7.102]$$

In the region $z > 0$, waves propagate only in the positive direction of the Oz axis, and no wave can propagate in the negative direction. Similarly, for $z < 0$, there are only waves propagating in the negative direction of the Oz axis. In addition to the boundary condition [7.101], the tangential component of $\vec{\mathbf{E}}$ must be continuous across the sheet (i.e. at $z = 0$):

$$\left.\begin{array}{l} H_{y,+}(t) - H_{y,-}(t) = -K_x(t) \\ Z\left[H_{y,+}(t) + H_{y,-}(t)\right] = 0 \end{array}\right\} \Longrightarrow H_{y,+}(t) = -H_{y,-}(t) = -\frac{1}{2}K_x(t) \quad [7.103]$$

and, consequently:

$$H_y(z,t) = \begin{cases} -\frac{1}{2}K_x\left(t - \frac{z}{v_\phi}\right) & z > 0 \\ \frac{1}{2}K_x\left(t + \frac{z}{v_\phi}\right) & z < 0 \end{cases}$$
$$E_x(z,t) = \begin{cases} -\frac{1}{2}ZK_x\left(t - \frac{z}{v_\phi}\right) & z > 0 \\ -\frac{1}{2}ZK_x\left(t + \frac{z}{v_\phi}\right) & z < 0 \end{cases} \quad [7.104]$$

Since the shape of the fields, in time and space, remains unchanged during propagation, linear dielectric media are said to be *non-dispersive*. Note also that the electric field at $z = 0$ is in the opposite direction to the current. The magnetic field is perpendicular to the current, but in the plane of the sheet and in opposite directions on either side. The power flux delivered by the sheet is:

$$\vec{\mathbf{\Pi}}(z,t) = \begin{cases} \frac{1}{4}ZK_x^2\left(t - \frac{z}{v_\phi}\right)\vec{\mathbf{k}} & z > 0 \\ -\frac{1}{4}ZK_x^2\left(t + \frac{z}{v_\phi}\right)\vec{\mathbf{k}} & z < 0 \end{cases} \quad [7.105]$$

The power fluxes measured on either side of the sheet are equal in magnitude but opposite in direction. If $K_x(t)$ varies sinusoidally with time, that is, it is of the form $\Re\left(K_0 e^{-\mathrm{i}\omega t}\right)$, then the 1D wave fields generated on each side of the sheet are written as:

$$H_y(z,t) = \begin{cases} \Re\left[-\frac{1}{2}K_0 e^{-\mathrm{i}\omega\left(t - \frac{z}{v_\phi}\right)}\right] & z > 0 \\ \Re\left[\frac{1}{2}K_0 e^{-\mathrm{i}\omega\left(t + \frac{z}{v_\phi}\right)}\right] & z < 0 \end{cases}$$
$$E_x(z,t) = \begin{cases} \Re\left[-\frac{1}{2}ZK_0 e^{-\mathrm{i}\omega\left(t - \frac{z}{v_\phi}\right)}\right] & z > 0 \\ \Re\left[-\frac{1}{2}ZK_0 e^{-\mathrm{i}\omega\left(t + \frac{z}{v_\phi}\right)}\right] & z < 0 \end{cases} \quad [7.106]$$

We observe that at a fixed t, the fields also vary sinusoidally with position. Therefore, it is useful to introduce the *wavenumber* $\mathsf{k} = \frac{2\pi}{\lambda} = \frac{\omega}{v_\phi} = \omega\sqrt{\varepsilon\mu}$, where λ is the *wavelength*. Similarly, at a fixed position, the fields vary sinusoidally with time, with a period $T = \frac{2\pi}{\omega} = \frac{1}{\nu}$, where ν is the *frequency* of the source. Thus, we obtain the relation $\lambda = v_\phi T$. The time-averaged densities of electric and magnetic energy are equal:

$$\langle w_e \rangle = \frac{\varepsilon}{4}\vec{\hat{\mathbf{E}}} \cdot \vec{\hat{\mathbf{E}}}^* = \frac{\mu}{16}|K_0|^2$$
$$\langle w_m \rangle = \frac{\mu}{4}\vec{\hat{\mathbf{H}}} \cdot \vec{\hat{\mathbf{H}}}^* = \frac{\mu}{16}|K_0|^2 \quad [7.107]$$

We also note that $\hat{p}_d = 0$ in both regions $z > 0$ and $z < 0$ (regions without sources), which shows that the complex Poynting vector has zero divergence. It follows that the time-averaged (electromagnetic) power flux must be constant throughout space:

$$\vec{\Pi} = \frac{1}{2}\vec{E} \times \vec{H}^* = \begin{cases} \frac{1}{8}Z\left|K_0\right|^2 \vec{k} & z > 0 \\ -\frac{1}{8}Z\left|K_0\right|^2 \vec{k} & z < 0 \end{cases} \Longrightarrow \langle\vec{\Pi}\rangle = \begin{cases} \frac{1}{8}Z\left|K_0\right|^2 \vec{k} & z > 0 \\ -\frac{1}{8}Z\left|K_0\right|^2 \vec{k} & z < 0 \end{cases} \quad [7.108]$$

EXAMPLE 7.4. (Doppler effect).– If the current sheet $\Re\left(K_0 e^{-i\omega t}\right)$ moves at constant velocity $v_0 \vec{k}$, as illustrated in Figure 7.5, the boundary conditions are no longer found in $z = 0$, but instead in $z = v_0 t$. The general solutions are:

$$H_y(z,t) = \begin{cases} \Re\left[H_{y,+} e^{-i\omega_+\left(t - \frac{z}{v_\phi}\right)}\right] & z > v_0 t \\ \Re\left[H_{y,-} e^{-i\omega_-\left(t + \frac{z}{v_\phi}\right)}\right] & z < v_0 t \end{cases}$$
$$E_x(z,t) = \begin{cases} \Re\left[Z H_{y,+} e^{-i\omega_+\left(t - \frac{z}{v_\phi}\right)}\right] & z > v_0 t \\ \Re\left[-Z H_{y,-} e^{-i\omega_-\left(t + \frac{z}{v_\phi}\right)}\right] & z < v_0 t \end{cases} \quad [7.109]$$

where the frequencies ω_+ and ω_- on each side of the sheet differ from one another, and from the source frequency ω. We assume that $\frac{v_0}{v_\phi} \ll 1$ so that relativistic effects can be neglected. The boundary conditions:

$$\left.\begin{array}{l} E_{x,+}(v_0 t) = E_{x,-}(v_0 t) \\ H_{y,+}(v_0 t) - H_{y,-}(v_0 t) = -K_x \end{array}\right\} \Longrightarrow \begin{cases} H_{y,+} e^{-i\omega_+\left(1 - \frac{v_0}{c}\right)t} \\ = -H_{y,-} e^{-i\omega_-\left(1 + \frac{v_0}{c}\right)t} \\ H_{y,+} e^{-i\omega_+\left(1 - \frac{v_0}{c}\right)t} \\ -H_{y,-} e^{-i\omega_-\left(1 + \frac{v_0}{c}\right)t} = -K_0 e^{-i\omega t} \end{cases} \quad [7.110]$$

must be satisfied for any t. Therefore, the exponential terms must match, which allows us to deduce the shifted frequencies on either side of the sheet:

$$\left.\begin{array}{l} \omega_+ = \dfrac{\omega}{1 - \frac{v_0}{c}} \simeq \omega\left(1 + \dfrac{v_0}{c}\right) \\ \omega_- = \dfrac{\omega}{1 + \frac{v_0}{c}} \simeq \omega\left(1 - \dfrac{v_0}{c}\right) \end{array}\right\} \Longrightarrow H_{y,+} = -H_{y,-} = -\frac{K_0}{2} \quad [7.111]$$

We conclude that when the source moves toward an observer, the frequency increases, and when it moves away from an observer, the frequency decreases.

These frequency shifts due to the motion of a source or an observer are known as *Doppler shifts* or the *Doppler effect.*

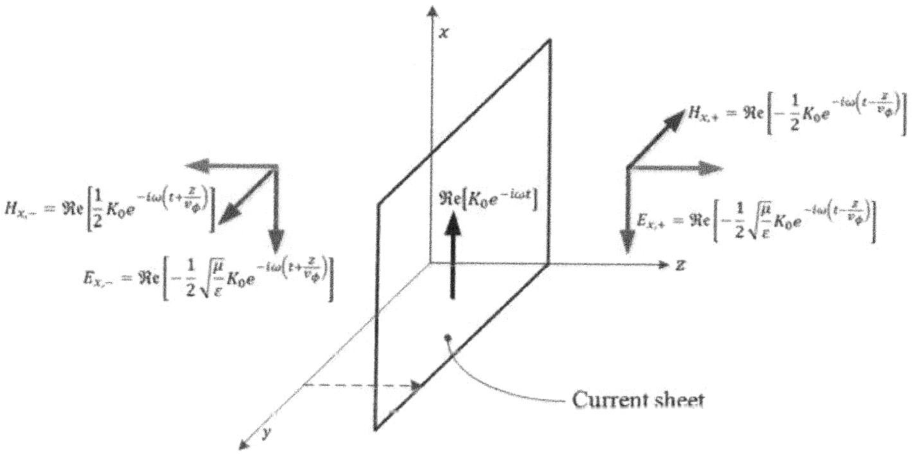

Figure 7.5. *Doppler shifts or Doppler effect. For a color version of this figure, see www.iste.co.uk/abdesselam/electrodynamics2.zip*

EXAMPLE 7.5. (Electromagnetic wave in a linear anisotropic medium).– Introducing anisotropy significantly complicates the equations governing field propagation. Instead of simply using $\vec{\mathbf{P}} = \chi\varepsilon_0\vec{\mathbf{E}}$, we must instead adopt the tensorial expression $P_i = \chi_{ij}\varepsilon_0 E^j$, where the electric susceptibility tensor is a matrix:

$$\overleftrightarrow{\chi} = \begin{pmatrix} \chi_{11} & \chi_{12} & \chi_{13} \\ \chi_{21} & \chi_{22} & \chi_{23} \\ \chi_{31} & \chi_{32} & \chi_{33} \end{pmatrix} \qquad [7.112]$$

In non-dispersive media, this tensor is symmetric, which means that the susceptibility matrix can be diagonalized relative to a system of principal axes. In that case, $\overleftrightarrow{\chi}$ has only diagonal elements χ_{11}, χ_{22} and χ_{33}, known as the *principal susceptibilities* (the principal susceptibilities correspond to the eigenvalues of the matrix [7.113]). As usual, the wave equation for the electric field is obtained by applying $\vec{\nabla} \times \left(\vec{\nabla} \times \vec{\mathbf{E}}\right)$ to the Maxwell equations, that is, $\vec{\nabla} \times \left(\vec{\nabla} \times \vec{\mathbf{E}}\right) = \vec{\nabla} \times \left(-\frac{\partial \vec{\mathbf{B}}}{\partial t}\right) = -\frac{\partial}{\partial t}\left(\mu_0\vec{\mathbf{J}} + \mu_0\frac{\partial \vec{\mathbf{P}}}{\partial t}\right)$. Note that magnetization effects are neglected here. In anisotropic dielectrics, the permittivity ε is not spatially uniform, so the term $\vec{\nabla} \cdot \vec{\mathbf{E}}$ can no longer be eliminated from the identity

$\overrightarrow{\nabla} \times \left(\overrightarrow{\nabla} \times \overrightarrow{\mathbf{E}}\right) = \overrightarrow{\nabla}\left(\overrightarrow{\nabla} \cdot \overrightarrow{\mathbf{E}}\right) - \Delta\overrightarrow{\mathbf{E}}$. Assuming $\overrightarrow{\mathbf{J}} = 0$ and substituting $\overrightarrow{\mathbf{D}} = \varepsilon_0 \overrightarrow{\mathbf{E}} + \overrightarrow{\mathbf{P}} = \varepsilon_0 \overrightarrow{\mathbf{E}} + \overleftrightarrow{\chi} \overline{\otimes} \overrightarrow{\mathbf{E}}$, we obtain the generalized wave equation:

$$\overrightarrow{\nabla}\left(\overrightarrow{\nabla} \cdot \overrightarrow{\mathbf{E}}\right) - \Delta\overrightarrow{\mathbf{E}} = -\frac{1}{c^2}\frac{\partial^2 \overrightarrow{\mathbf{E}}}{\partial t^2} - \frac{\overleftrightarrow{\chi}}{c^2} \overline{\otimes} \frac{\partial^2 \overrightarrow{\mathbf{E}}}{\partial t^2} \quad [7.113]$$

By choosing $\overrightarrow{\mathbf{E}} = \Re\left[\overrightarrow{\hat{\mathbf{E}}}_0 e^{\mathrm{i}\left(\omega t - \overrightarrow{\mathsf{k}} \cdot \overrightarrow{\mathbf{r}}\right)}\right]$ (plane wave), the previous wave equation yields, for the Cartesian component i:

$$\mathsf{k}_i\left(\overrightarrow{\mathsf{k}} \cdot \overrightarrow{\mathbf{E}}\right) - \mathsf{k}^2 E_i + \frac{\omega^2}{c^2}\left(\delta_{ij} + \chi_{ij}\right) E^j = 0 \quad [7.114]$$

In the system of principal axes (where the three principal susceptibilities are denoted as χ_{xx}, χ_{yy} and χ_{zz}), this becomes:

$$\begin{aligned}
&\left[-\mathsf{k}_y^2 - \mathsf{k}_z^2 + \frac{\omega^2}{c^2}\left(1 + \chi_{xx}\right)\right] E_x + \mathsf{k}_x \mathsf{k}_y E_y + \mathsf{k}_x \mathsf{k}_z E_z = 0 \\
&\mathsf{k}_y \mathsf{k}_x E_x + \left[-\mathsf{k}_x^2 - \mathsf{k}_z^2 + \frac{\omega^2}{c^2}\left(1 + \chi_{yy}\right)\right] E_y + \mathsf{k}_y \mathsf{k}_z E_z = 0 \\
&\mathsf{k}_z \mathsf{k}_x E_x + \mathsf{k}_z \mathsf{k}_y E_y + \left[-\mathsf{k}_x^2 - \mathsf{k}_y^2 + \frac{\omega^2}{c^2}\left(1 + \chi_{zz}\right)\right] E_z = 0
\end{aligned} \quad [7.115]$$

This system admits a non-trivial solution only if its determinant vanishes. Introducing the quantities $n_x = \sqrt{1 + \chi_{xx}}$, $n_y = \sqrt{1 + \chi_{yy}}$ and $n_z = \sqrt{1 + \chi_{zz}}$ called the *principal refractive indices*, we rewrite the system determinant condition as:

$$\begin{vmatrix}
\left[\frac{n_x^2\omega^2}{c^2} - \mathsf{k}_y^2 - \mathsf{k}_z^2\right] & \mathsf{k}_x \mathsf{k}_y & \mathsf{k}_x \mathsf{k}_z \\
\mathsf{k}_y \mathsf{k}_x & \left[\frac{n_y^2\omega^2}{c^2} - \mathsf{k}_x^2 - \mathsf{k}_z^2\right] & \mathsf{k}_y \mathsf{k}_z \\
\mathsf{k}_z \mathsf{k}_x & \mathsf{k}_z \mathsf{k}_y & \left[\frac{n_z^2\omega^2}{c^2} - \mathsf{k}_x^2 - \mathsf{k}_y^2\right]
\end{vmatrix} = 0 \quad [7.116]$$

For a given frequency ω, this equation defines a surface in the $\overrightarrow{\mathsf{k}}$-space. To simplify the analysis, we restrict $\overrightarrow{\mathsf{k}}$ to lie in a coordinate plane, say, the Oyz-plane. Then, $\mathsf{k}_x = 0$, and the characteristic equation reduces to:

$$\left[\frac{n_x^2\omega^2}{c^2} - \mathsf{k}_y^2 - \mathsf{k}_z^2\right]\left[\frac{n_y^2\omega^2}{c^2}\frac{n_z^2\omega^2}{c^2} - \mathsf{k}_y^2\frac{n_y^2\omega^2}{c^2} - \mathsf{k}_z^2\frac{n_z^2\omega^2}{c^2}\right] = 0 \quad [7.117]$$

This equation has two distinct roots: one corresponds to the ordinary wave vector, denoted as $\overrightarrow{\mathsf{k}}_o$ (*ordinary solution*), and the other corresponds to the extraordinary wave vector, denoted as $\overrightarrow{\mathsf{k}}_e$ (*extraordinary solution*). These roots satisfy the relations:

$$\mathsf{k}_{o,y}^2 + \mathsf{k}_{o,z}^2 = \frac{n_x^2\omega^2}{c^2} \qquad [7.118]$$

$$\frac{\mathsf{k}_{e,y}^2}{\frac{n_z^2\omega^2}{c^2}} + \frac{\mathsf{k}_{e,z}^2}{\frac{n_y^2\omega^2}{c^2}} = 1 \qquad [7.119]$$

We therefore conclude that when the wave vector $\overrightarrow{\mathsf{k}}$ lies in the plane Oyz, there are two possible solutions for k^2: (1) The first is simply $\left|\overrightarrow{\mathsf{k}}_o\right| = \frac{n_x\omega}{c}$. This ordinary root $\overrightarrow{\mathsf{k}}_o$ describes, in Oyz, a circle of radius $\frac{n_x\omega}{c}$; (2) the second, denoted $\overrightarrow{\mathsf{k}}_e$, varies with direction and describes an ellipse with semi-axes $\frac{n_z\omega}{c}$ along the Oy direction and $\frac{n_y\omega}{c}$ along the Oz direction. Let us now determine the eigenvectors (i.e. the polarizations of $\overrightarrow{\mathbf{E}}$) associated with these distinct values of $\overrightarrow{\mathsf{k}}$, by rewriting the set of equations for E_x, E_y and E_z (when $\overrightarrow{\mathsf{k}}$ is restricted to the Oyz plane):

$$\begin{pmatrix} \left[\frac{n_x^2\omega^2}{c^2} - \mathsf{k}_y^2 - \mathsf{k}_z^2\right] & 0 & 0 \\ 0 & \left[\frac{n_y^2\omega^2}{c^2} - \mathsf{k}_z^2\right] & \mathsf{k}_y\mathsf{k}_z \\ 0 & \mathsf{k}_z\mathsf{k}_y & \left[\frac{n_z^2\omega^2}{c^2} - \mathsf{k}_y^2\right] \end{pmatrix} \begin{pmatrix} E_x \\ E_y \\ E_z \end{pmatrix} = \begin{pmatrix} 0 \\ 0 \\ 0 \end{pmatrix} \qquad [7.120]$$

For the first eigenvalue, $\overrightarrow{\mathsf{k}}_o$, the corresponding eigenvector is $\overrightarrow{\mathbf{E}}_o = E_x\overrightarrow{\mathbf{i}}$, which means that the electric field is polarized along the Ox direction. This wave, called the *ordinary wave* (polarized along Ox), propagates in the Oyz plane with a constant phase velocity $v_\phi = \frac{c}{n_x}$ in all directions. The second eigenvalue, $\overrightarrow{\mathsf{k}}_e$, is associated with an eigenvector orthogonal to the first. Therefore, the electric field $\overrightarrow{\mathbf{E}}_e$ lies within the Oyz plane. The *extraordinary wave* has a phase velocity that varies depending on the direction of $\overrightarrow{\mathsf{k}}$. The polarizations of the waves are indicated by double-headed arrows placed on the $\overrightarrow{\mathsf{k}}$ vectors, as illustrated in Figure 7.6.

These results can be easily extended to other coordinate planes. In each coordinate plane, the tip of the vector $\overrightarrow{\mathsf{k}}$ describes a circle and an ellipse. If two of the principal refractive indices are equal, the tip of $\overrightarrow{\mathsf{k}}$ describes a sphere in 3D space and a revolution ellipsoid that touch at two points, as shown in Figure 7.7(a). In the direction defined by these two points, the two roots $\overrightarrow{\mathsf{k}}_o$ and $\overrightarrow{\mathsf{k}}_e$ are equal. A line passing through these points is called the *optical axis*, and a crystal with a single optical axis is said to be *uniaxial*. If all three indices are different, there will be two optical axes, as shown in Figure 7.7(b), and the crystal is said to be *biaxial*.

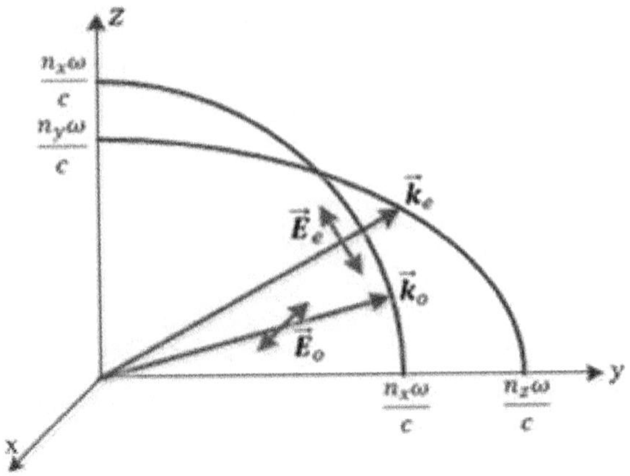

Figure 7.6. *If $\vec{k}$ is confined in Oyz: (a) waves polarized along the Ox direction propagate at the same speed in all directions; (b) waves with polarization parallel to the Oyz plane propagate at a speed that varies depending on the direction of $\vec{k}$. For a color version of this figure, see www.iste.co.uk/abdesselam/electrodynamics2.zip*

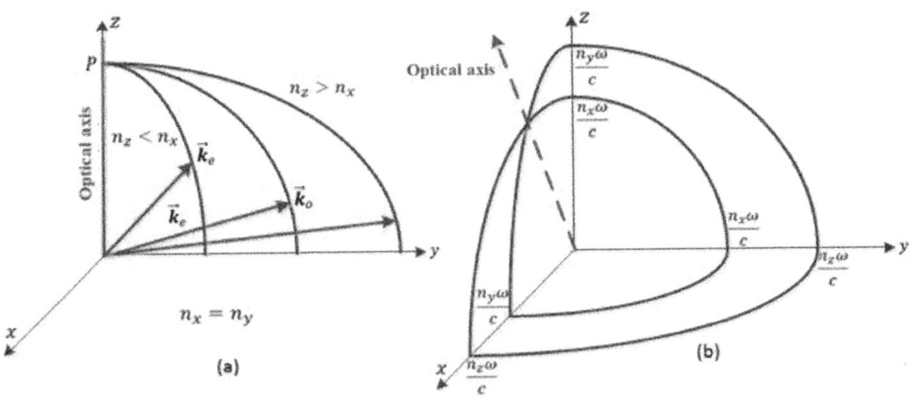

Figure 7.7. *(a) If $n_x = n_y$, the dielectric has a single optical axis along the Oz direction; (b) when the indices n_x, n_y and n_z are all different, the dielectric contains two distinct optical axes. For a color version of this figure, see www.iste.co.uk/abdesselam/electrodynamics2.zip*

7.4.2. *Electromagnetic wave in a conductive medium*

If we now consider an ohmic medium (with conductivity γ), the electric field $\vec{\mathbf{E}}(z,t)$ – propagating along the Oz axis and assumed, for simplicity, to have components in the Ox (transverse) and Oz (longitudinal) directions – will induce currents J_x and J_z in the same respective directions. These currents must be taken into account in the Maxwell–Ampère equation:

$$\begin{aligned} &\frac{\partial E_x}{\partial z}\vec{\mathbf{j}} = -\mu_0\frac{\partial H_y}{\partial t}\vec{\mathbf{j}} - \mu_0\frac{\partial H_z}{\partial t}\vec{\mathbf{k}} \\ &-\frac{\partial H_y}{\partial z}\vec{\mathbf{i}} = \left[J_x + \varepsilon_0\frac{\partial E_x}{\partial t}\right]\vec{\mathbf{i}} + \left[J_z + \varepsilon_0\frac{\partial E_z}{\partial t}\right]\vec{\mathbf{k}} \\ &\qquad = \left(\gamma E_x + \varepsilon_0\tfrac{\partial E_x}{\partial t}\right)\vec{\mathbf{i}} + \left[\gamma E_z + \varepsilon_0\tfrac{\partial E_z}{\partial t}\right]\vec{\mathbf{k}} \\ &\frac{\partial E_z}{\partial z} = \frac{\partial H_z}{\partial z} = 0 \end{aligned} \qquad [7.121]$$

The same reasoning applies when the electric field has a component along Oy (transverse) and a component along Oz (longitudinal). The solutions from section 7.4.1 are no longer valid due to the presence of the additional (conductivity) term, and therefore must be corrected. Equations [7.121] show that the component H_z remains constant in both time and space, and it will be set to zero. On the other hand, the component E_z is independent of z and varies with time according to:

$$E_z(z,t) = E_0 e^{-\gamma t/\varepsilon_0} \qquad [7.122]$$

This component quickly decays over time. It follows that no longitudinal electric field can persist in a conducting medium. In other words, the fields are purely transverse. Now, assume sinusoidal solutions for E_x and H_y in the form:

$$\begin{aligned} E_x(z,t) &= \Re\left[\hat{E}_0 e^{-\mathrm{i}(\omega t - \mathrm{k}z)}\right] \\ H_y(z,t) &= \Re\left[\hat{H}_0 e^{-\mathrm{i}(\omega t - \mathrm{k}z)}\right] \end{aligned} \qquad [7.123]$$

where $\hat{E}_0$ and $\hat{H}_0$ are complex amplitudes. The wavenumber k is also complex and will be determined by substituting [7.123] into [7.121]:

$$\begin{aligned} \mathrm{i}\mathrm{k}\hat{E}_0 &= \mathrm{i}\omega\mu_0\hat{H}_0 \\ \mathrm{i}\mathrm{k}\hat{H}_0 &= \mathrm{i}\omega\varepsilon_0\left(1 + \mathrm{i}\frac{\gamma}{\omega\varepsilon_0}\right)\hat{E}_0 \end{aligned} \qquad [7.124]$$

In reality, conductivity γ depends on angular frequency ω. According to the simplified Drude–Lorentz model, electrons are free to move inside the metal under the influence of an external electric field (without restoring forces), but they experience damping forces due to collisions. By solving the electron motion equation (see section 7.4.3), we find:

$$\gamma(\omega) = \frac{\gamma(0)}{1 - \mathrm{i}\omega\tau} \qquad [7.125]$$

where $\gamma(0) = \frac{ne^2\tau}{m}$, and τ^{-1} represents the electron collision frequency inside the conductor. Using the *plasma frequency* notation $\omega_p^2 = \frac{ne^2}{m\varepsilon_0}$, we can also write $\gamma(\omega) = \frac{\tau\omega_p^2\varepsilon_0}{1-\mathrm{i}\omega\tau}$.

To simplify the analysis, we introduce the *complex permittivity*:

$$\hat{\varepsilon} = \varepsilon_0\left(1 + \mathrm{i}\frac{\gamma(\omega)}{\omega\varepsilon_0}\right) \qquad [7.126]$$

This leads, from equation [7.126], to the relation $\frac{\hat{E}_0}{\hat{H}_0} = \frac{\omega\mu_0}{\mathsf{k}} = \frac{\mathsf{k}}{\omega\hat{\varepsilon}}$, which implies:

$$\mathsf{k}^2 = \omega^2\mu_0\hat{\varepsilon} = \omega^2\mu_0\varepsilon_0\left(1 + \mathrm{i}\frac{\gamma(\omega)}{\omega\varepsilon_0}\right) = \frac{\omega^2}{c^2}\left(1 + \mathrm{i}\frac{\gamma(\omega)}{\omega\varepsilon_0}\right) \qquad [7.127]$$

This is the *dispersion equation* in a conducting medium. Using equation [7.125], we can express the complex permittivity as:

$$\begin{aligned}
\hat{\varepsilon} &= \varepsilon_0\left[1 - \frac{\gamma(0)\,\tau}{\varepsilon_0(1+\omega^2\tau^2)} + \mathrm{i}\frac{\gamma(0)}{\omega\varepsilon_0(1+\omega^2\tau^2)}\right] \\
&= \varepsilon_0\left[1 - \frac{\omega_p^2\tau^2}{1+\omega^2\tau^2} + \mathrm{i}\frac{\tau\omega_p^2}{\omega(1+\omega^2\tau^2)}\right] \\
\mathsf{k}^2 &= \frac{\omega^2}{c^2}\left[1 - \frac{\gamma(0)\,\tau}{\varepsilon_0(1+\omega^2\tau^2)} + \mathrm{i}\frac{\gamma(0)}{\omega\varepsilon_0(1+\omega^2\tau^2)}\right] \\
&= \frac{\omega^2}{c^2}\left[1 - \frac{\omega_p^2\tau^2}{1+\omega^2\tau^2} + \mathrm{i}\frac{\tau\omega_p^2}{\omega(1+\omega^2\tau^2)}\right]
\end{aligned} \qquad [7.128]$$

At high frequencies ($\omega\tau \gg 1$), we can take approximately $\hat{\varepsilon} = \varepsilon_0\left(1 - \frac{\omega_p^2}{\omega^2}\right)$ and $\mathsf{k}^2 = \frac{\omega^2}{c^2}\left(1 - \frac{\omega_p^2}{\omega^2}\right)$. The conducting medium therefore behaves as a plasma (see section 7.4.3). At low frequencies ($\omega\tau \ll 1$), we approximate $\gamma(\omega) \approx \gamma(0)$, giving:

$$\begin{aligned}
\hat{\varepsilon} &= \varepsilon_0\left[1 + \mathrm{i}\frac{\gamma(0)}{\varepsilon_0\omega}\right] \\
\mathsf{k}^2 &= \frac{\omega^2}{c^2}\left[1 + \mathrm{i}\frac{\gamma(0)}{\varepsilon_0\omega}\right]
\end{aligned} \qquad [7.129]$$

These expressions arise directly from equation [7.127]. We observe the following limiting cases:

1) *Low-conductivity medium*: if the conductivity is very low, that is, if $\frac{\gamma(0)}{\omega\varepsilon_0} \ll 1$, equation [7.129] becomes:

$$\mathsf{k} = \pm\frac{\omega}{c}\left(1 + \mathrm{i}\frac{\gamma(0)}{2\omega\varepsilon_0}\right) = \pm\left(\frac{\omega}{c} + \mathrm{i}\frac{\gamma(0)}{2}\sqrt{\frac{\mu_0}{\varepsilon_0}}\right) \qquad [7.130]$$

Introducing $Z_0 = sqrt\frac{\mu_0}{\varepsilon_0}$, we then obtain:

$$H_y(z,t) = \begin{cases} \Re\left[\hat{H}_0 e^{-\frac{\gamma(0)Z_0 z}{2}} e^{-\mathrm{i}\omega\left(t-\frac{z}{c}\right)}\right] & z>0 \\ \Re\left[\hat{H}_0 e^{\frac{\gamma(0)Z_0 z}{2}} e^{-\mathrm{i}\omega\left(t+\frac{z}{c}\right)}\right] & z<0 \end{cases}$$
$$E_x(z,t) = \begin{cases} \Re\left[\hat{H}_0 Z_0\left(1-\mathrm{i}\frac{\gamma(0)}{2\varepsilon_0\omega}\right) e^{-\frac{\gamma(0)Z_0 z}{2}} e^{-\mathrm{i}\omega\left(t-\frac{z}{c}\right)}\right] & z>0 \\ \Re\left[-\hat{H}_0 Z_0\left(1-\mathrm{i}\frac{\gamma(0)}{2\varepsilon_0\omega}\right) e^{\frac{\gamma(0)Z_0 z}{2}} e^{-\mathrm{i}\omega\left(t+\frac{z}{c}\right)}\right] & z<0 \end{cases} \quad [7.131]$$

where the positive (respectively, negative) sign corresponds to a wave propagating in the positive (respectively, negative) Oz direction. We see that the imaginary part k_i of k causes an exponential attenuation of the fields from the surface of the conductor due to energy losses within the medium, as illustrated in Figure 7.8. Since the conductivity $\gamma(0)$ is small, we conclude that the wave is only slightly damped as it propagates. The wave will therefore penetrate deeper into the conductor to a distance of the order of $\frac{2}{\gamma(0)Z_0}$ (after several oscillations).

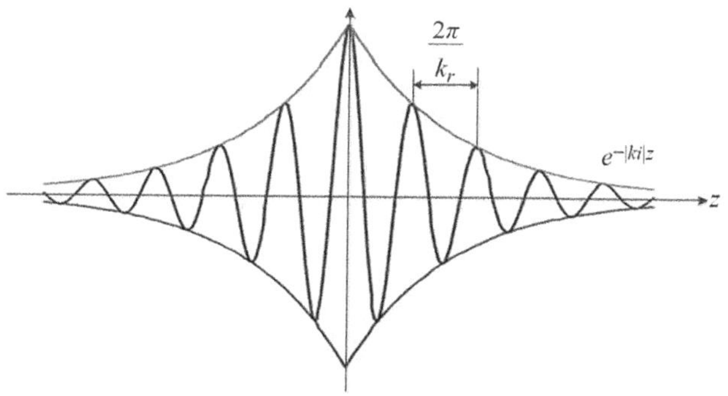

Figure 7.8. *Attenuation of a wave in a weakly conductive ohmic medium*

2) *Highly conductive medium*: if the conductivity is very high, that is, $\frac{\gamma(0)}{\omega\varepsilon_0} \gg 1$, we obtain approximately $\mathrm{k}^2 \simeq \mathrm{i}\omega\mu_0\gamma(0)$ and, consequently:

$$\mathrm{k} = \pm\frac{1+\mathrm{i}}{\delta} \quad [7.132]$$

with:

$$\delta = \sqrt{\frac{2}{\omega\mu_0\gamma(0)}} \quad [7.133]$$

It follows that:

$$H_y(z,t) = \begin{cases} \Re\left[\hat{H}_0 e^{-\frac{z}{\delta}} e^{-\mathrm{i}\left(\omega t - \frac{z}{\delta}\right)}\right] & z > 0 \\ \Re\left[\hat{H}_0 e^{\frac{z}{\delta}} e^{-\mathrm{i}\left(\omega t + \frac{z}{\delta}\right)}\right] & z < 0 \end{cases}$$
$$E_x(z,t) = \begin{cases} \Re\left[(1-\mathrm{i})\sqrt{\frac{\omega\mu_0}{2\gamma(0)}}\hat{H}_0 e^{-\frac{z}{\delta}} e^{-\mathrm{i}\left(\omega t - \frac{z}{\delta}\right)}\right] & z > 0 \\ \Re\left[-(1-\mathrm{i})\sqrt{\frac{\omega\mu_0}{2\gamma(0)}}\hat{H}_0 e^{\frac{z}{\delta}} e^{-\mathrm{i}\left(\omega t + \frac{z}{\delta}\right)}\right] & z < 0 \end{cases} \quad [7.134]$$

The fields decrease rapidly from the surface of the conductor. At $z = \delta$, its amplitude is reduced by a factor $\frac{1}{e}$. δ is called *skin depth*. We conclude that, in the sinusoidal regime, the electromagnetic wave remains localized in a thin layer, on the conductor surface, whose thickness decreases with increasing frequency, as shown in Figure 7.9.

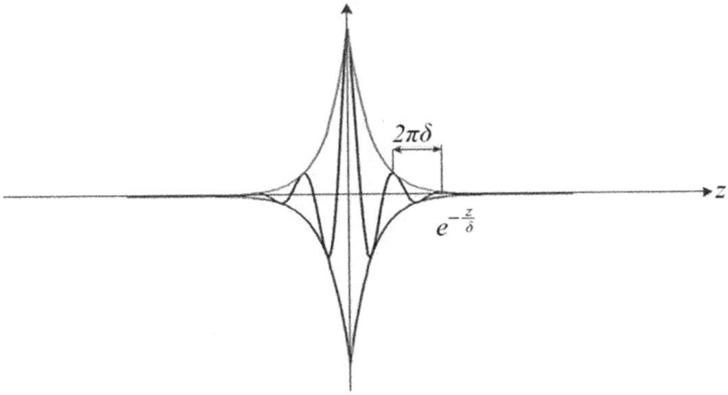

Figure 7.9. *For a highly conductive medium, the wave is rapidly attenuated over a distance δ, called the skin depth*

This is the *skin effect*.

EXAMPLE 7.6. (General case).– To solve equation [7.130] exactly, we separate the real and imaginary parts of k:

$$\mathsf{k} = \mathsf{k}_r + \mathrm{i}\mathsf{k}_i \implies \begin{cases} \mathsf{k}_r^2 - \mathsf{k}_i^2 = \frac{\omega^2}{c^2} \\ \mathsf{k}_r\mathsf{k}_i = \frac{1}{2}\omega\mu_0\gamma(0) \end{cases} \implies \begin{cases} \mathsf{k}_r^4 - \frac{\omega^2}{c^2}\mathsf{k}_r^2 - \frac{1}{4}\omega^2\mu_0^2\gamma^2(0) = 0 \\ \mathsf{k}_i = \frac{\omega\mu_0\gamma(0)}{2\mathsf{k}_r} \end{cases}$$

[7.135]

The first equation of [7.135] admits the solution $\mathsf{k}_r^2 = \frac{\omega^2}{2c^2}\left[1+\sqrt{1+\frac{\gamma^2(0)}{\varepsilon_0^2\omega^2}}\right] \equiv \frac{\omega^2\mathsf{n}^2}{c^2}$, where $\mathsf{n}^2 = \frac{1}{2}\left(1+\sqrt{1+\frac{\gamma^2(0)}{\varepsilon_0^2\omega^2}}\right)$. The negative solution is eliminated. Therefore, it follows that $\mathsf{k}_r = \pm\frac{\mathsf{n}}{c}\omega$ and $\mathsf{k}_i = \pm\frac{\gamma(0)Z_0}{2\mathsf{n}}$. Finally, the fields can be expressed (we consider here only waves propagating in the positive direction of the Oz axis) as:

$$
\begin{aligned}
H_y(z,t) &= \Re\left[\hat{H}_0 e^{-\frac{\gamma(0)Z_0 z}{2\mathsf{n}}} e^{-i\omega\left(t-\frac{\mathsf{n}}{c}z\right)}\right] \\
E_x(z,t) &= \Re\left[Z_0\frac{\hat{H}_0}{\mathsf{n}}\frac{1-\mathrm{i}\frac{\gamma(0)}{2\varepsilon_0\mathsf{n}^2\omega}}{1+\frac{\gamma^2(0)}{4\varepsilon_0^2\mathsf{n}^4\omega^2}} e^{-\frac{\gamma(0)Z_0 z}{2\mathsf{n}}} e^{-\mathrm{i}\omega\left(t-\frac{\mathsf{n}}{c}z\right)}\right]
\end{aligned}
\quad [7.136]
$$

The complex Poynting vector, the average electric energy density, the average magnetic energy density and the average power dissipation density in the medium are given, respectively, by:

$$
\begin{aligned}
\overrightarrow{\mathbf{\Pi}}(z) &= \frac{Z_0\left|\hat{H}_0\right|^2}{2\mathsf{n}}\frac{1-\mathrm{i}\frac{\gamma(0)}{2\varepsilon_0\mathsf{n}^2\omega}}{1+\frac{\gamma^2(0)}{4\varepsilon_0^2\mathsf{n}^4\omega^2}} e^{-\frac{\gamma(0)Z_0 z}{\mathsf{n}}}\overrightarrow{\mathbf{k}} \\
\langle w_e\rangle &= \frac{\mu_0}{4\mathsf{n}^2}\frac{1}{1+\frac{\gamma^2(0)}{4\varepsilon_0^2\mathsf{n}^4\omega^2}}\left|\hat{H}_0\right|^2 e^{-\frac{\gamma(0)Z_0 z}{\mathsf{n}}} \\
\langle w_m\rangle &= \frac{\mu_0}{4}\left|\hat{H}_0\right|^2 e^{-\frac{\gamma(0)Z_0 z}{\mathsf{n}}} \\
\hat{p}_d &= \frac{\gamma(0)Z_0^2}{2\mathsf{n}^2}\frac{1}{1+\frac{\gamma^2(0)}{4\varepsilon_0^2\mathsf{n}^4\omega^2}}\left|\hat{H}_0\right|^2 e^{-\frac{\gamma(0)Z_0 z}{\mathsf{n}}}
\end{aligned}
\quad [7.137]
$$

We can easily verify that (1) the complex Poynting theorem [7.73] is indeed satisfied; (2) the phase velocity is given by $v_\phi = \frac{\omega}{\mathsf{k}_r} = \frac{c}{\mathsf{n}}$, which depends on ω, so there is dispersion; (3) time-averaged Poynting vector $\langle\overrightarrow{\mathbf{\Pi}}(z)\rangle = \frac{Z_0\left|\hat{H}_0\right|^2}{2\mathsf{n}}\frac{e^{-\frac{\gamma(0)Z_0 z}{\mathsf{n}}}}{1+\frac{\gamma^2(0)}{4\varepsilon_0^2\mathsf{n}^4\omega^2}}\overrightarrow{\mathbf{k}}$ decreases exponentially as the wave penetrates the medium, indicating energy dissipation due to Joule heating; and (4) the wave is completely attenuated over a distance on the order of $\delta = \frac{1}{\mathsf{k}_i} = \frac{2\mathsf{n}}{\gamma(0)Z_0}$.

7.4.3. *Propagation of an electromagnetic wave in a plasma*

Note that Ohm's law is only valid for frequencies much lower than the collision frequency of the mobile charge carriers (electrons) with the ions of the crystal lattice. According to the Drude–Lorentz conduction model, there is a number n of electrons per unit volume in a metal that are free to move under the influence of an external electric field, but are also subject to damping forces due to collisions. The equation of motion for an electron inside the metal is thus:

$$
m\frac{\partial\overrightarrow{\mathbf{v}}}{\partial t} + \frac{m}{\tau}\overrightarrow{\mathbf{v}} = -e\overrightarrow{\mathbf{E}}\left(\overrightarrow{\mathbf{r}},t\right) \quad [7.138]
$$

where $\frac{1}{\tau}$ is the collision frequency. If the displacement of the electron is very small compared to the wavelength of the oscillating field, we can approximate:

$$m\frac{\partial \overrightarrow{\mathbf{v}}}{\partial t}+\frac{m}{\tau}\overrightarrow{\mathbf{v}}\approx -e\overrightarrow{\mathbf{E}}_0 e^{-\mathrm{i}\omega t} \qquad [7.139]$$

where $\overrightarrow{\mathbf{E}}_0$ is the average field experienced by the electron. In steady-state conditions, this equation has for solution:

$$\overrightarrow{\mathbf{v}}=-\frac{e}{m\left(\frac{1}{\tau}-\mathrm{i}\omega\right)}\overrightarrow{\mathbf{E}}_0 e^{-\mathrm{i}\omega t}\Longrightarrow \overrightarrow{\mathbf{J}}=\frac{ne^2}{m\left(\frac{1}{\tau}-\mathrm{i}\omega\right)}\overrightarrow{\mathbf{E}} \qquad [7.140]$$

Thus, the conductivity is:

$$\gamma=\frac{ne^2}{m\left(\frac{1}{\tau}-\mathrm{i}\omega\right)} \qquad [7.141]$$

At low frequencies ($\omega\tau \ll 1$), the inertia of the electrons is negligible compared to the collision force. Consequently, we can use the approximation $\gamma\approx\frac{ne^2\tau}{m}$. At low frequencies, the conductivity is constant (i.e. independent of ω). This corresponds to Ohm's law. For frequencies much higher than the collision frequency ($\omega\tau \gg 1$), inertia dominates, and the current is governed by an equation of the form:

$$m\frac{\partial \overrightarrow{\mathbf{v}}}{\partial t}\approx -e\overrightarrow{\mathbf{E}}\Longrightarrow \frac{\partial \overrightarrow{\mathbf{J}}}{\partial t}=\omega_p^2\varepsilon_0\overrightarrow{\mathbf{E}} \qquad [7.142]$$

where $\omega_p=\sqrt{\frac{e^2n}{m\varepsilon_0}}$ is the *plasma frequency*. This law applies to high-frequency waves to electromagnetic waves propagating in many dielectrics, and also to superconducting media, where the collision frequency is zero. For sinusoidal variations, equation [7.142] becomes:

$$\hat{J}_x=\mathrm{i}\frac{\varepsilon_0\omega_p^2}{\omega}\hat{E}_x \qquad [7.143]$$

with $E_x(z,t)=\Re\left[\hat{E}_x e^{-\mathrm{i}(\omega t-\mathrm{k}z)}\right]$ and $J_x(z,t)=\Re\left[\hat{J}_x e^{-\mathrm{i}(\omega t-\mathrm{k}z)}\right]$. In this regime, the conductivity becomes purely imaginary:

$$\gamma_{\text{plasma}}=\mathrm{i}\frac{\varepsilon_0\omega_p^2}{\omega}=\mathrm{i}\frac{ne^2}{m\omega} \qquad [7.144]$$

Let us provide an alternative derivation of the law [7.142]. To reproduce this law, consider a plasma (i.e. an electrically neutral ensemble of ions and electrons). If n_i and n_e denote the equilibrium number densities of ions (with charge $+e$) and electrons (with charge $-e$), respectively, then $n_i=n_e=n$. In the presence of an electric field, the electrons oscillate relative to the essentially immobile ions, whose inertia is much greater. Suppose that the electrons undergo a collective displacement $u(x,t)$ along the Ox-axis. Electrons originally at position x move to $x+u(x)$, while those at $x+dx$ move to $x+dx+u(x)+\frac{\partial u}{\partial x}dx$. This displacement causes a variation in the electron density, denoted δn_e, as illustrated in Figure 7.10.

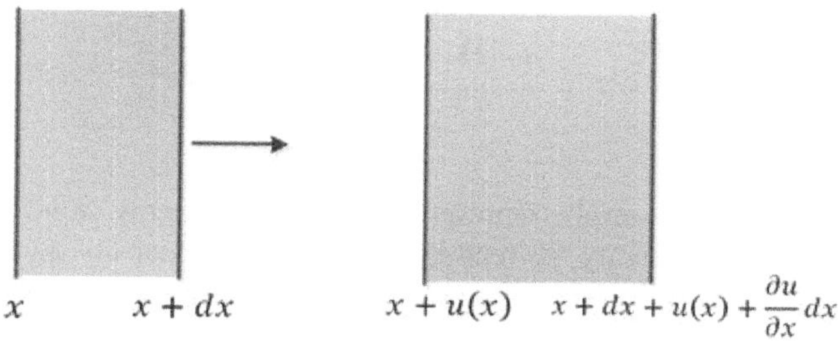

Figure 7.10. *Motion of a plasma wave*

In fact, a slice of thickness dx and cross-sectional area S experiences a variation in volume $d\tau = S\frac{\partial u}{\partial x}dx$. The conservation of charge also allows us to write $n_e S dx = (n_e+\delta n_e) S\left[dx+\frac{\partial u}{\partial x}dx\right]$, which gives $\delta n_e = -n_e\frac{\partial u}{\partial x}$. A volume charge density thus appears within the plasma:

$$\rho = n_i e - (n_e+\delta n_e)e = -e\delta n_e = en_e\frac{\partial u}{\partial x} = en\frac{\partial u}{\partial x} \qquad [7.145]$$

which is related to the local electric field through the Maxwell–Gauss equation $\frac{\partial E_x}{\partial x} = \frac{\rho}{\varepsilon_0} = \frac{en}{\varepsilon_0}\frac{\partial u}{\partial x}$. Integrating this expression gives:

$$E_x = \frac{en}{\varepsilon_0}u \qquad [7.146]$$

with the boundary condition $E_x = 0$ when $u = 0$, implying a zero constant of integration. The force experienced by the electron at its new position is $F_u = -\frac{ne^2}{\varepsilon_0}u$. This is a restoring force proportional to the displacement u of the electron. The corresponding equation of motion for a displaced electron is:

$$m\frac{\partial^2 u}{\partial t^2} + \frac{ne^2}{\varepsilon_0}u = 0 \qquad [7.147]$$

This is the *equation of an oscillatory motion with frequency* $\omega_p = \sqrt{\frac{ne^2}{\varepsilon_0 m}}$, which is called the *plasma frequency*. It is a characteristic quantity of plasma. The motion of the electrons gives rise to a conduction current $J_x = -ne\frac{\partial u}{\partial t}$. Finally, it comes that:

$$\frac{\partial J_x}{\partial t} = -ne\frac{\partial^2 u}{\partial t^2} = \frac{n^2e^3}{m\varepsilon_0}u = \omega_p^2\varepsilon_0 E_x \qquad [7.148]$$

For this medium, the Poynting theorem takes the form $\vec{\nabla} \cdot \vec{\Pi} + \frac{\partial}{\partial t}\left(\frac{\varepsilon_0 \left|\vec{\mathbf{E}}\right|^2}{2} + \frac{\mu_0 \left|\vec{\mathbf{H}}\right|^2}{2}\right) = -\vec{\mathbf{E}} \cdot \vec{\mathbf{J}} = -\frac{1}{\varepsilon_0 \omega_p^2} \vec{\mathbf{J}} \cdot \frac{\partial \vec{\mathbf{J}}}{\partial t} = -\frac{\partial}{\partial t}\left(\frac{\left|\vec{\mathbf{J}}\right|^2}{2\varepsilon_0 \omega_p^2}\right)$ and thus:

$$\vec{\nabla} \cdot \vec{\Pi} + \frac{\partial}{\partial t}\left(\frac{\varepsilon_0 \left|\vec{\mathbf{E}}\right|^2}{2} + \frac{\mu_0 \left|\vec{\mathbf{H}}\right|^2}{2} + \frac{\left|\vec{\mathbf{J}}\right|^2}{2\varepsilon_0 \omega_p^2}\right) = 0 \qquad [7.149]$$

This additional term simply represents the kinetic energy density of mobile electrons, since their velocity is related to the current density by the formula $\vec{\mathbf{J}} = -en\vec{\mathbf{v}}$, which implies:

$$\frac{\left|\vec{\mathbf{J}}\right|^2}{2\varepsilon_0 \omega_p^2} = \frac{1}{2} nm \left|\vec{\mathbf{v}}\right|^2 \qquad [7.150]$$

Let us now use equation [7.142] instead of Ohm's law in the Maxwell equations [7.121]:

$$\begin{aligned} \mathsf{i}\mathsf{k}\hat{E}_x &= \mathsf{i}\omega\mu_0 \hat{H}_y \\ \mathsf{i}\mathsf{k}\hat{H}_y &= \mathsf{i}\omega\varepsilon_0 \left(1 - \frac{\omega_p^2}{\omega^2}\right) \hat{E}_x \end{aligned} \qquad [7.151]$$

where we have used the expressions $E_x = \Re\left[\hat{E}_x e^{-\mathsf{i}(\omega t - \mathsf{k}z)}\right]$, $H_y = \Re\left[\hat{H}_y e^{-\mathsf{i}(\omega t - \mathsf{k}z)}\right]$ and $J_x = \Re\left[\hat{J}_x e^{-\mathsf{i}(\omega t - \mathsf{k}z)}\right] = \Re\left[\mathsf{i}\frac{\varepsilon_0 \omega_p^2}{\omega^2} \hat{E}_x e^{-\mathsf{i}(\omega t - \mathsf{k}z)}\right]$ have been used. Let us define the effective permittivity as:

$$\hat{\varepsilon} = \varepsilon_0 \left(1 - \frac{\omega_p^2}{\omega^2}\right) \qquad [7.152]$$

The system [7.151] immediately yields $\frac{\hat{E}_x}{\hat{H}_y} = \frac{\omega\mu_0}{\mathsf{k}} = \frac{\mathsf{k}}{\omega\hat{\varepsilon}}$ and consequently:

$$\mathsf{k}^2 = \omega^2 \mu_0 \hat{\varepsilon} = \frac{\omega^2 - \omega_p^2}{c^2} \qquad [7.153]$$

Expression [7.153] can also be obtained by inserting complex conductivity [7.144] into the dispersion equation [7.126]. Two cases are then possible (we will focus here on the wave propagating in the positive direction of the Oz axis; the wave propagating in the opposite direction is simply obtained by replacing z with $-z$):

– When $\omega < \omega_p$, it follows that $\mathsf{k}^2 < 0$. Therefore, k is purely imaginary. Taking $\mathsf{k} = \pm\mathrm{i}\frac{\sqrt{\omega_p^2-\omega^2}}{c}$, the wave takes the following form:

$$H_y(z,t) = \Re\left[\hat{H}_0 e^{-\mathrm{i}\omega t} e^{-\frac{\sqrt{\omega_p^2-\omega^2}}{c}z}\right]$$
$$E_x(z,t) = \Re\left[-\mathrm{i}\frac{\omega\mu_0 c}{\sqrt{\omega_p^2-\omega^2}}\hat{H}_0 e^{-\mathrm{i}\omega t} e^{-\frac{\sqrt{\omega_p^2-\omega^2}}{c}z}\right] \quad [7.154]$$

It follows that the wave no longer propagates. It becomes stationary with frequency ω. The wave behaves like an evanescent wave, whose amplitude decreases exponentially with z; it is absorbed over a characteristic thickness:

$$\delta = |\mathsf{k}|^{-1} = \frac{c}{\sqrt{\omega_p^2-\omega^2}} \quad [7.155]$$

– If $\omega > \omega_p$, then $\mathsf{k}^2 > 0$. Thus, $\mathsf{k} = \pm\frac{\sqrt{\omega^2-\omega_p^2}}{c}$ is real, and the wave expressions become:

$$H_y(z,t) = \Re\left[\hat{H}_0 e^{-\mathrm{i}\left(\omega t - \frac{\sqrt{\omega^2-\omega_p^2}}{c}z\right)}\right]$$
$$E_x(z,t) = \Re\left[\frac{\omega\mu_0 c}{\sqrt{\omega^2-\omega_p^2}}\hat{H}_0 e^{-\mathrm{i}\left(\omega t - \frac{\sqrt{\omega^2-\omega_p^2}}{c}z\right)}\right] \quad [7.156]$$

The wave is thus a traveling wave without any attenuation term. The dispersion relation can also be written as:

$$\omega = \sqrt{\omega_p^2 + c^2\mathsf{k}^2} \quad [7.157]$$

We conclude that the plasma behaves as a high-pass filter for incident waves. Plasma is transparent when $\hat{\varepsilon} > 0$, that is, when $\omega > \omega_p$. On the other hand, when $\omega < \omega_p$, an attenuation factor of the form $e^{-\frac{\sqrt{\omega_p^2-\omega^2}}{c}|z|}$ appears, and the wave ceases to propagate (it becomes stationary), as shown in Figure 7.11.

In other words, waves with $\lambda < \lambda_p$ propagate through the plasma without absorption. The condition $\hat{\varepsilon}(\omega_L) = 0$ defines the *cutoff frequency* ω_L. In our case, it is written as: $\hat{\varepsilon}(\omega_L) = \varepsilon\left(1 - \frac{\omega_p^2}{\omega_L^2}\right) = 0$, which gives $\omega_L = \omega_p$. Hence, the cutoff frequency is equal to the plasma frequency, that is, the frequency at which the electron gas undergoes longitudinal oscillations. This cutoff frequency of transverse electromagnetic waves corresponds to the longitudinal oscillation mode of the electron gas. Finally, the cutoff frequency provides a method for measuring plasma density. We can vary the frequency of a wave sent through the plasma and monitor its transmission. When the detected signal disappears, the wave is no longer transmitted. This corresponds to a density equal to $n = \frac{m\varepsilon_0\omega^2}{e^2}$.

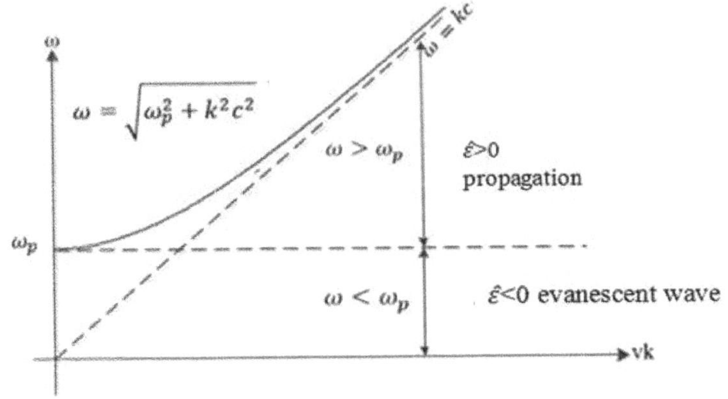

Figure 7.11. *Dispersion curve in a plasma*

7.4.4. *Propagation in a dielectric medium*

In a dielectric medium (without free charges and conduction currents), the Maxwell equations are written as:

$$\begin{aligned} \mathrm{i}k\hat{E}_x &= \mathrm{i}\mu_0\omega\hat{H}_y \\ \mathrm{i}k\hat{H}_y &= \mathrm{i}\varepsilon_0\omega\hat{E}_x + \mathrm{i}\omega\hat{P}_x = \mathrm{i}\varepsilon\omega\hat{E}_x \end{aligned} \qquad [7.158]$$

where $\hat{P}_x$ is the polarization. It is related to E_x by the relation $\hat{P}_x = \varepsilon_0\chi_e(\omega)\hat{E}_x$, where $\chi_e(\omega)$ is the complex electric susceptibility. According to the Drude–Lorentz model, the electric susceptibility and the permittivity of the medium are given by (see section 5.5.2):

$$\left\{\begin{aligned} \chi_e(\omega) &= \chi_e'(\omega) + \mathrm{i}\chi_e''(\omega) \\ &= \frac{\chi_e(0)\,\omega_0^2}{\omega_0^2 - \omega^2 - \mathrm{i}\gamma\omega} \\ \chi_e'(\omega) &= \frac{\chi_e(0)\left(\omega_0^2 - \omega^2\right)\omega_0^2}{\left(\omega_0^2 - \omega^2\right)^2 + \gamma^2\omega^2} \\ \chi_e''(\omega) &= \frac{\chi_e(0)\,\omega_0^2\Gamma\omega}{\left(\omega_0^2 - \omega^2\right)^2 + \gamma^2\omega^2} \end{aligned}\right. \Longrightarrow \left\{\begin{aligned} \varepsilon(\omega) &= \varepsilon_0\left(1 + \chi_e(\omega)\right) \\ &= \varepsilon_0\left(\varepsilon_r'(\omega) + \mathrm{i}\varepsilon''_r(\omega)\right) \\ \varepsilon_r'(\omega) &= 1 + \chi_e'(\omega) \\ \varepsilon''_r(\omega) &= \chi_e''(\omega) \end{aligned}\right. \qquad [7.159]$$

with $\chi_e(0) = \frac{\omega_p^2}{\omega_0^2} = \frac{ne^2}{m\varepsilon_0\omega^2}$. Here, ω_p is the plasma frequency, ω_0 the natural oscillation frequency of the electron within the atom, and γ the damping coefficient due to collisions in the crystal lattice. From equations [7.158], we immediately get $\frac{\hat{E}_x}{\hat{H}_y} = \frac{\mu_0\omega}{\mathsf{k}} = \frac{\mathsf{k}}{\varepsilon_0\omega}$, and consequently:

$$\mathsf{k}^2 = \mu_0\omega^2\varepsilon(\omega) = \frac{\omega^2}{c^2}\varepsilon_r(\omega) = \frac{\omega^2}{c^2}\left[1 + \chi_e(\omega)\right] \quad [7.160]$$

This is the dispersion relation in the medium. The wavenumber is therefore a complex, that is, $\mathsf{k} = \mathsf{k}_r + i\mathsf{k}_i$. Separating the real and imaginary parts in [7.160], we obtain $\mathsf{k}_r^2 - \mathsf{k}_i^2 = \frac{\varepsilon_r'\omega^2}{c^2}$ and $\mathsf{k}_i = \frac{\chi_e''\omega^2}{2c^2\mathsf{k}_r}$. Solving these yields $\mathsf{k}_r = \pm\frac{\omega}{c}\mathsf{n}(\omega)$ and $\mathsf{k}_i = \frac{\omega}{2c\mathsf{n}(\omega)}\chi_e''(\omega)$, where $\mathsf{n}(\omega) = \sqrt{\frac{1}{2}\left[1 + \chi_e'(\omega) + |1 + \chi_e(\omega)|\right]}$. $\mathsf{n}(\omega)$ is called the *refractive index of the medium*. We can also define a complex refractive index, as discussed in section 7.5.1 (both formulations are equivalent). Attenuation of the electric field during propagation (i.e. energy absorption) occurs if $\chi_e'' \neq 0$. To verify this, let us revisit the calculation for the Poynting vector, $\vec{\nabla} \cdot \vec{\Pi} = \frac{1}{\mu_0}\left[\vec{\mathbf{B}} \cdot \left(\vec{\nabla} \times \vec{\mathbf{E}}\right) - \vec{\mathbf{E}} \cdot \left(\vec{\nabla} \times \vec{\mathbf{B}}\right)\right] = \frac{1}{\mu_0}\left[-\vec{\mathbf{B}} \cdot \frac{\partial \vec{\mathbf{B}}}{\partial t} - \varepsilon_0\mu_0\vec{\mathbf{E}} \cdot \frac{\partial \vec{\mathbf{E}}}{\partial t} - \mu_0\vec{\mathbf{E}} \cdot \frac{\partial \vec{\mathbf{P}}}{\partial t}\right]$ and therefore:

$$\vec{\nabla} \cdot \vec{\Pi} + \frac{\partial}{\partial t}\left[\frac{\varepsilon_0}{2}\vec{\mathbf{E}}^2 + \frac{\vec{\mathbf{B}}^2}{2\mu_0}\right] + \vec{\mathbf{E}} \cdot \frac{\partial \vec{\mathbf{P}}}{\partial t} = 0 \quad [7.161]$$

The term $\vec{\mathbf{E}} \cdot \frac{\partial \vec{\mathbf{P}}}{\partial t} = 0$ represents the power received per unit volume. In fact, the force acting on an electron is $\vec{\mathbf{F}} = -e\vec{\mathbf{E}}$, which corresponds to the received power per unit volume $P_r = -ne\vec{\mathbf{E}} \cdot \frac{\partial \vec{\mathbf{r}}}{\partial t}$, where $\vec{\mathbf{r}}$ is the displacement of the electron. We recognize here the expression of the polarization $\vec{\mathbf{P}} = -ne\vec{\mathbf{r}}$, which leads to $P_r = \vec{\mathbf{E}} \cdot \frac{\partial \vec{\mathbf{P}}}{\partial t}$. Expressing the fields as $E_x = \Re\left[\hat{E}_x\right] = \Re\left[\hat{E}_0 e^{\mathrm{i}(\mathsf{k}_r z - \omega t)} e^{-\mathsf{k}_i z}\right]$ and $P_x = \Re\left[\hat{P}_x\right] = \Re\left[\varepsilon_0\left(\chi_e'(\omega) + \mathrm{i}\chi_e''(\omega)\right)\hat{E}_0 e^{\mathrm{i}(\mathsf{k}_r z - \omega t)} e^{-\mathsf{k}_i z}\right]$, we can derive the time-averaged power absorption over one period:

$$\langle\vec{\mathbf{E}} \cdot \frac{\partial \vec{\mathbf{P}}}{\partial t}\rangle = \frac{1}{2}\Re\left[\hat{E}_x\frac{\partial \hat{P}_x^*}{\partial t}\right] = \frac{1}{2}\Re\left[\mathrm{i}\omega\hat{E}_x\hat{P}_x^*\right] = \frac{1}{2}\varepsilon_0\left|\hat{E}_0\right|^2\omega\chi_e'' e^{-2\mathsf{k}_i z} \quad [7.162]$$

Therefore, absorption occurs if $\chi_e''(\omega) \neq 0$. In other words, energy absorption takes place when the polarization has a quadrature component with respect to the electric field. The general solutions for the electromagnetic fields are:

$$\begin{aligned} H_y(z,t) &= \Re\left[\hat{H}_0 e^{-\frac{\omega}{2c\mathsf{n}}\chi_e'' z} e^{-\mathrm{i}\omega\left(t - \frac{\mathsf{n}}{c}z\right)}\right] \\ E_x(z,t) &= \Re\left[\hat{H}_0 Z_0\left(\mathsf{n} + \mathrm{i}\frac{\chi_e''}{2\mathsf{n}}\right) e^{-\frac{\omega}{2c\mathsf{n}}\chi_e'' z} e^{-\mathrm{i}\omega\left(t - \frac{\mathsf{n}}{c}z\right)}\right] \end{aligned} \quad [7.163]$$

Let us now consider some specific cases:

1) At low frequencies, $\omega \ll \omega_0$ ($\omega \to 0$):

$$\begin{cases} \chi'_e = \chi_e(0) = \dfrac{\omega_p^2}{\omega_0^2} \\ \chi''_e = 0 \end{cases} \Longleftrightarrow \begin{cases} \varepsilon'_r = 1 + \dfrac{\omega_p^2}{\omega_0^2} \\ \varepsilon''_r = 0 \end{cases} \qquad [7.164]$$

At low frequencies, we observe that $\chi_e = \chi'_e$: χ_e is therefore real and positive. There is no phase shift between $\overrightarrow{\mathbf{P}}$ and $\overrightarrow{\mathbf{E}}$. There is no attenuation of the wave at low frequencies. Equations [7.163] then reduce to:

$$\begin{aligned} H_y(z,t) &= \Re\left[\hat{H}_0 e^{-\mathrm{i}\omega\left(t-\frac{\mathrm{n}}{c}z\right)}\right] \\ E_x(z,t) &= \Re\left[\hat{H}_0 Z_0 \mathrm{n} e^{-\mathrm{i}\omega\left(t-\frac{\mathrm{n}}{c}z\right)}\right] \end{aligned} \qquad [7.165]$$

where $\mathrm{n} = \sqrt{1 + \frac{\omega_p^2}{\omega_0^2}}$.

2) For $\omega = \omega_0$, we deduce:

$$\begin{cases} \chi'_e(\omega_0) = 0 \\ \chi''_e(\omega_0) = \dfrac{\omega_p^2}{\omega_0\gamma} \end{cases} \Longleftrightarrow \begin{cases} \varepsilon'_r(\omega_0) = 1 \\ \varepsilon''_r(\omega_0) = \dfrac{\omega_p^2}{\omega_0\gamma} \end{cases} \qquad [7.166]$$

We conclude that $\overrightarrow{\mathbf{P}}$ and $\overrightarrow{\mathbf{E}}$ are in quadrature, meaning they are phase-shifted by $\frac{\pi}{2}$. In this case, the fields are expressed as:

$$\begin{aligned} H_y(z,t) &= \Re\left[\hat{H}_0 e^{-\frac{\omega_p^2}{2c\mathrm{n}\gamma}z} e^{-\mathrm{i}\omega\left(t-\frac{\mathrm{n}}{c}z\right)}\right] \\ E_x(z,t) &= \Re\left[\hat{H}_0 Z_0\left(\mathrm{n} + \mathrm{i}\frac{\omega_p^2}{2\mathrm{n}\gamma\omega_0}\right) e^{-\frac{\omega_p^2}{2c\mathrm{n}\gamma}z} e^{-\mathrm{i}\omega\left(t-\frac{\mathrm{n}}{c}z\right)}\right] \end{aligned} \qquad [7.167]$$

with $\mathrm{n} = \sqrt{\frac{1}{2}\left(1 + \sqrt{1 + \frac{\omega_p^2}{\omega^2\gamma^2}}\right)}$. This is clearly seen that the electromagnetic wave is attenuated.

3) Near the resonance frequency ω_0 ($\omega \approx \omega_0$), we use the approximations $\omega_0^2 - \omega^2 = (\omega_0 + \omega)(\omega_0 - \omega) \approx 2\omega_0(\omega_0 - \omega)$ and $\gamma\omega \approx \gamma\omega_0$ to simplify the expressions of $\chi'_e(\omega)$ and $\chi''_e(\omega)$. Thus, it comes that:

$$\begin{aligned} \chi'_e(\omega) &\approx \frac{\omega_0\chi_e(0)}{2}\frac{\omega_0 - \omega}{(\omega_0 - \omega)^2 + \left(\frac{\gamma}{2}\right)^2} \\ \chi''_e(\omega) &\approx \frac{\omega_0\chi_e(0)}{2}\frac{\frac{\gamma}{2}}{(\omega_0 - \omega)^2 + \left(\frac{\gamma}{2}\right)^2} \end{aligned} \qquad [7.168]$$

We observe the following: (a) the function $\chi'_e(\omega)$ changes sign with $(\omega_0 - \omega)$, it is positive for $\omega < \omega_0$ and negative for $\omega > \omega_0$; (b) the function $\chi''_e(\omega)$ reaches its maximum at $\omega = \omega_0$, that is, $(\chi''_e)_{\max} = \chi''_e(\omega_0) = \frac{\chi_e(0)}{\gamma}$. We can therefore rewrite $\chi''_e(\omega)$ as:

$$\chi''_e(\omega) = \chi''_e(\omega_0)\frac{1}{1 + \frac{4}{\gamma^2}(\omega_0 - \omega)^2} \qquad [7.169]$$

The functions $\chi'_e(\omega)$ and $\chi''_e(\omega)$ near ω_0 are qualitatively represented in Figure 7.10, which indicates the frequencies $\omega_\pm = \omega_0 \pm \frac{\Gamma}{2}$ corresponding, respectively, to the extrema of $\chi'_e(\omega)$ on either side of ω_0, and defining the full width at half maximum (FWHM) of the bell-shaped curve for $\chi''_e(\omega)$. The function $\chi''_e(\omega)$ decreases very quickly around ω_0.

4) At high frequencies, $\omega \gg \omega_0$ $(\omega \to \infty)$, equations [7.159] show that both $\chi'_e(\infty) \to 0$ and $\chi''_e(\infty) \to 0$. More precisely, to estimate the order of magnitude of $\chi'_e(\omega)$ and $\chi''_e(\omega)$ at high frequencies, we can neglect all terms compared to ω^2 in their expressions. This gives:

$$\begin{aligned} \chi'_e(\omega) &\approx -\chi_e(0)\frac{\omega_0^2}{\omega^2} = -\frac{\omega_p^2}{\omega^2} \xrightarrow[\omega\to\infty]{} 0^- \\ \chi''_e(\omega) &\approx \chi_e(0)\frac{\omega_0^2\Gamma}{\omega^3} = \frac{\omega_p^2\Gamma}{\omega^3} \xrightarrow[\omega\to\infty]{} 0^+ \end{aligned} \qquad [7.170]$$

The quantity $\chi''_e(\omega)$ becomes negligible compared to $\chi'_e(\omega)$, so we approximate $\chi_e(\omega) \approx -\frac{\omega_p^2}{\omega^2}$. This indicates that the polarization $\overrightarrow{\mathbf{P}}$ and the electric field $\overrightarrow{\mathbf{E}}$ are therefore in phase opposition.

The previous analysis suggests that out resonance (i.e. far from ω_0), the imaginary part of $\chi_e(\omega)$ can be neglected, as it decreases rapidly around ω_0. In other words, we can assume that, far from resonance, the electron experiences only a weak damping force. We can therefore use the approximation: $\Gamma^2\omega^2 \ll (\omega_0^2 - \omega^2)^2$. Under this condition, the susceptibility becomes real:

$$\chi_e(\omega) \approx \chi_e(0)\frac{\omega_0^2}{\omega_0^2 - \omega^2} = \frac{\omega_p^2}{\omega_0^2 - \omega^2} \qquad [7.171]$$

We then say that the system is in a *transparency zone of the medium*. In this regime, a wave propagates without significant absorption. This phenomenon, or *transparency effect*, justifies the use of the term dielectric to describe certain media, which means that they allow a field to pass through them. Figure 7.12 shows the graphical representations of $\chi'_e(\omega)$ and $\varepsilon_r(\omega)$ far from the resonance frequency.

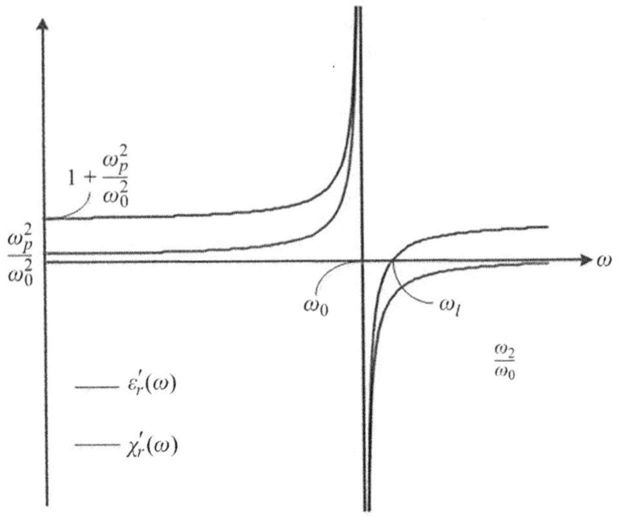

Figure 7.12. *Graphs of $\chi_e'(\omega)$ and $\varepsilon_r'(\omega)$ for $\omega \neq \omega_0$ ($\varepsilon_r'(\omega)$ is negative in the interval $]\omega_0, \omega_l[$ and positive elsewhere). For a color version of this figure, see www.iste.co.uk/abdesselam/electrodynamics2.zip*

We can see what follows:

a) The function $\varepsilon_r'(\omega)$ vanishes at the frequency ω_l (cutoff frequency). This frequency is determined by the condition $\varepsilon_r'(\omega_l) = 1 + \frac{\omega_p^2}{\omega_0^2-\omega_l^2} = 0$. Solving for ω_l gives:

$$\omega_l^2 = \omega_0^2 + \omega_p^2 \quad [7.172]$$

This relation shows that $\omega_l > \omega_0$.

b) The function $\varepsilon_r(\omega)$ is negative in the interval $]\omega_0, \omega_l[$. In this region, the dispersion relation [7.160] gives:

$$\begin{aligned} \mathsf{k}_r &= 0 \\ \mathsf{k}_i &= \pm\frac{\omega}{c}\sqrt{\frac{\omega_l^2-\omega^2}{\omega^2-\omega_0^2}} \end{aligned} \quad [7.173]$$

Consequently, the electromagnetic fields are expressed as:

$$\begin{aligned} H_y(z,t) &= \Re\left[H_0 e^{-\mathrm{i}\omega t} e^{-\frac{\omega}{c}\sqrt{\frac{\omega_l^2-\omega^2}{\omega^2-\omega_0^2}}z}\right] \\ E_x(z,t) &= \Re\left[-\mathrm{i}Z_0\sqrt{\frac{\omega^2-\omega_0^2}{\omega_l^2-\omega^2}}H_0 e^{-\mathrm{i}\omega t} e^{-\frac{\omega}{c}\sqrt{\frac{\omega_l^2-\omega^2}{\omega^2-\omega_0^2}}z}\right] \end{aligned} \quad [7.174]$$

There is therefore no wave propagation in this frequency range: the field becomes stationary and undergoes exponential attenuation inside the medium. This attenuation occurs over a skin depth defined by:

$$\delta = \frac{c}{\omega}\sqrt{\frac{\omega^2 - \omega_0^2}{\omega_l^2 - \omega^2}} \qquad [7.175]$$

c) If $\omega < \omega_0$ (respectively, $\omega > \omega_l$), then $\varepsilon_r = \varepsilon_r' > 1$ (respectively, $\varepsilon_r = \varepsilon_r' < 1$). In these regions, the dispersion relation [7.160] gives:

$$\mathsf{k} = \pm\frac{\omega}{c}\sqrt{\varepsilon_r} = \pm\frac{\omega}{c}\sqrt{\frac{\omega_l^2 - \omega^2}{\omega_0^2 - \omega^2}} \Longleftrightarrow \omega^4 - \omega^2\left(\mathsf{k}^2c^2 + \omega_l^2\right) + \mathsf{k}^2v^2\omega_0^2 = 0 \quad [7.176]$$

and, conversely:

$$\omega_\pm^2 = \frac{\mathsf{k}^2c^2 + \omega_l^2}{2} \pm \sqrt{\left(\frac{\mathsf{k}^2c^2 + \omega_l^2}{2}\right)^2 - \mathsf{k}^2c^2\omega_0^2} \qquad [7.177]$$

Thus, the fields take the following form:

$$\begin{aligned} H_y(z,t) &= \Re\left[\hat{H}_0 e^{-\mathrm{i}\omega\left(t - \frac{\omega}{c}\sqrt{\frac{\omega^2-\omega_l^2}{\omega^2-\omega_0^2}}z\right)}\right] \\ E_x(z,t) &= \Re\left[Z_0\sqrt{\frac{\omega^2 - \omega_0^2}{\omega^2 - \omega_l^2}}\hat{H}_0 e^{-\mathrm{i}\omega\left(t - \frac{\omega}{c}\sqrt{\frac{\omega^2-\omega_l^2}{\omega^2-\omega_0^2}}z\right)}\right] \end{aligned} \qquad [7.178]$$

We observe that the wave propagates without attenuation. Figure 7.13 summarizes the various behaviors of a dielectric medium subjected to electric polarization at different frequencies.

The phase velocity of the electromagnetic wave is equal to $v_\phi = \frac{c}{\mathsf{n}}$, where n is interpreted here as the *refractive index of the medium*. It is related to the relative permittivity through the relation:

$$\mathsf{n}^2 = \varepsilon_r = \frac{\omega_l^2 - \omega^2}{\omega_0^2 - \omega^2} \qquad [7.179]$$

For $\omega > \omega_l$ (respectively, $\omega < \omega_0$), we observe that $v_\phi > c$ (respectively, $v_\phi < c$). We can verify that $\frac{d\mathsf{n}^2}{d(\omega^2)} = \frac{\omega_p^2}{\left(\omega_0^2 - \omega^2\right)^2} > 0$, which shows that the refractive index increases with frequency. For $\omega < \omega_0$, we write $\mathsf{n}^2 = 1 + \frac{\omega_p^2}{\omega_0^2 - \omega^2} = 1 + \frac{\omega_p^2}{\omega_0^2}\frac{1}{1 - \frac{\omega^2}{\omega_0^2}} \approx$

$1 + \frac{\omega_p^2}{\omega_0^2} + \frac{\omega_p^2}{\omega_0^4}\omega^2$. Letting $A^2 = 1 + \frac{\omega_p^2}{\omega_0^2}$, $\omega = \frac{2\pi c}{\lambda_0}$ (where λ_0 is the wavelength of the signal in vacuum), and $B = \frac{(2\pi c)^2\omega_p^2}{\omega_0^4}$, we obtain $\mathsf{n}^2 = A^2 + \frac{B}{\lambda_0^2}$ and consequently:

$$\mathsf{n} = A\left(1 + \frac{B}{A^2\lambda_0^2}\right)^{1/2} \approx A + \frac{B}{2A\lambda_0^2} = A + \frac{C}{\lambda_0^2} \qquad [7.180]$$

Equation [7.180], known as the *Cauchy equation*, shows that the refractive index varies with wavelength.

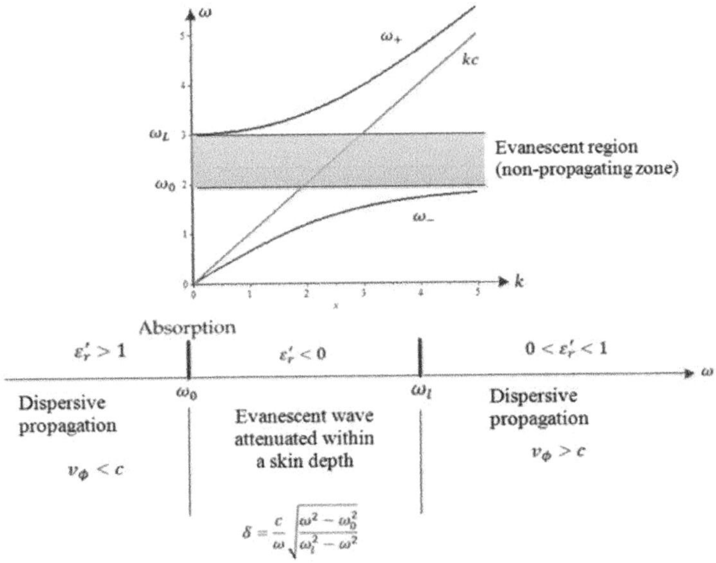

Figure 7.13. *Behavior of a dielectric at different frequencies (there is no wave propagation in* $]\omega_0, \omega_l[$*; the wave is attenuated over the skin depth* $\delta = \frac{c}{\omega}\sqrt{\frac{\omega^2-\omega_0^2}{\omega_l^2-\omega^2}}$*). For a color version of this figure, see www.iste.co.uk/abdesselam/electrodynamics2.zip*

7.5. Dispersive media

When the wavenumber k is not proportional to the frequency ω, the medium is said to be *dispersive*. As a result, an arbitrary (non-sinusoidal) wave changes shape during its propagation, because each of its Fourier components travels at a different speed. Consider a current sheet located at $z = 0$, carrying two identical currents flowing in the same direction, but with slightly different frequencies:

$$K(t) = K_0\left[\cos(\omega_0 + \Delta\omega)t + \cos(\omega_0 - \Delta\omega)t\right] = 2K_0\cos\Delta\omega t\cos\omega_0 t \qquad [7.181]$$

where $\Delta\omega \ll \omega_0$. $K(t)$ appears here as a time-dependent sinusoidal wave, called the *carrier wave*, with frequency ω_0 and whose amplitude is modulated by a slow-varying *envelope function* or *wave envelope* at frequency $\Delta\omega$. In a linear dielectric medium, such a wave packet would propagate away from the current sheet at a speed $\frac{1}{\sqrt{\varepsilon\mu}}$. However, in a dispersive medium, each component of the signal in equation [7.181] travels at a slightly different speed, because $\mathsf{k}(\omega)$ depends on ω. Since the two frequencies are very close to ω_0, we expand $\mathsf{k}(\omega)$ around ω_0: $\mathsf{k}(\omega_0 \pm \Delta\omega) = \mathsf{k}(\omega_0) \pm \left.\frac{d\mathsf{k}}{d\omega}\right|_{\omega_0} \Delta\omega$, where $\mathsf{k}(\omega_0)$ must be real to ensure propagation. For this wave packet propagating along the positive Oz direction, the electric field takes the following form:

$$\begin{aligned} E_x(z,t) &= \Re\left[\hat{E}_0 e^{-\mathrm{i}\left[(\omega_0+\Delta\omega)t-\left(k(\omega_0)+\left.\frac{dk}{d\omega}\right|_{\omega_0}\Delta\omega\right)z\right]}\right. \\ &\quad \left.+\hat{E}_0 e^{-\mathrm{i}\left[(\omega_0-\Delta\omega)t-\left(\mathsf{k}(\omega_0)-\left.\frac{d\mathsf{k}}{d\omega}\right|_{\omega_0}\Delta\omega\right)z\right]}\right] \\ &= \Re\left[\hat{E}_0 e^{-\mathrm{i}[\omega_0 t-\mathsf{k}(\omega_0)z]}\left\{e^{-\mathrm{i}\Delta\omega\left[t-\left.\frac{d\mathsf{k}}{d\omega}\right|_{\omega_0}z\right]}+e^{\mathrm{i}\Delta\omega\left[t-\left.\frac{d\mathsf{k}}{d\omega}\right|_{\omega_0}z\right]}\right\}\right] \\ &= 2\hat{E}_0\cos\left[\omega_0 t-\mathsf{k}(\omega_0)z\right]\cos\Delta\omega\left[t-\left.\frac{d\mathsf{k}}{d\omega}\right|_{\omega_0}z\right] \end{aligned} \quad [7.182]$$

where $\hat{E}_0$ is assumed to be real here for simplicity. This field is illustrated in Figure 7.14 as a function of z for a fixed t. The fast wave, with argument $\omega_0 t - \mathsf{k}(\omega_0)z$, propagates at the phase velocity $v_\phi = \frac{\omega_0}{\mathsf{k}(\omega_0)}$ through the envelope with argument $\Delta\omega\left[t - \left.\frac{d\mathsf{k}}{d\omega}\right|_{\omega_0} z\right]$, which modulates the amplitude. This envelope moves at the slower speed given by:

$$t - \left.\frac{d\mathsf{k}}{d\omega}\right|_{\omega_0} z = \text{cste} \iff \frac{dz}{dt} = v_g = \left.\frac{d\omega}{d\mathsf{k}}\right|_{\omega_0} \quad [7.183]$$

known as the *group velocity*. We conclude that a wave packet takes the form of a carrier wave propagating at the phase velocity $v_\phi = \frac{\omega_0}{\mathsf{k}(\omega_0)}$, whose amplitude is modulated by an envelope moving at the group velocity $v_g = \left.\frac{d\omega}{d\mathsf{k}}\right|_{\omega_0}$. If the constituent waves of the signal are concentrated near a central frequency ω_0, they remain closely grouped as they propagate through the dispersive medium, since they travel at nearly the same velocity v_g, which therefore represents the speed at which the wave packet and its energy propagate. The group velocity v_g also corresponds to the speed of information transmission, and it is always less than the speed of light in a vacuum. Hence, if $v_\phi \neq v_g$, the wave undergoes distortion – this is dispersion. However, if $\frac{d\omega}{d\mathsf{k}} = \frac{\omega}{\mathsf{k}}$, then there is a linear relation between ω and k, namely, $\omega = a \times \mathsf{k}$ (where a is a constant), which implies that the medium is *non-dispersive*.

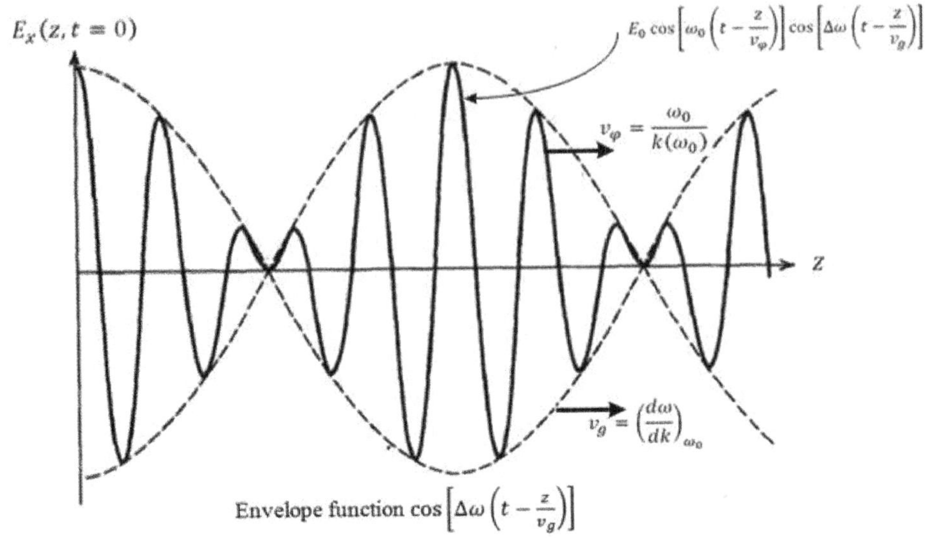

Figure 7.14. *Distortion of a non-sinusoidal wave as it propagates through a dispersive medium*

EXAMPLE 7.7.– For a wave propagating in a polarizable medium in the high-frequency limit ($\omega \gg \omega_0$), we have $\mathsf{n}^2 = 1 - \frac{\omega_p^2}{\omega^2}$. The refractive index, which remains real, becomes slightly less than 1. At such high frequencies, the phase velocity $v_\phi = \frac{c}{\mathsf{n}} > c$ is therefore greater than the speed of light in a vacuum. However, this velocity does not represent the actual speed at which a real signal propagates, associated with energy transport. Let us compute the group velocity: (1) differentiating n^2 with respect to ω, we get $\mathsf{n}\omega\frac{d\mathsf{n}}{d\omega} = \frac{\omega_p^2}{\omega^2}$; (2) since $v_g^{-1} = \frac{d\mathsf{k}}{d\omega}$ and $\mathsf{k} = \frac{\mathsf{n}\omega}{c}$, it follows that: $v_g^{-1} = \frac{\mathsf{n}}{c} + \frac{\omega}{c}\frac{d\mathsf{n}}{d\omega}$; (3) combining the above equations, we finally deduce $v_g^{-1} = \frac{\mathsf{n}}{c} + \frac{1}{c\mathsf{n}}\frac{\omega_p^2}{\omega^2} = \frac{1}{c\mathsf{n}}\left(\mathsf{n}^2 + \frac{\omega_p^2}{\omega^2}\right) = \frac{1}{c\mathsf{n}}$ and consequently:

$$v_g = c\mathsf{n} < c \qquad [7.184]$$

The real propagation speed of a signal at these frequencies is therefore less than the speed of light in a vacuum.

EXAMPLE 7.8.– For a wave propagating in a plasma in the high-frequency limit, we have $\omega^2 = \mathsf{k}^2c^2 + \omega_p^2$, which leads to $v_\phi = \frac{\omega}{\mathsf{k}}$ and $v_g = \frac{d\omega}{d\mathsf{k}} = \frac{\mathsf{k}}{\omega}c^2$. These velocities are meaningful only when k is real, that is, when $\omega > \omega_p$. In this limit, we observe that $v_\phi v_g = c^2$. ■

A general relation can be established between the group velocity and the phase velocity. We have $\frac{1}{v_g} = \frac{dk}{d\omega} = \frac{d}{d\omega}\left(\frac{\omega n}{c}\right) = \frac{n}{c} + \frac{\omega}{c}\frac{dn}{d\omega} = \frac{1}{v_\phi}\left(1 + \frac{\omega}{n}\frac{dn}{d\omega}\right)$ and since $\frac{\omega}{n}\frac{dn}{d\omega} = -\frac{\lambda_0}{n}\frac{dn}{d\lambda_0}$, we obtain the *Rayleigh equation*:

$$v_g = \frac{v_\phi}{1 - \frac{\lambda_0}{n}\frac{dn}{d\lambda_0}} \qquad [7.185]$$

This equation shows that if $\frac{dn}{d\lambda_0} < 0$ (regions of normal dispersion), then the group velocity is always less than the phase velocity. When $\frac{dn}{d\lambda_0} > 0$ (anomalous dispersion), Rayleigh's equation predicts a group velocity greater than the phase velocity. However, in such regions, the Rayleigh equation fails and must be modified due to the presence of absorption.

7.5.1. *Propagation in a dispersive and absorbent medium*

The study of dispersion and absorption phenomena, presented in section 7.4, shows that the electromagnetic properties of materials are, in fact, functions of frequency. This can be easily understood by considering that, at the microscopic scale, these phenomena result from energy exchanges between the elementary constituents of the medium and the photons of the electromagnetic signal. The behavior of the medium (i.e. the emergence of induced currents, whether due to polarization or conduction) results in a modification of wave propagation through an index $n(\omega)$, a generally complex quantity that characterizes the response of a medium to electromagnetic excitation – this is known as the complex refractive index. As seen previously, for the types of media considered here (dielectrics, conductors, plasmas, generally nonmagnetic), this amounts to replacing the vacuum permittivity ε_0 with $\varepsilon = \varepsilon_0\varepsilon_r$, where $\varepsilon_r = n^2(\omega)$. In a sinusoidal regime, the (macroscopic) Maxwell equations valid in an LIH medium then take the form:

$$\begin{aligned} &\vec{\nabla} \cdot \vec{\hat{\mathbf{E}}} = 0 \\ &\vec{\nabla} \cdot \vec{\hat{\mathbf{H}}} = 0 \\ &\vec{\nabla} \times \vec{\hat{\mathbf{E}}} = i\omega\mu_0 \vec{\hat{\mathbf{H}}} \\ &\vec{\nabla} \times \vec{\hat{\mathbf{H}}} = -i\omega\varepsilon_0 n^2(\omega)\,\vec{\hat{\mathbf{E}}} \end{aligned} \qquad [7.186]$$

As established in section 7.4.1, (1) from equations [7.186], we derive the wave equation for the electric field $\Delta\vec{\hat{\mathbf{E}}} + k^2\vec{\hat{\mathbf{E}}} = 0$, with $k = \frac{\omega n(\omega)}{c}$; (2) solving this equation yields two independent sets of solutions. Each set has a Poynting vector (power flux) oriented along the Oz direction. For the first set, the electric field (oriented along the Ox direction) is said to be linearly polarized along Ox; for the second set, the electric field is linearly polarized along the Oy direction (i.e. oriented along Oy). If both solutions are present simultaneously, the direction of the electric field varies with time.

Consider a progressive plane wave propagating along the Oz axis. The solution to equation [7.186] takes the form (the components $\hat{E}_x$ and $\hat{E}_y$ of the complex electric field differ by a phase ϕ):

$$\overrightarrow{\mathbf{E}} = \Re\left[E_{x,0}e^{-\mathrm{i}(\omega t-\mathrm{k}z)}\overrightarrow{\mathbf{i}} + E_{y,0}e^{-\mathrm{i}(\omega t-\mathrm{k}z+\phi)}\overrightarrow{\mathbf{j}}\right] \qquad [7.187]$$

(for simplicity, $E_{x,0}$ and $E_{y,0}$ are assumed to be real). From [7.186], we obtain:

$$\begin{aligned}\overrightarrow{\mathbf{H}} &= \Re\left[-\frac{\mathrm{i}}{\omega\mu_0}\overrightarrow{\nabla}\times\overrightarrow{\mathbf{E}}\right]\\ &= \Re\left[\mathrm{n}\varepsilon_0 c\left(-E_{y,0}e^{-\mathrm{i}(\omega t-\mathrm{k}z+\phi)}\overrightarrow{\mathbf{i}} + E_{x,0}e^{-\mathrm{i}(\omega t-\mathrm{k}z)}\overrightarrow{\mathbf{j}}\right)\right]\end{aligned} \qquad [7.188]$$

Let us write $\mathrm{k} = \mathrm{k}_r + \mathrm{i}\mathrm{k}_i$ and $\mathrm{n} = \mathrm{n}_r + \mathrm{i}\mathrm{n}_i$. The real fields are given by the real parts of the complex fields [7.187] and [7.188]. We see that:

$$\begin{aligned}\overrightarrow{\mathbf{E}} &= e^{-\mathrm{k}_i z}\left[E_{x,0}\cos\left(\omega t-\mathrm{k}_r z\right)\overrightarrow{\mathbf{i}} + E_{y,0}\cos\left(\omega t-\mathrm{k}_r z+\phi\right)\overrightarrow{\mathbf{j}}\right]\\ \overrightarrow{\mathbf{H}} &= c\varepsilon_0 e^{-\mathrm{k}_i z}\left\{-E_{y,0}\left[\mathrm{n}_r\cos\left(\omega t-\mathrm{k}_r z+\phi\right)+\mathrm{n}_i\sin\left(\omega t-\mathrm{k}_r z+\phi\right)\right]\overrightarrow{\mathbf{i}}+\right.\\ &\left.E_{x,0}\left[\mathrm{n}_r\cos\left(\omega t-\mathrm{k}_r z\right)+\mathrm{n}_i\sin\left(\omega t-\mathrm{k}_r z\right)\right]\overrightarrow{\mathbf{j}}\right\}\end{aligned} \qquad [7.189]$$

Thus, the corresponding Poynting vector is:

$$\begin{aligned}\overrightarrow{\mathbf{\Pi}} &= \left(E_x H_y - E_y H_x\right)\overrightarrow{\mathbf{k}} = c\varepsilon_0\overrightarrow{\mathbf{k}}e^{-2\mathrm{k}_i z}\left\{E_{x,0}^2\cos\left(\omega t-\mathrm{k}_r z\right)\left[\mathrm{n}_r\cos\left(\omega t-\mathrm{k}_r z\right)\right.\right.\\ &\left.+\mathrm{n}_i\sin\left(\omega t-\mathrm{k}_r z\right)\right] + E_{y,0}^2\cos\left(\omega t-\mathrm{k}_r z+\phi\right)\left[\mathrm{n}_r\cos\left(\omega t-\mathrm{k}_r z+\phi\right)\right.\\ &\left.\left.+\mathrm{n}_i\sin\left(\omega t-\mathrm{k}_r z+\phi\right)\right]\right\}\end{aligned} \qquad [7.190]$$

Its time-averaged value is:

$$\langle\overrightarrow{\mathbf{\Pi}}\rangle = \frac{c\mathrm{n}_r\varepsilon_0}{2}\overrightarrow{\mathbf{k}}e^{-2\mathrm{k}_i z}\left[E_{x,0}^2 + E_{y,0}^2\right] = \frac{1}{2}c\mathrm{n}_r\overrightarrow{\mathbf{k}}e^{-2\mathrm{k}_i z}\varepsilon_0\left|\overrightarrow{\mathbf{E}}\right|^2 \qquad [7.191]$$

In the case of an absorbing medium ($\mathrm{k}_i \neq 0$), the power transmitted per unit area in the direction of propagation decreases as the wave penetrates the medium.

We can eliminate the term $\omega t - \mathrm{k}_r z$ between the expressions of the components E_x and E_y by writing:

$$\begin{aligned}\cos\left(\omega t-\mathrm{k}_r z\right) &= \frac{E_x}{E_{x,0}e^{-\mathrm{k}_i z}}\\ \sin\left(\omega t-\mathrm{k}_r z\right) &= \frac{1}{\sin\phi}\left(\frac{E_x}{E_{x,0}e^{-\mathrm{k}_i z}}\cos\phi - \frac{E_y}{E_{y,0}e^{-\mathrm{k}_i z}}\right)\end{aligned} \qquad [7.192]$$

which yields:

$$\left(\frac{E_x}{E_{x,0}e^{-\mathrm{k}_i z}}\right)^2+\left(\frac{E_y}{E_{y,0}e^{-\mathrm{k}_i z}}\right)^2-2\frac{E_x}{E_{x,0}e^{-\mathrm{k}_i z}}\frac{E_y}{E_{y,0}e^{-\mathrm{k}_i z}}\cos\phi=\sin^2\phi \qquad [7.193]$$

This is the equation of an ellipse, as illustrated in Figure 7.15(a). As time progresses, the electric field describes an ellipse during each period – this is known as *elliptical polarization*. The area of the ellipse decreases exponentially as the wave penetrates the medium. There are two important special cases:

1) *Linear polarization*: if E_x and E_y are in phase, that is, $\phi = 0$, equation [7.193] reduces to $\left(\frac{E_x}{E_{x,0}e^{-\mathrm{k}_i z}}-\frac{E_y}{E_{y,0}e^{-\mathrm{k}_i z}}\right)^2=0$, which implies:

$$\tan\theta=\frac{E_y}{E_x}=\frac{E_{y,0}}{E_{x,0}} \qquad [7.194]$$

The electric field thus forms an angle θ with the Ox axis. The field amplitude oscillates over time along this line and attenuates as it penetrates the medium. The wave is linearly polarized, as shown in Figure 7.15(b).

2) *Circular polarization*: if $E_{x,0} = E_{y,0} = E_0$ and $\phi = \pm\frac{\pi}{2}$ (i.e. the components are in phase quadrature), we obtain the equation of a circle:

$$E_x^2+E_y^2=E_0^2e^{-2\mathrm{k}_i z} \qquad [7.195]$$

The tip of the electric field vector traces out a circle over the course of one period, with a radius that decreases exponentially as the wave propagates into the medium. If $\phi = \frac{\pi}{2}$, the electric field rotates clockwise. If $\phi = -\frac{\pi}{2}$, the rotation is counterclockwise. These cases are referred to, respectively, as right-handed and left-handed circular polarization for waves propagating in the positive Oz direction, as illustrated in Figures 7.15(c) and (d).

7.5.2. *Propagation in a transparent and inhomogeneous medium*

Consider a propagation in a dielectric medium of a monochromatic wave whose frequency lies outside the absorption bands of the medium. This medium is therefore transparent to the wave, which means that its refractive index is real. However, the medium may be inhomogeneous, in which case its refractive index depends on the spatial coordinates. For such a medium, the Maxwell equations are written as:

$$\begin{aligned}&\vec{\nabla}\cdot\vec{\hat{\mathbf{D}}}=0\\&\vec{\nabla}\cdot\vec{\hat{\mathbf{H}}}=0\\&\vec{\nabla}\times\vec{\hat{\mathbf{E}}}=\mathrm{i}\omega\mu_0\vec{\hat{\mathbf{H}}}\\&\vec{\nabla}\times\vec{\hat{\mathbf{H}}}=-\mathrm{i}\omega\varepsilon_0\mathrm{n}^2\vec{\hat{\mathbf{E}}}\end{aligned} \qquad [7.196]$$

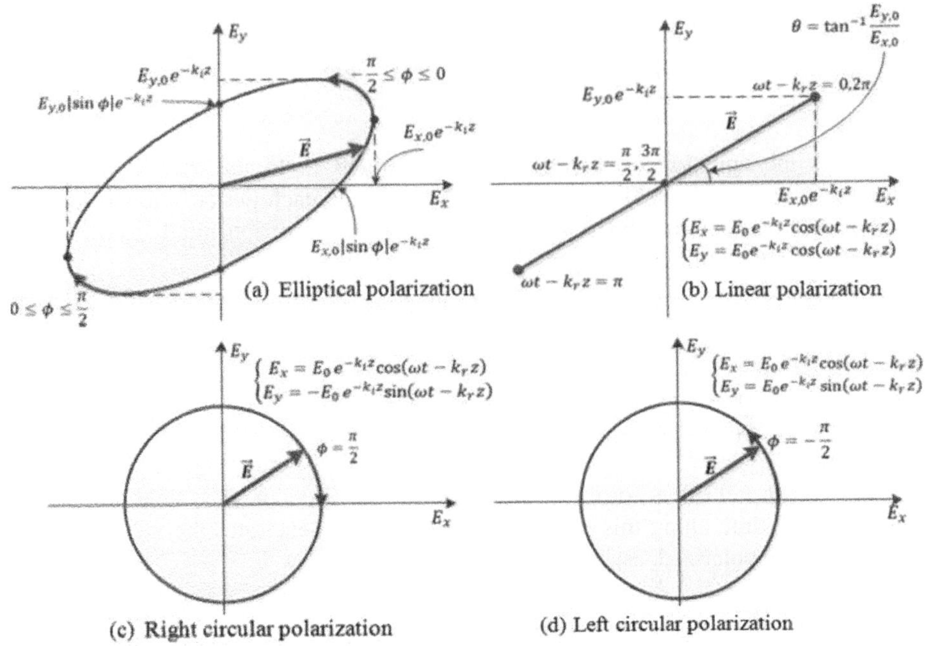

Figure 7.15. *Different polarizations of an electromagnetic wave in a dispersive and absorbent medium. For a color version of this figure, see www.iste.co.uk/abdesselam/electrodynamics2.zip*

The vector $\overrightarrow{\hat{\mathbf{D}}}$ is given here by $\overrightarrow{\hat{\mathbf{D}}} = \varepsilon_0 \mathrm{n}^2 \overrightarrow{\hat{\mathbf{E}}}$. If the refractive index is constant, the system [7.196] admits plane wave solutions. The fields take the form $\overrightarrow{\mathbf{E}}\left(\overrightarrow{\mathbf{r}}, t\right) = \Re\left[\overrightarrow{\hat{\mathbf{E}}}\left(\overrightarrow{\mathbf{r}}\right) e^{-\mathrm{i}\omega t}\right] = \Re\left[\overrightarrow{\hat{\mathbf{E}}}_0 e^{-\mathrm{i}\left(\omega t - \varphi\left(\overrightarrow{\mathbf{r}}\right)\right)}\right]$, where $\overrightarrow{\hat{\mathbf{E}}}_0$ is a constant complex vector. The phase $\varphi\left(\overrightarrow{\mathbf{r}}\right)$ is written as $\varphi\left(\overrightarrow{\mathbf{r}}\right) = \frac{\omega}{c} S_0\left(\overrightarrow{\mathbf{r}}\right) = \frac{2\pi}{\lambda_0} S_0\left(\overrightarrow{\mathbf{r}}\right)$, where $\overrightarrow{\mathrm{k}} = \frac{\omega \mathrm{n}}{c} \overrightarrow{\mathbf{T}} = \frac{2\pi}{\lambda_0} \mathrm{n} \overrightarrow{\mathbf{T}}$ is the wave vector, oriented in the direction of propagation defined by the unit vector $\overrightarrow{\mathbf{T}}$ (λ_0 being the wavelength in vacuum). The phase $\varphi\left(\overrightarrow{\mathbf{r}}\right)$ can also be written as $\varphi\left(\overrightarrow{\mathbf{r}}\right) = \frac{\omega}{c} S_0\left(\overrightarrow{\mathbf{r}}\right) = \frac{2\pi}{\lambda_0} S_0\left(\overrightarrow{\mathbf{r}}\right)$, where $S_0\left(\overrightarrow{\mathbf{r}}\right) = \mathrm{n} \overrightarrow{\mathbf{T}} \cdot \overrightarrow{\mathbf{r}}$. With this notation, the electric field $\overrightarrow{\mathbf{E}}\left(\overrightarrow{\mathbf{r}}, t\right)$ becomes:

$$\overrightarrow{\mathbf{E}}\left(\overrightarrow{\mathbf{r}}, t\right) = \Re\left[\overrightarrow{\hat{\mathbf{E}}}_0 e^{-\mathrm{i}\frac{2\pi}{\lambda_0}\left(ct - S_0\left(\overrightarrow{\mathbf{r}}\right)\right)}\right] \quad [7.197]$$

Now, if the refractive index n varies slowly with position over a distance on the order of λ_0, we seek solutions similar to [7.197], of the form:

$$\overrightarrow{\mathbf{E}}\left(\overrightarrow{\mathbf{r}}, t\right) = \Re\left[\overrightarrow{\hat{\mathbf{E}}}_0\left(\overrightarrow{\mathbf{r}}\right) e^{-\mathrm{i}\frac{2\pi}{\lambda_0}\left(ct - S\left(\overrightarrow{\mathbf{r}}\right)\right)}\right] \quad [7.198]$$

The amplitude $\overrightarrow{\hat{\mathbf{E}}}_0\left(\overrightarrow{\mathbf{r}}\right)$ (assumed to be real here for simplicity) is also a slowly varying function of position. It can be considered constant over a distance of the order of λ_0. This is the fundamental approximation of geometrical optics, also known as the *eikonal approximation*. Under this approximation, it is assumed that spatial variations of the refractive index, the amplitude and the function $S\left(\overrightarrow{\mathbf{r}}\right)$ occur over distances much larger than the wavelength. The function $S\left(\overrightarrow{\mathbf{r}}\right)$ is called the *eikonal function*, and it is worth noting that the surfaces where $S\left(\overrightarrow{\mathbf{r}}\right)$ is constant correspond to wavefronts. Substituting equation [7.198] into the first expression of system [7.196], we get $\overrightarrow{\nabla} \times \overrightarrow{\hat{\mathbf{E}}} = e^{\mathrm{i}\frac{2\pi}{\lambda_0}S(\overrightarrow{\mathbf{r}})}\overrightarrow{\nabla} \times \overrightarrow{\hat{\mathbf{E}}}_0 + \overrightarrow{\nabla}\left(e^{\mathrm{i}\frac{2\pi}{\lambda_0}S(\overrightarrow{\mathbf{r}})}\right) \times \overrightarrow{\hat{\mathbf{E}}}_0 = \mathrm{i}\omega\mu_0\overrightarrow{\hat{\mathbf{H}}}$, which gives $\overrightarrow{\hat{\mathbf{H}}} = \frac{1}{c\mu_0}e^{\mathrm{i}\frac{2\pi}{\lambda_0}S(\overrightarrow{\mathbf{r}})}\left[\overrightarrow{\nabla}S \times \overrightarrow{\hat{\mathbf{E}}}_0 + \frac{\lambda_0}{2\pi\mathrm{i}}\overrightarrow{\nabla} \times \overrightarrow{\hat{\mathbf{E}}}_0\right]$. Since $\overrightarrow{\hat{\mathbf{E}}}_0$ is approximately constant on a wavelength, the term $\frac{\lambda_0}{2\pi\mathrm{i}}\overrightarrow{\nabla} \times \overrightarrow{\hat{\mathbf{E}}}_0$ is nearly zero and can be neglected. Thus, it comes that:

$$\overrightarrow{\hat{\mathbf{H}}} = \overrightarrow{\hat{\mathbf{H}}}_0 e^{\mathrm{i}\frac{2\pi}{\lambda_0}S(\overrightarrow{\mathbf{r}})} = \frac{1}{c\mu_0}e^{\mathrm{i}\frac{2\pi}{\lambda_0}S(\overrightarrow{\mathbf{r}})}\overrightarrow{\nabla}S \times \overrightarrow{\hat{\mathbf{E}}}_0 = \frac{1}{c\mu_0}\overrightarrow{\nabla}S \times \overrightarrow{\hat{\mathbf{E}}} \qquad [7.199]$$

Conversely, let us substitute [7.198] and [7.199] into the fourth equation of [7.196]: $-\mathrm{i}\omega\varepsilon_0\mathrm{n}^2 e^{\mathrm{i}\frac{2\pi}{\lambda_0}S}\overrightarrow{\hat{\mathbf{E}}}_0 = \overrightarrow{\nabla} \times \overrightarrow{\hat{\mathbf{H}}} = \overrightarrow{\nabla} \times \left[\frac{1}{c\mu_0}e^{\mathrm{i}\frac{2\pi}{\lambda_0}S}\overrightarrow{\nabla}S \times \overrightarrow{\hat{\mathbf{E}}}_0\right] = \frac{1}{c\mu_0}\overrightarrow{\nabla}\left(e^{\mathrm{i}\frac{2\pi}{\lambda_0}S}\right) \times \left(\overrightarrow{\nabla}S \times \overrightarrow{\hat{\mathbf{E}}}_0\right) + \frac{1}{c\mu_0}e^{\mathrm{i}\frac{2\pi}{\lambda_0}S}\overrightarrow{\nabla} \times \left(\overrightarrow{\nabla}S \times \overrightarrow{\hat{\mathbf{E}}}_0\right)$. By simplifying, we obtain $-\mathrm{i}\mathrm{n}^2\overrightarrow{\hat{\mathbf{E}}}_0 = \mathrm{i}\overrightarrow{\nabla}S \times \left(\overrightarrow{\nabla}S \times \overrightarrow{\hat{\mathbf{E}}}_0\right) + \frac{c}{\omega}\overrightarrow{\nabla} \times \left(\overrightarrow{\nabla}S \times \overrightarrow{\hat{\mathbf{E}}}_0\right)$, which leads to:

$$-\mathrm{n}^2\overrightarrow{\hat{\mathbf{E}}}_0 = \overrightarrow{\nabla}S \times \left(\overrightarrow{\nabla}S \times \overrightarrow{\hat{\mathbf{E}}}_0\right) = \left(\overrightarrow{\nabla}S \cdot \overrightarrow{\hat{\mathbf{E}}}_0\right)\overrightarrow{\nabla}S - \left(\overrightarrow{\nabla}S\right)^2\overrightarrow{\hat{\mathbf{E}}}_0 \qquad [7.200]$$

$$0 = \overrightarrow{\nabla} \times \left(\overrightarrow{\nabla}S \times \overrightarrow{\hat{\mathbf{E}}}_0\right) \approx \left(\overrightarrow{\hat{\mathbf{E}}}_0 \cdot \overrightarrow{\nabla}\right)\overrightarrow{\nabla}S - \overrightarrow{\hat{\mathbf{E}}}_0(\Delta S) \qquad [7.201]$$

Thus, we arrive at the following equations:

$$\mathrm{n}^2 = \left(\overrightarrow{\nabla}S\right)^2 \qquad [7.202]$$

$$\overrightarrow{\hat{\mathbf{E}}} \cdot \overrightarrow{\nabla}S = 0 \qquad [7.203]$$

Formula [7.202] is known as the *eikonal equation*. Equations [7.199] and [7.203] show that the fields $\overrightarrow{\hat{\mathbf{E}}}$ and $\overrightarrow{\hat{\mathbf{H}}}$ are transverse. In fact, we also verify that $\overrightarrow{\hat{\mathbf{H}}} \cdot \overrightarrow{\nabla}S = 0$. Expression [7.203] can be confirmed starting from equation $\overrightarrow{\nabla} \cdot \left(n^2\overrightarrow{\hat{\mathbf{E}}}\right) = 0$: $0 =$

$$\vec{\nabla}\cdot\left(\mathrm{n}^2 e^{\mathrm{i}\frac{2\pi}{\lambda_0}S}\hat{\vec{\mathbf{E}}}_0\right)\approx\vec{\nabla}\left(\mathrm{n}^2 e^{\mathrm{i}\frac{2\pi}{\lambda_0}S}\right)\cdot\hat{\vec{\mathbf{E}}}_0=e^{\mathrm{i}\frac{2\pi}{\lambda_0}S}\left[\vec{\nabla}\left(\mathrm{n}^2\right)\cdot\hat{\vec{\mathbf{E}}}_0+\frac{2\pi\mathrm{i}}{\lambda_0}\mathrm{n}^2\vec{\nabla}S\cdot\hat{\vec{\mathbf{E}}}_0\right],$$

which implies $\vec{\nabla}S\cdot\hat{\vec{\mathbf{E}}}_0=0$ and $\vec{\nabla}\left(\mathrm{n}^2\right)\cdot\hat{\vec{\mathbf{E}}}_0=0$. Given relations [7.202] and [7.203], we verify that:

$$\hat{\vec{\mathbf{E}}}\simeq\frac{c\mu_0}{\mathrm{n}^2}\hat{\vec{\mathbf{H}}}\times\vec{\nabla}S \qquad [7.204]$$

An equation of the form $\varphi\left(\vec{\mathbf{r}}\right)=\text{const}$ defines a front wave. When following the wave as it propagates from a given front wave, its total phase $\omega t-\varphi\left(\vec{\mathbf{r}}\right)$ remains constant. If the wave passes through $\vec{\mathbf{r}}_1$ at time t_1 and through $\vec{\mathbf{r}}_2$ at time t_2, then:

$$\omega\left(t_2-t_1\right)=\varphi\left(\vec{\mathbf{r}}_2\right)-\varphi\left(\vec{\mathbf{r}}_1\right)=\frac{2\pi}{\lambda_0}\left(S\left(\vec{\mathbf{r}}_2\right)-S\left(\vec{\mathbf{r}}_1\right)\right) \qquad [7.205]$$

The quantity:

$$L_{12}=S\left(\vec{\mathbf{r}}_2\right)-S\left(\vec{\mathbf{r}}_1\right)=\int_1^2\vec{\nabla}S\cdot d\vec{\mathbf{r}} \qquad [7.206]$$

is called the *optical path* from $\vec{\mathbf{r}}_1$ to $\vec{\mathbf{r}}_2$. If $\vec{\mathbf{r}}_1$ and $\vec{\mathbf{r}}_2$ are infinitesimally close, then $\omega\left(t_2-t_1\right)\approx\vec{\nabla}S\cdot\overrightarrow{\mathbf{r_1r_2}}$. We conclude that to go from the wave surface defined by $\varphi_1=\varphi(\vec{\mathbf{r}}_1)$ to that defined by $\varphi_2=\varphi(\vec{\mathbf{r}}_2)$, the shortest path is along the field line of $\vec{\nabla}S$ passing through $\vec{\mathbf{r}}_1$. In other words, the wave propagates along the field lines of the function S (these field lines are orthogonal to the wave surfaces). The point $\vec{\mathbf{r}}_2$ actually reached by the wave is then uniquely defined: it is the point where the field line passing through $\vec{\mathbf{r}}_1$ intersects the wave surface φ_2. Assuming that the propagation speed of the wave at $\vec{\mathbf{r}}_1$ is known, this means that the travel time from $\vec{\mathbf{r}}_1$ to $\vec{\mathbf{r}}_2$ must be minimized. These results form the basis of *Fermat's principle*. We conclude that *the path actually taken by light to go from $\vec{\mathbf{r}}_1$ to $\vec{\mathbf{r}}_2$ is the one that makes the optical path extremal* [7.206].

In optics, light rays are therefore the trajectories orthogonal to the wavefronts (i.e. the field lines of $\vec{\nabla}S$). For a monochromatic wave, its wave vector at a point $\vec{\mathbf{r}}$ is:

$$\vec{\mathsf{k}}\left(\vec{\mathbf{r}}\right)=\frac{2\pi}{\lambda_0}\vec{\nabla}S=\frac{2\pi\mathrm{n}}{\lambda_0}\vec{\mathbf{T}} \qquad [7.207]$$

where $\vec{\mathbf{T}}$ is the unit vector tangent to the field line (since $\vec{\nabla}S=\mathrm{n}\vec{\mathbf{T}}$), called the *principal tangent*. The optical path then takes the form:

$$L_{12}=\int_1^2\mathrm{n}\left(\vec{\mathbf{r}}\right)ds \qquad [7.208]$$

where the integral is taken along the field line of $\vec{\nabla}S$ passing through $\vec{\mathbf{r}}_1$ and $\vec{\mathbf{r}}_2$. The arc element ds is related to displacements dx, dy and dz through equation

$ds = \sqrt{dx^2 + dy^2 + dz^2} = \sqrt{1 + y'^2 + z'^2}dx$, where y' and z' denote derivatives with respect to x. Thus, the optical path is written as:

$$L_{12} = \int_1^2 f\left(x, y\left(x\right), y'\left(x\right), z\left(x\right), z'\left(x\right)\right) dx \qquad [7.209]$$

with $f\left(x, y, y', z, z'\right) = n\left(x, y, z\right)\sqrt{1 + y'^2 + z'^2}$. The optical path is therefore a functional that depends on the shape of the path via the functions $y\left(x\right)$ and $z\left(x\right)$. Since the light ray corresponds to the path that makes the optical path stationary, we must apply the Euler–Lagrange equations $\frac{\partial f}{\partial y} = \frac{d}{dx}\left(\frac{\partial f}{\partial y'}\right)$ and $\frac{\partial f}{\partial z} = \frac{d}{dx}\left(\frac{\partial f}{\partial z'}\right)$, which leads to $\frac{\partial \mathrm{n}}{\partial y}\sqrt{1 + y'^2 + z'^2} = \frac{d}{dx}\left(\frac{\mathrm{n}y'}{\sqrt{1+y'^2+z'^2}}\right)$ and $\frac{\partial \mathrm{n}}{\partial z}\sqrt{1 + y'^2 + z'^2} = \frac{d}{dx}\left(\frac{\mathrm{n}z'}{\sqrt{1+y'^2+z'^2}}\right)$. Using the fact that $ds = \sqrt{1 + y'^2 + z'^2}, dx$, we obtain $\frac{\partial \mathrm{n}}{\partial y} = \frac{d}{ds}\left(\mathrm{n}\frac{dy}{ds}\right)$ and $\frac{\partial \mathrm{n}}{\partial z} = \frac{d}{ds}\left(\mathrm{n}\frac{dz}{ds}\right)$. A similar equation can be derived for x, $\frac{\partial \mathrm{n}}{\partial x} = \frac{d}{ds}\left(\mathrm{n}\frac{dx}{ds}\right)$, which can be obtained by expressing the optical path with y as the integration variable instead of x. These three partial differential equations provide the equation of the light ray in the medium. They can be compactly written in vector form as:

$$\vec{\nabla}\mathrm{n} = \frac{d}{ds}\left[\mathrm{n}\left(\frac{dx}{ds}\vec{\mathbf{i}} + \frac{dy}{ds}\vec{\mathbf{j}} + \frac{dz}{ds}\vec{\mathbf{k}}\right)\right] = \frac{d}{ds}\left(\mathrm{n}\vec{\mathbf{T}}\right) \qquad [7.210]$$

Taking into account the relations [7.199], [7.204] and [7.207], we see that in the eikonal approximation, the electromagnetic wave locally resembles a plane wave, whose local direction of propagation is given by $\vec{\nabla}S$, and whose local propagation velocity is:

$$\vec{\mathbf{v}} = \frac{c}{\mathrm{n}\left(\vec{\mathbf{r}}\right)}\vec{\mathbf{T}}\left(\vec{\mathbf{r}}\right) \qquad [7.211]$$

When the refractive index is constant, the light rays are straight lines. It is the inhomogeneity of the medium that causes the rays to bend. The vector form of the ray equation [7.210] can also be recovered as follows: consider a field line of $\vec{\nabla}S$, along which the position of a point is described by its arc length. The unit tangent vector at point $\vec{\mathbf{r}}$ on this curve is $\vec{\mathbf{T}} = \frac{d\vec{\mathbf{r}}}{ds}$, which implies $\vec{\nabla}S = \mathrm{n}\frac{d\vec{\mathbf{r}}}{ds}$. Differentiating this expression along the curve gives $d\left(\vec{\nabla}S\right) = \left(d\vec{\mathbf{r}}\cdot\vec{\nabla}\right)\left(\vec{\nabla}S\right) = ds\left(\vec{\mathbf{T}}\cdot\vec{\nabla}\right)\left(\vec{\nabla}S\right) = \frac{ds}{\mathrm{n}}\left(\vec{\nabla}S\cdot\vec{\nabla}\right)\left(\vec{\nabla}S\right) = \frac{ds}{\mathrm{n}}\sum_i\sum_j\frac{\partial S}{\partial x_i}\frac{\partial^2 S}{\partial x_i \partial x_j}\vec{\mathbf{e}}_j = \frac{ds}{\mathrm{n}}\sum_i \sum_j\frac{\partial S}{\partial x_i}\frac{\partial^2 S}{\partial x_j \partial x_i}\vec{\mathbf{e}}_j = \frac{ds}{2\mathrm{n}}\sum_i\sum_j\frac{\partial}{\partial x_j}\left(\frac{\partial S}{\partial x_i}\right)^2\vec{\mathbf{e}}_j = \frac{ds}{2\mathrm{n}}\vec{\nabla}\left(\vec{\nabla}S\right)^2 = \frac{ds}{2\mathrm{n}}\vec{\nabla}\left(\mathrm{n}^2\right) = ds\vec{\nabla}\mathrm{n}$. Hence, we obtain $\frac{d}{ds}\left(\mathrm{n}\frac{d\vec{\mathbf{r}}}{ds}\right) = \vec{\nabla}\mathrm{n}$. In particular, if n is constant, equation [7.210] simplifies to

$\frac{d^2\overrightarrow{\mathbf{r}}}{ds^2} = 0$, whose general solution is $\overrightarrow{\mathbf{r}} = \overrightarrow{\mathbf{a}}s + \overrightarrow{\mathbf{b}}$, where $\overrightarrow{\mathbf{a}}$ and $\overrightarrow{\mathbf{b}}$ are constant vectors. If $\overrightarrow{\mathbf{b}} \neq 0$, this solution represents a straight line (not passing through the origin) carried by the vector $\overrightarrow{\mathbf{a}}$; this is the case for plane waves. On the other hand, if $\overrightarrow{\mathbf{b}} = 0$, the curves obtained are straight lines passing through the origin. The corresponding eikonal is:

$$S\left(\overrightarrow{\mathbf{r}}\right) = S\left(r\right) = \pm \mathsf{n} r \qquad [7.212]$$

and therefore:

$$\overrightarrow{\mathbf{T}} = \pm\overrightarrow{\mathbf{e}}_r = \pm\left[\sin\theta\left(\cos\varphi\,\overrightarrow{\mathbf{i}} + \sin\theta\,\overrightarrow{\mathbf{j}}\right) + \cos\theta\,\overrightarrow{\mathbf{k}}\right] \qquad [7.213]$$

which corresponds to spherical waves. To obtain the radius of curvature of a light ray, we write $\overrightarrow{\nabla}\mathsf{n} = \frac{d\mathsf{n}}{ds}\overrightarrow{\mathbf{T}} + \mathsf{n}\frac{d\overrightarrow{\mathbf{T}}}{ds}$ and use the Frenet formula $\frac{d\overrightarrow{\mathbf{T}}}{ds} = \frac{1}{R}\overrightarrow{\mathbf{N}}$, where $\overrightarrow{\mathbf{N}}$ is the *principal normal*, a unit vector perpendicular to the curve at position $\overrightarrow{\mathbf{r}}$ (and thus perpendicular to $\overrightarrow{\mathbf{T}}$). This leads to:

$$\frac{\mathsf{n}}{R} = \overrightarrow{\mathbf{N}}\cdot\overrightarrow{\nabla}\mathsf{n} \qquad [7.214]$$

The gradient of the refractive index always points toward regions of higher index. In other words, the gradient vector is oriented in the direction of the concavity of the path.

7.6. Normal incidence on a perfect conductor

A plane wave, with an electric field oriented along the Ox direction, is normally incident on a perfect conductor located at $z = 0$, as shown in Figure 7.16. The presence of a perfect conductor will give rise to a reflected wave propagating in the negative Oz direction (denoted by $-Oz$). It is important to note that there is no electromagnetic field inside the perfect conductor. The incident fields on the conducting plane (propagating in the positive Oz direction, that is, along $+Oz$) are expressed as follows:

$$\begin{aligned}\overrightarrow{\mathbf{E}}_i\left(\overrightarrow{\mathbf{r}},t\right) &= \Re\left[\hat{E}_{0,i}e^{-\mathrm{i}(\omega t-\mathsf{k}z)}\overrightarrow{\mathbf{i}}\right]\\ \overrightarrow{\mathbf{H}}_i\left(\overrightarrow{\mathbf{r}},t\right) &= \Re\left[Z_0^{-1}\hat{E}_{0,i}e^{-\mathrm{i}(\omega t-\mathsf{k}z)}\overrightarrow{\mathbf{j}}\right]\end{aligned} \qquad [7.215]$$

Similarly, the reflected fields (propagating along the $-Oz$ direction) are given by:

$$\begin{aligned}\overrightarrow{\mathbf{E}}_r\left(\overrightarrow{\mathbf{r}},t\right) &= \Re\left[\hat{E}_{0,r}e^{-\mathrm{i}(\omega t+\mathsf{k}z)}\overrightarrow{\mathbf{i}}\right]\\ \overrightarrow{\mathbf{H}}_r\left(\overrightarrow{\mathbf{r}},t\right) &= \Re\left[-Z_0^{-1}\hat{E}_{0,r}e^{-\mathrm{i}(\omega t+\mathsf{k}z)}\overrightarrow{\mathbf{j}}\right]\end{aligned} \qquad [7.216]$$

where $\mathsf{k} = \frac{\omega}{c}$. The amplitude of the magnetic field (whether incident or reflected) is obtained by dividing the amplitude of the corresponding electric field by the impedance of the medium Z_0. The negative sign in front of the reflected magnetic field (for the wave traveling along the $-Oz$ direction) results from the fact that the Poynting vector $\vec{\mathbf{\Pi}}_r = \vec{\mathbf{E}}_r \times \vec{\mathbf{H}}_r$ of the reflected wave points in the $-Oz$ direction.

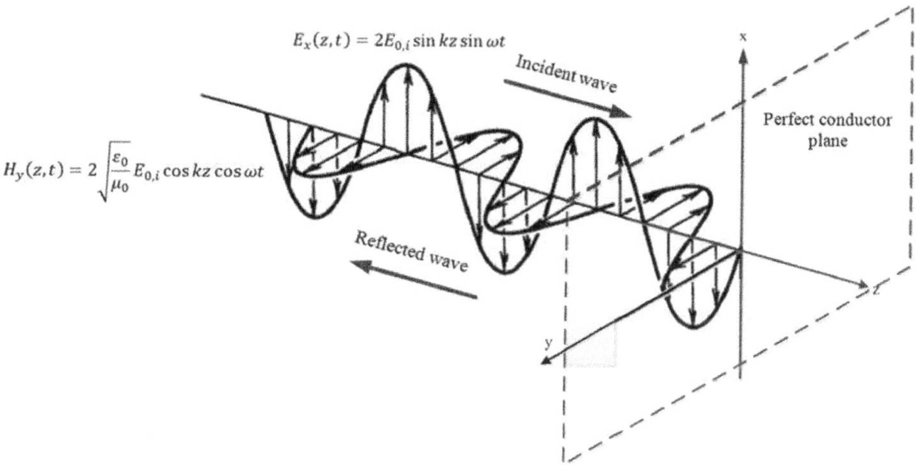

Figure 7.16. *Normal incidence of a plane wave on a perfect conductor*

Outside the perfect conducting plane, the resulting electric field (respectively, magnetic field) is simply the sum of the incident and reflected electric (respectively, magnetic) fields. The unknown field amplitude $E_{0,r}$ can be determined using the boundary condition at $z = 0$, where the tangential component of the total electric field $\vec{\mathbf{E}}$ must be continuous and thus zero at every point on the conductor's surface. This gives: $E_{0,i} + E_{0,r} = 0 \rightarrow E_{0,r} = -E_{0,i}$. The resulting fields are therefore:

$$\begin{aligned} E_x(z,t) &= E_i(z,t) + E_r(z,t) = \Re\left[\hat{E}_{0,i}\left(e^{\mathrm{i}kz} - e^{-\mathrm{i}kz}\right)e^{-\mathrm{i}\omega t}\right] \\ &= 2\hat{E}_{0,i} \sin \mathrm{k}z \sin \omega t \\ H_y(z,t) &= H_i(z,t) + H_r(z,t) = \Re\left[Z_0^{-1}\hat{E}_{0,i}\left(e^{\mathrm{i}kz} + e^{-\mathrm{i}kz}\right)e^{-\mathrm{i}\omega t}\right] \\ &= 2Z_0^{-1}\hat{E}_{0,i} \cos \mathrm{k}z \cos \omega t \end{aligned} \quad [7.217]$$

($\hat{E}_{0,i}$ is assumed here to be real for simplicity). From this, we observe that (1) the two solutions ([7.215] and [7.216]) propagating in opposite directions combine to form a standing wave. The resulting solution does not propagate (it is a spatially stationary sinusoidal wave with a time-varying amplitude); (2) the magnetic field is in phase quadrature (i.e. phase shifted by $\frac{\pi}{2}$) with respect to the electric field, both in time and space; (3) a surface current flows on the perfect conductor at $z = 0$, due to the discontinuity of the tangential component of $\vec{\mathbf{H}}$:

$$K_x = H_y(z = 0, t) = 2Z_0^{-1}\hat{E}_{0,i} \cos \mathrm{k}z \cos \omega t \quad [7.218]$$

(4) the perfect conducting plane will experience a time-averaged force per unit area (since the magnetic field is zero for $z = 0^+$) given by:

$$\begin{aligned}\vec{\mathbf{F}} &= \vec{\mathbf{K}} \times \mu_0 \frac{1}{2}\left[\vec{\mathbf{H}}\left(0^-, t\right) + \vec{\mathbf{H}}\left(0^+, t\right)\right] \\ &= \frac{1}{2}\mu_0 H_y^2\left(0^-, t\right)\vec{\mathbf{k}} = 2\varepsilon_0 E_{0,i}^2 \cos^2 \omega t\, \vec{\mathbf{k}}\end{aligned} \qquad [7.219]$$

This is known as the *radiation pressure.*

7.7. Normal incidence on a dielectric

7.7.1. *Lossless (perfect) dielectric*

Now, replace the perfect conductor with a lossless (perfect) dielectric of permittivity ε_2 and permeability μ_2, as shown in Figure 7.17. We then consider a plane wave (in the transparency domain, for simplicity) normally incident from a medium of permittivity ε_1 and permeability μ_1.

In addition to the incident and reflected (electric and magnetic) fields for $z < 0$, there are transmitted fields that propagate in the $+Oz$ direction within the dielectric medium for $z > 0$:

$$\begin{aligned}\vec{\mathbf{E}}_i(z,t) &= \Re\left[\hat{E}_{0,i}e^{-\mathrm{i}(\omega t - \mathrm{k}_1 z)}\vec{\mathbf{i}}\right] \\ \vec{\mathbf{H}}_i(z,t) &= \Re\left[Z_1^{-1}\hat{E}_{0,i}e^{-\mathrm{i}(\omega t - \mathrm{k}_1 z)}\vec{\mathbf{j}}\right] \\ \vec{\mathbf{E}}_r(z,t) &= \Re\left[\hat{E}_{0,r}e^{-\mathrm{i}(\omega t + \mathrm{k}_1 z)}\vec{\mathbf{i}}\right] \\ \vec{\mathbf{H}}_r(z,t) &= \Re\left[-Z_1^{-1}\hat{E}_{0,r}e^{-\mathrm{i}(\omega t + \mathrm{k}_1 z)}\vec{\mathbf{j}}\right] \\ \vec{\mathbf{E}}_t(z,t) &= \Re\left[\hat{E}_{0,t}e^{-\mathrm{i}(\omega t - \mathrm{k}_2 z)}\vec{\mathbf{i}}\right] \\ \vec{\mathbf{H}}_t(z,t) &= \Re\left[Z_2^{-1}\hat{E}_{0,t}e^{-\mathrm{i}(\omega t - \mathrm{k}_2 z)}\vec{\mathbf{j}}\right]\end{aligned} \qquad [7.220]$$

where $\mathrm{k}_1 = \omega\sqrt{\varepsilon_1\mu_1}$ and $\mathrm{k}_2 = \omega\sqrt{\varepsilon_2\mu_2}$. There is no wave propagating in the $-Oz$ direction in the region $z > 0$, since there is no boundary (on the right side – the second medium is semi-infinite) or source in this region. The quantities $\hat{E}_{0,r}$ and $\hat{E}_{0,t}$ can be determined from the continuity equations for the tangential components of $\vec{\mathbf{E}}$ and $\vec{\mathbf{H}}$ at $z = 0$:

$$\begin{aligned}&\hat{E}_{0,i} + \hat{E}_{0,r} = \hat{E}_{0,t} \\ &Z_1^{-1}\left(\hat{E}_{0,i} - \hat{E}_{0,r}\right) = Z_2^{-1}\hat{E}_{0,t}\end{aligned} \qquad [7.221]$$

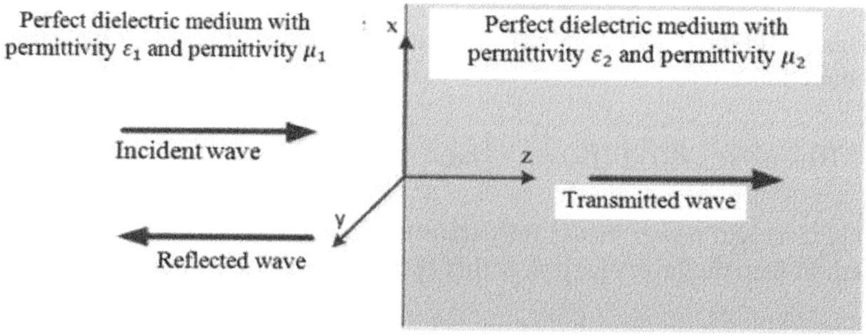

Figure 7.17. *Normal incidence of a plane wave on a lossless dielectric*

From these equations, the *reflection coefficient* R and the *transmission coefficient* T of the field can be calculated as follows:

$$\mathsf{R} = \frac{\hat{E}_{0,r}}{\hat{E}_{0,i}} = \frac{Z_2 - Z_1}{Z_2 + Z_1}$$
$$\mathsf{T} = \frac{\hat{E}_{0,t}}{\hat{E}_{0,i}} = \frac{2Z_2}{Z_2 + Z_1} \qquad [7.222]$$

These are the *Fresnel equations*. It is easily verified that:

$$1 + \mathsf{R} = \mathsf{T} \qquad [7.223]$$

If the two media have the same impedance, that is, $Z_2 = Z_1$, no reflected wave occurs.

The time average of the complex Poynting vector (average power over one period carried by one unit area by the total wave in the $+Oz$ direction) in the $z < 0$ region is equal to:

$$\begin{aligned}\langle \Pi_{z<0} \rangle &= \frac{1}{2}\Re\left[\hat{E}_x(z)\,\hat{H}_y^*(z)\right] \\ &= \frac{1}{2}Z_1^{-1}\Re\left[\left(\hat{E}_{0,i}e^{\mathrm{i}k_1 z} + \hat{E}_{0,r}e^{-\mathrm{i}k_1 z}\right)\left(\hat{E}_{0,i}^* e^{-\mathrm{i}k_1 z} - \hat{E}_{0,r}^* e^{\mathrm{i}k_1 z}\right)\right] \\ &= \frac{1}{2}Z_1^{-1}\left(\left|\hat{E}_{0,i}\right|^2 - \left|\hat{E}_{0,r}\right|^2\right) + \\ &\frac{1}{2}Z_1^{-1}\Re\left[\hat{E}_{0,r}\hat{E}_{0,i}^* e^{-2\mathrm{i}k_1 z} - \hat{E}_{0,r}^*\hat{E}_{0,i}e^{2\mathrm{i}k_1 z}\right]\end{aligned} \qquad [7.224]$$

The expression $\hat{E}_{0,r}\hat{E}_{0,i}^* e^{-2\mathrm{i}k_1 z} - \hat{E}_{0,r}^*\hat{E}_{0,i}e^{2\mathrm{i}k_1 z}$ is purely imaginary, which means that this term does not contribute to the real part. It follows that the power flux in the region $z < 0$ is equal to the difference between the power flux of the incident wave (propagating in the $+Oz$ direction) and that of the reflected wave (propagating in the $-Oz$ direction):

$$\langle \Pi_{z<0} \rangle = \frac{1}{2} Z_1^{-1} \left(\left| \hat{E}_{0,i} \right|^2 - \left| \hat{E}_{0,r} \right|^2 \right) = \frac{1}{2} Z_1^{-1} \left| \hat{E}_{0,i} \right|^2 \left(1 - \mathsf{R}^2 \right) \qquad [7.225]$$

This (average) power flux is fully transmitted into dielectric medium 2, as can be verified. In fact, the (average) power flux carried by the transmitted wave is:

$$\langle \Pi_{z>0} \rangle = \frac{1}{2} Z_2^{-1} \left| \hat{E}_{0,t} \right|^2 = \frac{1}{2} Z_2^{-1} \left| \hat{E}_{0,i} \right|^2 \mathsf{T}^2 = \frac{1}{2} Z_1^{-1} \left| \hat{E}_{0,i} \right|^2 \left(1 - \mathsf{R}^2 \right) \qquad [7.226]$$

7.7.2. Dielectric with losses (with ohmic conductivity)

Assume now that medium 2 has an ohmic conductivity $\gamma(0)$. To simplify, medium 1 is taken to be vacuum. The solutions [7.222] remain valid. It is sufficient to replace the permittivity ε_2 with the complex permittivity $\hat{\varepsilon}_2 = \varepsilon_0 \left(1 + i \frac{\gamma(0)}{\omega \varepsilon_0} \right)$. As a result, the impedance in medium 2 also becomes complex. The effects of losses can be easily analyzed in the limits of low and high conductivity:

a) *Low-conductivity medium*: if the conductivity is low, the impedance in medium 2 can be approximated as $\hat{Z}_2 \simeq Z_0 \left(1 - \frac{\mathrm{i}}{2} \frac{\gamma(0)}{\omega \varepsilon_0} \right)$. The reflection and transmission coefficients also become complex:

$$\begin{aligned} \mathsf{R} &\simeq -\frac{\mathrm{i}}{4} \frac{\gamma(0)}{\omega \varepsilon_0} \\ \mathsf{T} &\simeq 1 - \frac{\mathrm{i}}{4} \frac{\gamma(0)}{\omega \varepsilon_0} \end{aligned} \qquad [7.227]$$

The coefficients R and T thus introduce phase shifts between the fields.

b) *High-conductivity medium*: for high conductivities, the impedance in region 2 is approximately $\hat{Z}_2 \simeq \sqrt{\frac{\mu_0 \omega}{\mathrm{i}\gamma(0)}} = \frac{\omega \delta Z_0}{2c} (1 - \mathrm{i})$, where $\delta = \sqrt{\frac{2}{\omega \mu_0 \gamma(0)}}$ is the skin depth in medium 2. The reflection and transmission coefficients are therefore given by:

$$\begin{aligned} \mathsf{R} &\simeq -1 + \frac{\omega \delta}{c} (1 - \mathrm{i}) \\ \mathsf{T} &\simeq \frac{\omega \delta}{c} (1 - \mathrm{i}) \end{aligned} \qquad [7.228]$$

As $\gamma(0) \to \infty$ (i.e. $\delta \to 0$), we recover the limiting behavior $\mathsf{R} \to -1$ and $\mathsf{T} \to 0$, which is consistent with the results obtained in section 7.6.

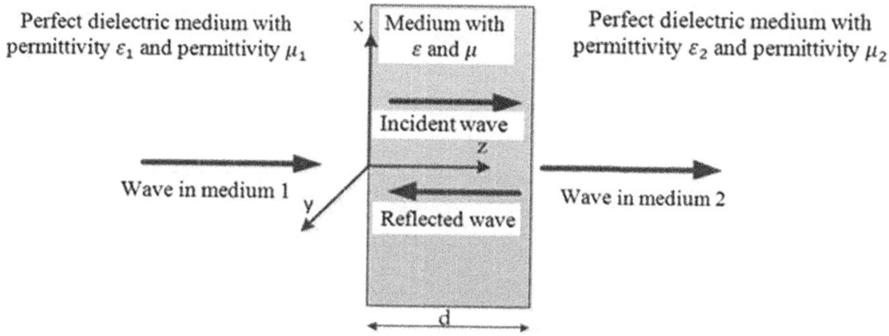

Figure 7.18. *Complete transmission of a signal through a coating*

7.7.3. Complete signal transmission through a coating

A thin, lossless dielectric layer (referred to here as a dielectric coating) of thickness d, permittivity ε and permeability μ is placed at the interface between two infinite half-spaces, ideal (non-dissipative) media with characteristics (ε_1, μ_1) and (ε_2, μ_2), respectively, as illustrated in Figure 7.18. Our goal is to determine the values of ε, μ and the thickness d that allow a wave propagating in the $+Oz$ direction to be completely transmitted through the coating from region 1 to region 2. In order for the wave to be transmitted into region 2 without reflection, no reflected field must exist in region 1. In each region, the electromagnetic fields take the following form:

$$\begin{aligned}
\vec{\mathbf{E}}_1(z,t) &= \Re\left[\hat{E}_{0,1}e^{-\mathrm{i}(\omega t-\mathrm{k}_1 z)}\vec{\mathbf{i}}\right]\\
\vec{\mathbf{H}}_1(z,t) &= \Re\left[Z_1^{-1}\hat{E}_{0,1}e^{-\mathrm{i}(\omega t-\mathrm{k}_1 z)}\vec{\mathbf{j}}\right]\\
\vec{\mathbf{E}}_+(z,t) &= \Re\left[\hat{E}_{0,+}e^{-\mathrm{i}(\omega t-\mathrm{k}z)}\vec{\mathbf{i}}\right]\\
\vec{\mathbf{H}}_+(z,t) &= \Re\left[Z^{-1}\hat{E}_{0,+}e^{-\mathrm{i}(\omega t-\mathrm{k}z)}\vec{\mathbf{j}}\right]\\
\vec{\mathbf{E}}_-(z,t) &= \Re\left[\hat{E}_{0,-}e^{-\mathrm{i}(\omega t+\mathrm{k}z)}\vec{\mathbf{i}}\right]\\
\vec{\mathbf{H}}_-(z,t) &= \Re\left[-Z^{-1}\hat{E}_{0,-}e^{-\mathrm{i}(\omega t+\mathrm{k}z)}\vec{\mathbf{j}}\right]\\
\vec{\mathbf{E}}_2(z,t) &= \Re\left[\hat{E}_{0,2}e^{-\mathrm{i}(\omega t-\mathrm{k}_2 z)}\vec{\mathbf{i}}\right]\\
\vec{\mathbf{H}}_2(z,t) &= \Re\left[Z_2^{-1}\hat{E}_{0,2}e^{-\mathrm{i}(\omega t-\mathrm{k}_2 z)}\vec{\mathbf{j}}\right]
\end{aligned} \qquad [7.229]$$

where $\mathrm{k}_1 = \omega\sqrt{\varepsilon_1\mu_1}$, $\mathrm{k} = \omega\sqrt{\varepsilon\mu}$ and $\mathrm{k}_2 = \omega\sqrt{\varepsilon_2\mu_2}$. The continuity of the tangential components of the electric and magnetic fields $\vec{\mathbf{E}}$ and $\vec{\mathbf{H}}$ at $z = 0$ and $z = d$ imposes the following constraints:

$$\begin{aligned}
&\hat{E}_{0,1} = \hat{E}_{0,+} + \hat{E}_{0,-} \\
&Z_1^{-1}\hat{E}_{0,1} = Z^{-1}\left(E_{0,+} - \hat{E}_{0,-}\right) \\
&\hat{E}_{0,+}e^{\mathrm{i}kd} + \hat{E}_{0,-}e^{-\mathrm{i}kd} = \hat{E}_{0,2}e^{\mathrm{i}k_2 d} \\
&Z^{-1}\left(\hat{E}_{0,+}e^{\mathrm{i}kd} - \hat{E}_{0,-}e^{-\mathrm{i}kd}\right) = Z_2^{-1}\hat{E}_{0,2}e^{\mathrm{i}k_2 d}
\end{aligned} \qquad [7.230]$$

and consequently:

$$\begin{aligned}
\hat{E}_{0,+} &= \frac{1}{2}\left(1 + Z_1^{-1}Z\right)\hat{E}_{0,1} \\
\hat{E}_{0,-} &= \frac{1}{2}\left(1 - Z_1^{-1}Z\right)\hat{E}_{0,1} \\
\hat{E}_{0,2} &= e^{-\mathrm{i}k_2 d}\left(\hat{E}_{0,+}e^{\mathrm{i}kd} + \hat{E}_{0,-}e^{-\mathrm{i}kd}\right) \\
&= Z_2 Z^{-1}e^{-\mathrm{i}k_2 d}\left(\hat{E}_{0,+}e^{\mathrm{i}kd} - \hat{E}_{0,-}e^{-\mathrm{i}kd}\right)
\end{aligned} \qquad [7.231]$$

where $Z_1 = \sqrt{\frac{\mu_1}{\varepsilon_1}}$, $Z = \sqrt{\frac{\mu}{\varepsilon}}$ and $Z_2 = \sqrt{\frac{\mu_2}{\varepsilon_2}}$. The last equation clearly shows that $\hat{E}_{0,+}$ and $\hat{E}_{0,-}$ must satisfy $E_{0,+}e^{\mathrm{i}kd}\left(1 - Z_2Z^{-1}\right) + E_{0,-}e^{-\mathrm{i}kd}\left(1 + Z_2Z^{-1}\right) = 0$. Substituting the expressions for $\hat{E}_{0,+}$ and $\hat{E}_{0,-}$ from above and simplifying by $\hat{E}_{0,1}$, we obtain $\left(1 + Z_1^{-1}Z\right)\left(1 - Z_2Z^{-1}\right)e^{2\mathrm{i}kd} + \left(1 - Z_1Z^{-1}\right)$. Since this is a complex relation, both the real and imaginary parts must vanish independently. The imaginary part is zero only if the coating thickness d satisfies $2kd = n\pi \Rightarrow d = \frac{n\lambda}{4}$, $n \in \mathbb{N}^\star$. For the real part, we will have $\left(1 + Z_1^{-1}Z\right)\left(1 - Z_2Z^{-1}\right)(-1)^n + \left(1 - Z_1^{-1}Z\right)\left(1 + Z_2Z^{-1}\right) = 0$.

Thus, we distinguish two cases:

1) For even n ($n = 2m$), the only possible solution is:

$$Z_1 = Z_2 \qquad \left(d = \frac{m\lambda}{2},\ m \in \mathbb{N}^\star\right) \qquad [7.232]$$

which implies that media 1 and 2 are identical – the coating has no effect in this case.

2) If media 1 and 2 have different impedances, then n must be odd, that is, $n = 2m + 1$. We must take:

$$Z = \sqrt{Z_1 Z} \qquad \left(d = \frac{(2m+1)\lambda}{4},\ m \in \mathbb{N}\right) \qquad [7.233]$$

We conclude that if the coating has a thickness $d = \frac{(2m+1)\lambda}{4}$ and an impedance equal to the geometric mean of the adjacent media impedances, then all of the average power in region 1 is transmitted through the coating into region 2:

$$\begin{aligned}\langle\Pi_{z<0}\rangle &= \frac{1}{2}Z_1^{-1}\left|\hat{E}_{0,1}\right|^2 = \frac{1}{2}\Re\left[Z^{-1}\left(E_{0,+}e^{\mathrm{i}kz} + E_{0,-}e^{-\mathrm{i}kz}\right)\right.\\ &\quad\left.\times\left(E_{0,+}^{*}e^{-\mathrm{i}kz} - E_{0,-}^{*}e^{\mathrm{i}kz}\right)\right]\\ &= \underbrace{\frac{1}{2}Z^{-1}\left(\left|E_{0,+}\right|^2 - \left|E_{0,-}\right|^2\right)}_{\langle\Pi_{0<z<d}\rangle} = \frac{1}{2}Z_2^{-1}\left|E_{0,2}\right|^2 = \langle\Pi_{z>d}\rangle\end{aligned} \quad [7.234]$$

7.8. Oblique incidence of a plane wave

So far, we have only considered waves (propagating along the Oz direction) normally incident on planar interfaces. In sections 7.6 and 7.7, the electric field is polarized along Ox. For waves polarized along Oy, the same solutions remain valid, since both polarizations are parallel to the interface. We now consider a plane wave whose Poynting vector makes an angle θ with the Oz axis, as shown in Figure 7.19. The electric field is directed along Oy, but the magnetic field (which is perpendicular to both $\vec{\mathbf{E}}$ and $\vec{\Pi}$) has components along both Ox and Oz. The direction of the Poynting vector, denoted here as Oz', is related to the axes Ox and Oz by the transformation $z' = x\sin\theta + z\cos\theta$. Consequently, the phase term of the wave takes the form $\omega t - \mathsf{k}z' = \omega t - \mathsf{k}_x x - \mathsf{k}_z z$, where $\mathsf{k}_x = \mathsf{k}\sin\theta$ and $\mathsf{k}_z = \mathsf{k}\cos\theta$ $(\mathsf{k} = \omega\sqrt{\varepsilon\mu})$. This allows us to write the fields as:

$$\begin{aligned}\vec{\mathbf{E}} &= \Re\left[\hat{E}_0 e^{-\mathrm{i}(\omega t - \mathsf{k}_x x - \mathsf{k}_z z)}\vec{\mathbf{j}}\right]\\ \vec{\mathbf{H}} &= \Re\left[Z^{-1}\hat{E}_0 e^{-\mathrm{i}(\omega t - \mathsf{k}_x x - \mathsf{k}_z z)}\left(-\cos\theta\,\vec{\mathbf{i}} + \sin\theta\,\vec{\mathbf{k}}\right)\right]\end{aligned} \quad [7.235]$$

The spatial dependence of the fields can therefore be expressed in the form $e^{i\vec{\mathsf{k}}\cdot\vec{r}}$, where the wavenumber can be considered as a vector $\vec{\mathsf{k}} = \mathsf{k}_x\vec{\mathbf{i}} + \mathsf{k}_z\vec{\mathbf{k}}$. These equations can be easily generalized to 3D. The magnitude of the wave vector $\vec{\mathsf{k}}$ is $\omega\sqrt{\varepsilon\mu}$, and its direction coincides with that of the Poynting vector:

$$\langle\vec{\Pi}\rangle = \frac{1}{2}\Re\left(\vec{\mathbf{E}} \times \vec{\mathbf{H}}^{*}\right) = \frac{1}{2}Z^{-1}\left|E_0\right|^2\left(\cos\theta\,\vec{\mathbf{k}} + \sin\theta\,\vec{\mathbf{i}}\right) = \frac{1}{2}\frac{\left|E_0\right|^2}{\omega\mu}\vec{\mathsf{k}} \quad [7.236]$$

The wave in [7.235] is called a *TE* (*transverse electric*) *wave*: the electric field $\vec{\mathbf{E}}$ is directed transversely to the plane of incidence, while the magnetic excitation field $\vec{\mathbf{H}}$ lies in the plane of incidence Oxz, as shown in Figure 7.19(a). On the other hand, in the case of a *TM* (*transverse magnetic*) *wave*, the magnetic excitation field $\vec{\mathbf{H}}$ is transverse (i.e. normal to the plane of incidence), and the electric field $\vec{\mathbf{E}}$ lies within the plane of incidence, as shown in Figure 7.19(b). A TM wave is described by the following fields:

$$\begin{aligned}\vec{\mathbf{E}} &= \Re\left[\hat{E}_0 e^{-\mathrm{i}(\omega t - \mathrm{k}_x x - \mathrm{k}_z z)}\left(\cos\theta\,\vec{\mathbf{i}} - \sin\theta\,\vec{\mathbf{k}}\right)\right]\\ \vec{\mathbf{H}} &= \Re\left[Z^{-1}\hat{E}_0 e^{-\mathrm{i}(\omega t - \mathrm{k}_x x - \mathrm{k}_z z)}\vec{\mathbf{j}}\right]\\ \langle\vec{\mathbf{\Pi}}\rangle &= \frac{1}{2}Z^{-1}\left|E_0\right|^2\left(\cos\theta\,\vec{\mathbf{k}} + \sin\theta\,\vec{\mathbf{i}}\right) = \frac{1}{2}\frac{\left|E_0\right|^2}{\omega\mu}\vec{\mathrm{k}}\end{aligned} \qquad [7.237]$$

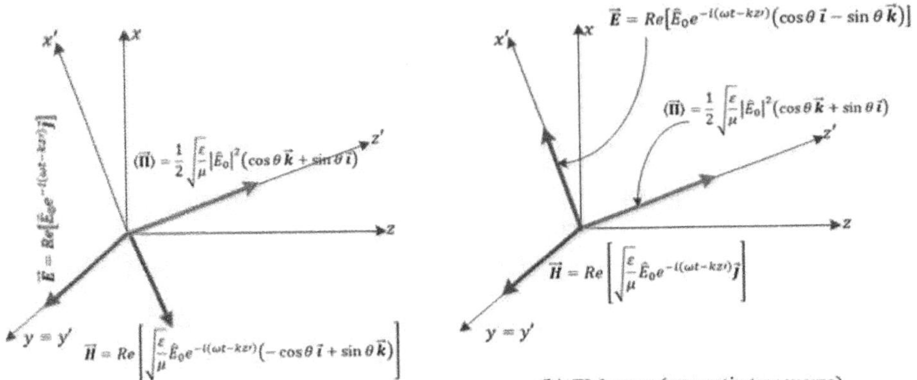

(a) TE wave (electric transverse) (b) TM wave (magnetic transverse)

Figure 7.19. *TE (transverse electric) and TM (transverse magnetic) waves. For a color version of this figure, see www.iste.co.uk/abdesselam/electrodynamics2.zip*

Any polarized wave can be viewed as a superposition of a TE and a TM waves (a property known as the decomposition of polarized waves). Using this decomposition, a linearly polarized wave in an arbitrary direction can be written as:

$$\begin{aligned}\vec{\mathbf{E}}(x,z,t) &= \Re\left\{\left[\hat{E}_{TM}\left(\cos\theta\,\vec{\mathbf{i}} - \sin\theta\,\vec{\mathbf{k}}\right) + \hat{E}_{TE}\,\vec{\mathbf{j}}\right]\right.\\ &\qquad\left. e^{-\mathrm{i}(\omega t - \mathrm{k}_x x - \mathrm{k}_z z)}\right\}\\ \vec{\mathbf{H}}(x,z,t) &= \Re\left\{Z^{-1}\left[\hat{E}_{TM}\,\vec{\mathbf{j}} - \hat{E}_{TE}\left(\cos\theta\,\vec{\mathbf{i}} - \sin\theta\,\vec{\mathbf{k}}\right)\right]\right.\\ &\qquad\left. e^{-\mathrm{i}(\omega t - \mathrm{k}_x x - \mathrm{k}_z z)}\right\}\end{aligned} \qquad [7.238]$$

To account for attenuation effects, we generalize the fields as follows:

$$\begin{aligned}\vec{\mathbf{E}} &= \Re\left[\vec{\hat{\mathbf{E}}}_0 e^{-i\omega t} e^{-\vec{\gamma}\cdot\vec{\mathbf{r}}}\right] = \Re\left[\vec{\hat{\mathbf{E}}}_0 e^{-\mathrm{i}\left(\omega t - \vec{\mathrm{k}}\cdot\vec{\mathbf{r}}\right)} e^{-\vec{\alpha}\cdot\vec{\mathbf{r}}}\right]\\ \vec{\mathbf{H}} &= \Re\left[\vec{\hat{\mathbf{H}}}_0 e^{-i\omega t} e^{-\vec{\alpha}\cdot\vec{\mathbf{r}}}\right] = \Re\left[\vec{\hat{\mathbf{H}}}_0 e^{-\mathrm{i}\left(\omega t - \vec{\mathrm{k}}\cdot\vec{\mathbf{r}}\right)} e^{-\vec{\alpha}\cdot\vec{\mathbf{r}}}\right]\end{aligned} \qquad [7.239]$$

where $\vec{\gamma} = \vec{\alpha} - i\vec{k} = \gamma_x\vec{\mathbf{i}} + \gamma_y\vec{\mathbf{j}} + \gamma_z\vec{\mathbf{k}}$ is called the *complex propagation vector*. The vector $\vec{\alpha}$ accounts for the attenuation phenomena observed in nature, while $\vec{k}$ is related to wave propagation. If $\vec{\alpha}$, the solutions are referred to as *non-uniform plane waves*.

The form of the fields in [7.239] suggests replacing the differential operator $\vec{\nabla}$ in Maxwell's equations with the vector operator $-\vec{\gamma}$, since all spatial derivatives act only on the exponential term $e^{-\vec{\gamma}\cdot\vec{\mathbf{r}}}$. In fact, we have $\vec{\nabla}e^{-\vec{\gamma}\cdot\vec{\mathbf{r}}} = -\vec{\gamma}e^{-\vec{\gamma}\cdot\vec{\mathbf{r}}}$. Therefore, in the absence of sources (perfect dielectric medium), the Maxwell equations become:

$$\begin{aligned} &-\vec{\gamma}\times\vec{\hat{\mathbf{E}}}_0 = \mathrm{i}\omega\mu\vec{\hat{\mathbf{H}}}_0 \\ &-\vec{\gamma}\times\vec{\hat{\mathbf{H}}}_0 = -\mathrm{i}\omega\varepsilon\vec{\hat{\mathbf{E}}}_0 \\ &-\vec{\gamma}\cdot\varepsilon\vec{\hat{\mathbf{E}}}_0 = 0 \\ &-\vec{\gamma}\cdot\mu\vec{\hat{\mathbf{H}}}_0 = 0 \end{aligned} \qquad [7.240]$$

The last two equations of [7.240] indicate that $\vec{\gamma}$ is orthogonal to both $\vec{\hat{\mathbf{E}}}_0$ and $\vec{\hat{\mathbf{H}}}_0$. To decouple the first two equations, we write $\vec{\gamma}\times\left(\vec{\gamma}\times\vec{\hat{\mathbf{E}}}_0\right) = -\mathrm{i}\omega\mu\left(\vec{\gamma}\times\vec{\hat{\mathbf{H}}}_0\right)$ $= \omega^2\varepsilon\mu\vec{\hat{\mathbf{E}}}_0$ and $\vec{\gamma}\times\left(\vec{\gamma}\times\vec{\hat{\mathbf{H}}}_0\right) = \mathrm{i}\omega\varepsilon\left(\vec{\gamma}\times\vec{\hat{\mathbf{E}}}_0\right) = \omega^2\varepsilon\mu\vec{\hat{\mathbf{H}}}_0$. Now, using the vector identity $\vec{\gamma}\times\left(\vec{\gamma}\times\vec{\hat{\mathbf{E}}}_0\right) = \vec{\gamma}\left(\vec{\gamma}\cdot\vec{\hat{\mathbf{E}}}_0\right) - (\vec{\gamma}\cdot\vec{\gamma})\vec{\hat{\mathbf{E}}}_0 = -(\vec{\gamma}\cdot\vec{\gamma})\vec{\hat{\mathbf{E}}}_0$ and $\vec{\gamma}\times\left(\vec{\gamma}\times\vec{\hat{\mathbf{H}}}_0\right) = \vec{\gamma}\left(\vec{\gamma}\cdot\vec{\hat{\mathbf{H}}}_0\right) - (\vec{\gamma}\cdot\vec{\gamma})\vec{\hat{\mathbf{H}}}_0 = -(\vec{\gamma}\cdot\vec{\gamma})\vec{\hat{\mathbf{H}}}_0$, since $\vec{\gamma}\cdot\vec{\hat{\mathbf{E}}}_0 = \vec{\gamma}\cdot\vec{\hat{\mathbf{H}}}_0 = 0$, we deduce $(\vec{\gamma}\cdot\vec{\gamma})\vec{\hat{\mathbf{E}}}_0 = -\omega^2\varepsilon\mu\vec{\hat{\mathbf{E}}}_0$ and $(\vec{\gamma}\cdot\vec{\gamma})\vec{\hat{\mathbf{H}}}_0 = -\omega^2\varepsilon\mu\vec{\hat{\mathbf{H}}}_0$.

The dispersion relation then becomes:

$$\vec{\gamma}\cdot\vec{\gamma} = \alpha^2 - \mathrm{k}^2 - 2\mathrm{i}\vec{\alpha}\cdot\vec{\mathrm{k}} = -\omega^2\varepsilon\mu \qquad [7.241]$$

When ε and μ are real (perfect dielectric), we obtain the following solution:

$$\begin{aligned} &\alpha^2 - \mathrm{k}^2 = -\omega^2\varepsilon\mu \\ &\vec{\alpha}\cdot\vec{\mathrm{k}} = 0 \end{aligned} \qquad [7.242]$$

If $\vec{\alpha} = 0$, equation [7.242] reduces to $\mathrm{k}^2 = \omega^2\varepsilon\mu$. The second equation in system [7.242] shows that the attenuation effect, represented here by $\vec{\alpha}$, is perpendicular to the propagation direction, represented by $\overrightarrow{k}$. To compute the time-averaged power flux transported by the fields [7.245], we use equations [7.240]:

$$\begin{aligned}\langle\overrightarrow{\Pi}\rangle &= \frac{1}{2}\Re\left[\overrightarrow{\hat{E}}_0 \times \overrightarrow{\hat{H}}_0^*\right] = \frac{1}{2}\Re\left[\overrightarrow{\hat{E}}_0 \times \left(\frac{\overrightarrow{\gamma}^* \times \overrightarrow{\hat{E}}_0^*}{i\omega\mu}\right)\right] \\ &= \frac{1}{2}\Re\left[\frac{\overrightarrow{\gamma}^*\left|\overrightarrow{\hat{E}}_0\right|^2 - \overrightarrow{\hat{E}}_0^*\left(\overrightarrow{\gamma}^*\cdot\overrightarrow{\hat{E}}_0\right)}{i\omega\mu}\right] \\ &= \frac{1}{2}\frac{\overrightarrow{k}}{\omega\mu}\left|\overrightarrow{\hat{E}}_0\right|^2 - \frac{1}{2}\Re\left[\frac{\overrightarrow{\hat{E}}_0^*\left(\overrightarrow{\gamma}^*\cdot\overrightarrow{\hat{E}}_0\right)}{i\omega\mu}\right] \\ \langle\overrightarrow{\Pi}\rangle &= \frac{1}{2}\Re\left[\overrightarrow{\hat{E}}_0 \times \overrightarrow{\hat{H}}_0^*\right] = \frac{1}{2}\Re\left[\left(\frac{\overrightarrow{\gamma} \times \overrightarrow{\hat{H}}_0}{i\omega\varepsilon}\right) \times \overrightarrow{\hat{H}}_0^*\right] \\ &= \frac{1}{2}\Re\left[\frac{-\overrightarrow{\gamma}\left|\overrightarrow{\hat{H}}_0\right|^2 + \overrightarrow{\hat{H}}_0\left(\overrightarrow{\gamma}\cdot\overrightarrow{\hat{H}}_0^*\right)}{i\omega\varepsilon}\right] \\ &= \frac{1}{2}\frac{\overrightarrow{k}}{\omega\varepsilon}\left|\overrightarrow{\hat{H}}_0\right|^2 + \frac{1}{2}\Re\left[\frac{\overrightarrow{\hat{H}}_0\left(\overrightarrow{\gamma}\cdot\overrightarrow{\hat{H}}_0^*\right)}{i\omega\varepsilon}\right]\end{aligned} \quad [7.243]$$

These two expressions are equivalent. When $\overrightarrow{\mathbf{E}}_0$ is simultaneously perpendicular to both $\overrightarrow{\alpha}$ and $\overrightarrow{k}$ (in other words, when $\overrightarrow{\gamma}^* \cdot \overrightarrow{\mathbf{E}}_0 = 0$), the first equation (of [7.243]) reduces to $\langle\overrightarrow{\Pi}\rangle = \frac{1}{2}\frac{\overrightarrow{k}}{\omega\mu}\left|\overrightarrow{\mathbf{E}}_0\right|^2$. The time-averaged power flow is always in the direction of the wave vector $\overrightarrow{k}$. This corresponds to the TE mode. Similarly, when $\overrightarrow{\gamma}^\star \cdot \overrightarrow{\mathbf{H}}_0 = 0$, the second expression gives $\langle\overrightarrow{\Pi}\rangle = \frac{1}{2}\frac{\overrightarrow{k}}{\omega\varepsilon}\left|\overrightarrow{\mathbf{H}}_0\right|^2$ (always in the direction of $\overrightarrow{k}$). This is the *TM mode*. This analysis is also valid for dissipative (ohmic) media, provided that the permittivity ε is replaced by the complex permittivity $\hat{\varepsilon}$.

EXAMPLE 7.9.– Let us examine non-uniform (plane wave) solutions with $\overrightarrow{\alpha} \neq 0$, generated by a current sheet located in the plane $z = 0$, carrying a surface current given by:

$$K_x = K_0 \cos(\omega t - \mathsf{k}_x x) = \Re\left[K_0 e^{-\mathrm{i}(\omega t - \mathsf{k}_x x)}\right] \quad [7.244]$$

This current, oriented along the Ox axis, gives rise to a magnetic excitation field directed along the Oy axis. Since the system is invariant along the Oy axis, the solutions must take the following form:

$$\overrightarrow{\mathbf{H}} = \begin{cases} \Re\left[\hat{H}_{0,+}\overrightarrow{\mathbf{j}}\, e^{-\mathrm{i}\omega t} e^{-\overrightarrow{\gamma}_+ \cdot \overrightarrow{\mathbf{r}}}\right] & z > 0 \\ \Re\left[\hat{H}_{0,-}\overrightarrow{\mathbf{j}}\, e^{-\mathrm{i}\omega t} e^{-\overrightarrow{\gamma}_- \cdot \overrightarrow{\mathbf{r}}}\right] & z < 0 \end{cases}$$
$$\overrightarrow{\mathbf{E}} = \begin{cases} \Re\left[\dfrac{\overrightarrow{\gamma}_+ \times \hat{H}_{0,+}\overrightarrow{\mathbf{j}}}{\mathrm{i}\omega\varepsilon} e^{-\mathrm{i}\omega t} e^{-\overrightarrow{\gamma}_+ \cdot \overrightarrow{\mathbf{r}}}\right] & z > 0 \\ \Re\left[\dfrac{\overrightarrow{\gamma}_- \times \hat{H}_{0,-}\overrightarrow{\mathbf{j}}}{\mathrm{i}\omega\varepsilon} e^{-\mathrm{i}\omega t} e^{-\overrightarrow{\gamma}_- \cdot \overrightarrow{\mathbf{r}}}\right] & z < 0 \end{cases} \quad [7.245]$$

where $\overrightarrow{\gamma}_+$ and $\overrightarrow{\gamma}_-$ are the complex propagation vectors on either side of the current sheet, given by $\overrightarrow{\gamma}_+ = \gamma_{+,x}\overrightarrow{\mathbf{i}} + \gamma_{+,z}\overrightarrow{\mathbf{k}}$ and $\overrightarrow{\gamma}_- = \gamma_{-,x}\overrightarrow{\mathbf{i}} + \gamma_{-,z}\overrightarrow{\mathbf{k}}$. At $z = 0$, the tangential component of $\overrightarrow{\mathbf{H}}$ must satisfy the discontinuity condition $-\hat{H}_{0,+}e^{-\gamma_{+,x}x} + \hat{H}_{0,-}e^{-\gamma_{-,x}x} = K_0 e^{\mathrm{i}\mathsf{k}_x x}$, which implies:

$$-\gamma_{+,x} = -\gamma_{-,x} = \mathrm{i}\mathsf{k}_x \quad [7.246]$$

The remaining components, $\gamma_{+,z}$ and $\gamma_{-,x}$, can then be determined using the dispersion relation [7.241], $\gamma_x^2 + \gamma_z^2 = -\omega^2\varepsilon\mu$, which yields $\gamma_z = \pm\sqrt{\mathsf{k}_x^2 - \omega^2\varepsilon\mu}$. This leads to the following observations.

1) If $\mathsf{k}_x^2 < \omega^2\varepsilon\mu$, then γ_z is purely imaginary and corresponds to wave propagation (this is a uniform plane wave). The components $\gamma_{+,z}$ and $\gamma_{-,z}$ are therefore equal to $\gamma_{+,z} = -\gamma_{-,z} = -\mathrm{i}\sqrt{\omega^2\varepsilon\mu - \mathsf{k}_x^2}$, since the waves must propagate away from the current sheet on both sides.

2) If $\mathsf{k}_x^2 > \omega^2\varepsilon\mu$, then γ_z is purely real and corresponds to attenuation along the direction of Oz (this is a non-uniform plane wave). The components $\gamma_{+,z}$ and $\gamma_{-,z}$ have opposite signs, $\gamma_{+,z} = -\gamma_{-,z} = \sqrt{\mathsf{k}_x^2 - \omega^2\varepsilon\mu}$, because the wave must decay away from the current sheet on both sides. In particular, when $\omega = 0$, we obtain $\gamma_z = \pm\mathsf{k}$.

3) Continuity of the tangential component of $\overrightarrow{\mathbf{E}}$ requires $\gamma_{+,z}\hat{H}_{0,+} = \gamma_{-,z}\hat{H}_{0,-}$, which implies:

$$\hat{H}_{0,-} = -\hat{H}_{0,+} = \frac{K_0}{2} \quad [7.247]$$

4) When $\mathsf{k}_x = 0$, we recover the solution of example 7.3. Increasing k_x generates waves whose Poynting vectors lie along $\mathsf{k}_x \overrightarrow{\mathbf{i}} \pm \mathsf{k}_z \overrightarrow{\mathbf{k}}$. For $\mathsf{k}_x^2 = \omega^2 \varepsilon \mu$, we have $\mathsf{k}_z = 0$, so the Poynting vector is purely along Ox with no z-dependence. An increase in k_z beyond $\omega\sqrt{\varepsilon\mu}$, γ_z becomes real again, and the field decreases with z.

7.8.1. *Oblique incidence on a perfect conductor*

7.8.1.1. *Electric field parallel to the interface (TE wave)*

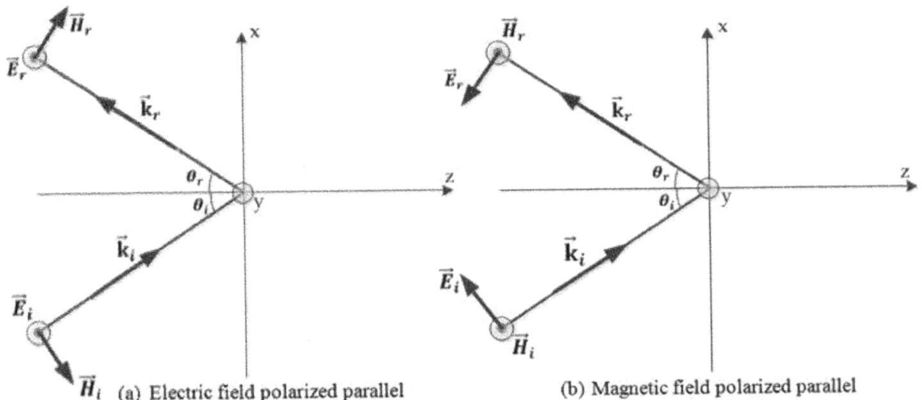

(a) Electric field polarized parallel to the interface

(b) Magnetic field polarized parallel to the interface

Figure 7.20. *Plane wave with oblique incidence on a perfect conductor*

Consider a uniform plane wave incident on a perfect conductor, with its Poynting vector making an angle θ_i with the normal to the interface (see Figure 7.20(a)). The electric field is parallel to the interface, while the magnetic excitation field has components along the Ox and Oz axes:

$$\begin{aligned} \overrightarrow{\mathbf{E}}_i &= \Re\left[\hat{E}_{0,i} e^{-\mathrm{i}(\omega t - \mathsf{k}_{i,x} x - \mathsf{k}_{i,z} z)} \overrightarrow{\mathbf{j}}\right] \\ \overrightarrow{\mathbf{H}}_i &= \Re\left[Z^{-1} \hat{E}_{0,i}\left(-\cos\theta_i \overrightarrow{\mathbf{i}} + \sin\theta_i \overrightarrow{\mathbf{k}}\right) e^{-\mathrm{i}(\omega t - \mathsf{k}_{i,x} x - \mathsf{k}_{i,z} z)}\right] \end{aligned} \quad [7.248]$$

where $\mathsf{k}_{i,x} = \mathsf{k}\sin\theta_i$, $\mathsf{k}_{i,z} = \mathsf{k}\cos\theta_i$ and $k = \omega\sqrt{\varepsilon\mu}$. No transmitted field exists inside the perfect conductor. Instead, there is a reflected field whose Poynting vector makes an angle θ_r with the normal. The reflected electric field remains along Oy, and the reflected magnetic field (perpendicular to $\overrightarrow{\mathbf{E}}_r$ and $\overrightarrow{\mathbf{\Pi}}_r$) lies in the Oxz plane (Figure 7.20(a)):

$$\begin{aligned} \overrightarrow{\mathbf{E}}_r &= \Re\left[\hat{E}_{0,r} e^{-\mathrm{i}(\omega t - \mathsf{k}_{r,x} x + \mathsf{k}_{r,z} z)} \overrightarrow{\mathbf{j}}\right] \\ \overrightarrow{\mathbf{H}}_r &= \Re\left[Z^{-1} \hat{E}_{0,r}\left(\cos\theta_r \overrightarrow{\mathbf{i}} + \sin\theta_r \overrightarrow{\mathbf{k}}\right) e^{-\mathrm{i}(\omega t - \mathsf{k}_{r,x} x + \mathsf{k}_{r,z} z)}\right] \end{aligned} \quad [7.249]$$

where $k_{r,x} = k\sin\theta_r$ and $k_{r,z} = k\cos\theta_r$. To determine the reflection angle θ_r and the reflected amplitude $\hat{E}_{0,r}$, apply the boundary conditions at $z = 0$: continuity of the tangential component of $\vec{\mathbf{E}}$ and the normal component of $\vec{\mathbf{B}}$. Since there is no field inside the conductor, at $z = 0$, these reduce to:

$$\begin{aligned}&\hat{E}_{0,i}e^{ik_{i,x}x} + \hat{E}_{0,r}e^{ik_{r,x}x} = 0\\&Z^{-1}\left(\hat{E}_{0,i}\sin\theta_i e^{ik_{i,x}x} + \hat{E}_{0,r}\sin\theta_r e^{ik_{r,x}x}\right) = 0\end{aligned} \qquad [7.250]$$

These equations are valid for any x, which means that $k_{i,x} = k_{r,x}$. Hence, we obtain:

$$\begin{aligned}&\hat{E}_{0,r} = -\hat{E}_{0,i}\\&\theta_i = \theta_r = \theta\end{aligned} \qquad [7.251]$$

The angles of incidence and reflection are therefore equal. The resulting fields are:

$$\begin{aligned}\vec{\mathbf{E}} &= \Re\left[\hat{E}_{0,i}\left(e^{ik_z z} - e^{-ik_z z}\right)e^{-i(\omega t - k_x x)}\vec{\mathbf{j}}\right]\\&= 2\hat{E}_{0,i}\sin k_z z\sin(\omega t - k_x x)\,\vec{\mathbf{j}}\\\vec{\mathbf{H}} &= \Re\left[Z^{-1}\hat{E}_{0,i}\left[\cos\theta\left(-e^{ik_z z} - e^{-ik_z z}\right)\vec{\mathbf{i}}\right.\right.\\&\quad\left.\left.+\sin\theta\left(e^{ik_z z} - e^{-ik_z z}\right)\vec{\mathbf{k}}\right]e^{-i(\omega t - k_x x)}\right]\\&= 2Z^{-1}\hat{E}_{0,i}\left[-\cos\theta\cos k_z z\cos(\omega t - k_x x)\,\vec{\mathbf{i}}\right.\\&\quad\left.+\sin\theta\sin k_z z\sin(\omega t - k_x x)\,\vec{\mathbf{k}}\right]\end{aligned} \qquad [7.252]$$

($\hat{E}_{0,i}$ is taken to be real here, for simplicity). From equations [7.252], we can now calculate the time-averaged power flux:

$$\begin{aligned}\langle\vec{\mathbf{\Pi}}\rangle = \langle\vec{\mathbf{E}}\times\vec{\mathbf{H}}\rangle &= 4Z^{-1}\hat{E}_{0,i}^2\sin\theta\sin^2 k_z z\langle\sin^2(\omega t - k_x x)\rangle\,\vec{\mathbf{i}}\\&= 2Z^{-1}\hat{E}_{0,i}^2\sin\theta\sin^2 k_z z\,\vec{\mathbf{i}}\end{aligned} \qquad [7.253]$$

This power flux is parallel to the interface (along Ox) and varies with z. To find the surface current flowing along Oy at $z = 0$, use the discontinuity in the tangential component of $\vec{\mathbf{H}}$:

$$K_y = -H_x(z = 0) = 2Z^{-1}\hat{E}_{0,i}\cos\theta\cos(\omega t - k_x x) \qquad [7.254]$$

7.8.1.2. *Magnetic excitation field parallel to the interface (TM wave)*

When the magnetic excitation field $\vec{\mathbf{H}}$ is parallel to the conducting interface, as in Figure 7.20(b), the incident and reflected fields are written as:

$$\begin{aligned}\vec{\mathbf{E}}_i &= \Re\left[\hat{E}_{0,i}\left(\cos\theta_i\,\vec{\mathbf{i}} - \sin\theta_i\,\vec{\mathbf{k}}\right)e^{-i(\omega t - k_{i,x}x - k_{i,z}z)}\right]\\\vec{\mathbf{H}}_i &= \Re\left[Z^{-1}\hat{E}_{0,i}e^{-i(\omega t - k_{i,x}x - k_{i,z}z)}\,\vec{\mathbf{j}}\right]\\\vec{\mathbf{E}}_r &= \Re\left[\hat{E}_{0,r}\left(-\cos\theta_r\,\vec{\mathbf{i}} - \sin\theta_r\,\vec{\mathbf{k}}\right)e^{-i(\omega t - k_{r,x}x + k_{r,z}z)}\right]\\\vec{\mathbf{H}}_r &= \Re\left[Z^{-1}\hat{E}_{0,r}e^{-i(\omega t - k_{r,x}x + k_{r,z}z)}\,\vec{\mathbf{j}}\right]\end{aligned} \qquad [7.255]$$

where $\mathsf{k}_{i,x} = \mathsf{k}\sin\theta_i$, $\mathsf{k}_{i,z} = \mathsf{k}\cos\theta_i$, $\mathsf{k}_{r,x} = \mathsf{k}\sin\theta_r$, $\mathsf{k}_{r,z} = \mathsf{k}\cos\theta_r$ and $k = \omega\sqrt{\varepsilon\mu}$. The continuity of the tangential component of the electric field $\vec{\mathbf{E}}$, at the interface $z = 0$, requires $\hat{E}_{0,i}\cos\theta_i\,\vec{\mathbf{i}}\,e^{\mathrm{i}\mathsf{k}_{i,x}x} - \hat{E}_{0,r}\cos\theta_i\,\vec{\mathbf{i}}\,e^{\mathrm{i}\mathsf{k}_{r,x}x} = 0$. There is no normal component of $\vec{\mathbf{B}}$ here. Since this must hold for all x, it follows that $\mathsf{k}_{i,x} = \mathsf{k}_{r,x}$. Thus, we find $\theta_i = \theta_r = \theta$ and $\hat{E}_{0,r} = -\hat{E}_{0,i}$. We again see that the angles of incidence and reflection are equal. Therefore, the resultant $\vec{\mathbf{E}}$ and $\vec{\mathbf{H}}$ fields are:

$$\begin{aligned}
\vec{\mathbf{E}} &= \Re\Big[E_{0,i}\Big[\cos\theta\left(e^{\mathrm{i}\mathsf{k}_z z} - e^{-i\mathsf{k}_z z}\right)\vec{\mathbf{i}} - \\
&\quad \sin\theta\left(e^{\mathrm{i}\mathsf{k}_z z} + e^{-\mathrm{i}\mathsf{k}_z z}\right)\vec{\mathbf{k}}\Big]e^{-\mathrm{i}(\omega t - \mathsf{k}_x x)}\Big] \\
&= 2E_{0,i}\Big[\cos\theta\sin\mathsf{k}_z z\sin(\omega t - \mathsf{k}_x x)\,\vec{\mathbf{i}} - \\
&\quad \sin\theta\cos\mathsf{k}_z z\cos(\omega t - \mathsf{k}_x x)\,\vec{\mathbf{k}}\Big] \\
\vec{\mathbf{H}} &= \Re\Big[Z^{-1}E_{0,i}\left(e^{\mathrm{i}\mathsf{k}_z z} + e^{-\mathrm{i}\mathsf{k}_z z}\right)e^{-\mathrm{i}(\omega t - \mathsf{k}_x x)}\vec{\mathbf{j}}\Big] \\
&= 2Z^{-1}E_{0,i}\cos\mathsf{k}_z z\cos(\omega t - \mathsf{k}_x x)\,\vec{\mathbf{j}}
\end{aligned} \qquad [7.256]$$

and, consequently:

$$\begin{aligned}
\langle\vec{\mathbf{\Pi}}\rangle &= 4Z^{-1}E_{0,i}^2\sin\theta\cos^2\mathsf{k}_z z\langle\cos^2(\omega t - \mathsf{k}_x x)\rangle\,\vec{\mathbf{i}} \\
&= 2Z^{-1}E_{0,i}^2\sin\theta\cos^2\mathsf{k}_z z\,\vec{\mathbf{i}}
\end{aligned} \qquad [7.257]$$

The average power flux, for this polarization, is also directed along the Ox direction. The surface current flowing over the interface $z = 0$ is given by the tangential component of $\vec{\mathbf{H}}$:

$$K_x(t,x) = H_y(z=0) = 2Z^{-1}E_{0,i}\cos(\omega t - \mathsf{k}_x x) \qquad [7.258]$$

For the surface charge, it is obtained from the normal component of the electric field $\vec{\mathbf{E}}$:

$$\sigma(t,x) = -\varepsilon E_z(z=0) = 2\varepsilon E_{0,i}\sin\theta\cos(\omega t - \mathsf{k}_x x) \qquad [7.259]$$

Note that the surface current $K_x(t,x)$ and the surface charge $\sigma(t,x)$ are related by the conservation equation (as can be verified):

$$\frac{\partial K_x}{\partial x} + \frac{\partial\sigma}{\partial t} = 0 \qquad [7.260]$$

7.8.2. *Oblique incidence on a perfect dielectric*

7.8.2.1. *Electric field parallel to the interface (TE wave)*

By replacing the perfect conductor with a lossless (perfect) dielectric, transmitted fields are observed inside the dielectric, in addition to the incident and reflected fields, as shown in Figure 7.21(a).

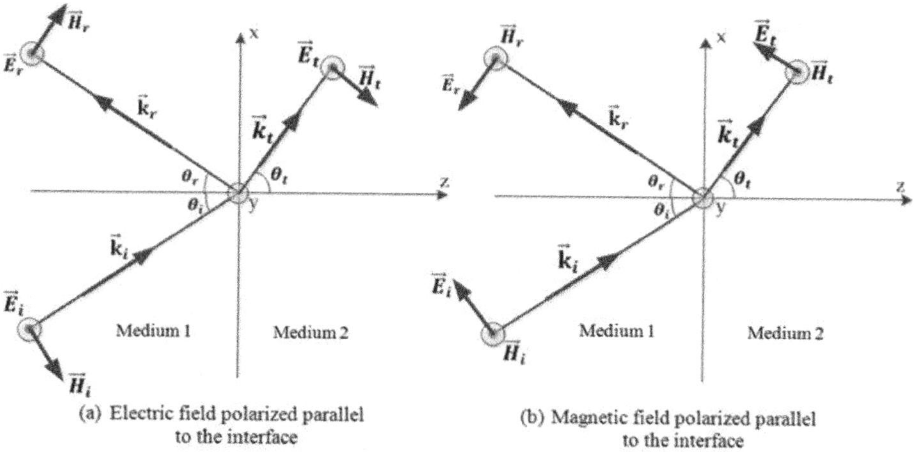

Figure 7.21. *Uniform plane wave incident obliquely on a perfect dielectric*

Suppose that the incident wave has an electric field polarized parallel to the interface. In each region, the fields are expressed as follows:

$$\begin{aligned}
\vec{\mathbf{E}}_i &= \Re\left[\hat{E}_{0,i} e^{-\mathrm{i}(\omega t - \mathrm{k}_{i,x}x - \mathrm{k}_{i,z}z)} \vec{\mathbf{j}}\right] \\
\vec{\mathbf{H}}_i &= \Re\left[Z_1^{-1}\hat{E}_{0,i}\left(-\cos\theta_i\,\vec{\mathbf{i}} + \sin\theta_i\,\vec{\mathbf{k}}\right) e^{-\mathrm{i}(\omega t - \mathrm{k}_{i,x}x - \mathrm{k}_{i,z}z)}\right] \\
\vec{\mathbf{E}}_r &= \Re\left[\hat{E}_{0,r} e^{-\mathrm{i}(\omega t - \mathrm{k}_{r,x}x + \mathrm{k}_{r,z}z)} \vec{\mathbf{j}}\right] \\
\vec{\mathbf{H}}_r &= \Re\left[Z_1^{-1}\hat{E}_{0,r}\left(\cos\theta_r\,\vec{\mathbf{i}} + \sin\theta_r\,\vec{\mathbf{k}}\right) e^{-\mathrm{i}(\omega t - \mathrm{k}_{r,x}x + \mathrm{k}_{r,z}z)}\right] \\
\vec{\mathbf{E}}_t &= \Re\left[\hat{E}_{0,t} e^{-\mathrm{i}(\omega t - \mathrm{k}_{t,x}x - \mathrm{k}_{t,z}z)} \vec{\mathbf{j}}\right] \\
\vec{\mathbf{H}}_t &= \Re\left[Z_2^{-1}\hat{E}_{0,t}\left(-\cos\theta_t\,\vec{\mathbf{i}} + \sin\theta_t\,\vec{\mathbf{k}}\right) e^{-\mathrm{i}(\omega t - \mathrm{k}_{t,x}x - \mathrm{k}_{t,z}z)}\right]
\end{aligned} \qquad [7.261]$$

where θ_i, θ_r and θ_t are the angles of incidence, reflection and transmission with respect to the normal. In each region, the wavenumbers are $\mathrm{k}_{i,x} = \mathrm{k}_1 \sin\theta_i$, $\mathrm{k}_{i,z} = \mathrm{k}_1\cos\theta_i$, $\mathrm{k}_{r,x} = \mathrm{k}_1\sin\theta_r$, $\mathrm{k}_{r,z} = \mathrm{k}_1\cos\theta_r$, $\mathrm{k}_{t,x} = \mathrm{k}_2\sin\theta_t$, $\mathrm{k}_{t,z} = \mathrm{k}_2\cos\theta_t$, $\mathrm{k}_1 = \omega\sqrt{\varepsilon_1\mu_1}$ and $\mathrm{k}_2 = \omega\sqrt{\varepsilon_2\mu_2}$. The angles (θ_r, θ_t) and the amplitudes $\left(\hat{E}_{0,r}, \hat{E}_{0,t}\right)$ in [7.261] are determined from the continuity equations for the tangential components of $\vec{\mathbf{E}}$ and $\vec{\mathbf{H}}$ at the interface $z = 0$:

$$\begin{aligned}
&\hat{E}_{0,i}e^{\mathrm{i}\mathrm{k}_{i,x}x} + \hat{E}_{0,r}e^{\mathrm{i}\mathrm{k}_{r,x}x} = \hat{E}_{0,t}e^{\mathrm{i}\mathrm{k}_{t,x}x} \\
&Z^{-1}\left[-\hat{E}_{0,i}\cos\theta_i e^{\mathrm{i}\mathrm{k}_{i,x}x} + \hat{E}_{0,r}\cos\theta_r e^{\mathrm{i}\mathrm{k}_{r,x}x}\right] = -Z_2^{-1}\hat{E}_{0,t}\cos\theta_t e^{\mathrm{i}\mathrm{k}_{t,x}x}
\end{aligned} \qquad [7.262]$$

These equations are satisfied for all x, which implies that $\mathsf{k}_{i,x} = \mathsf{k}_{r,x} = \mathsf{k}_{t,x}$ and, consequently, $\mathsf{k}_1 \sin\theta_i = \mathsf{k}_1 \sin\theta_r = \mathsf{k}_2 \sin\theta_t$. Thus, we obtain:

$$\theta_r = \theta_i, \qquad \sin\theta_t = \frac{\mathsf{k}_1}{\mathsf{k}_2}\sin\theta_i = \frac{n_1}{n_2}\sin\theta_i \tag{7.263}$$

where n_1 and n_2 are the refractive indices of media 1 and 2, respectively. We observe that:

1) The angles of incidence and reflection are equal.

2) The transmission angle obeys a more complicated relationship, known as the *Snell–Descartes law*, which relates the sines of the angles θ_i and θ_t.

3) If $n_1 > n_2$ (respectively, $n_1 < n_2$), then $\theta_t > \theta_i$ (respectively, $\theta_t < \theta_i$), as illustrated in Figure 7.22.

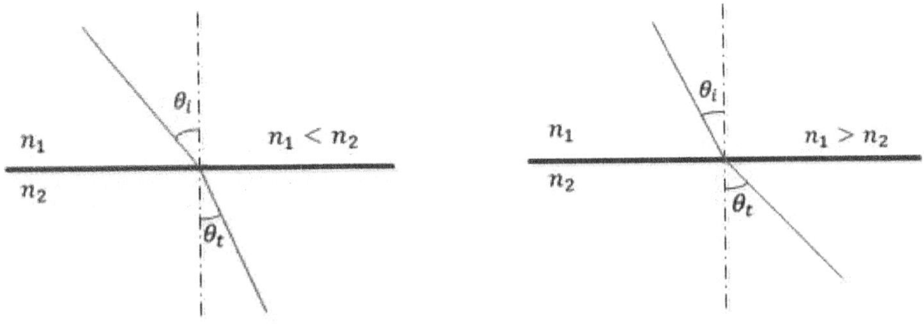

Figure 7.22. *When $n_1 < n_2$ (respectively, $n_1 > n_2$), the transmitted wave is closer to (respectively, farther from) the normal direction to the interface. For a color version of this figure, see www.iste.co.uk/abdesselam/electrodynamics2.zip*

4) The normal component of $\vec{\mathbf{B}}$ is conserved on both sides of the interface $z = 0$. The continuity of $\left(\vec{\mathbf{B}}\right)_\perp$ is expressed by the equation $\sqrt{\varepsilon_1\mu_1}\left(\hat{E}_{0,i} + \hat{E}_{0,r}\right)\sin\theta_i = \sqrt{\varepsilon_2\mu_2}\hat{E}_{0,t}\sin\theta_t$. Using the Snell–Descartes law, it is clear that $\hat{E}_{0,i} + \hat{E}_{0,r} = \hat{E}_{0,t}$.

For the amplitudes $\left(\hat{E}_{0,r}, \hat{E}_{0,r}\right)$, the reflection coefficient R_{TE} and the transmission coefficient T_{TE} (for a TE wave), we obtain (by solving [7.262]):

$$\begin{cases} \hat{E}_{0,r} = \dfrac{Z_2\cos\theta_i - Z_1\cos\theta_t}{Z_2\cos\theta_i + Z_1\cos\theta_t}\hat{E}_{0,i} \\ \hat{E}_{0,t} = \dfrac{2Z_2\cos\theta_i}{Z_2\cos\theta_i + Z_1\cos\theta_t}\hat{E}_{0,i} \end{cases} \Longleftrightarrow \begin{cases} R_{TE} = \dfrac{\hat{E}_{0,r}}{\hat{E}_{0,i}} = \dfrac{Z_2\cos\theta_i - Z_1\cos\theta_t}{Z_2\cos\theta_i + Z_1\cos\theta_t} \\ T_{TE} = \dfrac{\hat{E}_{0,t}}{\hat{E}_{0,i}} = \dfrac{2Z_2\cos\theta_i}{Z_2\cos\theta_i + Z_1\cos\theta_t} \end{cases} \tag{7.264}$$

7.8.2.2. *Magnetic excitation field parallel to the interface (TM wave)*

For this polarization, as shown in Figure 7.22(b), the fields are:

$$\begin{aligned}
\vec{\mathbf{E}}_i &= \Re\left[\hat{E}_{0,i}\left(\cos\theta_i\,\vec{\mathbf{i}} - \sin\theta_i\,\vec{\mathbf{k}}\right)e^{-\mathrm{i}(\omega t - \mathsf{k}_{i,x}x - \mathsf{k}_{i,z}z)}\right]\\
\vec{\mathbf{H}}_i &= \Re\left[Z_1^{-1}\hat{E}_{0,i}e^{-\mathrm{i}(\omega t - \mathsf{k}_{i,x}x - \mathsf{k}_{i,z}z)}\vec{\mathbf{j}}\right]\\
\vec{\mathbf{E}}_r &= \Re\left[\hat{E}_{0,r}\left(-\cos\theta_r\,\vec{\mathbf{i}} - \sin\theta_r\,\vec{\mathbf{k}}\right)e^{-\mathrm{i}(\omega t - \mathsf{k}_{r,x}x + \mathsf{k}_{r,z}z)}\right]\\
\vec{\mathbf{H}}_r &= \Re\left[Z_1^{-1}\hat{E}_{0,r}e^{-\mathrm{i}(\omega t - \mathsf{k}_{r,x}x + \mathsf{k}_{r,z}z)}\vec{\mathbf{j}}\right]\\
\vec{\mathbf{E}}_t &= \Re\left[\hat{E}_{0,t}\left(\cos\theta_t\,\vec{\mathbf{i}} - \sin\theta_t\,\vec{\mathbf{k}}\right)e^{-\mathrm{i}(\omega t - \mathsf{k}_{t,x}x - \mathsf{k}_{t,z}z)}\right]\\
\vec{\mathbf{H}}_t &= \Re\left[Z_2^{-1}\hat{E}_{0,t}e^{-\mathrm{i}(\omega t - \mathsf{k}_{t,x}x - \mathsf{k}_{t,z}z)}\vec{\mathbf{j}}\right]
\end{aligned} \qquad [7.265]$$

The continuity of the tangential components of $\vec{\mathbf{E}}$ and $\vec{\mathbf{H}}$ at the interface $z = 0$ requires that:

$$\begin{aligned}
&\hat{E}_{0,i}\cos\theta_i e^{\mathrm{i}\mathsf{k}_{i,x}x} - \hat{E}_{0,r}\cos\theta_r e^{\mathrm{i}\mathsf{k}_{r,x}x} = \hat{E}_{0,t}\cos\theta_t e^{\mathrm{i}\mathsf{k}_{t,x}x}\\
&Z_1^{-1}\left[\hat{E}_{0,i}e^{\mathrm{i}\mathsf{k}_{i,x}x} + \hat{E}_{0,r}e^{\mathrm{i}\mathsf{k}_{r,x}x}\right] = Z_2^{-1}\hat{E}_{0,t}e^{\mathrm{i}\mathsf{k}_{t,x}x}
\end{aligned} \qquad [7.266]$$

Once again, we see that $\mathsf{k}_{i,x} = \mathsf{k}_{r,x} = \mathsf{k}_{t,x}$; therefore, equations [7.263] remain valid. The Snell–Descartes law and the equality between the angles of incidence and reflection are independent of the polarization. Note that it is the normal component of $\vec{\mathbf{D}}$ that is now conserved at the interface, $\varepsilon_1\hat{E}_{0,i}\sin\theta_i e^{\mathrm{i}\mathsf{k}_{i,x}x} + \varepsilon_1\hat{E}_{0,r}\sin\theta_r e^{\mathrm{i}\mathsf{k}_{r,x}x} = \varepsilon_2\hat{E}_{0,t}\sin\theta_t e^{\mathrm{i}\mathsf{k}_{t,x}x}$, due to the Snell–Descartes law. To obtain the reflection coefficient R_{TM} and the transmission coefficient T_{TE}, we solve the system [7.266]:

$$\begin{cases}\hat{E}_{0,r} = \dfrac{Z_1\cos\theta_i - Z_2\cos\theta_t}{Z_1\cos\theta_i + Z_2\cos\theta_t}\hat{E}_{0,i}\\[2ex] \hat{E}_{0,t} = \dfrac{2Z_2\cos\theta_i}{Z_1\cos\theta_i + Z_2\cos\theta_t}\hat{E}_{0,i}\end{cases} \iff \begin{cases}R_{TM} = \dfrac{Z_1\cos\theta_i - Z_2\cos\theta_t}{Z_1\cos\theta_i + Z_2\cos\theta_t}\\[2ex] T_{TM} = \dfrac{2Z_2\cos\theta_i}{Z_1\cos\theta_i + Z_2\cos\theta_t}\end{cases} \qquad [7.267]$$

7.8.2.3. *Brewster angle*

If we consider TE polarization, we observe that the reflected wave vanishes if the reflection coefficient is zero, that is, $R_{TE} = 0$. This condition is satisfied when:

$$Z_2\cos\theta_i = Z_1\cos\theta_t \qquad [7.268]$$

This angle is called the *Brewster angle*. By squaring [7.268] and using the Snell–Descartes law, we obtain the following expression for the Brewster angle:

$$\sin^2\theta_B = \frac{1 - \frac{\varepsilon_2}{\varepsilon_1}\frac{\mu_1}{\mu_2}}{1 - \left(\frac{\mu_1}{\mu_2}\right)^2} \qquad [7.269]$$

This equation does not always admit real solutions. For example, if $\mu_1 = \mu_2$ and $\varepsilon_1 \neq \varepsilon_2$, then the right-hand side becomes infinite and the equation has no solution. To obtain real solutions, the right-hand side of [7.269] must lie between 0 and 1. This requires (1) if $\mu_1 < \mu_2$ (respectively, $\mu_1 > \mu_2$), then $\frac{\mu_1}{\mu_2} < \frac{\varepsilon_2}{\varepsilon_1} < \frac{\mu_2}{\mu_1}$ (respectively, $\frac{\mu_2}{\mu_1} < \frac{\varepsilon_2}{\varepsilon_1} < \frac{\mu_1}{\mu_2}$); (2) in particular, a Brewster angle does exist when $\mu_1 \neq \mu_2$ and $\varepsilon_1 = \varepsilon_2$, in which case:

$$\sin^2\theta_B = \frac{1}{1+\frac{\mu_1}{\mu_2}} \qquad \Longleftrightarrow \qquad \tan\theta_B = \sqrt{\frac{\mu_2}{\mu_1}} \tag{7.270}$$

It can be verified that $\theta_B + \theta_t = \frac{\pi}{2}$. In fact, using [7.263], [7.268] and [7.270], it comes that:

$$\begin{aligned} \cos(\theta_B+\theta_t) &= \cos\theta_B\cos\theta_t - \sin\theta_B\sin\theta_t = \cos^2\theta_B\sqrt{\frac{\mu_2}{\mu_1}} - \sin^2\theta_B\sqrt{\frac{\mu_1}{\mu_2}} \\ &= \sqrt{\frac{\mu_2}{\mu_1}} - \sin^2\theta_B\left(\sqrt{\frac{\mu_1}{\mu_2}}+\sqrt{\frac{\mu_2}{\mu_1}}\right) = 0 \end{aligned} \tag{7.271}$$

We thus conclude that if a wave with arbitrary polarization strikes an interface at the Brewster angle, the reflected wave will be linearly polarized. Moreover, the reflected and transmitted waves are perpendicular to each other.

For TM polarization, the reflected wave disappears if condition $R_{TM} = 0$ is met, which implies:

$$Z_1\cos\theta_i = Z_2\cos\theta_t \tag{7.272}$$

By combining this condition with the Snell–Descartes law, we obtain the Brewster angle:

$$\sin^2\theta_B = \frac{1-\frac{\varepsilon_1\mu_2}{\varepsilon_2\mu_1}}{1-\left(\frac{\varepsilon_1}{\varepsilon_2}\right)^2} \tag{7.273}$$

To have real solutions, the following must hold: (1) if $\varepsilon_1 < \varepsilon_2$ (respectively, $\varepsilon_1 > \varepsilon_2$), then $\frac{\varepsilon_1}{\varepsilon_2} < \frac{\mu_2}{\mu_1} < \frac{\varepsilon_2}{\varepsilon_1}$ (respectively, $\frac{\varepsilon_2}{\varepsilon_1} < \frac{\mu_2}{\mu_1} < \frac{\varepsilon_1}{\varepsilon_2}$); (2) when $\mu_1 = \mu_2$ and $\varepsilon_1 \neq \varepsilon_2$, a real solution exists:

$$\sin^2\theta_B = \frac{1}{1+\frac{\varepsilon_1}{\varepsilon_2}} \qquad \Longleftrightarrow \qquad \tan\theta_B = \sqrt{\frac{\varepsilon_2}{\varepsilon_1}} \tag{7.274}$$

For this Brewster angle, it is easy to verify that $\theta_B + \theta_t = \frac{\pi}{2}$. Indeed, we have:

$$\begin{aligned} \cos(\theta_B+\theta_t) &= \cos\theta_B\cos\theta_t - \sin\theta_B\sin\theta_t \\ &= \cos^2\theta_B\sqrt{\frac{\varepsilon_2}{\varepsilon_1}} - \sin^2\theta_B\sqrt{\frac{\varepsilon_1}{\varepsilon_2}} \\ &= \sqrt{\frac{\varepsilon_2}{\varepsilon_1}} - \sin^2\theta_B\left(\sqrt{\frac{\varepsilon_1}{\varepsilon_2}}+\sqrt{\frac{\varepsilon_2}{\varepsilon_1}}\right) = 0 \end{aligned} \tag{7.275}$$

7.8.2.4. *Maximum angle of incidence and critical angle*

The Snell–Descartes law shows that if $n_1 < n_2$ (respectively, $n_1 > n_2$), then $\theta_t < \theta_i$ (respectively, $\theta_t > \theta_i$). Thus, when the refractive index of medium 1 is smaller than that of medium 2, the transmission angle is smaller than the angle of incidence. However, when the refractive index of medium 1 is higher, the angle of the transmitted wave becomes larger. Therefore, depending on the case, the angles θ_i and θ_t will be limited by a critical value that cannot be exceeded, as illustrated in Figure 7.23.

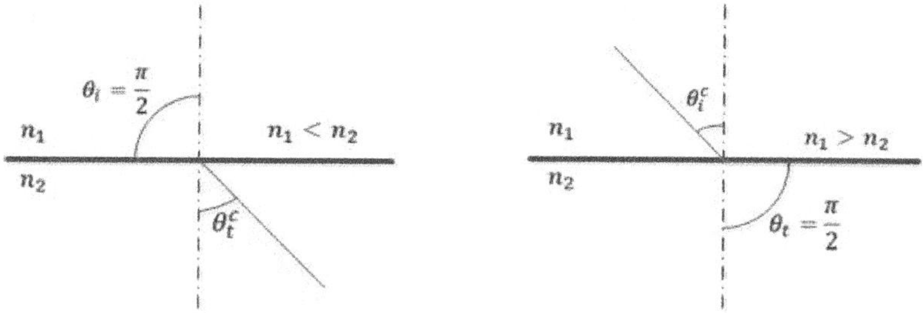

Figure 7.23. *Maximum angle of incidence and critical angle. For a color version of this figure, see www.iste.co.uk/abdesselam/electrodynamics2.zip*

– If $\mathrm{n}_1 < \mathrm{n}_2$, then there exists a critical angle θ_t^c such that $\theta_i = \frac{\pi}{2}$ and $\sin\theta_t^c = \frac{\mathrm{n}_1}{\mathrm{n}_2}$. In this case, regardless of the angle of incidence of the wave coming from medium 1, the wave will be transmitted in medium 2 and will propagate within it with a transmission angle smaller than θ_t^c.

– If $\mathrm{n}_1 > \mathrm{n}_2$, then there exists a critical angle θ_i^c such that $\theta_t = \frac{\pi}{2}$ and $\sin\theta_i^c = \frac{\mathrm{n}_2}{\mathrm{n}_1}$. In this second case, when the angle of incidence exceeds the critical angle θ_i^c, the transmitted wave does not enter the second medium at all but rather propagates along the interface. The transmission coefficient becomes zero, while the reflection coefficient equals 1. When $\theta_i = \theta_c$, the wavenumber $\mathrm{k}_{t,z}$ is zero. For an incidence angle smaller than θ_i^c, the quantity $\mathrm{k}_{t,z}$ is real. However, for an angle greater than θ_i^c, there is no real angle θ_t that satisfies the Snell–Descartes law. However, this law must always be verified to satisfy the boundary conditions at the $z = 0$ interface. Therefore, the angle θ_t becomes complex (while still satisfying the Snell–Descartes law). Although $\sin\theta_t$ remains a real number greater than 1 ($\sin\theta_t = \frac{\mathrm{n}_1}{\mathrm{n}_2}\sin\theta_i > \frac{\mathrm{n}_1}{\mathrm{n}_2}\sin\theta_i^c > 1$), $\cos\theta_t$ becomes purely imaginary, since $\cos\theta_t = \sqrt{1-\sin^2\theta_t}$ (and the term under the square root is negative). This therefore makes $\mathrm{k}_{t,z}$ purely imaginary. We then write $\mathrm{k}_{t,z} = \mathrm{k}_2\cos\theta_t = +i\alpha$ (the positive sign is chosen to represent evanescent waves in region 2):

$$\begin{aligned}\vec{\mathbf{E}}_t &= \Re\left[\hat{E}_{0,t}e^{-i(\omega t-\mathrm{k}_{t,x}x)}e^{-\alpha z}\vec{\mathbf{j}}\right]\\ \vec{\mathbf{H}}_t &= \Re\left[Z_2^{-1}\hat{E}_{0,t}\left(-\cos\theta_t\vec{\mathbf{i}}+\sin\theta_t\vec{\mathbf{k}}\right)e^{-i(\omega t-\mathrm{k}_{t,x}x)}e^{-\alpha z}\right]\end{aligned} \quad [7.276]$$

We can see that describing the wave behavior for incidence angles greater than the critical angle θ_i^c requires the introduction of an evanescent wave in region 2. It is easy to show that there is no power flux in region 2 along the Oz direction, since the time-averaged complex Poynting vector component is zero:

$$\langle \Pi_z \rangle = -\frac{1}{2}\Re\left[E_y H_x^*\right] = -\frac{1}{2}\Re\left[Z_2^{-1}\left|E_{0,t}\right|^2 (-\cos\theta_t)^* e^{-2\alpha z}\right] = 0 \quad [7.277]$$

because $\cos\theta_t$ is purely imaginary (so the quantity inside the brackets is purely imaginary). We conclude that the time-averaged incident power is completely reflected at the interface. It is also worth noting that, even though the time-averaged transmitted power inside the dielectric is zero, non-zero fields still exist and decay exponentially with distance.

7.8.2.5. *Reflection and transmission coefficients*

To simplify, we assume $\mu_1 = \mu_2 = \mu$. The reflection and transmission coefficients can also be expressed in terms of the refractive indices n_1 and n_2:

$$\begin{aligned}
R_{TE} &= \frac{Z_2\cos\theta_i - Z_1\cos\theta_t}{Z_2\cos\theta_i + Z_1\cos\theta_t} = \frac{\mathrm{n}_1\cos\theta_i - \mathrm{n}_2\cos\theta_t}{\mathrm{n}_1\cos\theta_i + \mathrm{n}_2\cos\theta_t} \\
T_{TE} &= \frac{2Z_2\cos\theta_i}{Z_2\cos\theta_i + Z_1\cos\theta_t} = \frac{2\mathrm{n}_1\cos\theta_i}{\mathrm{n}_1\cos\theta_i + \mathrm{n}_2\cos\theta_t} \\
R_{TM} &= \frac{Z_1\cos\theta_i - Z_2\cos\theta_t}{Z_1\cos\theta_i + Z_2\cos\theta_t} = \frac{\mathrm{n}_2\cos\theta_i - \mathrm{n}_1\cos\theta_t}{\mathrm{n}_2\cos\theta_i + \mathrm{n}_1\cos\theta_t} \\
T_{TM} &= \frac{2Z_2\cos\theta_i}{Z_1\cos\theta_i + Z_2\cos\theta_t} = \frac{2\mathrm{n}_1\cos\theta_i}{\mathrm{n}_2\cos\theta_i + \mathrm{n}_1\cos\theta_t}
\end{aligned} \quad [7.278]$$

Using the Snell–Descartes law, we can rewrite the Fresnel equations [7.278] to express them in terms of the angle of incidence θ_i:

$$\begin{aligned}
R_{TE} &= \frac{\cos\theta_i - \sqrt{\left(\frac{\mathrm{n}_2}{\mathrm{n}_1}\right)^2 - \sin^2\theta_i}}{\cos\theta_i + \sqrt{\left(\frac{\mathrm{n}_2}{\mathrm{n}_1}\right)^2 - \sin^2\theta_i}} \\
T_{TE} &= \frac{2\cos\theta_i}{\cos\theta_i + \sqrt{\left(\frac{\mathrm{n}_2}{\mathrm{n}_1}\right)^2 - \sin^2\theta_i}} = 1 + R_{TE} \\
R_{TM} &= \frac{\left(\frac{\mathrm{n}_2}{\mathrm{n}_1}\right)^2\cos\theta_i - \sqrt{\left(\frac{\mathrm{n}_2}{\mathrm{n}_1}\right)^2 - \sin^2\theta_i}}{\left(\frac{\mathrm{n}_2}{\mathrm{n}_1}\right)^2\cos\theta_i + \sqrt{\left(\frac{\mathrm{n}_2}{\mathrm{n}_1}\right)^2 - \sin^2\theta_i}} \\
T_{TM} &= \frac{2\left(\frac{\mathrm{n}_2}{\mathrm{n}_1}\right)\cos\theta_i}{\left(\frac{\mathrm{n}_2}{\mathrm{n}_1}\right)^2\cos\theta_i + \sqrt{\left(\frac{\mathrm{n}_2}{\mathrm{n}_1}\right)^2 - \sin^2\theta_i}} = \frac{\mathrm{n}_1}{\mathrm{n}_2}\left(1 + R_{TM}\right)
\end{aligned} \quad [7.279]$$

It is also possible to rewrite the Fresnel equations as follows:

$$\begin{aligned}
R_{TE} &= \frac{\cos\theta_i - \frac{n_2}{n_1}\sin\theta_i}{\cos\theta_i + \frac{n_2}{n_1}\sin\theta_i} = \frac{\cos\theta_i - \frac{\sin\theta_i}{\sin\theta_t}\sin\theta_i}{\cos\theta_i + \frac{\sin\theta_i}{\sin\theta_t}\sin\theta_i} = -\frac{\sin(\theta_i - \theta_t)}{\sin(\theta_i + \theta_t)} \\
T_{TE} &= \frac{2\cos\theta_i\sin\theta_t}{\sin(\theta_i + \theta_t)} = 1 - \frac{\sin(\theta_i - \theta_t)}{\sin(\theta_i + \theta_t)} = 1 + R_{TE}, \\
R_{TM} &= \frac{\tan(\theta_i - \theta_t)}{\tan(\theta_i + \theta_t)} \\
T_{TM} &= \frac{\tan(\theta_i + \theta_t)}{\cos(\theta_i - \theta_t)} - \frac{\tan(\theta_i - \theta_t)}{\cos(\theta_i + \theta_t)} = \frac{\sin\theta_t}{\sin\theta_i}(1 + R_{TM})
\end{aligned} \quad [7.280]$$

Let us summarize the results obtained for the TE waves:

– If $n_1 < n_2$, then $\theta_t < \theta_i$. The wave is always transmitted into medium 2, regardless of the angle of incidence. The reflection coefficient R_{TE} is always real and negative, since $0 < \theta_i - \theta_t < \frac{\pi}{2}$. We therefore conclude that the TE wave undergoes a phase shift of π upon reflection. At normal incidence (i.e. when $\theta_i = 0$), we get the value $R_{TE}(0) = \frac{n_1 - n_2}{n_1 + n_2} < 0$. For a grazing incidence ($\theta_i = \frac{\pi}{2}$), we find $R_{TE}\left(\frac{\pi}{2}\right) = -1$. As θ_i increases from 0 to $\frac{\pi}{2}$, the magnitude of R_{TE} continuously increases from $\frac{n_2 - n_1}{n_2 + n_1}$ to 1, as shown in Figure 7.24.

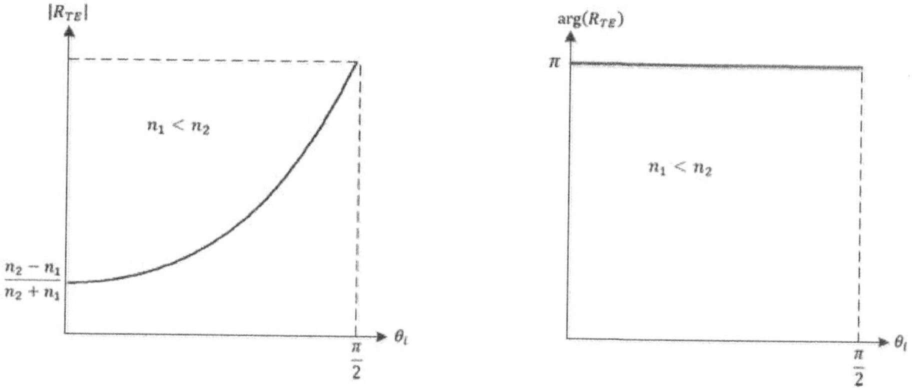

Figure 7.24. *Magnitude and phase of the reflection coefficient as a function of the angle of incidence for a TE wave with $n_1 < n_2$. For a color version of this figure, see www.iste.co.uk/abdesselam/electrodynamics2.zip*

– If $n_1 > n_2$ and $\theta_i < \theta_i^c$, then $\theta_t > \theta_i$ and $0 < \theta_t - \theta_i < \frac{\pi}{2}$; so $\sin(\theta_t - \theta_i) > 0$. Consequently, R_{TE} is real and positive. The reflected wave is in phase with the incident wave. Figure 7.25 shows that as θ_i increases from 0 to θ_i^c, the magnitude of R_{TE} increases from $\frac{n_1 - n_2}{n_1 + n_2}$ (value obtained for a normal incident wave, that is, for $\theta_i = 0$) to 1 (value given by the incidence limit θ_i^c – for this value, the angle θ_t takes the value $\frac{\pi}{2}$).

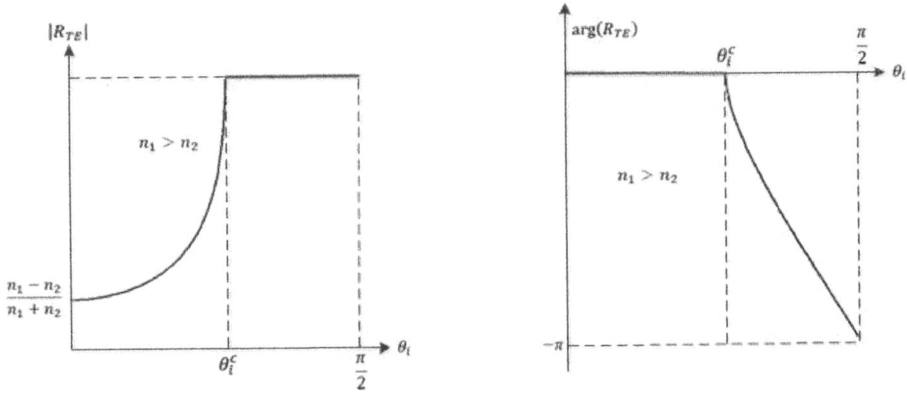

Figure 7.25. *Magnitude and phase of the reflection coefficient as a function of the angle of incidence for a TE wave with* $n_1 > n_2$. *For a color version of this figure, see www.iste.co.uk/abdesselam/electrodynamics2.zip*

– If $n_1 > n_2$ and $\theta_i > \theta_i^c$, then θ_t becomes complex (the transmitted wave is inhomogeneous). The reflection and transmission coefficients are written as:

$$R_{TE} = \frac{\cos\theta_i - i\sqrt{\sin^2\theta_i - \left(\frac{n_2}{n_1}\right)^2}}{\cos\theta_i + i\sqrt{\sin^2\theta_i - \left(\frac{n_2}{n_1}\right)^2}} = \frac{\cos\theta_i - i\sqrt{\sin^2\theta_i - \sin^2\theta_i^c}}{\cos\theta_i + i\sqrt{\sin^2\theta_i - \sin^2\theta_i^c}}$$
$$T_{TE} = \frac{2\cos\theta_i}{\cos\theta_i + i\sqrt{\sin^2\theta_i - \left(\frac{n_2}{n_1}\right)^2}} = \frac{2\cos\theta_i}{\cos\theta_i + i\sqrt{\sin^2\theta_i - \sin^2\theta_i^c}} \qquad [7.281]$$

In this area, the modulus of R_{TE} is 1, while its argument is:

$$\arg(R_{TE}) = -2\tan^{-1}\frac{\sqrt{\sin^2\theta_i - \sin^2\theta_i^c}}{\cos\theta_i} \qquad [7.282]$$

Thus, the reflected wave is out of phase with the incident wave by the angle $\arg(R_{TE})$. If $\theta_i = \theta_i^c$, then $\arg(R_{TE}) = 0$ and if $\theta_i = \frac{\pi}{2}$, we get $\arg(R_{TE}) = \pi$. The variations of the functions $|R_{TE}|$ and $\arg(R_{TE})$ in the area $\left[\theta_i^c, \frac{\pi}{2}\right]$ are illustrated in Figure 7.25.

Let us summarize the results obtained for TM waves:

– $n_1 < n_2 \Longrightarrow \theta_t < \theta_i$ (the wave is always transmitted in medium 2): the reflection coefficient R_{TM} is real, continuous, positive in the interval $[0, \theta_B[$, where θ_B is the Brewster angle, zero at angle θ_B and negative in $]\theta_B, \frac{\pi}{2}]$. Thus, the reflected and incident waves are in phase in the $[0, \theta_B[$ zone and in phase opposition in the $]\theta_B, \frac{\pi}{2}]$ zone. We can therefore see that the function $|R_{TM}|$ is decreasing in $[0, \theta_B[$, null in θ_B and increasing in $]\theta_B, \frac{\pi}{2}]$, as illustrated in Figure 7.26.

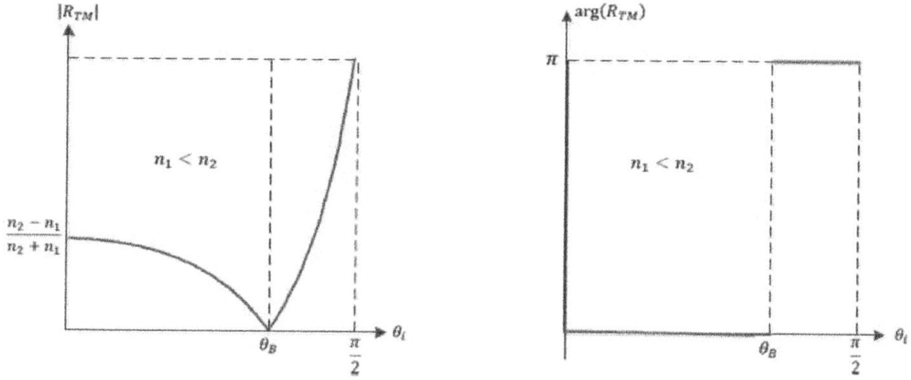

Figure 7.26. *Magnitude and phase of the reflection coefficient as a function of the angle of incidence for a TM wave with* $n_1 < n_2$. *For a color version of this figure, see www.iste.co.uk/abdesselam/electrodynamics2.zip*

– $n_1 > n_2$ and $\theta_i < \theta_i^c \Longrightarrow \theta_t > \theta_i$ (the wave passes partially through medium 2). In this zone, the reflection coefficient R_{TM} is negative in $[0, \theta_B[$, zero in θ_B and positive in $]\theta_B, \theta_i^c]$. The functions $|R_{TM}|$ and $\arg(R_{TM})$ are illustrated in Figure 7.27.

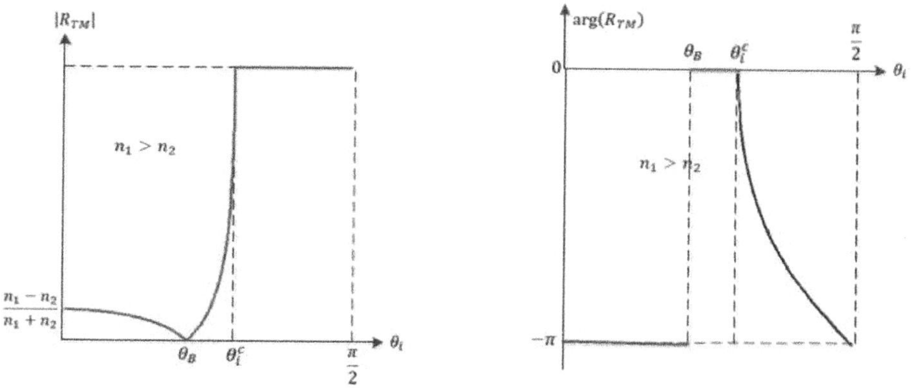

Figure 7.27. *Magnitude and phase of the reflection coefficient as a function of the angle of incidence for a TM wave with* $n_1 > n_2$. *For a color version of this figure, see www.iste.co.uk/abdesselam/electrodynamics2.zip*

– $n_1 > n_2$ and $\theta_i > \theta_i^c \Longrightarrow \theta_t$ is complex (the wave is completely reflected). The reflection coefficient is written as:

$$R_{TM} = \frac{\left(\frac{n_2}{n_1}\right)^2 \cos\theta_i - i\sqrt{\sin^2\theta_i - \left(\frac{n_2}{n_1}\right)^2}}{\left(\frac{n_2}{n_1}\right)^2 \cos\theta_i + i\sqrt{\sin^2\theta_i - \left(\frac{n_2}{n_1}\right)^2}} \qquad [7.283]$$

and, consequently:

$$|R_{TM}| = 1$$
$$\arg(R_{TM}) = -2\tan^{-1}\frac{\sqrt{\sin^2\theta_i - \left(\frac{n_2}{n_1}\right)^2}}{\left(\frac{n_2}{n_1}\right)^2 \cos\theta_i} \qquad [7.284]$$

The functions $|R_{TM}|$ and $\arg(R_{TM})$, in this zone, are shown in Figure 7.27.

References

Abbel, M.L. and Braselton, J.P. (2005). *Maple by Example*. Elsevier Academic Press, Cambridge.

Abdesselam, B. (2007). *Introduction à la mécanique analytique*. Ellipses, Paris.

Ait-Ameur, K. (2016). *Physique, les rayonnements electromagnétiques : résumé de cours et problèmes corrigés*. Ellipses, Paris.

Alastuey, A., Margo, M., Pujol, P. (2008). *Physique et outils mathématiques : méthodes et examples*. EDP Sciences/CNRS Editions, Paris.

Alfvèn, H. (1950). *Cosmical Electrodynamics*. Oxford University Press, Oxford.

Arfken, G. (1985). *Mathematical Methods for Physicists*. Academic Press, Cambridge.

Arnold, V.I. (1980). *Ordinary Differential Equations*. MIT Press, Cambridge.

Baker, B.B. and Copson, E.E. (2003). *Mathematical Theory of Huygens's Principle*. American Mathematical Society, Providence.

Bérest, P. (1997). *Calcul des variations : application à la mécanique et à la physique*. Ellipses, Paris.

Berthon, A. (2011). *Electromagnétisme-équations intégrales de l'électromagnétisme, applications aux problèmes axisymétriques*. Ellipses, Paris.

Bitsadze, A.V. (1980). *Equations of Mathematical Physics*. Mir Publishers, Moscow.

Bitsadze, A.V. and Kalinichenko, D.F. (1980). *A Collection of Problems on the Equations of Mathematical Physics*. Mir Publishers, Moscow.

Bok, J. and Hulin-Jung, N. (1975). *Ondes électromagnétiques relativité*. Hermann, Paris.

Boyce, W.E. and Diprima, R.C. (2009). *Elementary Equations and Boundary Value Problems*. John Wiley and Sons, New York.

Brewster, H.D. (2010). *Electromagnetism*. Oxford Book Company, Oxford.

Brillouin, L. (1960). *Wave Propagation and Group Velocity*. Academic Press, Cambridge.

Butkov, E. (1968). *Mathematical Physics*. Addison-Wesley, Boston.

Cahn, S.B. and Nadgorny, B.E. (1994). *A Guide to Physics Problems. Part 1: Mechanics, Relativity and Electrodynamics*. Kluwer Academic Publishers, Alphen aan den Rijn.

Cardone, F. and Mignani, R. (2004). *Energy and Geometry: An Introduction to Deformed Special Relativity*. World Scientific, Singapore.

Cohen-Tannoudji, C. (2004). *Atoms in Electromagnetic Fields*. World Scientific, Singapore.

Cordier, D. (2007). *Cours de physique – électromagnétisme – 2. Phénomènes d'induction et ondes électromagnétiques*. Dunod, Paris.

Cornille, P. (2003). *Advanced Electromagnetism and Vacuum Physics*. World Scientific, Singapore.

Coulomb, C.A. (1884). *Collection de mémoires relatifs à la physique 3*. Gauthier-Villars, Paris.

Courant, R. and Hilbert, D. (1989). *Methods of Mathematical Physics*. John Wiley and Sons, New York.

D'Auria, R. and Trigiante, M. (2016). *From Special Relativity to Feynman Diagrams: A Course of Theoretical Particle Physics for Beginners*. Springer, Berlin.

Debye, P. (1945). *Polar Molecules*. Dover Publications, New York.

Dirac, P.A.M. (1958). *The Principles of Quantum Mechanics*. Oxford University Press, Oxford.

Durand, E. (1953). *Electrostatique et magnétostatique*. Masson et Cie, Paris.

Edminister, J.A. (1979). *Schaum's Outline of Theory and Problems of Electromagnetics*. McGraw-Hill, New York.

Einstein, A., Lorentz, H.A., Minkowski, H., Weyl, H. (1952). *The Principle of Relativity*. Dover Publications, New York.

Eom, H.J. (2004). *Electromagnetic Wave Theory of Boundary-Value Problems: An Advanced Course on Analytical Methods*. Springer, Berlin.

Faraday, M. (1965). *Experimental Researches in Electricity*. Dover Publications, New York.

Feynman, R.P., Leighton, R.B., Sands, M. (1979). *Le cours de physique de Feynman – électromagnétisme 1 et 2*. Inter-Editions, Paris.

Fitzpatrick, R. (2008). *Maxwell's Equations and the Principles of Electromagnetism*. Infinity Science Press LLC, Sydney.

Foot, C.J. (2005). *Atomic Physics*. Oxford University Press, Oxford.

Friedman, B. (2011). *Principles and Techniques of Applied Mathematics*. Dover Publications, New York.

Friedrich, H. (2004). *Theoretical Atomic Physics*. Springer, Berlin.

Fröhlich, H. (1949). *Theory of Dielectrics*. Oxford University Press, Oxford.

Fujimoto, M. (2007). *Physics of Classical Electromagnetism*. Springer, Berlin.

Funaro, D. (2008). *Electromagnetism and the Structure of Matter*. World Scientific, Singapore.

Goldstein, H. (1950). *Classical Mechanics*. Addison Wesley, Boston.

Grimes, D.M. and Grimes, G.A. (2004). *The Electromagnetic Origin of Quantum Theory of Light*. World Scientific, Singapore.

Guru, B. and Hiziroglu, H.R. (2009). *Electromagnetic Fields Theory Fundamentals*. Cambridge University Press, Cambridge.

Hairer, E. and Wanner, G. (2006). Analyse complexe et séries de Fourier. Thesis, Université de Genève, Geneva.

Halpern, A. and Erlbach, E. (1998). *Theory and Problems of Beginning Physics II: Waves, Electromagntism, Optics and Modern Physics*. McGraw-Hill, New York.

Heitler, W. (2010). *Quantum Theory of Radiation*. Dover Publications, New York.

Henley, E.M. and Garcia, A. (2007). *Subatomic Physics*. World Scientific, Singapore.

Hertz, H. (1962). *Electric Waves: Being Researches on the Propagation of Electric Action with Finite Velocity Through Space*. Dover Publications, New York.

Hoffmann, J.D. (2001). *Numerical Methods for Engineers and Scientists*. Marcel Dekker, New York.

Iwanenko, D. and Sokolow, A. (1949). *Klassische fieltheorie*. Akademic-Verlag, Berlin.

Jackson, J.D. (1962). *Classical Electrodynamics*. John Wiley and Sons, New York.

Jeans, J.H. (2009). *Mathematical Theory of Electricity and Magnetism*. Cambridge University Press, Cambridge.

Jeffereys, H. and Jeffreys, B.S. (2000). *Methods of Mathematical Physics*. Cambridge University Press, Cambridge.

Johnson, W.R. (2007). *Atomic Structure Theory. Lectures on Atomic Physics*. Springer, Berlin.

Jordan, E.C. and Balman K.G. (1968). *Electromagnetic Waves and Radiating Systems*. Prentice-Hall, New Jersey.

Kamal, A.A. (2010). *1000 Solved Problems in Modern Physic*. Springer, Berlin.

Kay, D.C. (2011). *Theory and Problems of Tensor Calculus*. McGraw-Hill, New York.

Kibble, T.W.B. and Berkshire, F.H. (2004). *Classical Mechanics*. Imperial College Press, London.

King, R.W.P. and Wu, T.T. (1959). *Scattering and Diffraction of Waves*. Harvard University Press, Cambridge.

Landau, L.D. and Lifchitz, E.M. (1962). *Classical Theory of Fields*. Addison-Wesley, Boston.

Landau, L.D., Lifshitz, E.M., Pitaevskii, L.P. (1984). *Electrodynamics of Continuous Media*. Addison-Wesley, Boston.

Le Bella, C. (1998). *Des phénomènes critiques aux champs de jauge : une introduction aux méthodes et aux applications de la théorie quantique des champs*. EDP Sciences, Les Ulis.

Lichnerowicz, A. (1955). *Théories relativistes de la gravitation et de l'électromagnétisme, relativité générale et théories unitaires*. Masson et Cie, Paris.

Lorentz, H.A. (1952). *Theory of Electrons*. Dover Publications, New York.

Mathey, P. (2017). *Electromagnétisme : ondes et propagation guidée, cours, 130 QCM et exercices corrigés, exemples d'applications*. Dunod, Paris.

Maxwell, J.C. (1954). *Treatise on Electricity and Magnetism*. Dover Publications, New York.

Moliton, A. (2004). *Applications de l'électromagnétisme dans les milieux matériels*. Hermès–Lavoisier, Paris.

Morin, D. (2008). *Introduction to Classical Mechanics with Problems and Solutions*. Cambridge University Press, Cambridge.

Neff, H.P. (1991). *Introductory Electromagnetics*. John Wiley and Sons, New York.

Nikiforov, A. and Ouvarov, V. (1983). *Fonctions spéciales de la physique mathématique*. Mir Publishers, Moscow.

Nogarède, B. (2005). *Electrodynamique appliquée : bases et principes physiques de l'électrotechnique – cours et exercices corrigés licence 3ème année, master, écoles d'ingénieurs*. Dunod, Paris.

Okun, L.B. (2008). *Energy and Mass in Relativity Theory*. World Scientific, Singapore.

Panofsky, W. and Philips, M. (1955). *Classical Electricity and Magnetism*. Addison-Wesley, Boston.

Pauli, W. (1981). *Theory of Relativity*. Dover Publications, New York.

Pérez, J.P.H., Carles, R., Fleckinger, R. (2020). *Electromagnétisme : fondements et applications, avec 300 exercices et problèmes résolus*, 4th edition. Dunod, Paris.

Provost, J.P. and Vallée, J.P. (2011). *Les maths en physique, la physique à travers le filtre des mathématiques avec éléments d'analyse numérique, cours et applications*. Dunod, Paris.

Raab, R.E. and de Lange, O.L. (2004). *Multipole Theory in Electromagnetism: Classical, Quantum, and Symmetry Aspects, with Applications*. Oxford Science Publications, Oxford.

Rax, J.M. (2017). *Electromagnétisme : milieux, structures et énergie*. De Boeck, Brussels.

Reed, M. and Barry, S. (1980). *Methods of Modern Mathematical Physics*. Academic Press, Cambridge.

Reitz, J.R., Milford, F.J., Christy, R.W. (2008). *Foundations of Electromagnetic Theory*. Addison-Wesley, Boston.

Reuse, F.A. (2012). *Electrodynamique : 125 exercices avec solutions – bachelor, licence, classes prépas*. PPUR, Lausanne.

Riley, K.F., Hobson, M.P., Bence, S.J. (2012). *Mathematical Methods for Physics and Engineering*. Cambridge University Press, Cambridge.

Rudnicki, K. (2000). *Gravitation, Electromagnetism and Cosmology: Toward a New Synthesis*. Apeiron, Montreal.

Samarskii, A.A. and Vabishchevich, P.N. (2007). *Numerical Methods for Solving Inverse Problems of Mathematical Physics*. De Gruyter, Berlin.

Schultz, B. (1980). *Geometrical Methods of Mathematical Physics*. Cambridge University Press, Cambridge.

Sénéchal, D. (2007). Electromagnétisme avancé. Thesis, Université de Sherbrooke, Sherbrooke.

Sommerfeld, A. (1952). *Electrodynamics*. Academic Press, Cambridge.

Spiegel, M.R. (1968). *Mathematical Handbook of Formulas and Tables*. McGraw-Hill, New York.

Spiegel, M.R. (1974). *Fourier Analysis with Applications to Boundary Value Problems*. McGraw-Hill, New York.

Thirring, W. (1997). *Classical Mathematical Physics, Dynamical Systems and Field Theories*, 3rd edition. Springer, Berlin.

Van Bladel, J. (2007). *Electromagnetic Fields*, 2nd edition. John Wiley and Sons, New York.

Vanderlinde, J. (2005). *Classical Electromagnetic Theory*, 2nd edition. Kluwer Academic Publishers, Dordrecht.

Verweij, M.D., Van Den Berg, P.M., Blok H. (2006). *Electromagnetic Waves: An Introductory Course*, 2nd edition. VSSD, Delft.

Wrede, R.C. and Spiegel, M.R. (2010). *Advanced Calculus*. McGraw-Hill, New York.

Zhang, M. (1979). *Electromagnetic Field Theory: A Problem Solving Approach*. John Wiley and Sons, New York.

Index

D, E

F

G, H

K, L

M

O, P

Q, R

S

T

V

Other titles from

ISTE

in

Waves

2025

ABDESSELAM Boucif
Introduction to Classical Electrodynamics 1

2024

BERNARD Jean-Michel L.
Advanced Studies in the Mathematical Theory of Scattering (Waves and Scattering Set – Volume 3)

JUVÉ Daniel, GALLAND Marie-Annick, CLAIR Vincent
Acoustics of Fluid Media 1: Principles and Applications

RÉVEILLAC Jean-Michel
Synthesizers and Subtractive Synthesis 1: Theory and Overview
Synthesizers and Subtractive Synthesis 2: Application and Practice

2022

GONTRAND Christian
Electromagnetism: Links to Special Relativity

LACAZE Pierre Camille, LACROIX Jean-Christophe
Nanotechnology and Nanomaterials for Energy

RÉVEILLAC Jean-Michel
Recording and Voice Processing 1: History and Generalities
Recording and Voice Processing 2: Working in the Studio

ROYER Daniel, VALIER-BRASIER Tony
Elastic Waves in Solids 1: Propagation

SAKHO Ibrahima
Nuclear Physics 1: Nuclear Deexcitations, Spontaneous Nuclear Reactions

VALIER-BRASIER Tony, ROYER Daniel
Elastic Waves in Solids 2: Radiation, Scattering, Generation

2021

DAHOO Pierre-Richard, LAKHLIFI Azzedine
Infrared Spectroscopy of Symmetric and Spherical Top Molecules for Space Observation 1
(Infrared Spectroscopy Set – Volume 3)
Infrared Spectroscopy of Symmetric and Spherical Top Molecules for Space Observation 2
(Infrared Spectroscopy Set – Volume 4)

2020

DANIELE Vito G., LOMBARDI Guido
Scattering and Diffraction by Wedges 1: The Wiener-Hopf Solution - Advanced Applications
(Waves and Scattering Set – Volume 1)
Scattering and Diffraction by Wedges 2: The Wiener-Hopf Solution - Advanced Applications
(Waves and Scattering Set – Volume 2)

SAKHO Ibrahima
Introduction to Quantum Mechanics 2: Wave-Corpuscle, Quantization & Schrödinger's Equation

2019

BERTRAND Pierre, DEL SARTO Daniele, GHIZZO Alain
The Vlasov Equation 1: History and General Properties

DAHOO Pierre-Richard, LAKHLIFI Azzedine
Infrared Spectroscopy of Triatomics for Space Observation
(Infrared Spectroscopy Set – Volume 2)

RÉVEILLAC Jean-Michel
Electronic Music Machines: The New Musical Instruments

ROMERO-GARCÍA Vicente, HLADKY-HENNION Anne-Christine
Fundamentals and Applications of Acoustic Metamaterials: From Seismic to Radio Frequency
(Metamaterials Applied to Waves Set – Volume 1)

SAKHO Ibrahima
Introduction to Quantum Mechanics 1: Thermal Radiation and Experimental Facts Regarding the Quantization of Matter

2018

SAKHO Ibrahima
Screening Constant by Unit Nuclear Charge Method: Description and Application to the Photoionization of Atomic Systems

2017

DAHOO Pierre-Richard, LAKHLIFI Azzedine
Infrared Spectroscopy of Diatomics for Space Observation
(Infrared Spectroscopy Set – Volume 1)

PARET Dominique, HUON Jean-Paul
Secure Connected Objects

PARET Dominque, SIBONY Serge
Musical Techniques: Frequencies and Harmony

RÉVEILLAC Jean-Michel
Analog and Digital Sound Processing

STAEBLER Patrick
Human Exposure to Electromagnetic Fields

2016

ANSELMET Fabien, MATTEI Pierre-Olivier
Acoustics, Aeroacoustics and Vibrations

BAUDRAND Henri, TITAOUINE Mohammed, RAVEU Nathalie
The Wave Concept in Electromagnetism and Circuits: Theory and Applications

PARET Dominique
Antennas Designs for NFC Devices

PARET Dominique
Design Constraints for NFC Devices

WIART Joe
Radio-Frequency Human Exposure Assessment

2015

PICART Pascal
New Techniques in Digital Holography

2014

APPRIOU Alain
Uncertainty Theories and Multisensor Data Fusion

JARRY Pierre, BENEAT Jacques N.
RF and Microwave Electromagnetism

LAHEURTE Jean-Marc
UHF RFID Technologies for Identification and Traceability

REMY Jean-Gabriel, LETAMENDIA Charlotte
LTE Standards

REMY Jean-Gabriel, LETAMENDIA Charlotte
LTE Services

SAVAUX Vincent, LOUËT Yves
MMSE-based Algorithm for Joint Signal Detection, Channel and Noise Variance Estimation for OFDM Systems

THOMAS Jean-Hugh, YAAKOUBI Nourdin
New Sensors and Processing Chain

TING Michael
Molecular Imaging in Nano MRI

VALIÈRE Jean-Christophe
Acoustic Particle Velocity Measurements using Laser: Principles, Signal Processing and Applications

VANBÉSIEN Olivier, CENTENO Emmanuel
Dispersion Engineering for Integrated Nanophotonics

2013

BENMAMMAR Badr, AMRAOUI Asma
Radio Resource Allocation and Dynamic Spectrum Access

BOURLIER Christophe, PINEL Nicolas, KUBICKÉ Gildas
Method of Moments for 2D Scattering Problems: Basic Concepts and Applications

GOURE Jean-Pierre
Optics in Instruments: Applications in Biology and Medicine

LAZAROV Andon, KOSTADINOV Todor Pavlov
Bistatic SAR/GISAR/FISAR Theory Algorithms and Program Implementation

LHEURETTE Eric
Metamaterials and Wave Control

PINEL Nicolas, BOURLIER Christophe
Electromagnetic Wave Scattering from Random Rough Surfaces: Asymptotic Models

SHINOHARA Naoki
Wireless Power Transfer via Radiowaves

TERRE Michel, PISCHELLA Mylène, VIVIER Emmanuelle
Wireless Telecommunication Systems

2012

BÉCHERRAWY Tamer
Electromagnetism: Maxwell Equations Wave Propagation and Emission

BOUCHET Olivier
Wireless Optical Communications

BOUTIER Alain
Laser Metrology in Fluid Mechanics: Granulometry, Temperature and Concentration Measurements

LALAUZE René
Chemical Sensors and Biosensors

LE MENN Marc
Instrumentation and Metrology in Oceanography

PICART Pascal, LI Jun-chang
Digital Holography

SAGUET Pierre
Numerical Analysis in Electromagnetics: The TLM Method

2011

BÉCHERRAWY Tamer
Mechanical and Electromagnetic Vibrations and Waves

BESNIER Philippe, DÉMOULIN Bernard
Electromagnetic Reverberation Chambers

GOURE Jean-Pierre
Optics in Instruments

GROUS Ammar
Applied Metrology for Manufacturing Engineering

LE CHEVALIER François, LESSELIER Dominique, STARAJ Robert
Non-standard Antennas

2010

BEGAUD Xavier
Ultra Wide Band Antennas

GUNIOT Vincent
Wave Propagation in Fluids: Second Edition – Models and Numerical Techniques

MARAGE Jean-Paul, MORI Yvon
Sonar and Underwater Acoustics

2009

BOUDRIOUA Azzedine
Photonic Waveguides

BRUNEAU Michel, POTEL Catherine
Materials and Acoustics Handbook

DE FORNEL Frédérique, FAVENNEC Pierre-Noël
Measurements using Optic and RF Waves

FRENCH COLLEGE OF METROLOGY
Transverse Disciplines in Metrology

2008

CHARRUAU Stéphane
Electromagnetism and Interconnections: Advanced Mathematical Tools for Computer-aided Simulation

FILIPPI Paul J.T.
Vibrations and Acoustic Radiation of Thin Structures

LALAUZE René
Physical Chemistry of Solid-Gas Interfaces

2007

KUNDU Tribikram
Advanced Ultrasonic Methods for Material and Structure Inspection

PLACKO Dominique
Fundamentals of Instrumentation and Measurement

RIPKA Pavel, TIPEK Alois
Modern Sensors Handbook

2006

BALAGEAS Daniel, FRITZEN Claus-Peter, GÜEMES Alfredo
Structural Health Monitoring

BOUCHET Olivier, SIZUN Hervé, BOISROBERT Christian, DE FORNEL Frédérique, FAVENNEC Pierre-Noël
Free-Space Optics

BRUNEAU Michel, SCELO Thomas
Fundamentals of Acoustics

FRENCH COLLEGE OF METROLOGY
Metrology in Industry

GUILLAUME Philippe
Music and Acoustics

GUYADER Jean-Louis
Vibration in Continuous Media

RIGNEAULT Hervé, LOURTIOZ Jean-Michel, DELALANDE Claude, LEVENSON Juan Ariel
Nanophotonics

Printed and bound by CPI Group (UK) Ltd, Croydon, CR0 4YY

13/09/2026

14961011-0008